ACOUSTICS *and* AUDIO TECHNOLOGY

Third Edition

Mendel Kleiner

ISBN: 978-1-60427-052-5

Printed and bound in the U.S.A. Printed on acid-free paper.

10 9 8 7 6 5 4 3 2 1

Library of Congress Cataloging-in-Publication Data

Kleiner, Mendel, 1946-
Acoustics and audio technology / by Mendel Kleiner. -- 3rd ed.
p. cm.
Includes bibliographical references and index.
ISBN 978-1-60427-052-5 (pbk. : alk. paper)
1. Architectural acoustics. 2. Sound--Recording and reproducing--Equipment and supplies. 3. Hearing. I. Title.
NA2800.K559 2011
620.2--dc22

2011002859

Direct all inquiries to J. Ross Publishing, Inc., 5765 N. Andrews Way, Fort Lauderdale, FL 33309.

Phone: (954) 727-9333
Fax: (561) 892-0700
Web: www.jrosspub.com

CONTENTS

PREFACE

The purpose of this book, *Acoustics and Audio Engineering*, is to help you learn:

- The physical background to and mathematical treatment of sound propagation. These topics are treated in chapters on the fundamentals of acoustics and theoretical room acoustics.
- The properties of human hearing. This topic is discussed in chapters on hearing and speech and on the spatial aspects of room acoustics.
- The generation and radiation of sound as well as noise control. These topics are treated in chapters on sound and vibration isolation.
- The technologies used for pickup, recording, and reproduction of sound in various environments. These topics are treated in chapters on microphones, loudspeakers, and headphones.

In Chapter 1, the theory of sound propagation in fluids such as gases is discussed for the linear case. You will study the wave equation and its solutions as well as the concepts of sound pressure, particle velocity, and impedance. The physical background to *acoustical components* is discussed and the concept of acoustical circuits. The chapter also deals with sound propagation in layered media that is of interest in the cases of sound propagation outdoors and in water.

Chapter 2 discusses the $j\omega$-method for solving acoustic problems, various measures of sound, filter methods used to separate spectral components, and the concept of level.

Human hearing is one of the most important reasons to study audio and acoustics. Chapter 3 discusses the hearing mechanism and the properties of human hearing such as its sensitivity, bandwidth, and so on. The chapter also includes material on the effect of noise on hearing, hearing damage, and hearing related measures of sound. The concept of binaural hearing is introduced; binaural hearing is of great importance in everyday situations. Finally, the chapter is concluded by a section on the properties of voice and speech. Many communication systems are based on transmission of speech, and it is, therefore, important to be familiar with its properties.

Room acoustics is the subject of interest in Chapter 4. Alternative approaches to the study of room acoustics are introduced: geometrical acoustics, energy balance or statistical acoustics, and physical acoustics. Geometrical acoustics is of great importance for the practical treatment of time related properties in the propagation of transient signals. In practice, geometrical acoustics is being used in the forms of ray tracing and image source modeling. Statistical acoustics is also a *high-frequency* approach to solving room acoustics problems but looks at the conditions in the room for the steady state case and assumes that the signal emitted into the room has sufficient frequency bandwidth and that the room is sufficiently reverberant. In contrast to geometrical acoustics, it cannot be used for analysis of sound propagation outdoors. Physical acoustics uses the approach of finding a solution to the wave equation for the boundary conditions of the room for some sound source and reveals some of the shortcomings of the previously mentioned methods. The practical use of physical acoustics is limited except for understanding the basic principles of sound propagation and resonance unless one uses numerical methods to solve the equations. Such methods may be applied in both the time and frequency domains.

The way that human hearing analyzes the spatial properties of sound fields is the topic of Chapter 5. To be able to work effectively with room acoustical planning, it is necessary to understand the behavior of human hearing regarding how hearing analyzes the complex sound fields that are set up by direct sound and the many reflections off the room boundaries and how we perceive these sound fields. The chapter also discusses some useful measures for describing room acoustical quality.

The application of room acoustic planning is discussed in Chapter 6 and is focused on ways to make sure that a design results in good acoustical conditions in rooms for music performance, theaters, auditoria, and studios. Good room acoustical conditions include low noise levels so that listening and recording can be done without disturbing noise. Some noise criteria are discussed as well as the influence of noise on speech intelligibility.

Chapter 7 includes material on sound absorbers, diffusers, and reflectors that are tools used on the plain room surfaces to achieve the desired room acoustical conditions. Sound absorption is discussed with regard to sound-absorptive materials and discrete sound absorbers, such as membrane and other resonance absorbers. The subject of sound barriers is discussed as well since sound barriers are often used in offices along with sound-absorptive treatment. Sound barriers are of great importance in reducing outdoor and indoor noise levels in urban areas.

Wave propagation in solids is characterized by the possibility of having vibration propagation by many wave types besides longitudinal waves. The reason for this is the presence of shear in solids. Most of the discussion in Chapter 8 is devoted to one particular wave type—bending waves. Bending waves in plates are of great practical importance because they couple easily to waves in the surrounding air. Consequently, the attenuation of bending waves by various techniques is of great importance, such as in the construction of loudspeaker units, loudspeaker boxes, household appliances, cars, and many other devices.

The radiation of waves from vibrating structures is the subject of Chapter 9. Sound generation by turbulence is also included in the material in this chapter because its sound is generated by fast air flows, for example, wind around vehicles, microphones, and in loudspeaker ports.

To enjoy musical performance and high fidelity sound reproduction, it is necessary to reduce background noise levels so that the noise is almost inaudible. Since noise affects human ability to sleep and concentrate, it is necessary to have low noise levels in work places and homes, for example. Chapter 10 deals with the subject of sound insulation. The properties of various sound insulating constructions are discussed, as well as the influence of leaks and flanking transmission. The impact noise properties of floors are also discussed.

The concept of vibration isolation is discussed in Chapter 11. Classical one-dimensional vibration isolation theory is introduced as is the concept of insertion loss to allow calculation of the vibrational properties of simple dynamical systems. Mechanical mobility and impedance considerations allow a more wide frequency range treatment of the subject and are also treated.

Audio technology starts being treated in Chapter 12. This chapter is devoted to an overview of various microphone types, their construction, electromechanical and acoustical properties. Interest is primarily focused on electrodynamic and electrostatic microphones since these are the most common types for professional recording of speech and music. There are various ways microphones can be designed for desired directional properties. Since microphones are so important in the measurement of sound, this is a very important chapter.

Gramophone cartridges are covered in Chapter 13. Different types of pickups and their dynamic properties are treated as are the fundamentals of gramophone recording technology.

Chapter 14 is devoted to the subject of loudspeakers. Most loudspeakers are composed of a driver unit and a loudspeaker box. The loudspeaker box can be designed in many ways and has a large influence on the characteristics of the complete loudspeaker. Various designs of boxes and how these designs are interrelated is discussed. The chapter focuses mainly on electrodynamic motor units since most commercial loudspeakers use such units. The interaction between the loudspeaker and the room as well as the influence of the listening environment are also a part of this chapter.

Headphones and earphones are personal listening devices; these are treated in Chapter 15. They can be thought of as small loudspeakers that are placed close to or inside the ear canal. Various types of constructions are discussed such as electrodynamic, piezoelectric, and electrostatic headphones. Other differences between headphones and loudspeakers are also pointed out.

Digital sound reproduction is the subject of Chapter 16. Many errors can be introduced in the digitizing the representation of sound. The influence of sampling frequency, amplitude resolution and other factors on the reproduced sound is discussed. Examples are given of digital signal processing in audio and measurement. Such digital signal processing is important in the design of various types of codecs used, for example, for MP3, AAC, and other digital audio data storage and transmission systems.

The final chapter of the book deals with sound reproduction system aspects—mono, stereo, surround sound, binaural sound pickup, and reproduction. The chapter also gives an introduction to basic audio measurements using instruments and to subjective testing of audio using listening tests.

This book is intended as an Acoustics 101 for students of Sound and Vibration as well as students of electrical and electronic engineering, civil and mechanical engineering, computer science, signals and systems, and engineering physics. The course requires knowledge of basic engineering mathematics and physics. A solutions manual accompanies the book.

The author thanks all his colleagues at the Division of Applied Acoustics for their constant encouragement and interest, as well as views on the contents. The author especially wants to thank past students, particularly MSc. Jonah Sacks and MSc. Jennifer Martin, as well as Dr. Peter Svensson, Dr. Bengt-Inge Dalenbäck, Dr. Per Sjösten, and Dr. Rendell R. Torres for making the author observe and correct a number of omissions, errors, and other shortcomings. Thanks are also due to Samuel Kleiner for solving many Mathematica related issues.

Gothenburg, Sweden, October 2010.
Mendel Kleiner

INTRODUCTION

ABOUT ACOUSTICS AND AUDIO TECHNOLOGY

The subject areas of acoustics and audio technology span a wide range of subareas and are of interest for many technical systems. Professionals, such as architects, town planners, civil engineers, doctors, biologists, oceanographers, electronics engineers, and computer scientists, as well as specialists in digital signal processing, are in contact with audio and acoustics in their work. The interest in the subjects is self-evident for radio, television, and other media professionals. Acoustical and audio engineering touch on many other areas of knowledge as indicated in Figure i.1.

AUDIO ENGINEERING

Knowledge of audio engineering technology is useful in dealing with recording, storage, and playback of sound such as speech, music, and other signals. The design of musical instruments, samplers, and synthesizers can be done better and with more success with acoustics theory, technology, design, and measurement. Computer games and virtual environments of many kinds can be rendered more efficiently and with better results, for enjoyment and presence, by using proper acoustics theory and audio technology.

Audio and acoustic engineering has one challenge, however. The final determination of quality is determined by listeners whose ability to hear features varies considerably between listeners and for a particular listener also with time, previous exposure to sound, general conditions, memory, among others.

Since audio engineering is a commercially important field, it is natural to think that over the years there would have been a development of adequate measurement techniques. In a sense, that is correct. By measurement of traditionally evaluated audio equipment properties, such as frequency response,

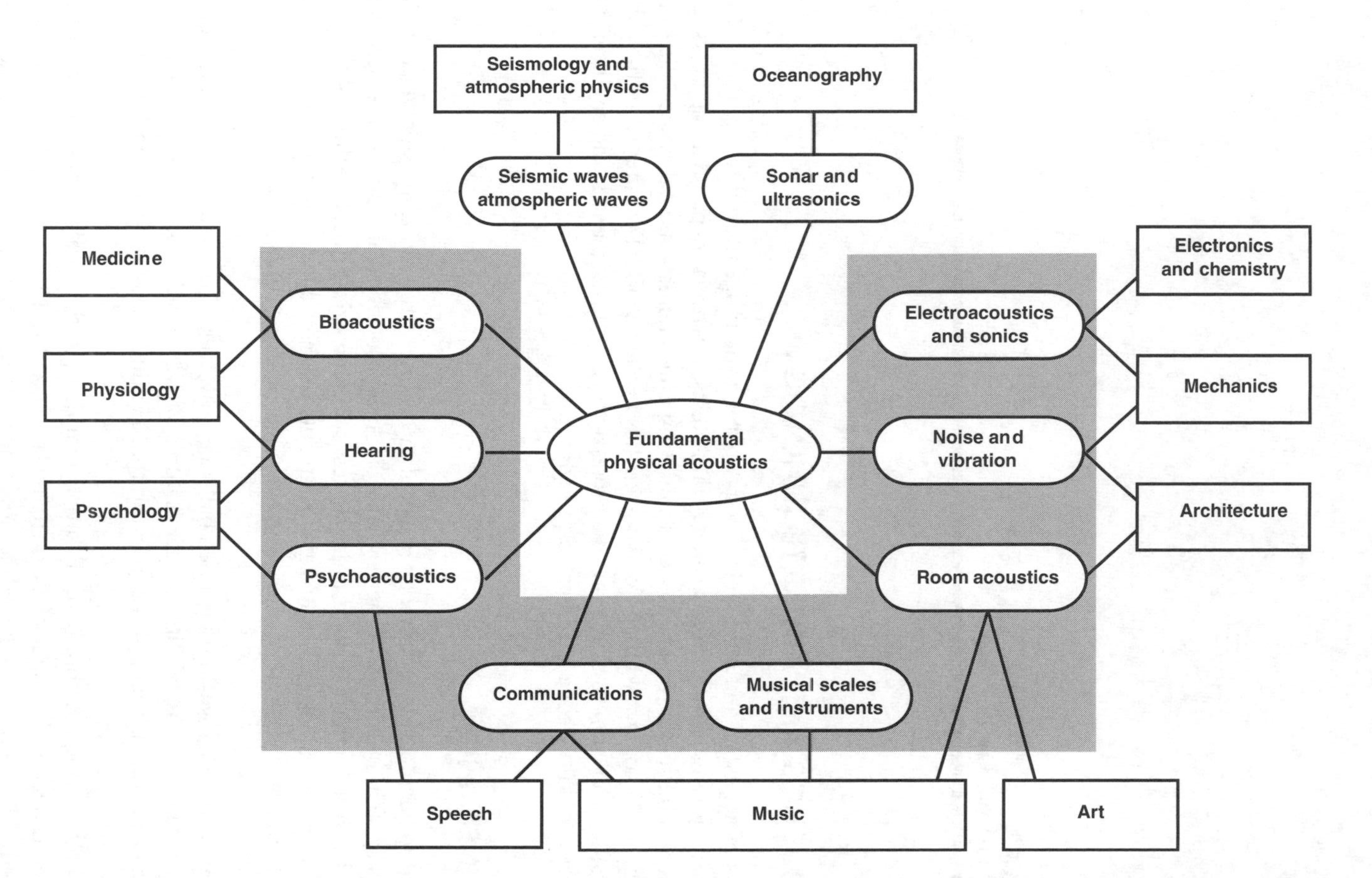

Figure i.1 The interdisciplinary world of acoustics: areas on gray background are of particular interest to the audio and acoustics professional (developed from Lindsay's "Wheel of Acoustics", Ref. i.22)

signal-to-noise ratio, harmonic and other forms of nonlinear distortion, much progress has been made. For high-end audio, however, the situation is quite unsatisfactory; it is difficult to separate marketing hype and blind faith from reality. One problem is that many audio equipment properties only become apparent after prolonged listening over weeks or months. This makes listening tests such as pair comparison tests unsatisfactory.

At the same time, these circumstances make audio engineering a quite fascinating field because there is always the possibility that your design is truly better than anyone else's; you just have to prove it. As Beranek writes, "It has been remarked that if one selects his own components, builds his own enclosure, and is convinced that he has made a wise choice of design, then his own loudspeaker sounds better to him than does anyone else's loudspeaker." (See Reference i.1.)

This book can only give a short introduction to the field; the reader is recommended to go on by reading the references. The books by Beranek and Olson (see References i.1-2) are classics in basic audio engineering; they are unique in their coverage and proper engineering treatment of the field. Recent books, such as the ones by Benson and Ballou (see References i.3-4), contain summaries by many specialists in the field. More specialized books on loudspeakers are the ones by Geddes, Borwick, and Colloms, ranging from science by way of engineering to subjectivism (see References i.5-7). Studio microphones are well covered in Gayford's book (see Reference i.8). An area not discussed at all in this book is audio electronics; the reader is advised to start by reading Linsley Hood's interesting book (see Reference i.9). Studio engineering and sound reinforcement engineering are fields that combine architectural acoustics with audio engineering; the books by Woram and Ahnert & Steffen cover these areas (see References i.10-11). The digitization of audio has introduced many new possibilities to manipulate sounds as well as many new types of effects that may or may not be audible; the book by Watkinson is an interesting starting guide to this field (see Reference i.12) as is the one by Pohlmann (see Reference i.23).

The Audio Engineering Society (http://www.aes.org) publishes the *Journal of Audio Engineering* and arranges many meetings and conferences that usually are documented by proceedings. It is the premier organization for audio engineering knowledge dissemination.

ACOUSTICS

With knowledge in acoustics, we wish to be able to design indoor and outdoor venues for public performance that are ideal for their intended purpose, for example, for speech, drama theater and musical performance. We want to design the venues so that they are excellent for listeners, speakers, singers and musicians, and others. We also want to help achieve a sustainable society, where noise exposure is voluntary and the acoustic climate agreeable to us.

It is virtually impossible to compare the acoustical* quality of rooms in an objective way, since human memory is short, and there are many variables outside the control of the scientist. It is a field where subjective judgment by single individuals, such as music critics, plays an inordinate role. Both

*Acoustic is used to modify terms that designate an object, or physical characteristics, associated with sound waves.

*Acoustical is used when the term being qualified does not designate explicitly something that has such properties, dimensions, or physical characteristics.

orchestras and audiences are influenced by the visual appearance of the hall, prior reports, and reputation, etc. Binaural sound rendering offers an avenue toward the goal of objective comparison but is not yet sufficiently perfected. Little objective research has been devoted to small room acoustics, such as that of home listening environments, cars, and even studios.

This book focuses on architectural acoustics, while discussing building acoustics and vibration isolation engineering on a small scale. The book by Raichel (see Reference i.13) covers many areas of engineering acoustics while still at an easy undergraduate level. A classic introduction to engineering acoustics is the book by Kinsler & Frey (see Reference i.14). A book discussing acoustics from the viewpoint of noise control is the one by Beranek (see Reference i.15). In modern product sound quality thinking, quiet is, however, not synonymous with quality; mechanical products are supposed to use appropriate sound to enhance their quality appearance—the car door slam or closing sound is but one example.

Architectural acoustics is another area where quality is difficult to assess. Many texts have been written on architectural acoustics, mostly from the viewpoint of the design of large auditoria, and then mostly with the audience in mind. The standard books are the ones by Kuttruff, Ando, Beranek, Barron, Egan, and Long (see References i.16-19, i.24). The books by Doelle and Cavenaugh & Wilkes look at architectural acoustics from the viewpoint of the architect (see References i.20-21).

The Acoustical Society of America (http://asa.aip.org) and the European Acoustics Association (http://www.eaa-fenestra.org) offer much information on acoustics through their journals and their conferences and meetings. For noise control, the International Institute of Noise Control Engineering is the premier information source (http://www.i-ince.org).

ABOUT THE AUTHOR

Mendel Kleiner is professor of acoustics at Chalmers University of Technology, Gothenburg, Sweden, in charge of the Chalmers Room Acoustics Group since 1989. Dr. Kleiner obtained his Ph.D. in architectural acoustics in 1978. He was professor of architectural acoustics at Rensselaer Polytechnic Institute, Troy, New York from 2003 to 2005. Kleiner is responsible for teaching room acoustics, audio technology, electroacoustics, and ultrasonics in the Chalmers master program on Sound and Vibration (http://www.ta.chalmers.se/intro/index.php). Professor Kleiner has brought his Chalmers group to the international research front of predictive room acoustics calculation, audible simulation, and 3-D sound. He returned to Chalmers in 2005 to continue leading the Chalmers Room Acoustics Group. He has more than 50 publications, has presented more than 110 papers and keynote lectures, has led courses at international conferences on acoustics and noise control, and organized an international conference on acoustics. Kleiner's main research areas are computer simulation of room acoustics, electro-acoustic reverberation enhancement systems, room acoustics of auditoria, sound and vibration measurement technology, product sound quality, and psychoacoustics. He is the author of *Audio Technology and Acoustics* published by the Division of Applied Acoustics, Chalmers and coauthor and editor of *Worship Space Acoustics* published by J. Ross Publishing, Ft. Lauderdale, FL. Kleiner is a Fellow of the Acoustical Society of America, Chair for the Audio Engineering Society's Technical Committee on Acoustics and Sound Reinforcement and its Standards Committee on Acoustics.

List of Symbols

The SI (metric) system of units is used in the book.

Upper case letters

A equivalent sound absorption area (m^2S metric sabin)
B bandwidth (Hz), magnetic field strength (Wb/m^2)
B' bending stiffness for sheets (Nm)
C capacitance (F)
C_A acoustic capacitance, acoustic compliance (m^3/Pa)
D damping, attenuation (dB), directivity index (dB), directivity factor
E modulus of elasticity (Pa), energy (J)
F force (N), directivity function
G shear modulus (N/m^2), power spectrum density (W/m^2)
I intensity (W/m^2), moment of inertia (m^4)
K_c coincidence number (m/s)
L level (dB)
L_I sound intensity level (dB). Normal reference level $1 \cdot 10^{-12}$ W/m^2
L_p sound pressure level (dB). Normal reference level $2 \cdot 10^{-5}$ Pa
L_u velocity level (dB). Normal reference levels are $1 \cdot 10^{-9}$m/s or $5 \cdot 10^{-8}$ m/s
L_W sound power level (dB). Normal reference level $1 \cdot 10^{-12}$ W
M mass, moment (Nm), molecular weight (kg)
M_A acoustic mass (kg/m^4)
N number of modes
P energy density (J/m^3)
Q volume flow (m^3)

R transmission loss (dB), viscous damping, resistance (Ns/m) (resistive part of impedance), gas constant, normalized autocorrelation
R_A acoustic resistance (Ns/m^5)
S surface area (m^2)
T reverberation time (s), absolute temperature (K), period (s)
U volume velocity (m^3/s)
V volume (m^3), potential energy (J)
W power (W)
X reactance (reactive part of impedance)
Z impedance (sound field impedance, Pa · s/m, acoustical impedance, Pa · s/m^3; mechanical impedance, Ns/m)

Lower case letters

a acceleration (m/s^2)
c speed of sound (m/s)
f frequency (Hz)
j imaginary unit, $j = \sqrt{-1}$
k wave number (m^{-1}), spring stiffness constant (N/m), Boltzmann's constant
m mass (kg), molecular damping coefficient (1/m)
m'' mass per unit area (kg/m^2)
n modal density (1/Hz)
p sound pressure (Pa)
q volume velocity (m^3/s)
r reflection coefficient, radius (m)
s radiation factor
t time (s)
u particle velocity (m/s)
u fluid velocity (m/s)
v velocity (m/s)

Greek letters

α absorption coefficient
β angle
γ standing wave ratio
δ damping constant (m^{-1})
ζ z-component of displacement (m), normalized impedance
η y-component of displacement (m), loss factor, efficiency
κ ratio between the specific heat at constant pressure and constant volume, $\kappa = C_p/C_V$
λ wavelength (m)
ν Poisson's ratio
ε extension

ξ x-component of displacement (m), relative standard deviation
ρ density (kg/m^3)
σ tension (Pa)
τ shear stress (Pa), transmission factor
φ angle
θ angle, temperature (°C)
ψ phase angle
ω angular frequency, $2\pi f$ (radians/s)
Δ difference
Λ constant

General symbols

$\hat{x}$ maximum value or peak value
$\bar{x}$ average of x over time
$\langle x \rangle$ average of x over space
$\tilde{x}$ rms value of x
$\underline{x}$ underline indicates that x is a complex quantity

Certain indices

′ per unit length
″ per unit area
0 normal condition, resonance, perpendicular to
b refraction, bending
c coincidence, critical
d diffuse
g limit
i incident
m average
r reflected
t transmitted
A acoustic, A-weighting
B bending wave, B-weighting
C C-weighting
L longitudinal
M mechanical, receiver
S transmitter
T transversal

Upper case letters are usually used to indicate amplitude (usually *A*, *B*) and number (usually *N*, *M*, *Q*).

Lower case letters are used for length and distance (usually *a*, *b*, *c*, *h*, *l*, *r*, *s*, *t*) and number (usually *l*, *m*, *n*, *q*).

This book has free material available for download from the Web Added Value™ resource center at *www.jrosspub.com*

At J. Ross Publishing we are committed to providing today's professional with practical, hands-on tools that enhance the learning experience and give readers an opportunity to apply what they have learned. That is why we offer free ancillary materials available for download on this book and all participating Web Added Value™ publications. These online resources may include interactive versions of material that appears in the book or supplemental templates, worksheets, models, plans, case studies, proposals, spreadsheets, and assessment tools, among other things. Whenever you see the WAV™ symbol in any of our publications, it means bonus materials accompany the book and are available from the Web Added Value™ Download Resource Center at www.jrosspub.com.

Downloads for *Acoustics and Audio Technology, Third Edition*, consist of a solutions manual for instructors to end-of-chapter problems.

Sound 1

1.1 INTRODUCTION

We are familiar with the fact that a mechanical system consisting of an inter-coupled mass and spring can store mechanical energy in the forms of potential and kinetic energy. If the system is excited by an outside force, the system will respond by motion. When the force is eliminated, the energy stored will cause the system to oscillate—the system will move in a periodic way. For the simple system mentioned, the free motion will occur at one frequency only; we call this the resonance frequency. We say that the system is resonating, and unless there is damping in the system, the system will go on oscillating with the same amplitude forever. Damping always results when energy is removed from the oscillating system.

Friction and radiation are two processes by which energy can be removed from the vibrating system. Friction, also internal, leads to energy being transformed from motion into heat. Radiation of sound waves is an important way in which a mechanical system may lose energy. However, energy may also be lost by the mechanical system generating movement of air in close vicinity to the system if there are damping mechanisms that affect the air motion.

1.2 THE WAVE EQUATION

We use the expression *sound waves* to characterize wave motion in gases and fluids, such as air or water. Such waves are *longitudinal*, since fluids cannot exhibit shear motion. Longitudinal waves are characterized by the oscillation of the wave motion being in the direction of the *wave propagation*. The direction of the wave propagation is determined by the geometrical and vibrational properties of the radiating mechanical system as well as by the surrounding fluid and its boundaries.

Sound waves have *spatial extension*; the period of distance until the wave repeats its pattern of vibration is called the *wavelength* of sound. The wavelength depends both on the *speed of sound* in the medium and the *frequency of oscillation*. The basic governing equations lead to a differential equation, the *wave equation*. The wave equation describes the physics of wave motion, it defines the relationship between the spatial and temporal behavior of sound. The speed depends on the physical properties of the medium, such as the gas mixture. The wave equation also describes the propagation properties of sound waves.

The wave equation is derived from using three equations that each describes a particular property of the fluid. These equations are the *equation of motion*, the *equation of continuity*, and an equation describing the *thermodynamic properties* of the medium.

In the derivation of the wave equation, we assume the gas to be *elastic* and *continuous*. The small amount of gas considered in the derivation of the wave equation may be regarded as a *volume element* large enough to contain many molecules but still small enough for the *acoustical quantities* of interest to be constant throughout the element. Of course, in reality, molecules move in and out of the element, but the assumption is valid from a statistical point of view if the element is large enough.

Acoustic waves are characterized by quantities such as:

- *Sound pressure* is the dynamic excess pressure over the atmospheric static pressure in a small volume.
- *Particle velocity* is the average velocity of the gas molecules in a small volume.

Our derivation of the wave equation requires several things:

- The medium is considered to be *at rest*.
- The sound pressure must be small enough compared to the static pressure for the medium to be considered *linear*. In practice, the maximum sound pressure is assumed to be less than 0.001 times the static pressure.
- There can be no heat exchange within the medium; that is, no heat flows into or away from the volume element. The process is said to be *adiabatic*.
- There should be no losses in the medium. The medium does not exhibit *viscosity* or other phenomena leading to damping.
- The medium is *homogenous*—the effects of gravitation are not considered.

In many practical cases, it is necessary to take both losses, movement, and inhomogeneity, into account, such as when considering wave propagation over long distances, indoors, or outdoors. *Wind* and *temperature variations* cause the wave to change direction. *Internal losses* in the medium result in attenuation of the wave.

Plane Waves

Mathematically, it is relatively easy to derive the wave equation for wave motion in a one-dimensional medium. Let us consider nonviscous wave propagation in a gas, or in a gas mixture such as air, in a narrow, rigid tube at frequencies low enough for the wave propagation in the tube to be an approximation to one-dimensional wave propagation.

The Equation of Continuity

The *equation of continuity* expresses the relationship between density changes and volume changes in a volume element. The mass of the volume is considered constant. Assume that the volume element is V_0, which is trapped between the two planes at x and $x + dx$ in a rectangular canal having a cross section area of $dydz$ as shown in Figure 1.1, that is, $V_0 = dxdydz$. The particle is assumed to be displaced for some reason.

The volume element's displacement at x is ξ, and its displacement at $x + dx$ is $\xi + (\partial\xi/\partial x)dx$. The derivative of x with respect to time, $\partial\xi/\partial t$, is called the *particle velocity* and is denoted u_x. Note that the

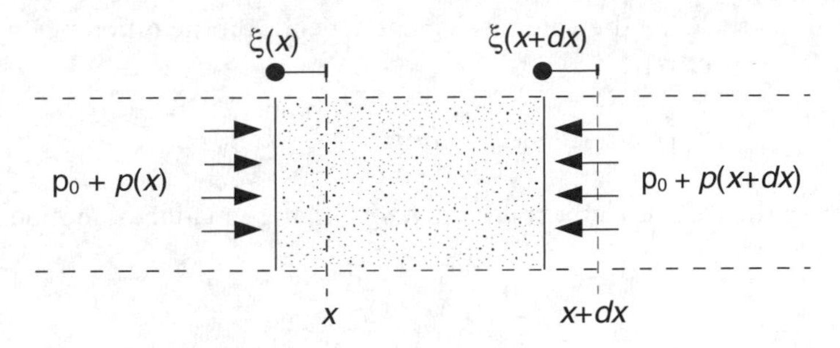

Figure 1.1 The small volume element under consideration. p_0 is the static pressure.

particle velocity is a *vectorial* quantity. The difference in displacement multiplied by the cross-sectional area *dydz* will be the change of volume ∂V, of the original volume V_0:

$$\partial V = \frac{\partial \xi}{\partial x} dx\, dy\, dz = \frac{\partial \xi}{\partial x} V_0 \tag{1.1}$$

Since the mass M_0 of the volume element will be unchanged, its density $\rho = M_0/V$ must change, leading to:

$$\frac{\partial \rho}{\partial V} = \frac{-M_0}{V^2} \quad \rightarrow \left[\frac{\partial \rho}{\partial V}\right]_{V=V_0} = \frac{-M_0}{V_0^2} \tag{1.2}$$

When we combine Equations 1.1 and 1.2, we find that the change in the element's density is inversely proportional to its relative elongation, expressed as:

$$\frac{\partial \rho}{\rho_0} = -\frac{\partial \xi}{\partial x} \tag{1.3}$$

When we calculate the derivative of this relationship with respect to time, we obtain the equation of continuity:

$$\frac{\partial \rho}{\partial t} = -\rho_0 \frac{\partial u_x}{\partial x} \tag{1.4}$$

For the three-dimensional case, we write the equation of continuity:

$$\frac{\partial \rho}{\partial t} = -\rho_0 \operatorname{div} \vec{u} \tag{1.5}$$

where $\operatorname{div} \vec{u} = \partial u_x / \partial x$ for a one-dimensional problem.

The Equation of Motion

Newton's equation of motion states that the force F needed to accelerate a mass m at an acceleration of a is $F = ma$. Of course, this applies also to the mass of the gas in the volume element under consideration in our case. Again, consider the situation shown in Figure 1.1. We denote the static pressure by p_0 and the sound pressure by p.

The force needed to accelerate the volume element depends on the difference in pressure between the planes x and $x + dx$ respectively:

$$\left[p(x) - p(x+dx)\right] dy\, dz = -\frac{\partial p}{\partial x} dx\, dy\, dz \tag{1.6}$$

Since the acceleration of the volume element is written $\partial\xi^2/\partial t^2$, the equation of motion can be written as:

$$\rho_0 \frac{\partial^2 \xi}{\partial t^2} = -\frac{\partial p}{\partial x} \tag{1.7}$$

Usually it is more practical to express the movement of the volume element by its velocity, $\partial\xi/\partial t = u_x$, instead of by its acceleration, which allows us to express the equation of motion as:

$$\rho_0 \frac{\partial u_x}{\partial t} = -\frac{\partial p}{\partial x} \tag{1.8}$$

For the three-dimensional case, it is expressed:

$$\rho_0 \frac{\partial \vec{u}}{\partial t} = -\operatorname{grad} p \tag{1.9}$$

where $\operatorname{grad} p = \partial u_x / \partial x$ for a one-dimensional problem.

Thermodynamic Properties

To analyze the dynamic properties of the gas, we also need to study the relationship between instantaneous density changes and sound pressure at a *temperature* T in the medium. If the system is adiabatic, the relationship between sound pressure and instantaneous volume is described by Poisson's equation:

$$(p_0 + p)V^{\kappa} = \text{constant} \tag{1.10}$$

Here κ is the ratio of specific heats at constant pressure and volume respectively; that is, $\kappa = C_p/C_v$. We also know that since the mass of the volume element is unchanged:

$$V(\rho_0 + \rho) = \text{constant} \tag{1.11}$$

Combining Equations 1.10 and 1.11, we obtain the partial derivatives as:

$$\frac{\partial p}{\partial V} = -\kappa \frac{p_0}{\rho_0} \quad \text{and} \quad \frac{\partial \rho}{\partial V} = -\frac{\rho_0}{V_0} \quad \text{i.e.,} \tag{1.12}$$

$$\frac{\partial p}{\partial \rho} = \kappa \frac{p_0}{\rho_0} \tag{1.13}$$

The *equation of state* for a gas is:

$$\frac{p_0}{\rho_0} = \frac{RT}{M} \tag{1.14}$$

where T is the temperature, M is the molecular weight of the gas or gas mixture, and R is the universal gas constant. Insertion of Equation 1.12 into Equation 1.11 gives the desired result:

$$\frac{\partial p}{\partial \rho} = \frac{\kappa RT}{M} \tag{1.15}$$

The Wave Equation

Combining Equations 1.4, 1.8, and 1.15, we obtain the wave equation for plane waves as expressed in sound pressure p:

$$\frac{\partial^2 p}{\partial x^2} - \frac{1}{c^2}\frac{\partial^2 p}{\partial t^2} = 0 \tag{1.16}$$

The wave equation may, of course, be expressed equally well in particle velocity, u. The quantity c depends on the thermodynamic properties of the gas:

$$c^2 = \frac{\kappa RT}{M} = \kappa\frac{p_0}{\rho_0} \tag{1.17}$$

For the three-dimensional case, the wave equation is written as:

$$\nabla^2 p - \frac{1}{c^2}\frac{\partial^2 p}{\partial t^2} = 0 \tag{1.18}$$

where $\nabla^2 p = \frac{\partial^2 p}{\partial x^2} + \frac{\partial^2 p}{\partial y^2} + \frac{\partial^2 p}{\partial z^2}$ in a Cartesian coordiante system.

1.3 SOLUTIONS TO THE WAVE EQUATION

One way of finding solutions to an equation is to try various solutions to see if they satisfy the equation. If we use this approach in investigating the possible solutions to the wave equation (see Reference 1.16), a reasonable guess to mathematically describe waves is a general pair of functions such as:

$$p(x,t) = f_1(tc - x) + f_2(tc + x) \tag{1.19}$$

where f_1 and f_2 are arbitrary functions, required to have continuous derivatives of the first and second order. Testing this solution, we find that it satisfies the wave equation. Studying the properties of the solution, we see that the shape of the functions remains unchanged for variations of the variables x and t, but they are displaced in x according to the value of t. The relationship between the time variable t and the space variable x is $x = tc$. The shape of the wave moves with the speed c, which we call the *propagation speed of sound.*

We note from Equation 1.17 that the propagation speed in a gas depends only on temperature. Under normal conditions—for example, in air—close to a temperature of 20°C, the propagation speed is:

$$c_0 = \sqrt{\kappa\frac{p_0}{\rho_0}} \approx 331.29 + 0.606 t_{°C} \quad [m/s] \tag{1.20}$$

where $t_{°C}$ is the temperature of the air expressed in degrees Celsius. At a temperature of 20°C, the speed of sound is $c_0 \approx 343.4$ m/s.

Since the shape of the wave does not change as it moves, we understand from Fourier theory that all frequency components, building up the shape of the wave, propagate at the same velocity. Sound propagation in gases is fairly independent of frequency, but frequency-dependent phenomena such as damping is discussed later. The term *nondispersive* is used to describe the fact that the propagation speed is frequency independent.

Stationary, One-dimensional Solutions for Cartesian Coordinates

In the rest of the discussion of the physical principles of sound and vibration, we will only study *continuous sinusoidal sound and vibration*. We know that according to the Fourier theorem, any physical wave shape may be obtained by suitable superposition of sines and cosines. It is practical to use the transform method called the *jω-method* to describe time variation as well as various properties of sound and media. The variables of sound pressure and particle velocity, for example, then become complex variables, which greatly simplifies the mathematics. The physical variables of instantaneous sound pressure and particle velocity are given by the real part of the complex quantities—designated by an underscore—for example $\underline{p}$, in equations. Our use of the *jω*-method is discussed in Chapter 2, and more information in general on the method can be found in Reference 1.3.

Using the *jω*-method, we can rewrite the wave equation as written in Equation 1.16 to:

$$\frac{\partial^2 \underline{p}}{\partial x^2} + k^2 \underline{p} = 0 \tag{1.21}$$

Here k is the *wave number*, sometimes called the *propagation constant*. The relationships between the wave number, the *speed* of sound *c*, *frequency f*, and *wavelength* λ, are given by $k = \omega/c = 2\pi f/c = 2\pi/\lambda$.

Using the *jω*-method, we can write, as a special case, a possible solution to the wave equation as written in Equation 1.21 to:

$$p(x, \omega, t) = \hat{p}_+ e^{j(\omega t - kx)} \tag{1.22}$$

This equation describes a one-dimensional sinusoidal wave propagating in the direction of positive *x*. The magnitude of the sound pressure $p(x,t)$ is denoted $\hat{p}_+$. Generally, we do not explicitly write the time variation when we use this transform method, instead we write $\underline{p}(x)$. The general solution to the wave equation in the one-dimensional case also has to feature a wave propagating in the direction of negative *x* and is written as:

$$p(x, k) = \hat{p}_+ e^{-j(kx + \beta_+)} + \hat{p}_- e^{j(kx + \beta_-)} \tag{1.23}$$

where β_- and β_+ are case-dependent phase constants.

Stationary, One-dimensional Solutions for Spherical Coordinates

Using the formulation of the wave Equation 1.18, one can write the general solutions for the case of *spherical symmetry* as:

$$\underline{p}(r, k) = \frac{A_+ e^{-j(kr + \beta_+)}}{r} + \frac{A_- e^{j(kr + \beta_-)}}{r} \tag{1.24}$$

where A_+ and A_- are the wave amplitudes at 1 m distance, for the outward and inward moving waves respectively, and r is the radius from the center of origin. The angles β_- and β_+ are case-dependent constants.

For the case of spherical symmetry, we are generally only interested in the wave that moves outward, in the direction of positive *r*. Using the equation of motion, we can show that the particle velocity of such an expanding wave is given by:

$$\underline{u}(r, k) = \frac{A_+}{\rho_0 c_0 r}\left(1 + \frac{1}{jkr}\right) e^{-jkr} \tag{1.25}$$

We now have an additional term $1/jkr$. This implies that there will be a phase difference between $\underline{p}$ and $\underline{u}$ that will be dependent on the value of kr; that is, on frequency and distance to origin. The phase difference will be largest when $kr \ll 1$; that is, when the frequency is low and/or the distance to the center is small.

For the *near-field region*, where $kr \ll 1$, the radial particle velocity is described by:

$$\underline{u}(r,k) \approx \frac{A_+}{j\rho_0 c_0 k r^2} e^{-jkr} \tag{1.26}$$

We see that there is a phase difference of 90° between particle velocity and sound pressure; the particle velocity lags behind the sound pressure. Another observation of importance is that the amplitude of the particle velocity u_r increases proportionally to $1/r^2$ as the distance to the origin is diminished.

For the *far-field region*, where $kr \gg 1$, the radial particle velocity is described by:

$$\underline{u}(r,k) \approx \frac{A_+}{\rho_0 c_0 r} e^{-jkr} \tag{1.27}$$

In this outer region, particle velocity and sound pressure are in-phase. The amplitude of both quantities also depends on distance in the same way in the far-field region.

To radiate an ideal spherically symmetrical wave, a radiator would have to feature a spherical surface moving at the same radial velocity at all points. To be a *point source*, such a radiator would be infinitely small. We realize that it is not physically possible to have such a radiator. However, we use the terms *point source* and *monopole* to label radiators that, from our point of observation, behave as if they were ideally small and radially radiating. One way of approximating such radiators, under certain conditions, is to use small loudspeakers.

A *loudspeaker* usually consists of a *loudspeaker box* and a *loudspeaker driver* (motor). If we have a loudspeaker box designed in such a way that only one side of the loudspeaker driver's diaphragm is facing the exterior of the box, and if the *dimensions* of the loudspeaker box, d, are much smaller than the wavelength of sound being generated—that is, $kd < 0.1$—then the sound field radiated by the loudspeaker will be a fairly good approximation to a spherical sound field at a sufficiently large distance.

It is important to realize that close to the loudspeaker the sound field will not have the characteristics of a true, spherical sound source. Figure 1.2 shows a loudspeaker box using a dodecahedron arrangement of loudspeaker drivers to approximately achieve omnidirectional characteristics. Typically, the deviations start to become large when $ka > 3$, where a is the *radius* of the dodecahedron.

One might think that it would be easy to simulate a spherical source by using a large number of small loudspeakers mounted on a sphere. However, because of the finite difference in distance between the loudspeaker diaphragms as well as the fact that these usually have a conical shape and are not a continuous part of the spherical surface, we find that the radiation pattern is not ideally spherical at frequencies where the distance between loudspeakers or the depth of the cones are larger than a small fraction of the wavelength.

A different way of studying the properties of a radiator is to study its *impulse response* in various directions. If the surface of the spherical radiator could move with infinite acceleration for a short time, the radiated sound pressure would be a *Dirac pulse*, since sound pressure is proportional to the volume acceleration of the sound source. Only if the source has the radiation properties of an ideal monopole can we have a Dirac-type sound pressure response in all directions.

Notice that only a truly spherical source can radiate having this ideal *geometrical* impulse response. The geometrical impulse response of a dodecahedron-shaped loudspeaker, for example, having ideal

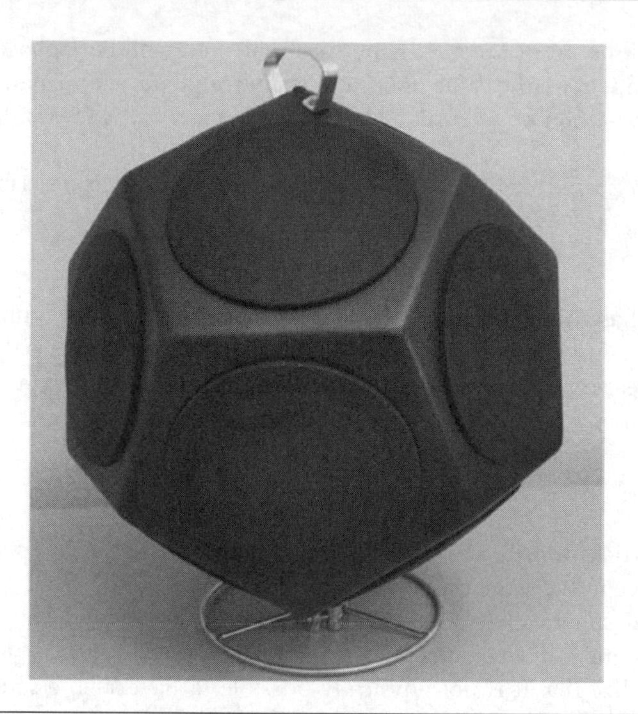

Figure 1.2 A loudspeaker using a dodecahedron arrangement of drivers to approximate the omnidirectional sound radiation characteristics of a monopole. (Photo by Mendel Kleiner.)

loudspeaker drivers—all in-phase, on all its flat surfaces—will feature impulse response contributions due to edge diffraction. Correspondingly, the frequency response of the loudspeaker will feature frequency response irregularities.

1.4 IMPEDANCE

The ratio between two interdependent quantities—such as force and linear velocity—at some point in a system is called *impedance*, $\underline{Z}$. The inverse of impedance is usually called mobility, $\underline{Y}$.

Both impedance and mobility are generally frequency-dependent, complex quantities, characterized by a real and an imaginary part, or, if we prefer, by magnitude and phase. The real part of impedance is usually called *resistance*, and its imaginary part is called *reactance*.

Definitions of Impedance Used in Acoustics

Because ratios between various sound field quantities often play an important role in problem-solving processes in acoustics, some impedances have been defined in various ways. Note that each impedance is characterized by its own unit or combination of units. A summary of the impedance relationships is shown in Table 1.1.

Characteristic impedance, $\underline{Z}_0$, is the ratio of sound pressure to particle velocity in an infinite plane wave. The unit of characteristic impedance is Rayl; that is, Ns/m^3. The characteristic impedance is usu-

Table 1.1 Summary of the impedance relationships.

Multiply by / to obtain	Z_M	Z_S	Z_A
Z_M	1	S	S^2
Z_S	1/S	1	S
Z_A	$1/S^2$	1/S	1

ally resistive if there are no propagation losses in the medium, and equal to the product of the density of the medium and the propagation speed of sound in the medium; that is, $Z_0 = \rho_0 c_0$.

Sound field impedance, $\underline{Z}_S$, is the ratio of sound pressure to particle velocity in the reference direction at a chosen point in the medium. Sometimes the term *specific acoustical impedance* is used for this quantity. The unit of sound field impedance is Ns/m^3.

Acoustical impedance, $\underline{Z}_A$, is the ratio of sound pressure to volume velocity in the normal direction to a reference surface in an acoustic system. Volume velocity is the product of the normal direction of the particle velocity and the surface area under consideration. The particle velocity is then considered constant regarding amplitude and phase over the surface. The unit of acoustical impedance is the *acoustical ohm*; that is, Ns/m^5.

Mechanical impedance, $\underline{Z}_M$, is the ratio of force to velocity at a chosen point in a mechanical system. The unit of mechanical impedance is the *mechanical ohm*; that is, Ns/m.

Characteristic and Sound Field Impedances in a Plane Wave

Using Equation 1.8, we find the sound field impedance for a plane wave propagating in the direction of positive x:

$$\frac{\underline{p}_+}{\underline{u}_{x+}} = \rho_0 c_0 = Z_0 \tag{1.28}$$

We see in this case that the sound field impedance is the same as the characteristic impedance of the medium. Obviously, the sound pressure and the particle velocity are *in-phase*.

In the same way, we obtain the sound field impedance for a plane wave propagating in the direction of negative x:

$$\frac{\underline{p}_-}{\underline{u}_{x-}} = -\rho_0 c_0 = -Z_0 \tag{1.29}$$

The minus sign in Equation 1.29 is a result of our choice of always regarding positive particle velocity as being in the direction of positive x, even for the wave that propagates in the negative x direction.

Sound Field Impedance in a Spherical Wave

The sound field impedance in a spherical wave varies according to the distance to the origin. Using Equation 1.25, we can show that the sound field impedance of a spherical, expanding wave varies with distance to the origin r as:

$$\underline{Z}_S(r,k) = \frac{\underline{p}(r,k)}{\underline{u}_r(r,k)} = Z_0 \frac{jkr}{1+jkr} = Z_0 \frac{1}{\frac{1}{jkr}+1} \tag{1.30}$$

In the *near-field* region ($kr \ll 1$), for low frequencies and/or small distances to the origin, the sound field impedance of the spherical wave is nearly pure reactance and has a *mass-type* character. That means that the reactance is positive and increases proportionally to frequency:

$$\underline{Z}_S \approx jkrZ_0 = j\frac{\omega}{c_0} r\rho_0 c_0 = j\omega\rho_0 r \quad ; \quad kr \ll 1 \tag{1.31}$$

If the sound field impedance is primarily reactance, as it is close to a small sound source, the sound field is said to be a *reactive near-field.*

In the *far-field* region ($kr \gg 1$), that is, for high frequencies and/or large distances to the origin, the sound field impedance of the spherical wave will be almost real and equal to the characteristic impedance of the medium:

$$\underline{Z}_S \approx Z_0 \quad ; \quad kr \gg 1 \tag{1.32}$$

Note that, in the far-field region, the sound field impedance of an expanding spherical wave is the same as that for an infinite plane wave.

1.5 SOUND INTENSITY

Wave propagation is characterized by transport of energy in the direction of the wave. The energy is transported at a velocity called the *group velocity* c_g. If the medium is *dispersive*, that is, the group velocity varies by frequency, the group velocity will be different from the *phase velocity* c_{ph}—that is, the velocity needed to always see the same phase in the wave. For sound waves in air, the difference between group velocity and phase velocity is usually negligible.

Sound intensity is a measure of the rate of energy transport per unit area in the wave. The sound intensity is a vectorial quantity and is directed along the direction of particle velocity:

$$\vec{I} = \frac{1}{2}\mathrm{Re}\left[\underline{p}\vec{\underline{u}}\right] \tag{1.33}$$

We know for an infinite plane wave in the positive x-direction that sound pressure and particle velocity will be in-phase, which results in:

$$I_x = \tilde{p}\tilde{u}_x = \frac{\tilde{p}^2}{Z_0} = Z_0\tilde{u}^2 \tag{1.34}$$

In this text, the ~ sign is used on top of letters to denote that the *root of the mean square value*—often called the *rms value*—is intended.

The sound intensity in a spherical wave is:

$$I_r(r) = \mathrm{Re}[\underline{Z}_S]\tilde{u}_r^2 \tag{1.35}$$

where

$$\mathrm{Re}[\underline{Z}_S] = Z_0\frac{(kr)^2}{1+(kr)^2} \tag{1.36}$$

We note that, according to Equation 1.35, the intensity in the sound field depends on the variables k and r. For a spherical sound source, the sound power radiated, for a given value of surface normal velocity u_0 and wave number k, will be much less for a small source than for a large source.

1.6 SOUND POWER

We can calculate the *sound power* being transported by a sound field by integrating the component of sound intensity normal to the surface over the surface area in question.

Consider a plane wave of intensity I [W/m^2] perpendicularly incident on a surface of area S. The intensity of the plane wave is given by Equation 1.34:

$$W = IS = \tilde{p}\tilde{u}S = \frac{\tilde{p}^2}{Z_0}S = Z_0\tilde{u}^2 S \tag{1.37}$$

Under a short time dt, the energy $Wdt = Sc_0t$ will be traveling through the surface. This energy is confined to a volume $Sdx = Sc_0dt$. Consequently, the energy density P in this volume must be:

$$P = \frac{\text{energy}}{\text{volume}} = \frac{I}{c_0} = \frac{\tilde{p}^2}{\rho_0 c_0^2} \tag{1.38}$$

Equation 1.32 shows us that since the sound field impedances for a plane wave and a spherical wave under the condition $kr \gg 1$ are nearly the same, this expression can be used to calculate the sound power in a spherical wave at significant distances and/or high frequencies.

The radial sound intensity at a distance r from the origin of a spherical wave depends on the radiated sound power W as:

$$I_r(r) = \frac{W}{4\pi r^2} \tag{1.39}$$

The intensity of sound in a spherical wave diminishes as $1/r^2$ and the sound pressure as $1/r$ as a function of distance from the point source. We call this the *distance law* for the geometrical sound attenuation of a point source.

Using Equations 1.35–1.36, we can determine the radiated power of a small spherical sound source having a radius of a:

$$W = Z_0 \frac{(ka)^2}{1+(ka)^2} \frac{\tilde{U}^2}{4\pi a^2} \tag{1.40}$$

Here, of course, the *volume velocity* is $\tilde{U} = 4\pi a^2 \tilde{u}$. That is:

$$W \approx \frac{\pi \rho_0}{c_0} f^2 \tilde{U}^2 \quad \text{if } ka \ll 1 \tag{1.41}$$

1.7 PROPAGATION LOSSES

During the propagation of sound away from a small sound source, the amplitude and intensity of the wave are reduced not only because power is spread over a larger area—*geometrical attenuation*—but also because of various loss mechanisms that turn the sound energy in the sound wave into heat. These losses are known as excess absorption. The excess absorption is usually accounted for by simply introducing a damping term in the expression for the intensity as a function of propagation distance x:

$$I(x) \propto e^{-mx} \tag{1.42}$$

that is:

$$\tilde{p}(x) \propto e^{-mx/2} \tag{1.43}$$

where m is the attenuation coefficient.

The losses in sound propagation in air are due to *heat conduction* (not perfectly adiabatic processes), *viscous losses*, and *relaxation* phenomena. The relaxation phenomena are responsible for the major part of the losses at normal conditions. The attenuation depends both on *temperature* and *relative humidity*.

The attenuation due to heat conduction and viscous losses increases proportionally to frequency squared. The relaxation causes an attenuation that peaks at certain frequencies. The relaxation process involves excitation of the gas molecules to other energy states, for example, from translational energy to vibrational/rotational energy. The relaxation time depends on the time that the molecule remains in the excited state. Maximum attenuation will occur when the period of oscillation in the sound wave is approximately the same as the relaxation time.

Figure 1.3 shows curves of the frequency-dependent attenuation of sound propagation in air at a temperature of 20°C for some values of the relative humidity (see also Figure 4.9). Using these curves, it is possible to calculate corresponding values for the attenuation coefficient m. Notice that the gas mixture of air involves several relaxation processes.

The standard ISO 9613-1:1993 (E) can be used to calculate the attenuation of sound in air for general combinations of temperature and relative humidity.

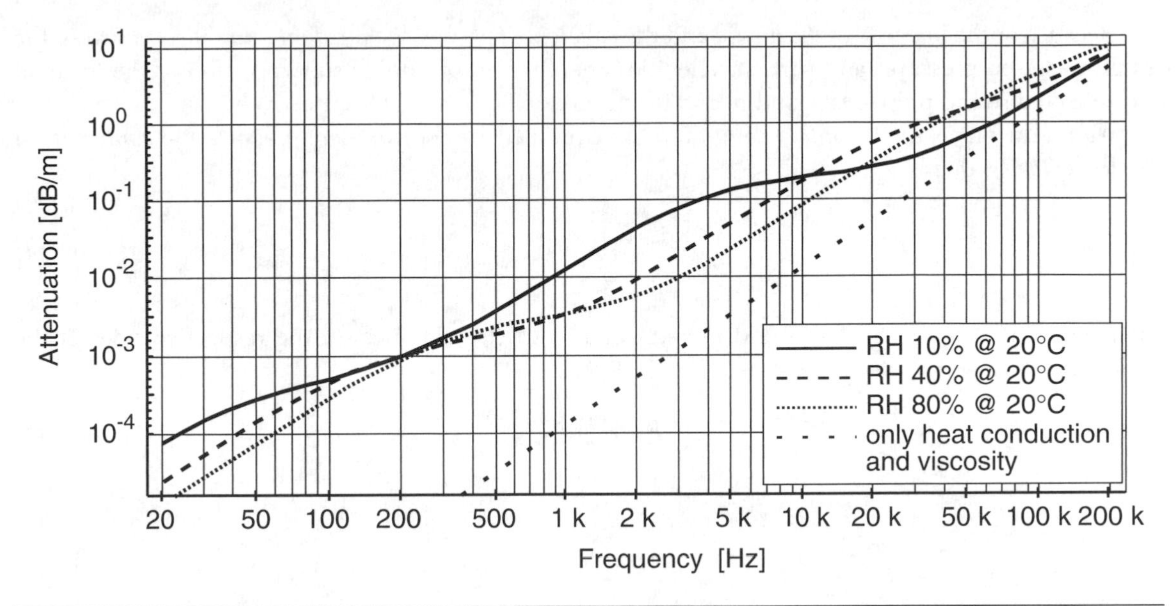

Figure 1.3 Attenuation in dB/m for sound propagation in air as a function of frequency at a temperature of 20°C and with relative humidity, RH, as a parameter. (After Reference 1.2) The graph can be used to calculate the attenuation coefficient m since the attenuation L in dB over a distance x is $L \approx 4.3\,mx$.

1.8 REFLECTION AND TRANSMISSION AT BOUNDARIES

When a sound wave is incident on the boundary between two media, or on a boundary where the characteristic impedance of a medium changes, part of the power in the incident wave may be reflected and some transmitted. The ratios of reflected and transmitted powers to that incident are determined by the characteristic impedances Z_{01} and Z_{02} at each side of the boundary as well as on the angles of incidence and transmission relative to the boundary.

If the boundary is not plane, but has a surface structure that has random variations in height, a major part of the reflection and transmission will be diffuse if the rms value of the height variations are of a size similar to the wavelength of the incident sound. By using special surface structures, one can obtain nearly perfect diffuse reflection (see Section 7.4). We will study the case of plane boundaries in this chapter.

Because of the *interference* between the incident sound wave and the reflected sound wave, note the *standing-wave phenomena* as the waves add up, in- or out-of-phase.

Normal Sound Incidence

Assume a plane wave incident on the plane impedance boundary at $x = 0$. Also, assume the angle of incidence to be 0° (*normal incidence*). The case of 90° incidence is called *grazing incidence*.

Because of the *linearity of the media* and the *coherence between the incident and reflected waves,* the resulting sound pressures and particle velocities are obtained by simple superposition of the incident and reflected sound pressures $\underline{p}_i$ and $\underline{p}_r$ and particle velocities $\underline{u}$ and $\underline{u}_r$, respectively.

Along the plane, the *boundary conditions* to be fulfilled are *continuity of pressure* and *continuity of particle velocity,* that is:

$$\left.\begin{aligned} \underline{p}_1 &= \underline{p}_2 \\ \underline{u}_1 &= \underline{u}_2 \end{aligned}\right\} \quad \text{at } x = 0 \tag{1.44}$$

where $\underline{p}_1$, $\underline{p}_2$ and $\underline{u}_1$, $\underline{u}_2$ are the sound pressures and particle velocities on the respective sides of the boundary, $x > 0$ and $x < 0$:

$$\underline{p}_1 = \left(\underline{p}_i + \underline{p}_r\right) \tag{1.45}$$

where

$$\underline{p}_i = \hat{p}_i e^{-jk_1 x} \tag{1.46}$$

$$\underline{p}_r = \hat{p}_r e^{jk_1 x} \tag{1.47}$$

$$\underline{p}_2 = \underline{p}_t = \hat{p}_t e^{-jk_2 x} \tag{1.48}$$

Because of the definition of sound field impedance, we find:

$$\underline{u}_1 = \left(\underline{u}_i + \underline{u}_r\right) = \frac{\underline{p}_i - \underline{p}_r}{\underline{Z}_{01}} \tag{1.49}$$

$$\underline{u}_2 = \underline{u}_t = \frac{\underline{p}_t}{\underline{Z}_{02}} \tag{1.50}$$

where index t means transmitted wave. If $\underline{u}_1$, $\underline{u}_2$, $\underline{p}_1$, and $\underline{p}_2$ are eliminated from these equations, we further find:

$$\underline{Z}_{01} \frac{\underline{p}_i + \underline{p}_r}{\underline{p}_i - \underline{p}_r} = \underline{Z}_{02} \qquad \text{at } x = 0 \tag{1.51}$$

The relationships between the sound pressures, in the respective waves, are usually expressed by a *pressure transmission coefficient* $\underline{t}$ and a *pressure reflection coefficient* $\underline{r}$ as:

$$\underline{t} = \frac{\underline{p}_t}{\underline{p}_i} = \frac{2\underline{Z}_{02}}{\underline{Z}_{02} + \underline{Z}_{01}} \tag{1.52}$$

$$\underline{r} = \frac{\underline{p}_r}{\underline{p}_i} = \frac{\underline{Z}_{02} - \underline{Z}_{01}}{\underline{Z}_{02} + \underline{Z}_{01}} \tag{1.53}$$

where index i means incident wave. These transmission and reflection coefficients are usually complex quantities. It is only when the incident wave is plane and the boundary plane is infinite that the reflection and transmission can be formulated in this simple way.

Associated with these reflection coefficients, we also define the *intensity transmission coefficient* τ and the *intensity reflection coefficient* γ:

$$\tau = \frac{I_t}{I_i} \tag{1.54}$$

$$\gamma = \frac{I_r}{I_i} \tag{1.55}$$

If there are no losses at the boundaries, the incident, reflected, and transmitted sound powers (W_i, W_r, and W_t) must be related as:

$$W_i = W_r + W_t \tag{1.56}$$

that is:

$$\gamma + \tau = 1 \tag{1.57}$$

From the side of wave incidence, it seems as if the energy of the transmitted wave is absorbed. One usually calls the quantity $(1 - \gamma)$ the sound-absorption coefficient, α; that is:

$$\alpha = 1 - |\underline{r}|^2 \tag{1.58}$$

Quasi-plane waves can be generated in long, straight tubes if the tube walls are hard and the tubes are wide enough, so that the influence of the viscosity of the medium can be neglected. The tubes must not be so wide as to allow crosswise sound propagation. Such tubes can be used as acoustical components to build various types of acoustical circuits, as described in Section 1.7.

Oblique Sound Incidence

Oblique incidence means that the angle of incidence of the sound is larger than zero, relative to the normal of the boundary. At oblique incidence, the reflection and transmission coefficients will be dependent on the angles of incidence and transmission. If the phase velocities in the two media are different, the transmitted wave will be *refracted.* Two cases will be studied: *normal reaction* and *Rayleigh type of reflection.*

Normal reaction (sometimes called *local reaction*) means that the sound wave transmitted into medium 2 propagates at right angles to the boundary. Normal reaction is accomplished when the sound—for example, due to a guiding structure—is forced to propagate in a particular direction. This is similar to when sound is transmitted into a porous absorber.

Rayleigh type of reflection (also known as *extended reaction*) means that the sound wave transmitted into medium 2 propagates at an angle to the boundary, determined by Snell's law (see Equation 1.6). Extended reaction typically occurs at boundaries between media in which free wave propagation is possible, for example, a large surface between water and air.

We will now study the conditions for a plane wave incident at an oblique angle as shown in Figure 1.4. Besides the requirement for continuity of sound pressure and particle velocity, according to Equation 1.44, it is also necessary for the three sound waves on either side of the boundary to propagate at the same phase velocity along the boundary. This means that the angle of reflection must be the same as the angle of incidence.

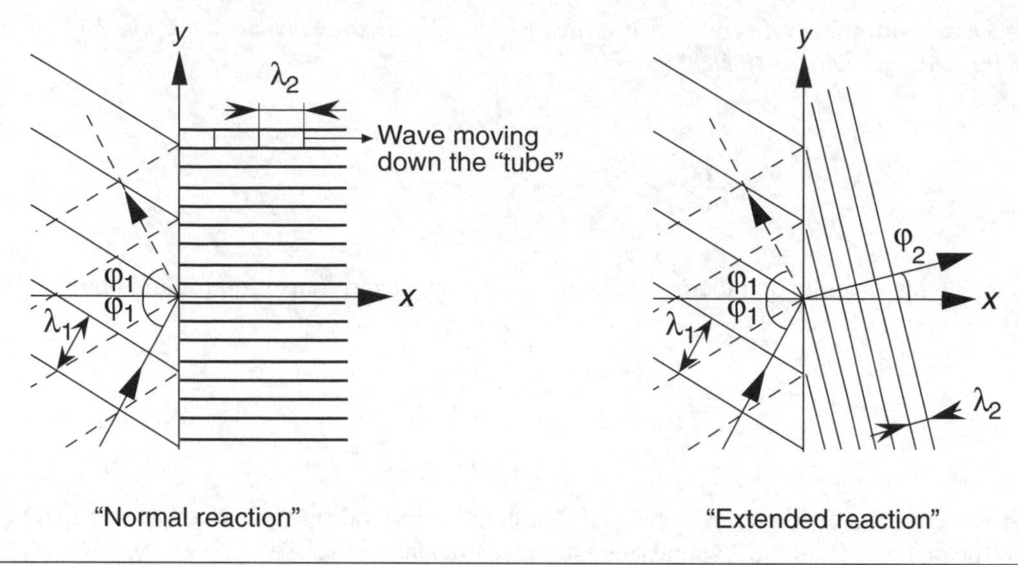

Figure 1.4 Oblique sound incidence on a plane impedance boundary.

The wave vector components in the y-direction must be the same for all three waves, since the wave numbers are based on the frequency of sound and the phase velocity. The wave number, which we have studied so far, is just the magnitude of the wave vector. In the present case, it is practical to regard the wave numbers for waves traveling at oblique angles as components of wave vectors in the x- and y-directions.

The angle of reflection is given by the equality of the wave vector components on either side in the y-direction, $k_{1y} = k_{2y}$; that is:

$$k_1 \sin(\varphi_1) = k_2 \sin(\varphi_2) \tag{1.59}$$

It can be reformulated to:

$$\frac{c_1}{\sin(\varphi_1)} = \frac{c_2}{\sin(\varphi_2)} \tag{1.60}$$

where c_1 and c_2 are the speed of sound in the two regions. This condition is known as *Snell's law*.

Normal Reaction

The obliquely incident sound wave can be formulated as:

$$\underline{p}_1 = \hat{p}_1 e^{-jk_1(x\cos(\varphi_1)+y\sin(\varphi_1))} \tag{1.61}$$

Since the reflection leads to the wave vector component in the x-direction of the reflected wave changing sign, one can write the resulting sound field in medium 1, using superposition, as:

$$\underline{p}_1 = \underline{p}_i + \underline{p}_r = \hat{p}_i e^{-jk_{1y}y}\left(e^{-jk_{1x}x} + \underline{r}e^{jk_{1x}x}\right) \tag{1.62}$$

where

$$k_{1x} = k_1 \cos(\varphi_1)$$
$$k_{1y} = k_1 \sin(\varphi_1) \quad (1.63)$$

The particle velocity can now be obtained using the definition of sound field impedance for a wave moving in the negative x-direction (see Equation 1.29):

$$\underline{u}_{1x} = \cos(\varphi_1)\frac{\hat{p}_i}{\underline{Z}_{01}} e^{-jk_{1y}y}\left(e^{-jk_{1x}x} - \underline{r}e^{jk_{1x}x}\right) \quad (1.64)$$

Let us now study the conditions on the side of medium 1. Remember that in the case of normal reaction studied now, the particle velocity in medium 2 is only in the x-direction; there is no propagation in the y-direction. The local complex sound field impedance on the side of medium 2 at the boundary $x = 0$ is written as:

$$\underline{Z}_{02} = \frac{\underline{p}_2}{\underline{u}_2} \quad (1.65)$$

Consequently, the boundary conditions for the particle velocity will be fulfilled if:

$$\frac{\underline{p}_1(0)}{\underline{u}_{1x}(0)} = \underline{Z}_{02} \quad (1.66)$$

This results in the pressure reflection coefficient $\underline{r}$:

$$\underline{r} = \frac{\underline{Z}_{02}\cos(\varphi_1) - \underline{Z}_{01}}{\underline{Z}_{02}\cos(\varphi_1) + \underline{Z}_{01}} \quad (1.67)$$

One notes that the reflection coefficient is dependent not only on the characteristic impedances of the two media but also on the angle of incidence (see Section 1.8).

Rayleigh Type of Reflection

In this case, the boundary condition with regard to the propagation direction is given by Snell's law. Of course, the regular requirements for sound pressure and particle velocity continuity across the boundary still apply.

Utilizing the same approach previously used, we obtain the reflection coefficient in the case of extended reaction as:

$$\underline{r} = \frac{\underline{Z}_{02}\cos(\varphi_1) - \underline{Z}_{01}\cos(\varphi_2)}{\underline{Z}_{02}\cos(\varphi_1) + \underline{Z}_{01}\cos(\varphi_2)} \quad (1.68)$$

The propagation velocities of sound in the two media influence the pressure reflection and transmission coefficients as given by Snell's law.

Note that:

$$\cos(\varphi_2) = \sqrt{1 - \left(\frac{c_2}{c_1}\sin(\varphi_1)\right)^2} \quad (1.69)$$

In some particularly interesting acoustical cases (for example, in some porous sound absorbers), the speed of sound in medium 2 is considerably smaller than in medium 1; that is, $c_2 \ll c_1$.

This means that $\cos(\varphi_2) \approx 1$, that is, $\varphi_2 \approx 0$ and, on the whole, independent of the angle of incidence φ_1. The propagation in medium 2 is then approximately equal to the case of normal reaction and the propagation in the y-direction is neglected.

When the angle of incidence φ_1 becomes close to 90°—grazing incidence—the pressure reflection coefficient $\underline{r}$ will become −1 if medium 2 has a finite characteristic impedance. This means that all energy will be reflected, and that the phase of the reflected wave will be inverted. In many situations, this causes the phase inverted wave canceling sound that is not reflected, leading to substantial cancellation of the sound pressure close to the sound-absorbing surface. Two examples of this cancellation at shallow angles of incidence are when sound from vehicles propagates nearly parallel to grassy ground and when sound from an orchestra propagates almost parallel to the audience seating area in an auditorium.

1.9 ACOUSTICAL COMPONENTS AND CIRCUITS

In many technical applications, one needs acoustic frequency response functions, or impedances, which vary in a predetermined way. Such applications can be found in acoustic measurement and excitation systems, and in transducer design, such as in the design of loudspeakers and microphones.

To obtain the desired response or impedance, it is practical to use electroacoustical analogies where the desired response is obtained by combination of discrete *acoustical* inductors, capacitors, and resistors. The circuits can then be analyzed either from the electrical or acoustical viewpoint (see Reference 1.1).

In this section, we discuss how, as long as the component size is small compared to the acoustic wavelength, this approach can be used to obtain acoustical equivalents of such electrical components. It is much easier, of course, to build discrete electrical components than acoustical components, because the speed of electromagnetic waves is approximately one million times greater than that of sound in air, making the wavelengths about one million times longer.

For the design of acoustical components to be practical, by the simple theory shown here, it is necessary to assume plane waves of sound. As discussed previously, it is possible to obtain approximations to plane waves in tubes over a certain frequency range. The frequency range is limited upward due to the possibility of oblique waves in the tubes at high frequencies. We will study hard-walled tubes.

In tubes having a rectangular cross section, the maximum distance between two parallel sides must be smaller than 0.5 λ. For tubes with a circular cross section, the diameter must be smaller than approximately 0.59 λ. In practice, it is wise to choose maximum cross section values less than 0.5 λ. The frequency range is limited downward because of the viscous layer at the tube walls. The layer becomes more important at low frequencies, and it prevents the particle velocity from being constant over the tube cross section. A commonly used limit is that the minimum dimension should be larger than 0.003 λ.

It is best to study the components using the idea of acoustical impedance, since the components can then be interconnected in the acoustical circuit as if they were electrical components. At the junctions, there will be continuity of pressure and volume velocity. In the acoustical circuits, the volume velocity is analogous to electrical current and pressure to electrical voltage. The pressure drop in an acoustical circuit is equivalent to the voltage drop in an electrical circuit.

A sound wave traveling down an infinite tube, with cross-sectional area S, sees a real acoustical impedance (*acoustical resistance*), which is $\rho c/S$. Obviously, making acoustical resistors by using infinite tubes is impractical and other ways have to be used. The design of pure acoustical resistance is difficult, however, and not studied here. In practice, one uses fine meshes or open-pore foam plastics and sintered materials (see Reference 1.1). In such acoustic resistors, one uses viscous losses to obtain the resistance.

If the tube is finite, the wave propagating in the tube will see changes to the acoustical impedance as the cross-sectional area is changed or if the tube end is closed or open. This will result in reflection of the sound wave and, consequently, standing waves in the tube. The local acoustical impedance corresponding to the sound field impedance discussed previously will then depend on the position in the tube.

By using abruptly close- and open-ended tubes, one can obtain *acoustical compliance components* and *acoustical mass components*, *acoustic inductors*, and *acoustic capacitors*. The compliance components will have impedance—*reactance*—proportional to $1/j\omega$, and the mass components will have impedances—*reactance*—proportional to $j\omega$.

Acoustical Compliance Components

Figure 1.5 shows a cut through a cylindrical tube, having a cross section area $S = \pi b^2$ and maximum and minimum dimensions corresponding to the limits mentioned previously.

Assume a plane wave propagating in the positive x-direction:

$$\underline{p}(x,k) = \hat{p}_+ e^{-jkx} \tag{1.70}$$

If the tube is truncated by a hard wall at $x = 0$, the particle velocity perpendicular to the wall must be zero, and the incident wave will be fully reflected by the discontinuity. The reflection coefficient will be 1 because of the hard wall, and the reflected wave is subject to the same impedance as the incoming wave (except for sign). Thus, the pressure amplitudes of the incoming and reflected waves must be equal.

Using superposition, the sound pressure in the tube at negative x coordinates can be written:

$$\underline{p}(x,k) = \hat{p}_+ \left(e^{-jkx} + e^{jkx}\right) \tag{1.71}$$

The particle velocity can similarly be written as:

$$\underline{u}_x(x,k) = \frac{\hat{p}_+}{Z_0}\left(e^{-jkx} - e^{jkx}\right) \tag{1.72}$$

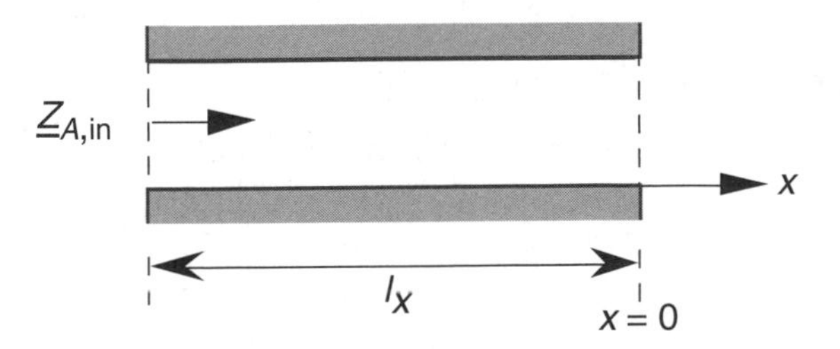

Figure 1.5 A short tube, open at both ends, to be used as an acoustical component.

By definition, volume velocity is the product of particle velocity by cross section area, $\underline{U} = \underline{u}S$. The acoustical impedance seen by the wave for $x < 0$ can then be written as:

$$\underline{Z}_A(x,k) = \frac{\underline{p}(x,k)}{S\underline{u}_x(x,k)} = \frac{Z_0}{S}\frac{e^{-jkx}+e^{jkx}}{e^{-jkx}-e^{jkx}} \tag{1.73}$$

It can also be written as:

$$\underline{Z}_A(x,k) = -\frac{Z_0}{jS}\cot(kx) = \frac{jZ_0}{S}\cot(kx) \tag{1.74}$$

This means that a wave that propagates along a tube, closed at the far end, at $x = -l_x$ sees an acoustical impedance, that is:

$$\underline{Z}_A(k)\Big|_{x=-l_x} = -j\frac{Z_0}{S}\cot(kl_x) \tag{1.75}$$

One notes that the acoustical impedance is *fully reactive* (no real part) so the sound pressure and particle velocity are 90° out-of-phase. Depending on the frequency of the sound and on the length of the tube l_x, the impedance will be either positive or negative.

We say that the tube is resonant for frequencies where its acoustical input impedance is zero. Resonance requires $\cot(kl_x) = 0$, which gives the conditions for resonance as $l_x = \lambda/4$, $3\lambda/4$, and so forth.

For the tube entrance to act as a discrete acoustical component, the tube length l_x must be short. A commonly used limit is $l_x < \lambda/8$—that is, $kl_x \ll 1$—that corresponds to a tube that is short compared to the wavelength. We can then write:

$$\underline{Z}_A(\omega)\Big|_{x=-l_x} \approx \frac{1}{j\omega C_A} \tag{1.76}$$

where C_A, the *acoustical compliance* (capacitance), is:

$$C_A = \frac{V}{\rho_0 c_0^2} \tag{1.77}$$

when $kl_x \ll 1$, $kb \ll 1$, and where $V = Sl_x$.

As long as the tube diameter and tube length are small compared to the wavelength of sound, it usually does not matter what the exact shape is; it is the confined volume V that is important. This property of C_A is used when designing loudspeaker boxes because the box can be given many shapes, as long as its dimensions are small compared to the maximum wavelength of sound.

Acoustical Mass Components

Again, consider the tube discussed previously and shown in Figure 1.5. Assume an incoming wave that can be formulated as in Equation 1.70.

Now assume the far end, at $x = 0$, to be open. The boundary condition will now be that there is pressure release; that is, the pressure at the far end is $p = 0$. This is not strictly correct since there is some reaction from the air outside the tube, but this is neglected here since we have assumed that $kb \ll 1$.

The reflection coefficient will, in this case, be −1, which is a result of a phase change in the reflected wave by 180°.

The sound pressure in the resulting sound field in the tube can then, in analogy to Equation 1.71, be written as:

$$\underline{p}_x(x,k) = \hat{p}_+\left(e^{-jkx} - e^{jkx}\right) \tag{1.78}$$

Correspondingly, the particle velocity can be written:

$$\underline{u}_x(x,k) = \frac{\hat{p}_+}{Z_0}\left(e^{-jkx} + e^{jkx}\right) \tag{1.79}$$

The acoustical impedance seen by the incoming wave at the tube entrance will then be:

$$\underline{Z}_A(x,k) = \frac{\underline{p}(x,k)}{S\underline{u}_x(x,k)} = \frac{Z_0}{S}\frac{e^{-jkx} - e^{jkx}}{e^{-jkx} + e^{jkx}} \tag{1.80}$$

In turn, it can be rewritten as:

$$\underline{Z}_A(x,k) = -\frac{jZ_0}{S}\tan(kx) \tag{1.81}$$

The acoustical input impedance Z_A for a narrow tube of length l_x can thus be written:

$$\underline{Z}_A(\omega)\Big|_{x=-l_x} = \frac{jZ_0}{S}\tan(kl_x) \tag{1.82}$$

As previously stated, we say that the tube is resonant for frequencies where its acoustical input impedance is zero. The impedance is zero when $\tan(kl_x) = 0$; that is, when the tube length is such that $l_x = n\lambda/2$ where $n = 1, 2, 3 \ldots$

Equation 1.82 can be simplified for the special case of narrow and short tubes. Using a series expansion of $\tan(kl_x)$, we can write the acoustical input impedance Z_A as:

$$\underline{Z}_A(\omega)\Big|_{x=-l_x} \approx \frac{jZ_0kl_x}{S} = j\omega\frac{\rho_0 l_x}{S} = j\omega\frac{\rho_0 l_x S}{S^2} \tag{1.83}$$

Since the magnitude of the impedance increases proportionally to frequency, and since the impedance has a positive reactance, it is called an acoustical mass impedance or acoustical inductance. The acoustical mass impedance M_A is:

$$M_A = \frac{\rho_0 l_x}{S} \tag{1.84}$$

So the acoustical input impedance at the entrance of the tube is:

$$\underline{Z}_A(\omega)\Big|_{x=-l_x} \approx j\omega M_A \tag{1.85}$$

when $kl_x \ll 1$ and $kb \ll 1$.

Since the air outside the far end of the tube is moved by the sound field, there must be some energy radiated. The energy radiation becomes smaller as the tube diameter is reduced. The air at the far end also behaves as an extra added mass, virtually extending the tube slightly. Both of these effects can be taken into account, approximately, by including the radiation impedance at the end of the tube in the calculation instead of putting the value of the impedance at the tube far end to zero when calculating the reflection coefficient. The acoustical mass component is usually not as *ideal* as the acoustical compliance component.

The extra length is handled by using an *end correction* when calculating the effective length of the tube. The end correction is primarily important when dealing with short tubes such as small holes in thin plates and sheets. One can show that the end correction for a circular opening, having a radius b, is approximately $0.78b$ long, resulting in an added acoustical mass of:

$$M_{A,\text{end corr}} \approx \frac{\rho_0}{4b} \tag{1.86}$$

If the circular tube is a hole through a plate, the acoustical mass of the tube (open at both ends) can be written as approximately:

$$M_{A,\text{end corr}} \approx \frac{\rho_0 l_x}{S} + \frac{\rho_0}{2b} \tag{1.87}$$

As in the case of the acoustical compliance, it is primarily the volume of the enclosed air that is important. The exact shape of the tube is unimportant, as long as its maximum cross-sectional dimensions are much smaller than a wavelength.

Acoustical Circuits

If one attaches two acoustical impedance elements to one another in a series as shown in Figure 1.6, the volume velocity $\underline{U}$ in both elements will be the same, and the total pressure over the added components will be $\underline{p} = \underline{p}_1 + \underline{p}_2$.

The total impedance $\underline{Z}_A$, obtained when series coupling acoustical impedance components, is written:

$$\underline{Z}_A = \sum_i \underline{Z}_{Ai} \tag{1.88}$$

When attaching two acoustical impedance components in parallel, as in Figure 1.7, the resulting impedance will be:

$$\frac{1}{\underline{Z}_A} = \sum_i \frac{1}{\underline{Z}_{Ai}} \tag{1.89}$$

The pressure drop will be the same over each element, but the volume velocity in each component will depend on the impedances of both components. It is also possible to introduce acoustic transformers and sources, for example, (see Reference 1.1).

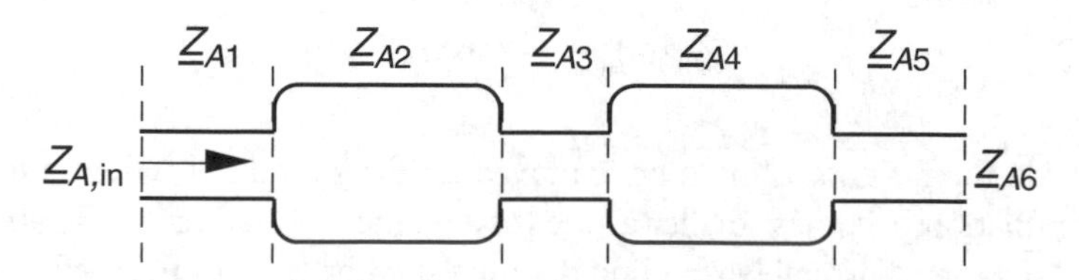

Figure 1.6 Some acoustical impedance elements in series.

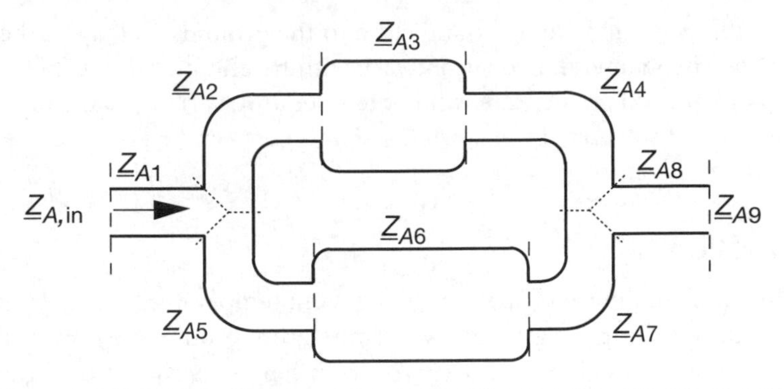

Figure 1.7 Acoustical impedance elements in parallel.

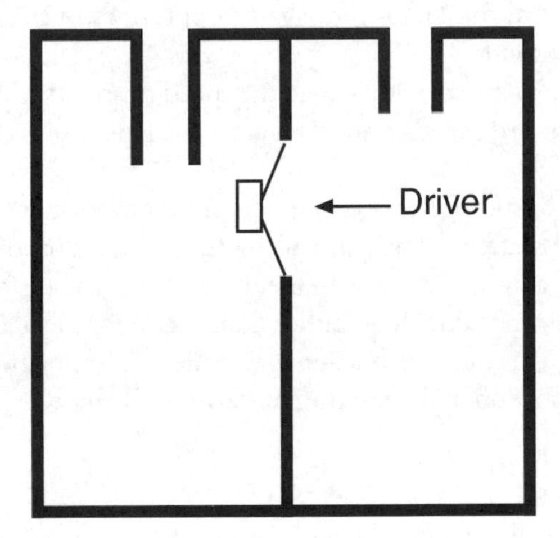

Figure 1.8 A bandpass loudspeaker box design often used for sub-bass loudspeakers and that has four acoustical impedance elements to obtain the desired acoustical filter characteristics.

Figure 1.8 shows a loudspeaker box design often used for sub-bass loudspeakers. It uses two acoustically resonant circuits on either side of the loudspeaker driver to improve the sound radiation from the driver. The calculation of the resonance frequencies uses the previously discussed compliance and mass components.

1.10 SOUND PROPAGATION IN INHOMOGENEOUS MEDIA

In many practical cases sound propagation in inhomogeneous media is characterized by varying temperature, density, and composition. Such problems are found when dealing with sound propagation outdoors and under water. Even sound propagation over long distances in rooms can be affected by air currents and temperature variations.

Outdoors most of the variation will be found close to the ground. As long as the air is at rest (no wind or turbulence) and the temperature does not vary with height, high-frequency sound will propagate linearly, like rays of light. When there are wind or temperature variations, the ray paths will become nonlinear. For some types of weather (inversion), the sound rays can be effectively reflected by a layer of air having a positive temperature gradient.

Temperature Variation

As shown by Equation 1.17, the speed of sound will vary with temperature. Usually the air temperature will vary with height but not along the ground. Two temperature profiles are shown in Figure 1.9.

Normally, the temperature of air at rest decreases with height as shown in Figure 1.9b. One calls the case of negative temperature gradient of −0.01°C/m the *normal temperature profile*. In the case of negative temperature gradients, the speed of sound also decreases with height, resulting in ray paths that bend upward. This is typical for sound propagation outdoors over grassy land on sunny summer afternoons, resulting in *sound shadow*.

When instead, temperature increases with height, the speed of sound will also increase with height, and the ray paths will be bent downward as indicated in Figure 1.9a. This case is called *inversion* and is typical of clear winter nights and mornings.

Sound shadow and the acoustic effects of inversion can be important for the resulting sound level at a far distance from the sound source. When measuring the attenuation of sound at low frequencies by distance outdoors, one typically finds that measurement results for the distance law show a level decrease of approximately −7 dB per distance doubling, rather than the theoretical decrease of −6 dB per distance doubling for free propagation. This difference is due to turbulence. In practice, there will also be additional interference due to ground reflection and scattering by various objects.

Wind

The (lateral) wind velocity u will usually vary with height due to air viscosity, ground roughness, and vegetation. Typically, the velocity varies with height z over the ground as approximately $u(z) \sim ln(z)$.

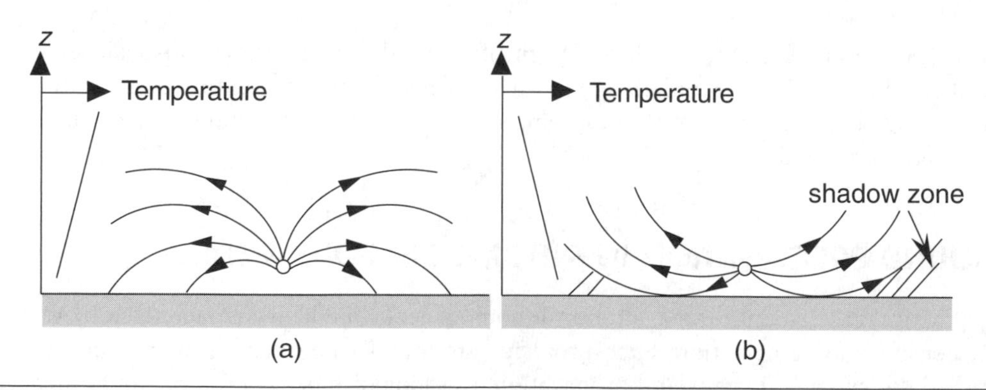

Figure 1.9 Sound propagation paths in the presence of temperature gradients above ground.

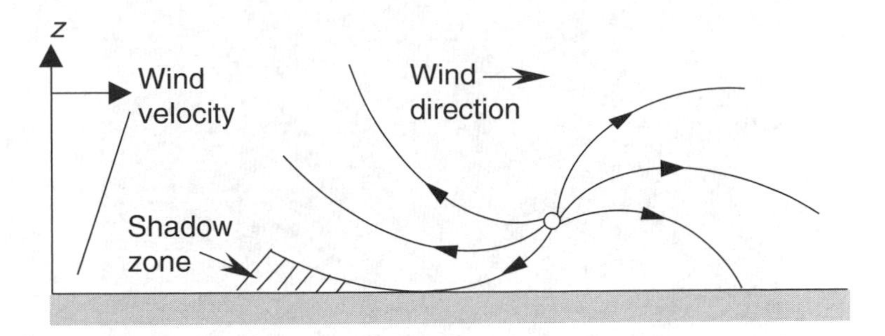

Figure 1.10 Influence of wind on the sound propagation paths.

In the case of tailwind, the sound ray paths will bend upward, whereas, in the case of headwind, they will bend downward as shown in Figure 1.10. The influence of wind and temperature gradients must be added when calculating the ray paths. Note that shadow zones also can occur in the case of tailwind, also shown in Figure 1.10.

One must also remember that the wind seldom is homogeneous—wind creates turbulence, resulting in vortices that scatter sound. The sky is *acoustically blue* because of the scattering of sound by such turbulence. In practice, the sound shadow in the case of negative temperature gradient seldom results in a level decrease of more than 20 dB, because of the scattering of sound from higher air layers due to turbulence.

The turbulence also introduces amplitude and frequency modulation to the sound, making it garbled and difficult to understand. It is important to consider this effect in the design of speech and music sound reinforcement systems to be used outdoors over long distance.

Radius of Curvature

One can show that the resulting ray paths, due to wind and temperature gradients, are approximately circular, having a certain radius of curvature. This applies also when both wind and temperature gradients are present at the same time. One can show that the radius of curvature for sound propagating close to the ground is given by:

$$r = \frac{20\sqrt{T}}{\frac{10}{\sqrt{T}}\frac{\partial T}{\partial z} + \frac{\partial u}{\partial z}} \tag{1.90}$$

In this formula, the speed of sound has been approximated by $c \approx 20\sqrt{T}$, and the wind speed has been assumed to be small compared to the speed of sound.

Figure 1.11 shows a high-intensity, high-directivity voice warning system used in a civil-defense application. Temperature and wind gradient effects can be quite noticeable in these applications, such as in the use of similar loudspeaker array technology for outdoor concert applications.

Figure 1.11 A high-directivity voice warning system using an array of toroidal pattern loudspeakers in a civil-defense application. Temperature and wind gradient effects can be noticeable in such applications involving communication outdoors over long distances. (Photo by Mendel Kleiner.)

1.11 DIPOLES AND QUADRUPOLES

A *dipole* is an elementary source that can be thought of as consisting of two point sources, monopoles, at a small distance b from one another (see Figure 1.12), oscillating at the same frequency but 180° out-of-phase. In the figure this property is indicated by plus and minus signs.

The sound pressure at point A at a distance r will be superposition of the pressures from each monopole, since the sources are assumed not to influence each other's radiation. Because of both positive and negative interference, the total sound pressure will vary, not only with distance but also with angle as in Fig. 1.12.

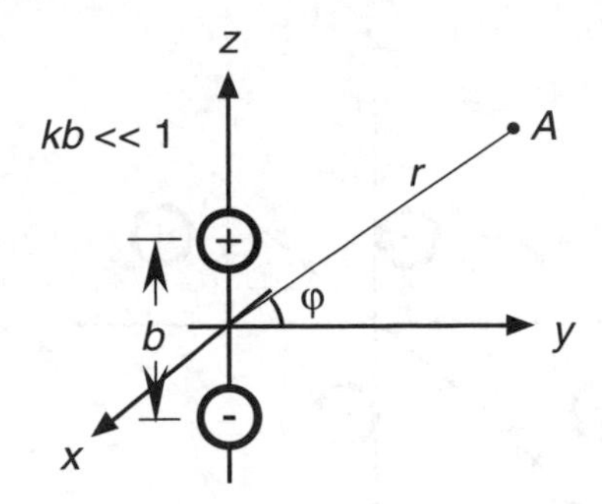

Figure 1.12 The coordinate system used in the analysis of the dipole. The two monopole sources (+/–) are on the *z* axis.

Assume that the sources are much closer than a wavelength and sum up the sound pressures. One then finds that the ratio between the sound pressure at point A from just the *plus* monopole, $\underline{p}_m$, and the sound pressure from the dipole, $\underline{p}_d$, can be written as:

$$\frac{\underline{p}_d}{\underline{p}_m} = \left(jkb + \frac{b}{r} \right) \sin(\varphi) \tag{1.91}$$

The sound power radiated by the dipole will be different from that radiated by a single monopole. Assume that each source acting as a free monopole, by itself, could radiate a power W_m. Calculating the sound power radiated by the dipole, one finds that the ratio between W_d and W_m is:

$$\frac{W_d}{W_m} = \frac{(kb)^2}{3} \tag{1.92}$$

where the *radiated power by a monopole* is:

$$W_m = \frac{\pi \rho_0}{c_0} f^2 \tilde{U}^2 \tag{1.93}$$

As noted, the dipole radiates much less energy than a monopole. The dipole cancellation is an important asset in many cases of noise control engineering. The cancellation is sometimes called *aerodynamic short-circuit* or *aerodynamic cancellation*. Some interesting situations when this phenomenon occurs are:

- The reduction of low frequency response when using a free suspended loudspeaker instead of using the same loudspeaker so that its rear side radiates into a closed box
- The reduced sound radiation from perforated metal sheets compared to solid metal sheets
- The weak sound radiation by small oscillating bodies, such as oscillating machine parts and rods
- Some cases of active noise control by interference

Quadrupoles are somewhat more complicated elementary sources than dipoles. One type of quadrupole is shown in Figure 1.13. This particular quadrupole consists of four monopoles, two in-phase and two out-of-phase at distances b_x and b_y in the $z = 0$ plane.

As in the case of the dipole, one can show that the effective sound power radiation for a quadrupole, W_q, is much smaller than for either of the free sources separately, W_m. One can show that the ratio W_q/W_m becomes:

$$\frac{W_q}{W_m} = \frac{(kb_x)^2 (kb_y)^2}{15} \tag{1.94}$$

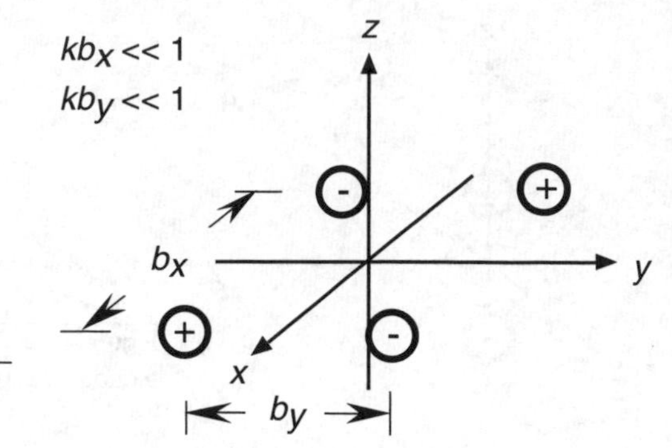

Figure 1.13 An example of a possible quadrupole configuration; all of the monopole sources are in the $z = 0$ plane.

The sound power radiation for a quadrupole is consequently a factor $(kb)^2/5$ less than that of the dipole studied earlier. The radiation pattern of the quadrupole will also be more complicated. Examples of quadrupole-type sound radiation can be found in the radiation of sound from plates and membranes, as well as in the sound generation mechanisms of turbulent flow.

1.12 PROBLEMS

(Note: Related problems can be found in Chapter 2)

1.1 A spherical, sinusoidal wave propagates away from a point source in air. The frequency is 1 kHz. The acoustic pressure amplitude at 1 m from the point source is 0.1 Pa. For air $\rho_0 = 1.20$ kg/m^3, $c_0 = 343$ m/s.

Tasks:

a. Plot the acoustic pressure amplitude and the particle velocity amplitude as functions of the distance from the source.
b. Plot the phase angle between acoustic pressure and particle velocity as a function of the distance from the source.
c. Graph the functions as functions of *kr*—that is, the product between wave number and distance.

1.2 The sound pressure in front of a wall varies with distance from the wall due to interference.

Task: Calculate the sound pressure increase, relative to the free field pressure, that one obtains when measuring the sound pressure level at a distance of 0.1 m in front of a flat, hard wall. The sound wave is perpendicularly incident to the wall.

1.3 The surface of a sphere, whose radius is 0.05 m, vibrates radially, and its volume velocity is frequency-independent. The amplitude of the surface vibration is 0.001 m at 100 Hz.

Task: Calculate the radiated sound power as a function of frequency.

1.4 A Kundt's tube (a.k.a., plane wave tube) can be used to measure the complex reflection coefficient of a material at normal incidence. The standing wave ratio γ is defined as:

$$\gamma = \frac{\tilde{p}_{max}}{\tilde{p}_{min}}$$

Due to exterior noise, it is usually difficult to determine γ accurately when $\tilde{p}_{min}$ is small. Consequently, it is also difficult to determine the sound-absorption coefficient α. By measuring the distance from the surface of the test sample facing the tube to the point where the first sound pressure level minimum is obtained, it is possible to obtain data for calculating the phase angle of the reflection coefficient.

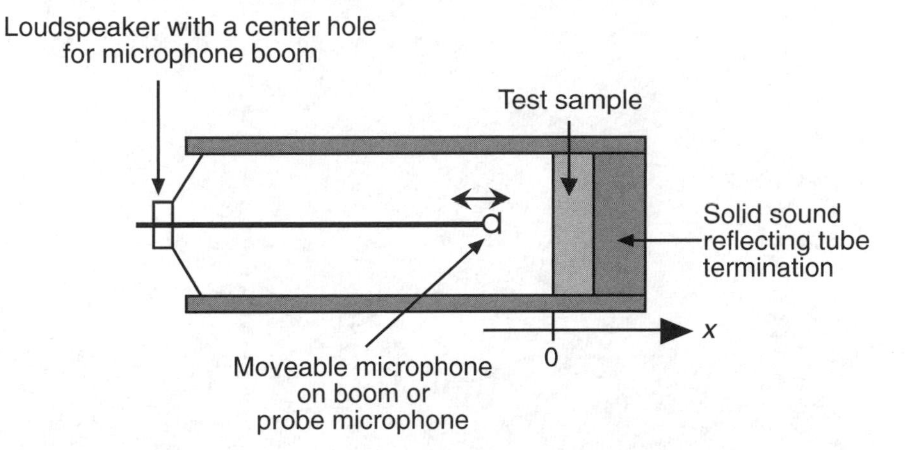

P1.4a The principle of the Kundt's tube equipment for measurement of the sound pressure reflection coefficient.

Tasks:

a. Derive an expression for the sound-absorption coefficient α of the material as a function of the standing wave ratio γ. Calculate the sound-absorption coefficient for γ = 10 ± 1.
b. Derive an expression for how the phase angle of the reflection coefficient can be calculated.

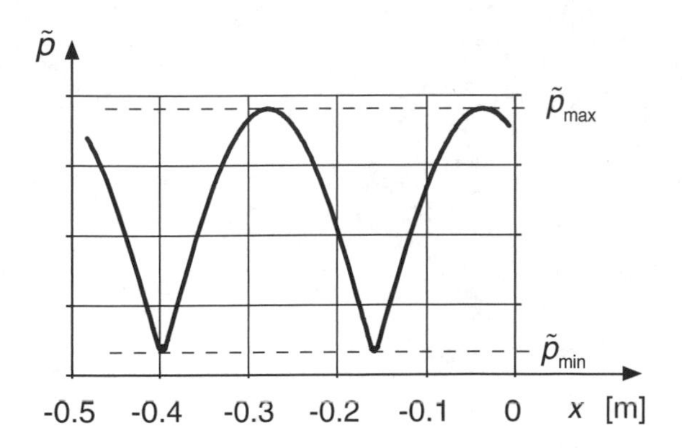

P1.4b Example of the measured rms sound pressure of a standing wave in a Kundt's tube. Note the locations of the peaks and dips of the sound pressure relative to *x* = 0.

Audio Signals 2

2.1 SPECTRUM AND TIME HISTORY

We need to study the properties of sound both from the technical and from the subjective viewpoints. Hearing acts as a complicated time-frequency analyzer that can also analyze patterns in sound and use cognition to determine the source identity, among other things. Technical analysis at this time cannot imitate most of these advanced properties. While time-frequency analysis, such as short-time Fourier analysis and wavelet analysis, can simulate some of the analysis features of hearing (see Chapter 17), we, in practice, primarily study the spectra of signals—that is, their power content as a function of frequency—using Fourier analysis. The signals, music, and noise are continuous or quasiperiodic from an analysis viewpoint (their spectra do not change much over the time of the analysis window). For transient signals or impulse responses, we may use the already mentioned short-time Fourier analysis with good results. References 2.1–2.4 are good introductions to the analysis and measurement of various sounds.

Measurement of audio system characteristics and qualities is discussed in Chapter 17. Chapters 3, 5, and 6 also discuss measurement aspects of room acoustics, while Chapter 10 discusses measurement in building acoustics.

2.2 SIGNALS AND THE $j\omega$-METHOD

The acoustic waves that we are interested in from the viewpoint of sound, such as noise, voice, and music, can be regarded as signals in time and space. We are accustomed to writing waves in this form:

$$p(x,t) = A\cos(2\pi ft - 2\pi x/\lambda) + B\sin(2\pi ft - 2\pi x/\lambda) \tag{2.1}$$

describing the shape of the signal at points in time and space (see Figure 2.1).

The signals picked up by the microphones apply to the particular measurement point in space and give a signal that is a function of time $s(t)$. A practical audio signal is usually finite and quasiperiodic. Assuming the period of the signal to be T_0, the signals can be represented by a sum (possibly infinite) of simple signals such as sine and/or cosine signals:

$$\begin{aligned} s(t) &= \sum_{n=0}^{\infty} A_n \cos(n\omega_0 t) + \sum_{n=0}^{\infty} B_n \sin(n\omega_0 t) \\ &= A_0 + \sum_{n=1}^{\infty} A_n \cos(n\omega_0 t) + \sum_{n=1}^{\infty} B_n \sin(n\omega_0 t) \end{aligned} \tag{2.2}$$

where

$$\omega_0 = \frac{2\pi}{T_0} \tag{2.3}$$

Since there is no steady state offset to be accounted for in audio, the coefficient $A_0 = 0$. The coefficients A_n and B_n are obtained by a form of correlation:

$$A_n = \frac{2}{T_0}\int_{-T_0/2}^{T_0/2} s(t)\cos(n\omega_0 t)\,dt \tag{2.4}$$

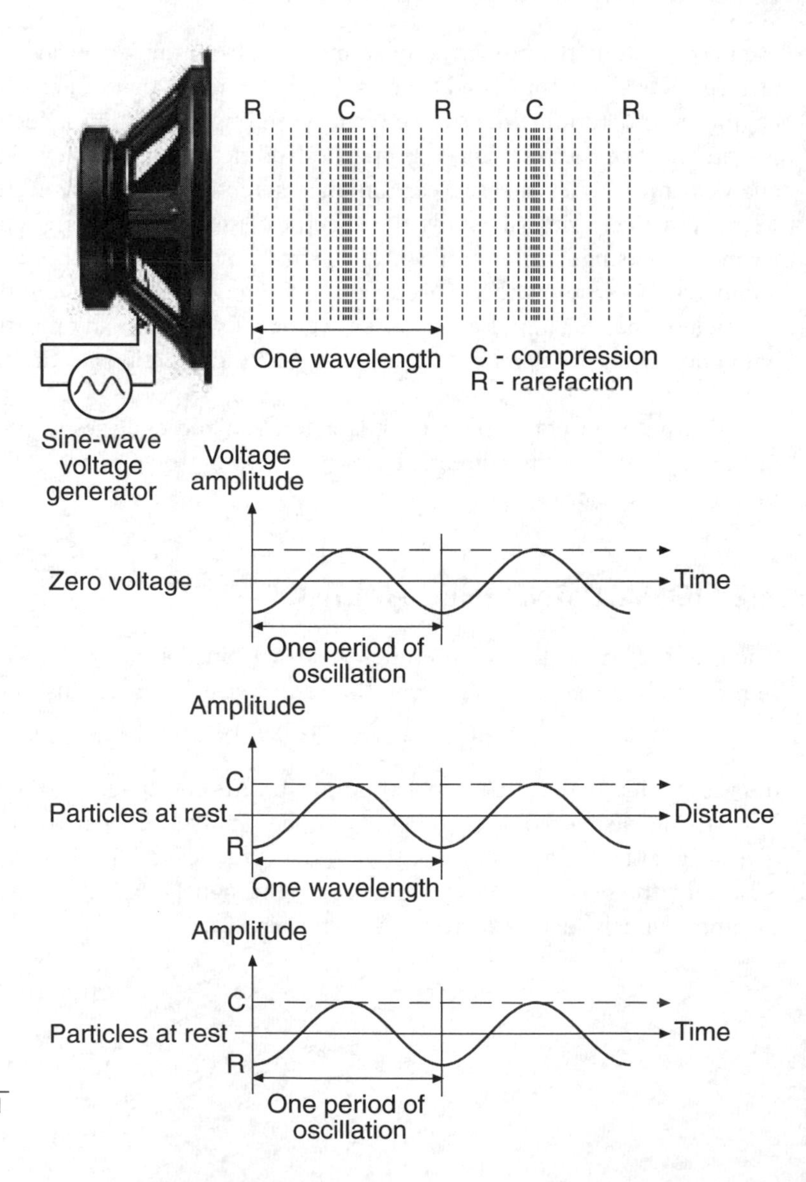

Figure 2.1 A wave represents a signal in time and space.

$$B_n = \frac{2}{T_0} \int_{-T_0/2}^{T_0/2} s(t)\sin(n\omega_0 t)\,dt \tag{2.5}$$

It is more convenient to use only cosines and a phase angle than to use the sum of sine and cosine because of the simpler mathematical handling:

$$\begin{aligned} s(t) &= \sum_{n=1}^{\infty} C_n \cos(n\omega_0 t - \phi_n) \\ C_n &= \sqrt{A_n^2 + B_n^2} \\ \phi_n &= \arctan(\frac{B_n}{A_n}) \end{aligned} \tag{2.6}$$

Using Euler's formula:

$$e^{j\omega t} = \cos(\omega t) + j\sin(\omega t) \tag{2.7}$$

we can make use of the simpler math afforded us. We use complex notation in the mathematical handling of the wave equation and its solutions but consider the real part of the exponential the actual real-world signal.

The coefficients C_n and angles φ_n are called the frequency domain representation of the time signal $s(t)$ or its spectrum. Phase is often of secondary interest in signal analysis, but the phase information can be used to analyze the delay of the signal in a transmission path. The reason for neglecting phase information is that, from the viewpoint of noise abatement, phase is of little interest because it is the energy exposure of the ear that is important for hearing loss.

In audio signal characterization and listening, phase is important. A transient sound and a noise sound may have the same energy but will appear entirely different to our hearing. It is sufficient to time mirror (reverse) a musical signal or speech and listen to it, to be convinced that phase is important; both the source and the reversed source signals have the same spectra.

The *jω-method* (pronounced the j-omega method), sometimes called the complex method, is the practical application of Fourier transform theory to the analysis of signals and linear systems. Since audio signals can be expanded into complex exponential series, each series term (frequency), can be analyzed separately as to how a transmission system affects it. Superposition (addition of the time history) of all relevant frequency terms with amplitude and phase then gives the time history of the output signal. Because the time function is always of the form:

$$e^{j\omega t} \tag{2.8}$$

It is not necessary to include the function in each equation. Instead, in this book, we use a shorthand notation and underline the variable to mark that it is has the time dependence given by Equation 2.8. We also use the underline to mark out other complex quantities and variables.

The ratio between two complex variables, such as pressure and particle velocity, in general will be complex as well but will not have the time component:

$$\underline{Z}_n = \frac{\underline{p}_n}{\underline{u}_n} = \frac{(A_n + jB_n)e^{j\omega t}}{(C_n + jD_n)e^{j\omega t}} = \frac{A_n + jB_n}{C_n + jD_n} \tag{2.9}$$

One of the most useful properties of the $j\omega$-method is that it represents differentation with respect to time by multiplication with $j\omega$ and integration with respect to time by division with $j\omega$. We have already used this property several times in Chapter 1.

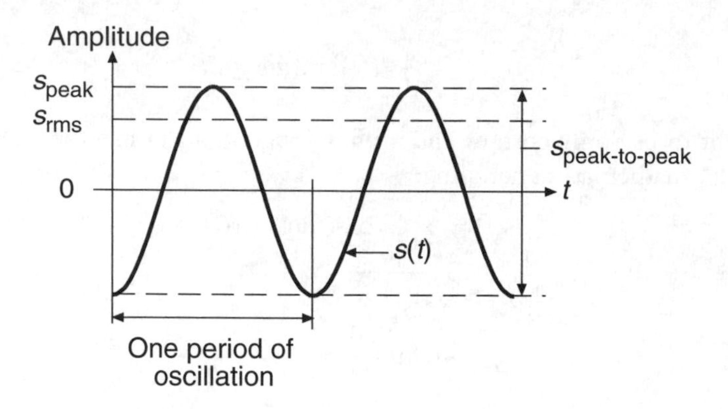

Figure 2.2 Characteristics of a sine-wave signal.

2.3 SINE-WAVE SIGNALS

Signals that can be described by just one frequency are called sine waves. Some terms used to characterize sine-wave signals are shown in Figure 2.2. A sinusoidal signal having a *peak value* of A has an *mean square value* (effective value) of $A^2/2$ and *root-mean-square (rms) value* of $A/\sqrt{2}$. The effective value is defined in Equation 2.21. The rms value of a sine-wave is approximately 71% of the peak value. The magnitude of the peak-to-rms ratio of a signal, sometimes called the *crest factor*, is an important quantity in the characterization of signals.

In this book, we write the rms value of variables using a tilde sign such as $\tilde{p}$ for rms sound pressure. The peak value is written $\hat{p}$.

2.4 NOISE

The term "noise" in colloquial language stands for a loud, unpleasant, undesired, or disturbing sound. In acoustics, the use of the term is more limited. While many use the term to describe the sound along a freeway, airport, or from a neighbor, the use of the word should really be restricted to signals (such as sound and voltage) that do not carry meaning. Thermal noise, the result of random physical processes, is such a signal. Figure 2.3 shows an example of such a noise signal. The peak value is nearly infinite but seldom occurs. In classical music, the crest factor is typically less than 10 and in speech less than 4.

The autocorrelation of a signal can be used to describe its quality of randomness. The normalized *autocorrelation function* for a signal $s(t)$ is defined by:

$$R_{ss}(\tau) = \frac{\int_{-\infty}^{\infty} s(t+\tau)s(t)\,dt}{\int_{-\infty}^{\infty} s(t)s(t)\,dt} \tag{2.10}$$

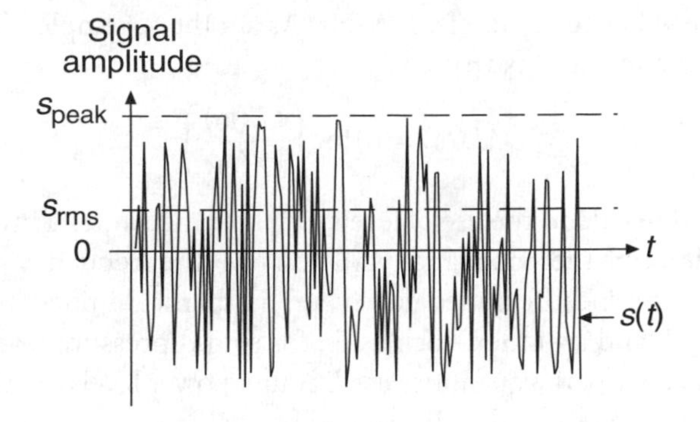

Figure 2.3 A noise signal.

The normalized autocorrelation $R(\tau)$ of thermal noise, often called *white noise*, is zero except for $\tau = 0$ (i.e., when the signal coincides identically with itself when $R(0) = 1$). Another characteristic of a white noise signal is that it has the same *spectral density* at any frequency. Spectral density is the power per Hz of the signal. Many other signals are characterized by an autocorrelation function like the one of white noise, for example, pink and red noise. These noise signals have the following power spectra $G(\omega)$:

$$\begin{aligned} G_{\text{white}}(\omega) &\propto 1 \\ G_{\text{pink}}(\omega) &\propto \frac{1}{\omega} \\ G_{\text{red}}(\omega) &\propto \frac{1}{\omega^2} \end{aligned} \tag{2.11}$$

The power spectrum describes the power for unit frequency. Long time segments of signals, such as most music and speech, also have properties that result in their normalized autocorrelation being almost zero for $\tau \neq 0$.

2.5 THE LEVEL CONCEPT

The psychophysical sensitivity of hearing is such that each increase of intensity by a certain ratio, over some intensity limit, is perceived approximately equally strong subjectively. This is a good reason to use logarithmic ratios to describe the subjective response.

Levels can be used both to describe differences between quantities and magnitudes of quantities. It is practical to use the internationally agreed reference values, when applicable, in calculating levels.

The concept of *level*, in acoustics, implies that one uses the logarithm between two values of a power related sound field property to describe their difference. Power-related quantities are sound intensity, sound power, energy, and energy density. Since sound pressure and particle velocity are related to sound intensity (and voltage and electric current to electric power), the use of levels can be applied to them as well. As an example, a power density level is written:

$$L(\omega) = 10 \log\left(\frac{G_1(\omega)}{G_0(\omega)}\right) \tag{2.12}$$

where $G_0(\omega)$ is a reference value, for example, 1 W/Hz. As another example, the difference between two power density levels $G_2(\omega)$ and $G_1(\omega)$ is written:

$$\Delta L(\omega) = 10\log\left(\frac{G_2(\omega)}{G_1(\omega)}\right) \tag{2.13}$$

Note also that it is practical to extend the use of levels to impedance and its inverse mobility as well—for example, to more easily describe the impedance variation over the frequency range. Sound field impedance is often referenced to the specific impedance of air, mechanical impedance to 1 Ns/m.

Levels are denoted by L and an index, such as L_p, for sound pressure level. Levels are expressed in units of decibel [dB] irrespective of the quantities used. Sound power levels are, at times, expressed in units of Bel. (One Bel is, of course, ten decibel.) This should be avoided.

As an application of the level idea, consider the use of level to describe the properties of attenuation of a plane wave by air or by an electric filter.

2.6 FILTERS AND FREQUENCY BANDS

In audio technology, as in acoustics, the primary frequency range of interest is usually that of normal hearing (20 Hz–20 kHz). In many cases, we need to know the property of the sound in narrower frequency ranges—for example, when analyzing noise or speech.

To remove uninteresting sound components, we use *analog* or *digital* filters. While analog filters could, in principle, be acoustical, we usually use electronic filters for analog filtering since this is more practical. Digital filtering can be done both by direct simulation of analog filters and by digital Fourier transform techniques. Digital filters have the advantage of higher stability but have the disadvantage of signal artifacts being created in the analog-to-digital and digital-to-analog conversion processes. Most digital filters are also not true *real-time filters*, in contrast to analog filters. This has important implications in many signal processing applications.

The *frequency response* of a filter is generally considered to be the output level minus the input level over the frequency range. The frequency response is obtained from the *transfer function* as 20 log($|H(\omega)|$):

$$\textit{Frequency response} = 20\log(|\underline{H}(f)|) + C \tag{2.14}$$

where C is a constant. In audio and acoustics, the constant is usually chosen so that the frequency response is zero dB at 1 kHz.

The transfer function operates on the input signal $\underline{S}_{in}(\omega)$ so that $\underline{S}_{in}(\omega)$ is transformed by the filter's transfer function $H(\omega)$ so that the output signal $\underline{S}_{out}(\omega)$ is:

$$\underline{S}_{out}(\omega) = \underline{H}(\omega)\underline{S}_{in}(\omega) \tag{2.15}$$

Filters are typically characterized by *pass bands* and *stop bands*, as shown in Figure 2.4. For each band, the filter will have a certain *attenuation* as a function of frequency. Typically, one does not want any attenuation in the pass band. The attenuation in the pass band is, however, usually characterized by *ripple*—small variations in attenuation. The pass band and the stop bands are characterized by their

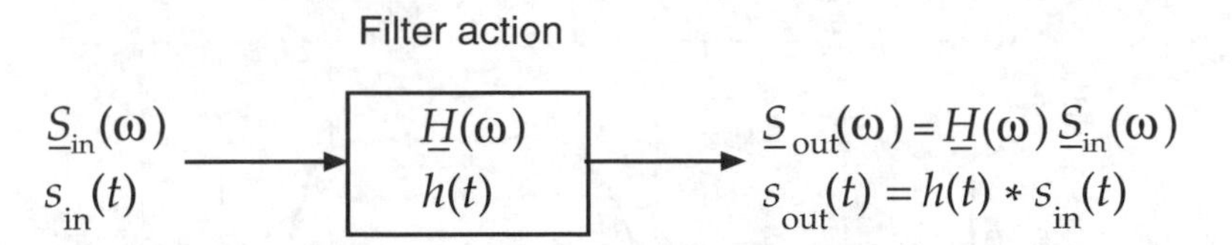

Figure 2.4 The filter operates on the signal $s_{in}(t)$ having a spectrum $S_{in}(\omega)$. The resulting signal spectrum is obtained by multiplication—at each frequency—of the input spectrum and the complex transfer function. The output time function is obtained by convolution of the input waveform by the filter's impulse response (see Equations 2.16 & 2.17).

start and *stop frequencies*, generally called *cutoff frequencies*, usually taken to be at the *−3 dB points of the filter* (power) *response* relative to the pass band average, or the filter's center frequency. Simple filters are *high-* and *low-pass filters*, so called because of their frequency rejection characteristics.

Practical filters do not have the sharp band edges shown in Figure 2.5. The frequency response of a practical analog commercial bandpass filter is illustrated in Figure 2.6. We note that the band edges have a finite *slope* (dB/Hz). Because of the finite slope, the filter will also let some signal through outside the pass band. This can have undesired influence on the accurate measurement of noise spectra, as an example.

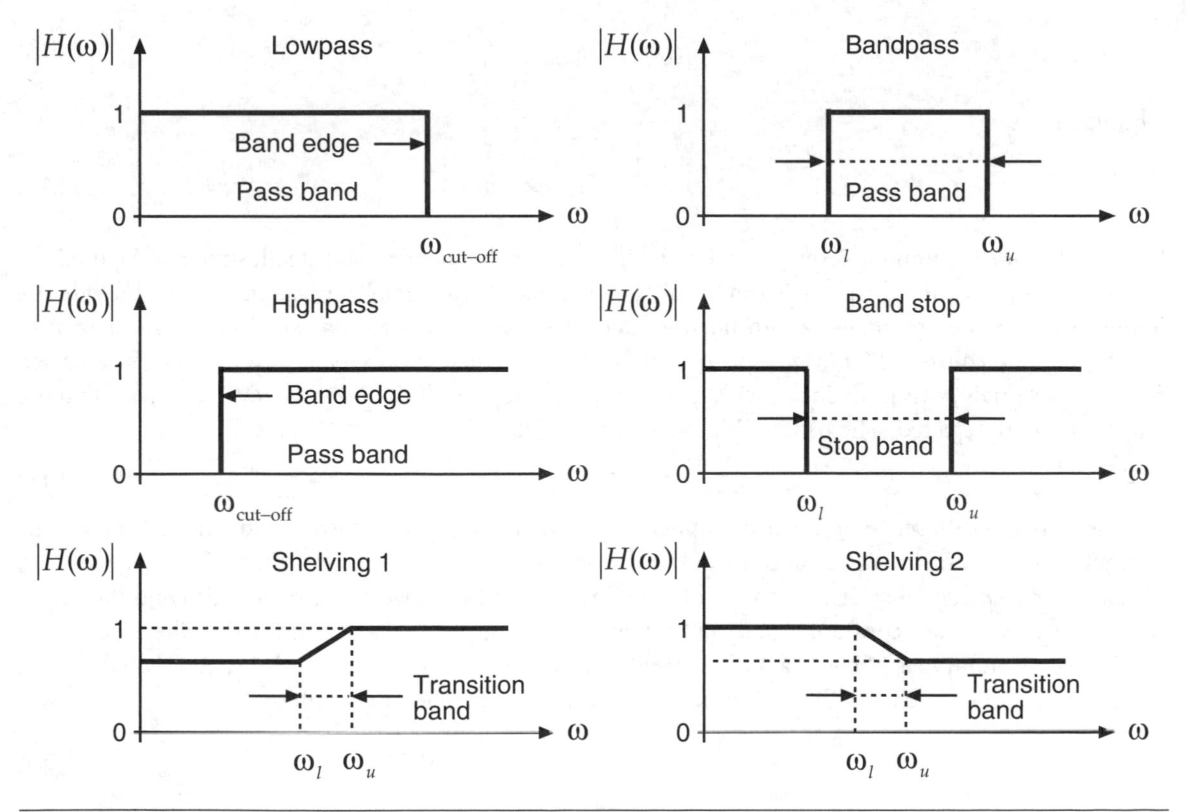

Figure 2.5 Some common filter types and their nominal frequency responses.

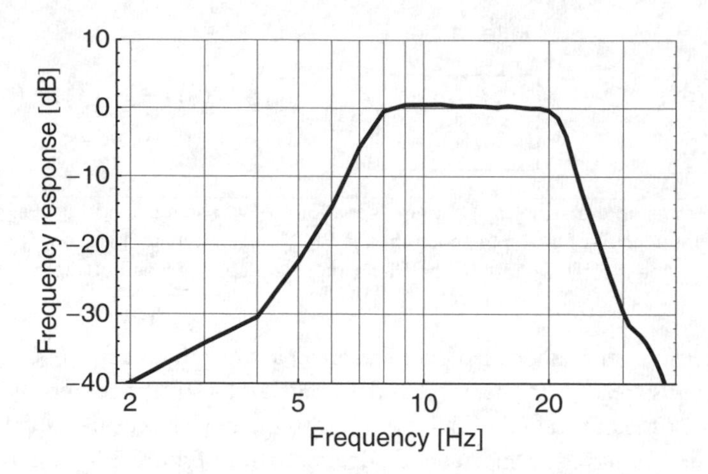

Figure 2.6 An example of the frequency response of a commercial analog octave band filter centered around 12.5 Hz.

It is also important to note that a filter's transfer function is related to its impulse response $h(t)$ by the Fourier transform:

$$\underline{H}(\omega) = \int_{-\infty}^{\infty} h(t)e^{-j\omega t}\, dt \tag{2.16}$$

And its inverse:

$$h(t) = \frac{1}{2\pi}\int_{-\infty}^{\infty} \underline{H}(\omega)e^{j\omega t}\, d\omega \tag{2.17}$$

Figure 2.7 shows the impulse response of the filter that has the frequency response illustrated in Figure 2.6.

The analysis of the frequency content of signals is called *spectral analysis*. One can subdivide the many forms of spectral analysis into *narrow-* and *wide-band* spectral analysis, into *constant relative bandwidth*, and *constant bandwidth* spectral analysis. In spectral analysis, *bandpass filters* are used. Such filters pass signals with frequencies with a certain frequency band, the *pass band*. Frequencies that are *out-of-band* are rejected. The bandwidth of a bandpass filter is:

$$B = \Delta f = f_u - f_l \tag{2.18}$$

That is, it is the difference between the upper cutoff frequency, f_u, and lower cutoff frequency, f_1. The geometrical mean of the upper and lower cutoff frequencies, the *geometrical mean frequency*, usually called the *center frequency*, is used to denote the filter. The ratio between the bandwidth and the center frequency, f_m, is always constant for any particular type of constant relative bandwidth filter.

Two common types of bandpass filters used in wide-band, relative bandwidth spectral analysis, are *third octave* and *octave* band filters. For octave band filters the following applies:

$$\frac{B}{f_m} = \frac{\Delta f}{f_m} = \frac{f_u - f_l}{f_m} = \sqrt{2} - \frac{1}{\sqrt{2}} \approx 0.71 \tag{2.19}$$

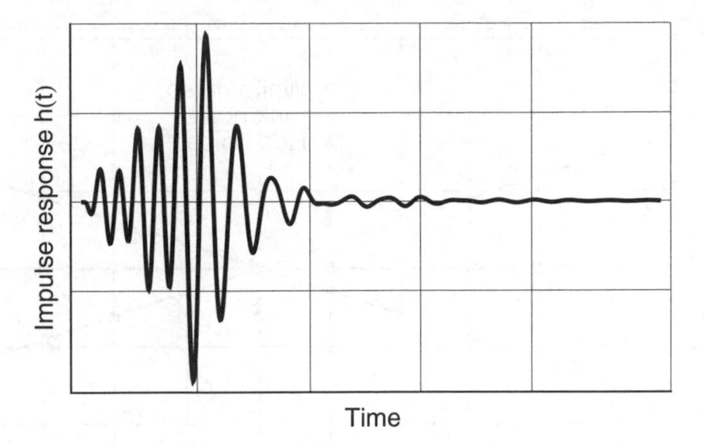

Figure 2.7 The impulse response of the octave band filter discussed in Figure 2.5. Note that low frequencies are more delayed by this filter than high frequencies.

Similarly, for third octave band filters:

$$\frac{B}{f_m} = \frac{\Delta f}{f_m} = \frac{f_u - f_l}{f_m} = \sqrt[6]{2} - \frac{1}{\sqrt[6]{2}} \approx 0.23 \qquad (2.20)$$

The center frequencies (in Hz) of octave- and third octave band filters have been internationally standardized as:

← . .160 200 **250** 315 400 **500** 630 800 **1000** 1250 1600 . . →

The numbers in bold style are the internationally agreed center frequencies of octave band filters (see the IEC 651 and ANSI S1.16 standards).

When analyzing sounds for information about tonal components, it is often practical to use filters having *constant bandwidth*. Such narrow-band, constant bandwidth filters for the audio frequency range are characterized by bandwidths as small as 0.3 Hz, in the case of analog filters, and by bandwidths almost arbitrarily small in the case of digital filters.

The level of a band-limited signal $L_{\Delta f}(f)$ having a bandwidth Δf is related to its spectrum level $L(f)$ as:

$$L_{\Delta f}(f) = L(f) + 10\log(\Delta f) \qquad (2.21)$$

The constant bandwidth filters will give a constant value when applied to a white noise signal. Since the bandwidths of the octave- and third octave band filters increase in proportion to frequency, the octave- and third octave band levels of white noise will also increase as shown for the octave band levels in Figure 2.8. Pink and red noise signals will have 3 dB/octave and 6 dB/octave lower-level slopes as shown in the figure. In the figure, the three noises were arbitrarily adjusted to have the same level in the 1 kHz octave band.

It should be noted that all real audio signals are low-pass filtered because of the high-frequency attenuation properties of air, microphones, loudspeakers, and amplifiers, among others.

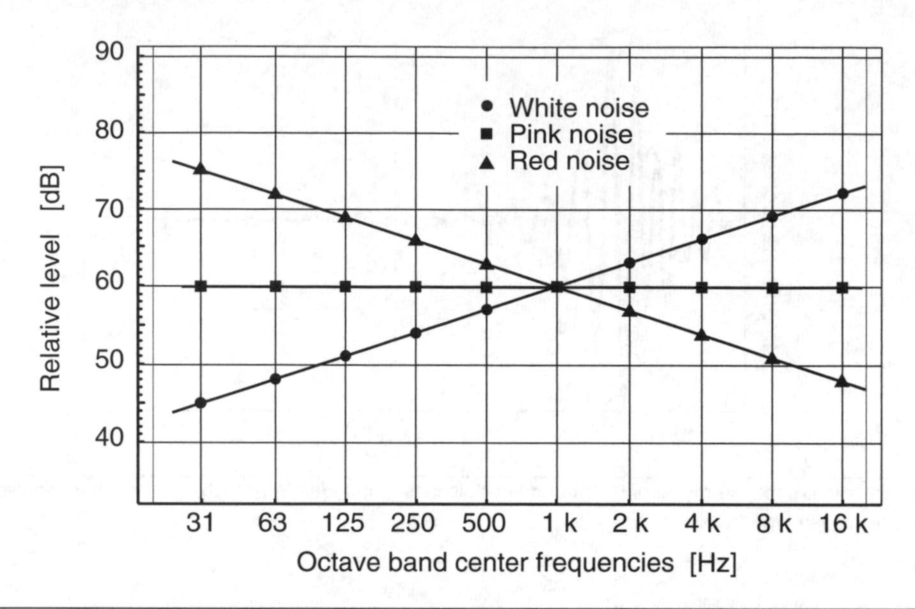

Figure 2.8 The octave band level value characteristics of white, pink, and red noise.

2.7 EFFECTIVE VALUE AND MEASUREMENT UNCERTAINTY

Any physical signal carries energy. The effective value is used to describe the power of a signal. The effective value or mean-square of a signal $\tilde{s}(t)$ is estimated as:

$$\tilde{s}^2 = \frac{1}{T}\int_0^T s^2(t)\,dt \tag{2.22}$$

where T is the integration time; generally a better estimate is obtained the longer the integration time.

A white noise signal has an amplitude that varies at random over time. One can show that the estimate of the rms value of bandpass filtered white noise has a *relative standard deviation* that is:

$$\xi_{ms} \approx \frac{1}{2\sqrt{BT}} \tag{2.23}$$

where B is the bandwidth in Hz of the noise and T is the time of integration for large vales of BT. Over small bandwidths, pink and red noises, as well as other noise-like signals, follow the same rule.

Since the variance determines the *confidence interval* of the estimated value, clearly to make good noise measurements, one must either use wide-band filters or long averaging times to obtain precise values. Figure 2.9 shows the 95% confidence intervals for a band-limited white noise signal as a function of the variable BT.

2.8 REFERENCE LEVELS

Hearing senses sound pressures in a range of approximately $.5 \cdot 10^{-5}$ Pa to 100 Pa. It is practical to use a *temperature* similar scale to describe the strength and relative strength of sounds since, as mentioned

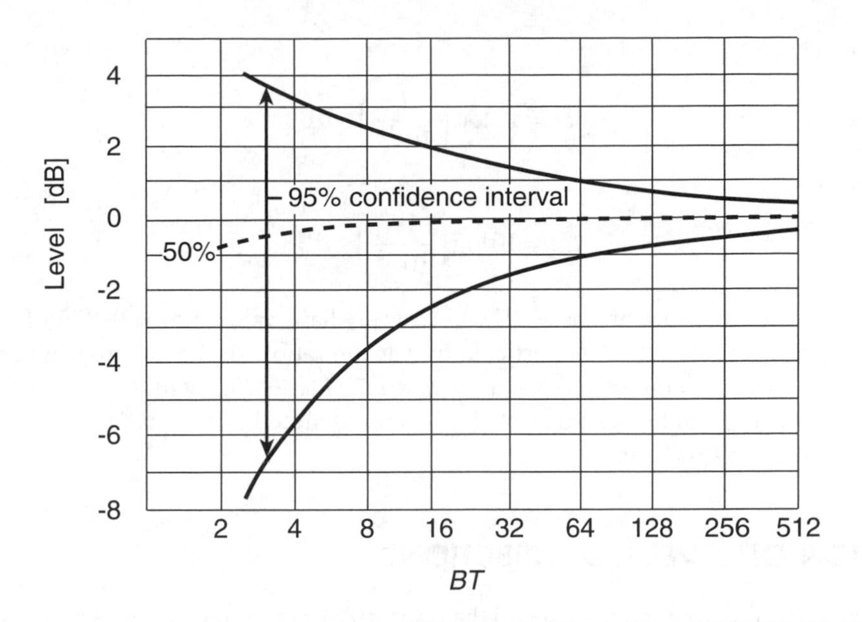

Figure 2.9 The confidence interval for the level of band-limited white noise as a function of the *BT*-product. (After Ref. 2.1)

previously, hearing perceives an increase of intensity by a certain ratio, over some intensity limit, as approximately equally strong subjectively over its sensitivity range.

For any scale to be practical, it needs a reference point or value. In acoustics, the level reference values for intensity and sound pressure are chosen so that related levels have the same size and the number 0 dB corresponds closely to the hearing threshold. When there is any risk of misunderstanding, one should always give the reference value. (Note, for example, that the reference values for sound pressure are different for sound in air and sound in water.)

Some common reference values for airborne sound are:

- Sound pressure (air) $p_{ref} = 2 \cdot 10^{-5}$ Pa
- Particle velocity (air) $u_{ref} = 10^{-9}$ m/s (sometimes $u_{ref} = 5 \cdot 10^{-8}$ m/s)
- Intensity $I_{ref} = 10^{-12}$ W/m^2
- Power $W_{ref} = 10^{-12}$ W

These values give the respective levels:

Sound pressure level

$$L_p = 20 \log\left(\frac{\tilde{p}}{2 \cdot 10^{-5}}\right) \text{ dB} \tag{2.24}$$

Particle velocity level

$$L_u = 20 \log\left(\frac{\tilde{u}}{10^{-9}}\right) \text{ dB} \tag{2.25}$$

Sound intensity level

$$L_I = 10\log\left(\frac{I}{10^{-12}}\right) \quad \text{dB} \tag{2.26}$$

Sound power level

$$L_W = 10\log\left(\frac{W}{10^{-12}}\right) \quad \text{dB} \tag{2.27}$$

With these reference values, the various levels will typically have values in the 0–140 dB range for airborne sound of practical interest. Note particularly that the ratios, in the case of sound pressure and particle velocity, are based on the *rms* (*root-mean-square*) *values of the quantities.*

In American usage, sound pressure level (L_p) is often denoted SPL, L_I is SIL, and L_W PWL. This convention is not used in this book.

2.9 ADDITION OF LEVEL CONTRIBUTIONS

It is important to note that one cannot just add the respective levels to obtain the resulting level of many sound sources acting at the same time. One must return to the *true* values of the respective sound field quantities and add those according to the rules applicable to each sound field quantity.

For sound power, one adds the power values before taking the logarithm. For sound intensity one adds the intensity values, but since intensity is a vectorial quantity, one must be careful with signs and directions.

For sound pressures and particle velocities, things are somewhat more complicated, as illustrated by the following example. It is practical to separate between two different cases, those of *correlated* and *uncorrelated* sounds. If the sounds are correlated—as, for example, the sound from the front and the back of a vibrating loudspeaker diaphragm, one must add the sound pressures directly before obtaining the root-mean-square value and dividing by the reference value.

Uncorrelated sounds are sounds for which the time weighted integral of the product of the sound pressures tends toward zero for sufficiently long integration times. Examples of uncorrelated sounds are the noises from two jets, two automobiles, and two voices. The sounds in different frequency bands of noise, music, and speech also tend to be uncorrelated and can be treated as such when calculating a sum over frequency bands.

For uncorrelated sounds, a simpler method is used. One can add the *effective value* (defined by Equation 2.22), of the sound pressure of each contributing sound source. Since the root mean square value, the *rms* value, is the square root of the effective value, we have:

$$\tilde{p}_{\text{tot}}^2 = \tilde{p}_1^2 + \tilde{p}_2^2 = p_{\text{ref}}^2\left(10^{L_{p1}/10} + 10^{L_{p2}/10}\right) \tag{2.28}$$

That is:

$$L_{p,\text{tot}} = 10\log\left(10^{L_{p1}/10} + 10^{L_{p2}/10}\right) \tag{2.29}$$

The resulting level, due to more than two sources, is obtained by simply repeating this procedure or by directly adding up all the effective values before taking the logarithm.

2.10 WEIGHTED SOUND PRESSURE LEVELS

It is often necessary to characterize sound in a way similar to how it is subjectively perceived. It is tradition to analyze the strength of signals by using so-called *A*, *B*, and *C filters* in the electronic filtering process before obtaining the effective values of the signals.

These filters are characterized by frequency dependence that is somewhat similar to that of human hearing in certain sound pressure ranges (see Section 3.3). Figure 3.13 shows the frequency response characteristics of these filters, which are defined by the IEC651 standard.

The levels measured using the filters are called *weighted levels*. The sound-pressure level measured using an *A filter* is expressed in units of dB and is usually called *sound level* and denoted L_{pA}. However, many still write the A-weighted sound level as L_p in units of dBA for extra clarity (as sometimes in this book). Corresponding indices are used when measuring using the B and C filters. Sometimes one writes L_{plin} to emphasize that no filter has been used in a measurement—the measurement value is said to be *linear* or *unweighted*.

2.11 EQUIVALENT LEVEL

Sometimes neither the sound level nor the mean sound level is sufficient to describe a sound having a time varying sound level. One might want to describe how the sound level changes over long time intervals. It is then possible to use the statistical distribution of sound level over a certain time interval—for example, 10:00 p.m. to 6:00 a.m. for night time values—and the averaged integrated energy value of the sound level over a certain time, the *equivalent level*.

The *energy-equivalent A-weighted sound pressure level*, in decibels, often referred to as simply the equivalent level $L_{Aeq,T}$, is defined as:

$$L_{Aeq,T} = 10\log\left(\frac{1}{T}\int_0^T \frac{\tilde{p}^2}{p_{ref}^2}dt\right) = 10\log\left(\frac{1}{T}\int_0^T 10^{L_p/10}\,dt\right) \tag{2.30}$$

where T is the measurement period (usually in hours), $p(t)$ is the short time averaged A-weighted rms sound pressure and $L_p(t)$ is the short time averaged A-weighted sound level.

Typical integration times are 8 and 24 hours for many types of environmental noise measurements. The equivalent level can also be estimated approximately from a statistical distribution of the A-weighted sound pressure level:

$$L_{Aeq,T} = 10\log\left(\frac{1}{T}\sum_i t_i \cdot 10^{L_{p,i}/10}\right) \tag{2.31}$$

where t_i is the time that the sound level has fallen between the limits of class *i*, and $L_{p,i}$ is the A-weighted sound level corresponding to the class-midpoint of class *i*. The level classes are usually 5 dB wide.

2.12 PROBLEMS

2.1 To calculate the total (unweighted) sound pressure level using *band-limited* data for noise, one usually assumes that the octave band signal components are uncorrelated. A certain spectrum has the following octave band sound pressure level data:

Octave band	63	125	250	500	1 k	2 k	4 k	8k	[Hz]
L_p	50	58	64	59	51	46	41	35	[dB] re 20 μPa

Task: Calculate, assuming the signal components uncorrelated, the linear sound pressure level.

2.2 Consider the octave band spectrum. One can often estimate the A-weighted sound pressure level value from the octave band that has the highest level.

Tasks:

a. Calculate the total A-weighted sound pressure level for the signal spectrum given in Problem 2.2.
b. Which octave bands dominate in calculating the resulting value?

2.3 Three sources of sound are radiating inside a room. The rms value of the sound pressure was measured for each source radiating alone, giving the following set of results:

Source l: $p_{1,\text{rms}} = 0.63$ Pa
Source 2: $p_{2,\text{rms}} = 0.11$ Pa
Source 3: $p_{3,\text{rms}} = 0.20$ Pa

Tasks:

a. Determine the sound pressure level at the measuring point when all three sources are active. Assume that the sources are uncorrelated.
b. Repeat Part a but assume that Source 1 has been modified so that $\tilde{p}^2_{1,rms}$ is reduced to half its value.

2.4

Task: Calculate the equivalent level over eight hours for a noise having the following sound level (L_{pA}) distribution over time: 60 dB for one hour, 80 dB for six hours, and 90 dB for one hour.

2.5 Noise measurements are often done using third octave band filters.

Task: In which third octave frequency bands do you expect a 60 Hz tone and its accompanying overtones to appear?

2.6 The diagram on the next page shows octave band spectra for two noise signals.

Tasks:

Calculate and plot (in the diagram):

a. The linear and total A-weighted sound pressure levels for the signals.
b. The sum of the spectra A and B, assuming that they are uncorrelated. Also calculate the total A-weighted sound pressure level for the sum of spectra A and B.
c. The A-weighted spectrum of the sound pressure level for signal A.

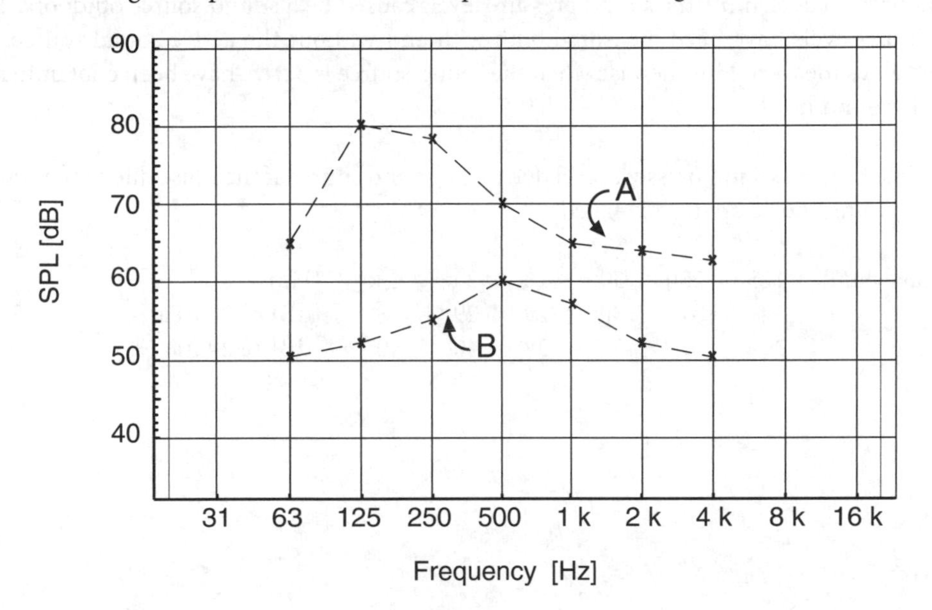

2.7 The graph in Figure 3.29 shows an example of pressure spectrum level (PSL), of speech at 5 cm distance from the lips. The PSL is the sound pressure level measured using an ideal bandpass filter 1 Hz wide.

Task: Calculate, using the graph, the approximate octave- and third octave band levels for speech at a distance of 1 m from the speaker's mouth.

2.8

Task: While measuring a particular noise in frequency bands 100 Hz wide, it is desired that the normalized variance should not exceed 0.3% (± 0.03 dB). Over what time is it necessary to integrate to obtain the effective value with this precision?

2.9 The sound pressure in front of a wall varies with distance from the wall due to interference.

Task: Plot the increase in sound pressure level as a function of frequency and calculate the values at the nominal octave band filter frequencies in the range of 63 Hz to 8 kHz.

2.10 The sphere surface vibrates radially and its volume velocity is frequency-independent. The sphere has a radius of 0.05 m. The amplitude of the surface vibration is 0.001 m at 100 Hz.

Task: Plot the sound power level as a function of frequency at the nominal octave band filter frequencies in the range of 125 Hz to 2 kHz.

2.11 You need to determine the sound pressure levels caused by a sound source outdoors. The sound pressure levels have been measured both with and without the active sound source. However, the values measured for the case when the sound source is active have been contaminated by the background noise.

Task: Calculate the *real* sound pressure level due to the sound source if the noise due to the sound source and the background noise are uncorrelated.

Octave band	125	250	500	1 k	2 k	4 k	[Hz]
$L_{p,\text{background only}}$	34	31	30	29	29	28	[dB] re 20 μPa
$L_{p,\text{with source}}$	37	37	37	38	40	40	[dB] re 20 μPa

Hearing and Voice

3

3.1 INTRODUCTION

Human hearing is an extremely sensitive mechanism for sensing acoustic signals. As we *listen*, we use a set of subsystems: *ears*, *auditory nerves*, and *brain*. Humans may be able to detect rms sound pressures as low as 5 μPa and to endure pressures as high as 100 Pa, a significant dynamic range for any transducer system. The frequency range extends from 20 Hz to 20 kHz for a *normal hearing* person.

The high frequencies have short wavelengths that allow good directional hearing. The directional hearing properties are further improved by *binaural hearing* and the associated correlation of the instantaneous signals from the two ears. Hearing also acts as an adaptive signal processor that gives us the possibility of detecting signal patterns and weak signals, even in noisy environments.

Although the physiological functions of the ear are well known, little is known about the signal processing properties of the auditory nerve and the brain. Using *psychoacoustic tests*, it is possible to check various hypotheses for the function of these systems. Such experiments and results are discussed later in this chapter.

Figure 3.1 shows a schematic representation of a psychoacoustic experiment. In listening, we perceive internally the effect of the *stimulus* on our ears, the *perception* formed depends on emotional status, prior experience, and so forth. When we need to output our experience, we have to use language—writing or speaking—or some help device such as a pointer or a musical instrument. All of these factors influence the response for any given stimulus. This makes it difficult to obtain exactly the same response from various listeners. Listener response will depend on such matters as the experimental layout, hearing threshold, experience, and personal ability to verbalize what is being perceived. This means that any test involving listening must be performed under well-thought-through, specified, and controlled circumstances to eliminate *bias error*. The test must also involve many listeners to reduce *random error*. In psychoacoustic tests involving spatial discrimination, it is often practical to use the *head-related coordinate system* shown in Figure 3.2.

3.2 THE COMPONENTS OF THE EAR

Functionally, the ear can be regarded as consisting of three parts: (1) *outer ear*, (2) *middle ear*, and (3) *inner ear*. Together these form a system for the conversion of sound pressure oscillations to associated nerve signals in the auditory nerve. A drawing showing the anatomy of the ear is shown in Figure 3.3.

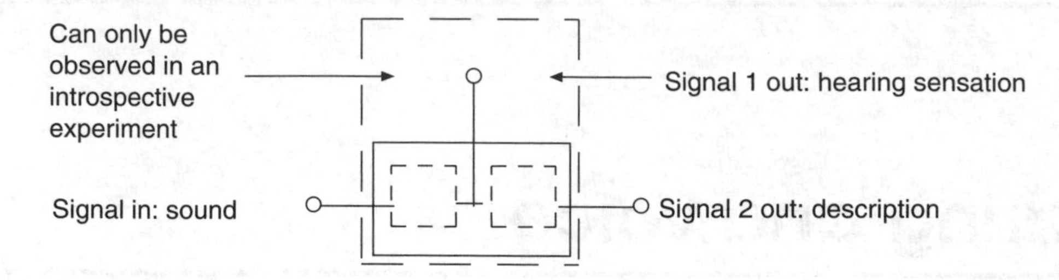

Figure 3.1 The psychophysical system.

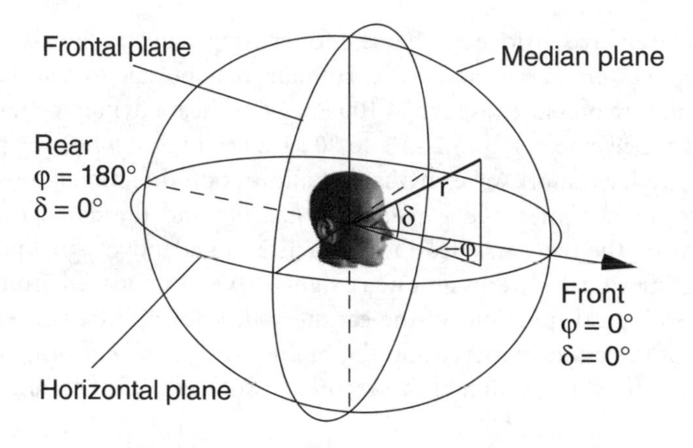

Figure 3.2 The head-related coordinate system.

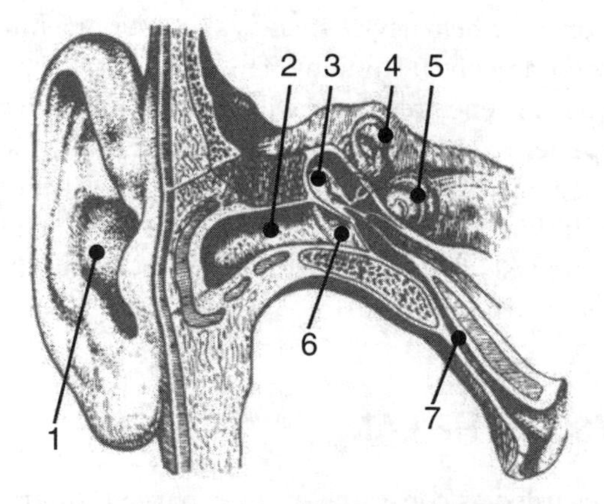

Figure 3.3 Anatomy of the ear: (1) pinna, (2) ear canal, (3) ossicles, (4) semicircular canals, (5) cochlea, (6) eardrum, and (7) eustachian tube.

The functional principles of the three parts are shown in Figure 3.4. The outer ear may be regarded as a signal collector or antenna, the middle ear as an impedance converter and the inner ear as a transducer from mechanical vibration to electrical signal.

The Outer Ear

The outer ear consists of the *eardrum*, the *ear canal*, and the outer, visible part of the ear, the *pinna*. The pinna contributes to many of the directional properties of hearing mentioned previously, but the head and the torso also influence the directional characteristics of hearing. A trained listener can quite easily discern from where a signal is coming, using only one ear. This is due to the changes in the *spectral content* of the signal that in turn are a result of the reflection and diffraction of sound around the head. These phenomena are angle-dependent and affect the sound primarily at frequencies above approximately 1 kHz. The curves shown in Figure 3.5 are examples of the frequency response in a pinna for various angles of sound incidence in the horizontal plane (sometimes called lateral angles). The curves show the magnitude of the *head-related transfer functions* (HRTFs). They are complex and relate to the sound pressure at the place of the head center. The HTRFs are quite individual, and human heads have small asymmetries between the left and right ears, influencing the localization ability.

The shape of the pinna is of great importance for our directional hearing. We can differentiate between sounds incident from the rear and front directions because of the shadow action of the pinna. At frequencies above 2 kHz, the level difference in sound from rear and front directions can exceed 10 dB.

Various *resonances* in the pinna are excited at frequencies typically above 3 kHz. The *modes* of these resonances are important for the directional hearing outside the horizontal plane, particularly at angles of incidence close to the median plane. Some examples of the behavior of these modes are shown in Figure 3.6. The excitation level of the various resonances depends on the angle of incidence of the sound.

Accurate aural localization of sound is important in many cases, such as hearing warning signals or optimizing speech intelligibility in noisy surroundings. Accurate localization is also important when

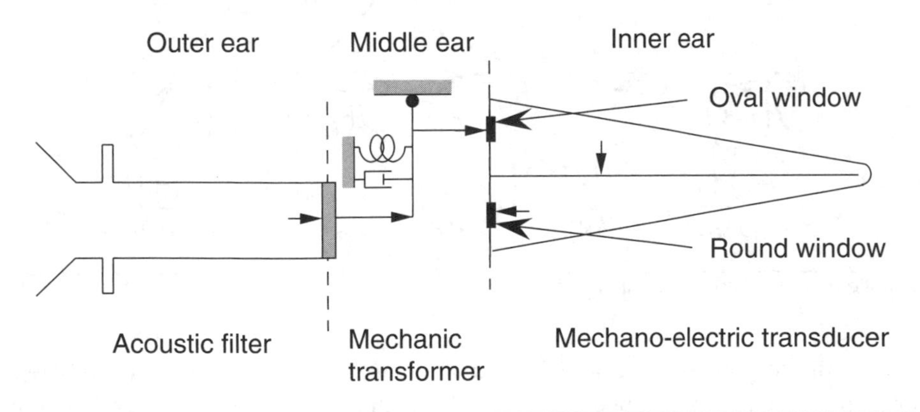

Figure 3.4 The functional components of the ear.

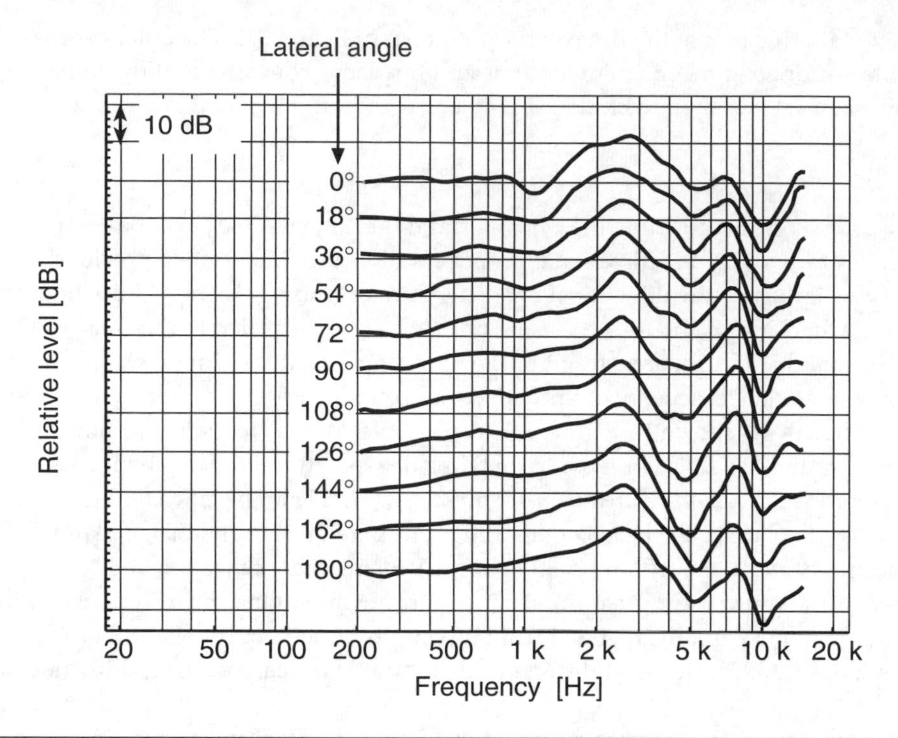

Figure 3.5 Frequency response for the head-related transfer functions from free field to the entrance of the ear canal as a function of the lateral angle φ in Figure 3.2. (After Ref. 3.2)

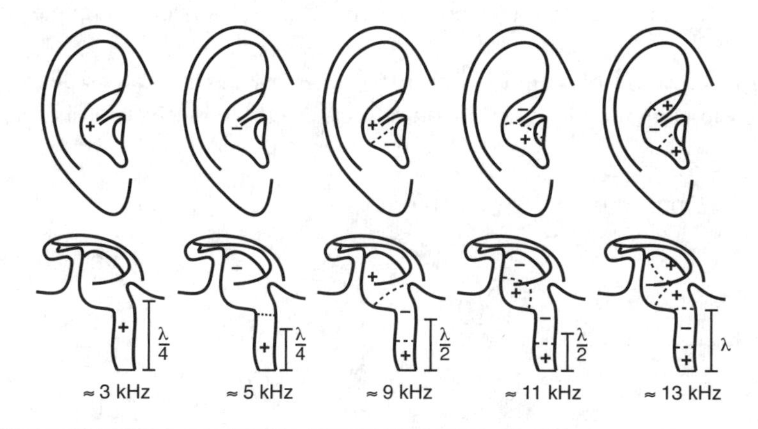

Figure 3.6 Acoustic modes in the pinna. (After Ref. 3.1)

simulating 3-D sound for games, virtual reality, and architecture. It is difficult to recognize the angle of incidence for sound arriving in the median plane. This is due to the lack of binaural cues as described later in this chapter. In the special case of sound incident in the median plane, hearing has to use memory and spectral cues to localize from which angle the sound is coming. The excitation of the resonances in the outer ear is important under such conditions.

As an example, the mode corresponding to the resonance at the frequency f_{03} in Figure 3.6 will be difficult to excite by sound from the frontal direction but will be easy to excite using sound coming from above. Modes at higher frequencies will behave analogously for various angles. The low frequency resonance f_{01} (at approximately 3 kHz) is of great importance for the impedance matching of the ear to the incident sound. The acoustic sensitivity of the ear is usually highest at this frequency. When the ear canal length is equal to $(1 + 2n)\lambda/4$ where λ is the wavelength of the sound, and n stands for a natural number, the ear canal will act as a quarter-wave transformer, and the impedance of the eardrum will appear small to the outside sound field.

The ear canal is essentially a tube, having an average length of 25 mm and an average diameter of 7 mm. The walls of the ear canal are acoustically hard. An example of the frequency response function of an ear canal is shown in Figure 3.7. The input impedance of the ear canal is of great importance in the design of earphones and headphones. The frequency response sensitivity increase of 10–20 dB also results in increased risk of hearing damage for sounds in this frequency range.

The eardrum is an elastic membrane separating the outer and middle ear sections. Most of the acoustic power at the eardrum will be transferred to the middle ear at medium frequencies.

The Middle Ear

The middle ear contains the three *middle ear bones*. These are ossicles and are the malleus, the incus, and the stapes. Normally, the middle ear is acoustically sealed from the outside except for the sound that enters through the vibrations of the eardrum (see Figure 3.8).

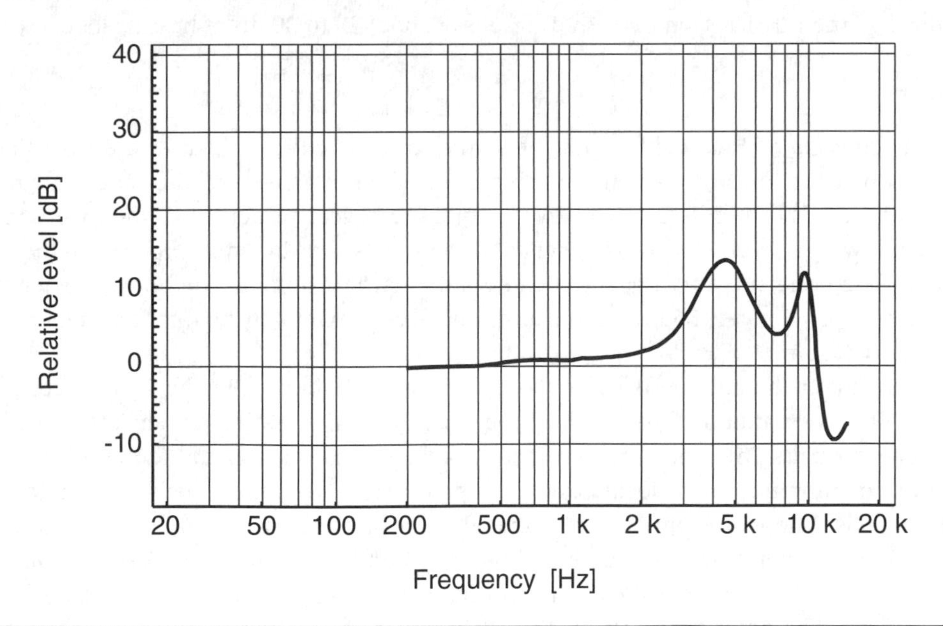

Figure 3.7 An example of the measured level difference between sound pressure at the eardrum and the entrance of the ear canal. (After Ref. 3.2)

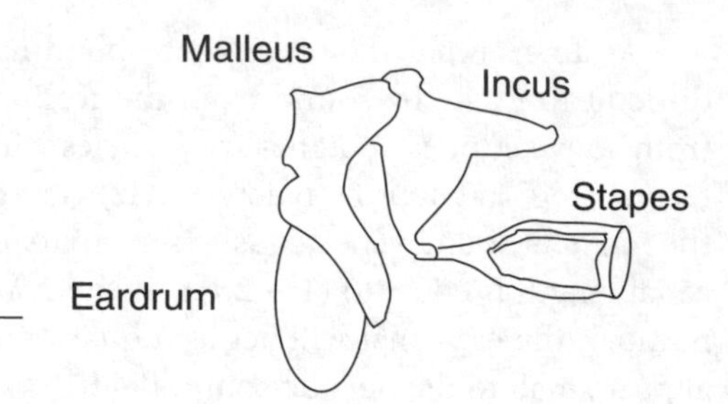

Figure 3.8 The ossicles attach to the eardrum and to the oval window of the cochlea.

To make it possible to equalize the static pressure, so that the eardrum will not be pre-tensioned, there is an opening, the Eustachian tube, leading to the mouth. The ossicles act as a mechanical impedance transformer to transfer the power from the eardrum to the inner ear in an optimal way. The muscles holding the ossicles can act as a form of ear protection. At sound pressure levels above approximately 70 dB, the *stapedius muscle* can change its mechanical properties and give approximately 20 dB of reduction in sound transmission. This is called the *acoustic reflex*. However, because of the reaction time required, more than 0.1 s, there is little protection against *transient sounds*.

Because of the mechanical transformation between large and small movement due to the lever action of the ossicles, and because of the area change, approximately 30:1, between the eardrum and the oval window (where the stapes attaches), one will have an impedance transformation between the high internal impedance of the inner ear and the low impedance at the eardrum. The impedance change reduces what would otherwise be approximately 30 dB attenuation to nearly zero. Measurements of hearing loss in people having had their ossicles removed show results of about 20 to 30 dB of hearing loss.

The Inner Ear

The inner ear consists of the cochlea and the semicircular canals. The latter are essential for our ability to keep balance but are normally not included when we speak of the inner ear or cochlea. The cochlea borders the middle ear with two membrane covered openings—the oval window and the round window. The oval window is attached to the stapes. As shown in Figure 3.3, the cochlea is spiral-shaped. The bone surrounding the cochlea is the hardest bone in the body. The length of the cochlea spiral is approximately 35 mm. It has a volume of approximately 50 mm^3 and a mean diameter of approximately 1.5 mm.

The physiology of the cochlea is shown in the section in Figure 3.9. The cochlea contains three canals filled with watery liquids. The canals of the cochlea are the vestibular canal, the cochlear duct, and the tympanic canals. The canals are joined at the helicotrema, the top end of the cochlea.

The basilar membrane, which separates the tympanic canal from the other canals, holds the inner and outer hair cells. There are approximately 12,000 outer hair cells, each holding some 100–200 stereocilia (hairs). The inner hair cells number about 3,500 and typically hold 50 stereocilia each. The movement of the stereocilia triggers the inner hair cells, which then send their signals to the brain down the auditory nerve. The outer hair cells play an important role as effector cells that are mechanically active and can change their length by acoustic stimulus. This results in an active feedback mechanism that amplifies the response of the ear, particularly close to the threshold of hearing. The system acts as

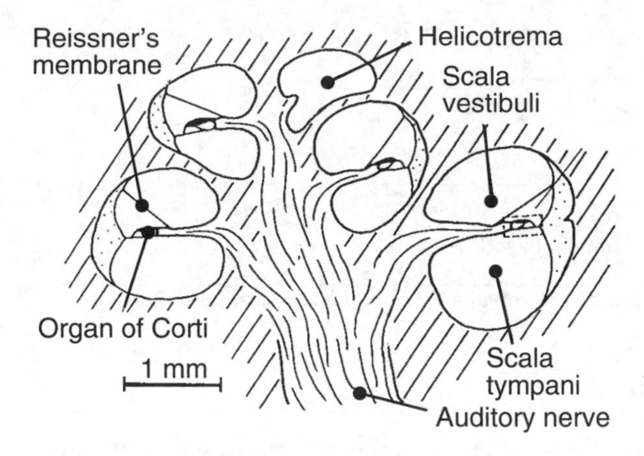

Figure 3.9 Physiology of the cochlea. The small rectangular area at the right is shown magnified in Figure 3.10.

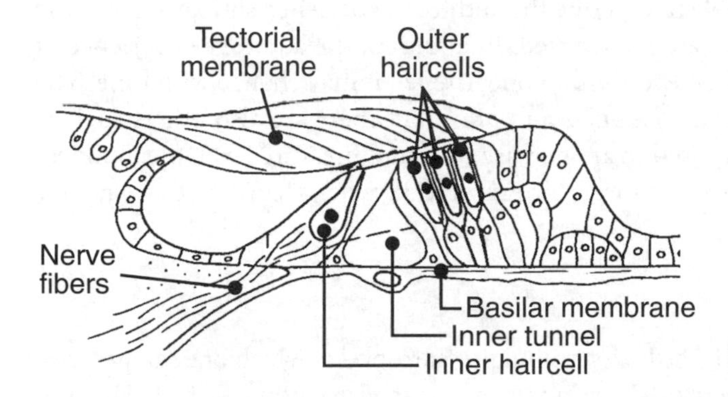

Figure 3.10 The organ of Corti.

a cochlear amplifier. One can note the active process by the phenomenon of otoacoustic emissions. If the ear is stimulated by a click, it will quickly respond by sending out sound. This forms the basis of a simple noninvasive test for hearing defects. The ear may also sometimes send out sound in the absence of exterior stimuli.

Figure 3.10 shows a section through the organ of Corti, indicated in Figure 3.9. One can see the arrangement of the basilar membrane, the hair cells, and the cilia. The upper part of the cilia attach to the tectorial membrane, which gives the arrangement extra moving leverage.

The movement of the basilar membrane is determined by the vibrations induced at the oval window by the sound incident on the ear. Since the speed of sound in the fluid is high, and the cochlea is short, the sound pressure at all points in the fluid will be approximately equal inside both the vestibular and tympanic canals respectively. The basilar membrane can be thought of as built of resonant sections that are highly damped. The sound pressure difference between the two canals will set the membrane in motion. The motion extends over a large area as illustrated in Figure 3.11, even with single frequency excitation. The motion of the basilar membrane is, however, not frequency selective enough to explain the frequency resolution of human hearing. Further signal processing takes place as the nerve signals propagate to the brain and in the brain.

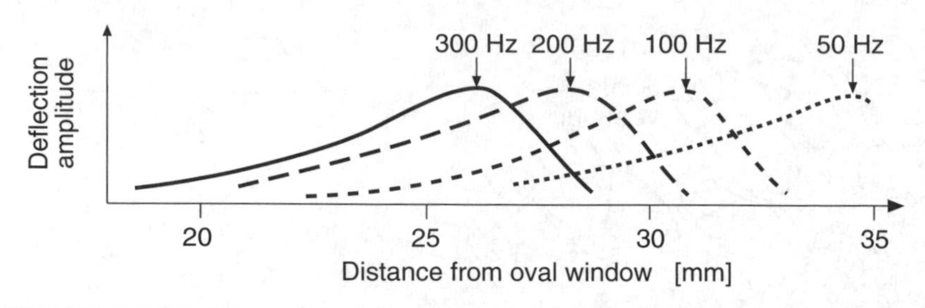

Figure 3.11 The envelope of the vibration amplitude of the basilar membrane for some frequencies. (After Ref. 3.4)

3.3 PSYCHOACOUSTIC DIMENSIONS

While it is possible to describe the physical properties of sound by using such terms as sound pressure, phase, frequency, and *duration*, it is not possible to describe the auditory experience similarly, since the various dimensions of the auditory experience are interrelated. In addition, the auditory experience of one listener is likely to be different from another. Even within one listener, it may change with time. The two most obvious psychoacoustic dimensions are *loudness* and *pitch*, but there are also others such as *roughness*, *sharpness*, and *timbre*. As one is exposed to an increase in sound pressure level, the loudness of the sound increases. The relationship is quite nonlinear; for example, it varies according to frequency and duration.

Loudness and Loudness Level

The *threshold of audibility* (hearing threshold) is the lowest sound pressure level at which one can perceive sounds. The threshold of audibility is usually measured over a frequency range extending from 20 Hz to 20 kHz for a plane-wave incident in the frontal direction of the listener. It is difficult to measure the hearing threshold correctly at low frequencies because of the difficulty of insulating the test chamber against other interfering low frequency sounds and loudspeaker or headphone nonlinear distortion. Different frequency response curves will be obtained for the threshold of audibility for different persons, but even different measurement procedures give different results. The threshold of audibility is more rigorously defined as the minimum perceptible free-field sound pressure level that can be detected at any frequency over the frequency range of hearing.

The threshold for a particular test person is in practice determined by subjecting the person to test tones at known sound pressure levels in an *anechoic chamber*. The sound field should be a plane-wave over a sufficiently large area. The background noise in the chamber has to be low enough so as not to interfere with the measurement. Since the threshold of audibility varies with signal duration, the test tone time has to be more than 0.3 s. A duration of 1 s is typical.

The threshold of audibility (the dotted curve marked MAF [minimum audible field]) is shown in Figure 3.12 and is frequency-dependent. The curves in the figure are from the ISO 226:2003 standard and are the measurement results obtained with the listeners facing a pure tone plane-wave in an anechoic chamber. The sound levels were measured at the listeners' head position without a listener present. The listeners were otologically normal young adults, ages 18 to 25 years.

The *Loudness Level* (LL) of a sound is the sound pressure level of a reference tone at 1 kHz that is perceived as equally loud as the test sound. The LL is expressed in the unit *phon*. The variation in sound

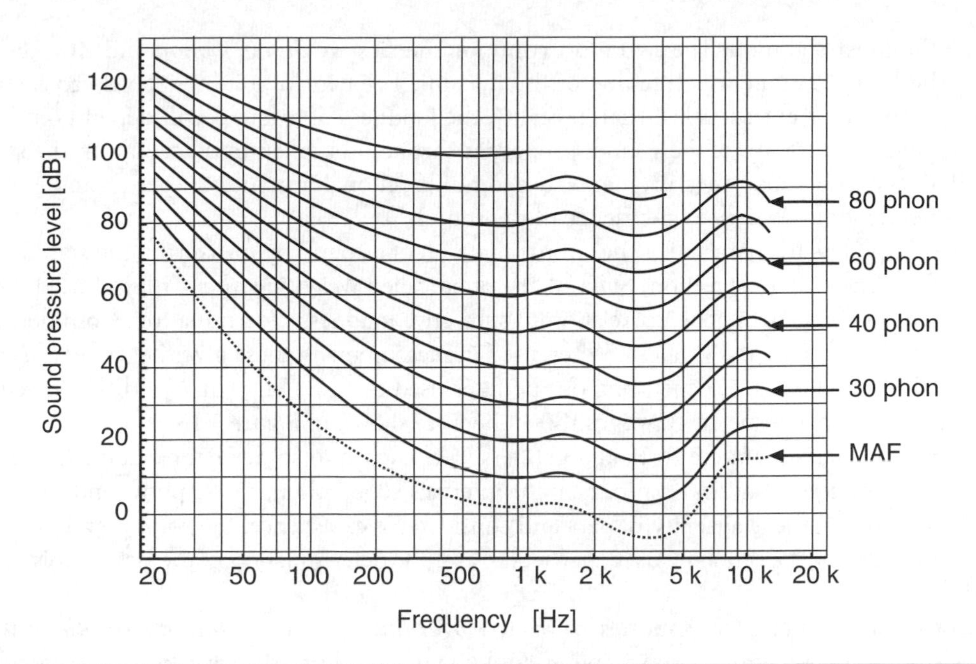

Figure 3.12 Equal loudness contours. The curve marked MAF illustrates the minimum audible sound field. (after ISO 226:2003)

pressure level for equal LL is described by *equal loudness contours*. These are measured using test persons in the same way as when determining the threshold of audibility, but in this case, the persons are asked to determine when the test tone at some frequency is equally loud as the reference tone. The standard deviations obtained in these types of psychoacoustic experiments are quite large, and the curves shown in Figure 3.12 represent averages of the equal loudness contours of many persons determined to have normal, undamaged hearing.

The equal loudness curves are very frequency-dependent at below 30 phon, particularly at low frequencies and low sound pressure levels. This is of great significance in the design of audio equipment since it makes it virtually impossible to have correct timbre in sound reproduction unless one listens at the same LL as was used in the actual recording. Only by using an adaptive frequency response equalization system that senses the LL (and has knowledge about the original level) can one have approximately the right timbre in sound reproduction. It is impossible to achieve this control by using what is known as *loudness compensation* in conventional amplifiers. A sound reproduction system playing louder than another does usually appears *better* in comparison, having seemingly stronger base and treble response. Because of this, it is always important to use the correct presentation sound levels in listening tests.

A trained listener has a threshold for *just noticeable differences* in sound pressure level of approximately 1 dB for pure tones and may sense down to about 0.3 dB deviations in octave bands in broadband spectra. At LL above 120 phon, the sound will be so strong that one can *feel* the sound, i.e., *threshold of feeling*. In the same way, there is a *threshold of pain* at approximately 140 phon.

It is useful to be able to measure the LL by instruments. Various methods have been derived to make this possible, both analog and digital. Today, digital methods based on various types of the spectral analysis are the most common. The two most common metrics are *Steven's phon* and *Zwicker's phon*, defined by the ISO 532 standard. Tracking of the dynamic behavior of loudness level can be difficult though. It is also important to realize that in reality, the loudness of a sound will depend on the angle of incidence and the duration of the sound, properties that are not taken into account in the standard. Conventional acoustic noise measurements use omnidirectional microphones, sensing the sound equally independent of the angle of incidence of the sound.

One usually prefers to characterize the approximate LL of a sound by measuring the sound pressure level with a frequency filter correction. Such filters, often called *weighting filters*, may, if used appropriately, yield measurements with good correlation to subjective loudness determination. Common weighting filters are so called A, B, and C filters. When used in measurement, these give the *sound level* reading expressed in dBA, dBB, or dBC depending on the filter used or as L_{pA}, L_{pB}, or L_{pC} [dB] as discussed in Chapter 2. The frequency response curves of these filters are shown in Figure 3.13.

The filter characteristics of the A, B, and C filters were chosen to somewhat resemble the sensitivity characteristics of the equal loudness contours in the intervals 20–55 phon, 55–85 phon, and 85–140 phon respectively. Because of the availability of data and considerable experience, the sound level expressed in dBA has become the most commonly used, particularly for the determination of the risk of noise-induced hearing loss.

As most sensory experiences, loudness is related *logarithmically* to *excitation strength*, in this case sound pressure, over some level. For example, a doubling of sound pressure at 1 kHz corresponds to an increase in LL of about 6 phon. An alternate way of measuring loudness is to measure using a metric, which has the property that an increase in loudness corresponds to a proportional increase in the metric value. Such a metric is *Loudness* (*N*), which is expressed in units of *sone*. By definition, 1 sone corresponds to a Loudness Level of a 40 phon, 1 kHz, reference tone. The Loudness value is equal to how many times stronger, or weaker, a sound is compared to this reference tone. Loudness can be measured

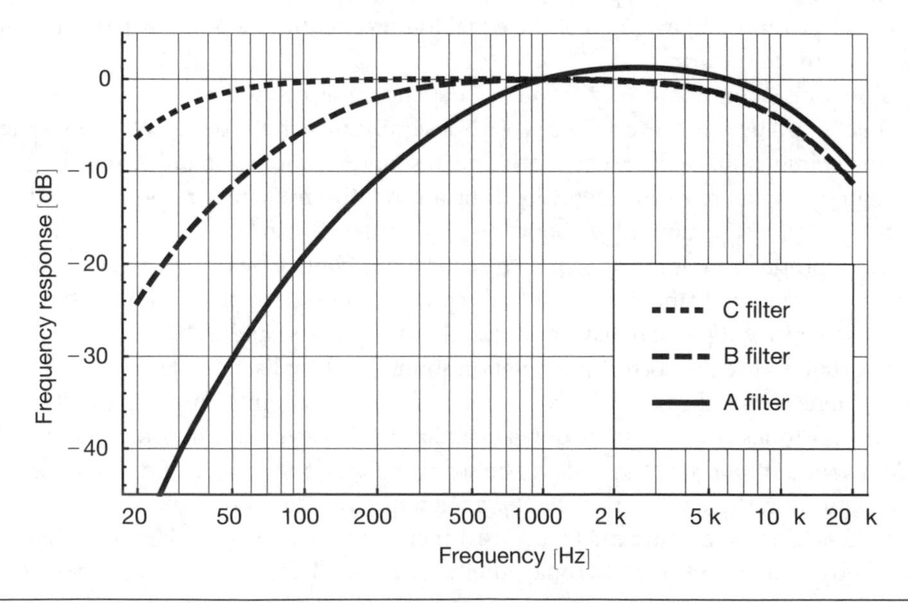

Figure 3.13 Attenuation characteristics for A, B, and C filters.

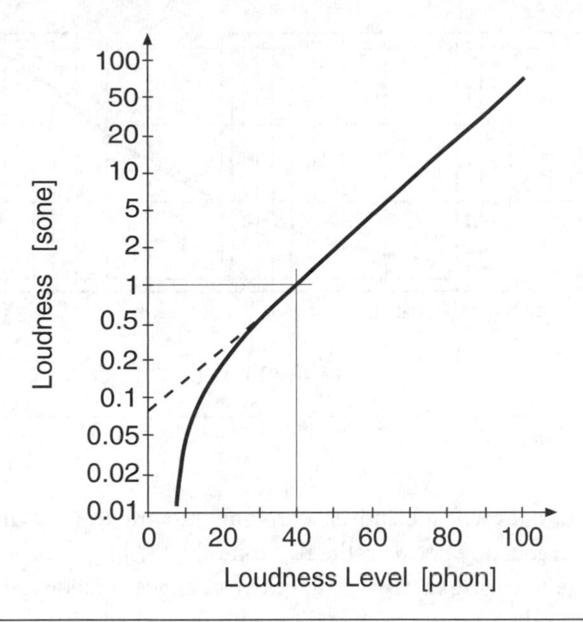

Figure 3.14 The nominal relationship between Loudness and Loudness Level at a frequency of 1 kHz. (After Ref. 3.3)

subjectively using a panel of listeners judging the loudness or by analog or digital means as with LL. The approximate relationship between Loudness and Loudness Level at 1 kHz is shown by the curve in Figure 3.14.

Pitch and Timbre

There are many subjective dimensions of sound other than loudness. They include *timbre*, the subjective perception of spectral content (frequency and balance between various parts in the spectrum) and *pitch*, the subjectively perceived *frequency* of a tone, the subjective *duration* of a sound, and the perception of the variation in time of sounds, such as *modulation* and *fluctuation*.

Timbre is defined as the attribute of sound that makes two sounds, having the same loudness and pitch, dissimilar. The timbre of a complex sound depends on its spectral characteristics. In audio equipment, various types of frequency selective filters are used to change the timbre of sounds, such as bass and treble filters.

Pitch is influenced by not only the frequency of the sound but also the intensity, the duration, and the spectral content. The presence of other harmonically related tones also influence pitch. Pitch is expressed in units of *mel*. A reference sine wave, having a frequency of 1 kHz, is defined as having a pitch of 1000 mel. One measures pitch by subjective comparison to this tone. Pitch does not vary linearly with frequency, as illustrated by the curve in Figure 3.15. The curve shows (ratio) pitch for various frequency tones heard at a loudness level of 60 phon.

Humans can sense static frequency deviations of 3 Hz or more at frequencies below approximately 0.5 kHz and of about 0.2–0.3 percent above, if the method of beats is not used. In the method of beats, one listens to two tones having frequencies f_1 and f_2 simultaneously, and, because of the limited frequency

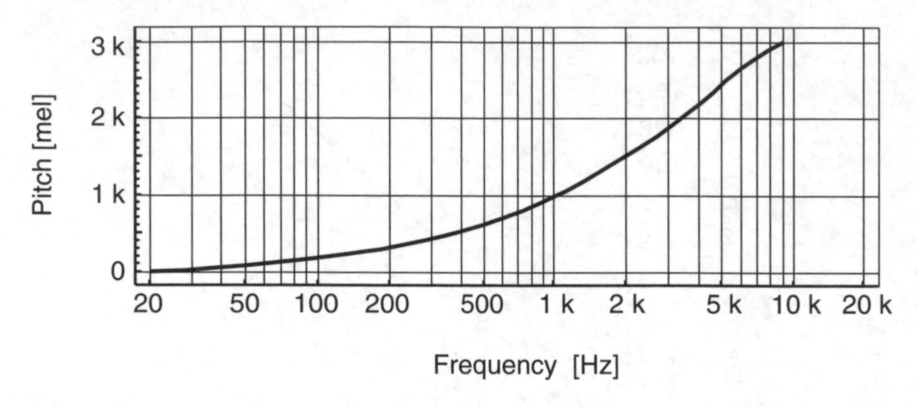

Figure 3.15 Pitch as a function of frequency (loudness level 60 phon). (After Ref. 3.3)

resolution of hearing, the two tones will subjectively combine into one tone having a frequency $(f_1 + f_2)/2$, modulated by the difference frequency between the two tones $|f_1 - f_2|$ if $|f_1 - f_2| < 10$ Hz.

Hearing is quite sensitive to various forms of repetitive changes, such as the amplitude or frequency of a tone. These changes are called *modulation*. The maximum sensitivity is found to modulation frequencies of about 4 Hz. Slow modulations having frequencies below 20 Hz are called *fluctuations*. For frequencies above approximately 20 Hz, the modulation subjectively changes character, and the modulation is perceived as *roughness*. The roughness sensitivity depends on the critical bandwidth. At 1 kHz the maximum roughness is heard for modulation frequencies of about 70 Hz (when the tone components are at the frequency band edges of the critical band).

An interesting phenomenon concerning pitch is virtual pitch. It is connected to a phenomenon called the *missing fundamental*. If one listens to a sound composed of a harmonic series of tones—for example, 200 Hz plus its first ten harmonics—the elimination of the fundamental (200 Hz) does not make much difference to the pitch. This is of great importance in audio engineering, since it allows even a loudspeaker with poor low frequency reproduction to have an *acceptable* low frequency response for many sounds.

3.4 EFFECTS OF EXCESSIVE SOUND EXPOSURE

Temporary and Permanent Threshold Shift

Excess sound exposure can damage various parts of the hearing mechanism, but even quite normal excitation can influence the sensitivity of hearing. We have noted earlier that the Stapedius muscle can become tensioned and in that way provide some hearing protection by reducing the vibration transmission by the ossicles. Short-time exposures, even below those that cause *noise-induced hearing loss*, may lead to so called *temporary threshold shift* (TTS). The recovery time depends on the sound pressure level and on the duration of the sound as illustrated in Figure 3.16. With exposure to a sufficiently high sound pressure level and long exposure times, hearing will not recover; the sound will have caused *permanent threshold shift* (PTS).

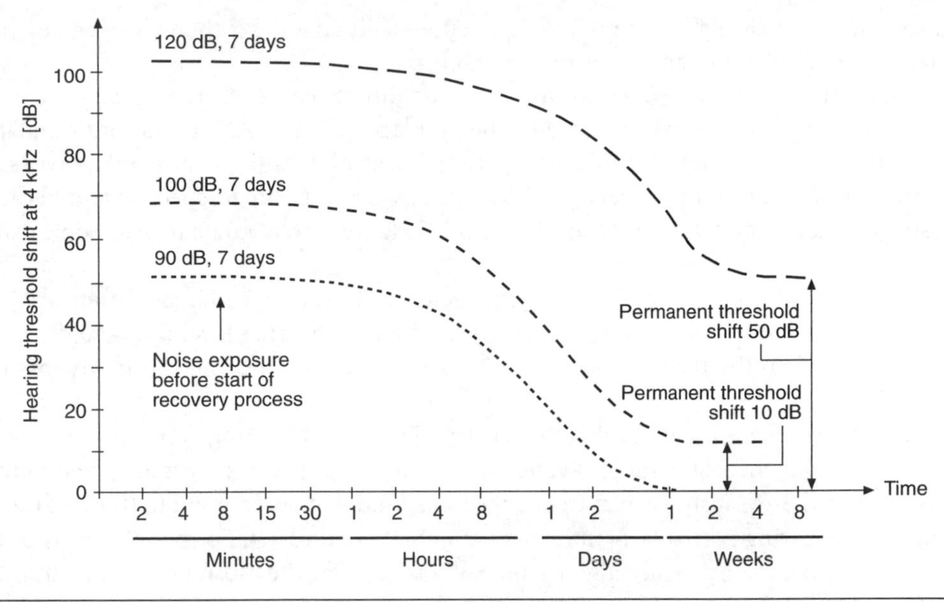

Figure 3.16 Examples of time dependence of recovery from TTS. (After Ref. 3.5)

Both TTS and PTS relate to phenomena in the inner ear, but PTS means that the sound exposure has led to permanent physical damage of the hair cells. The hair cells die, to some extent because the noise reduces their blood supply but also because of actual physical damage to the cilia. Such damage cannot be repaired, and the use of hearing aids can only restore some of the hearing abilities.

Hearing Impairment and Hearing Loss

The two main categories of hearing loss are *sensorineural hearing loss* that affects the cochlea or the auditory nerve and *conductive hearing loss* that affects the ear canal, tympanic membrane, or ossicles. In addition, there are cases of mixtures of these two categories, *mixed hearing loss.*

Sensorineural hearing loss is usually permanent and irreversible. Beside noise, it can be caused by poor blood supply or ototoxic chemicals. Since conductive hearing loss is of mechanical origin, it can often be compensated for by surgery in which the ossicles are replaced by a wire coupling the eardrum and the oval window. Conductive hearing loss is often the result of middle ear infections.

Prevention of noise-induced hearing loss is usually based on eliminating noise exposure, including less noisy equipment, workplace changes, use of ear defenders, or minimization of exposure time. Many studies indicate that hearing loss is linked to the *energy exposure,* that is, sound intensity and exposure time. This is the rationale for the use of the energy equivalent noise level $L_{\mathrm{Aeq},T}$ mentioned in Chapter 2.

Audiometry and Audiogram

In many countries, there is a requirement for yearly monitoring of an individual's hearing status using *audiometry* in an *audiometric testing program.* The purpose of such testing is to protect hearing sensitive employees so that their hearing is minimally impaired. Usually hearing is then checked at a number of specified frequencies, a method known as *tone audiometry*.

For each ear, the *pure tone audiogram* indicates the difference between a person's actual hearing ability and that accepted as normal, undamaged hearing according to ANSI and ISO standards, often expressed in dB HL (hearing level). Since occupational hearing loss (and most other types of noise-induced hearing loss) is caused by a variety of sounds having quite broad spectra, and because of the sound pressure amplification by the ear canal in the 3–4 kHz frequency range many people exhibit large hearing loss in this frequency region and above.

Tone audiometry average values for the frequencies .5, 1, 2, and 3 kHz are often used to define hearing loss. For adults mild hearing loss is defined as when the hearing level is 27–40 dB HL. A hearing aid is necessary when the the value exceeds 40 dBHL. Moderate hearing sensitivity loss is defined as 41–55 dB HL.

Figure 3.17 shows examples of hearing loss curves. The reason hearing is not completely damaged in the frequency range where it is most severely damaged is that the basilar membrane moves over a large area as discussed previously, even at frequencies of sound that correspond to those of the damaged hair cells. Impaired hearing particularly influences our ability to understand speech and to enjoy music. The consonants that consist of mainly high frequency sound suffer the most, whereas the fundamentals and harmonics of vocal sounds are less affected by hearing loss. However, in most western languages, the consonants are the major carriers of speech information.

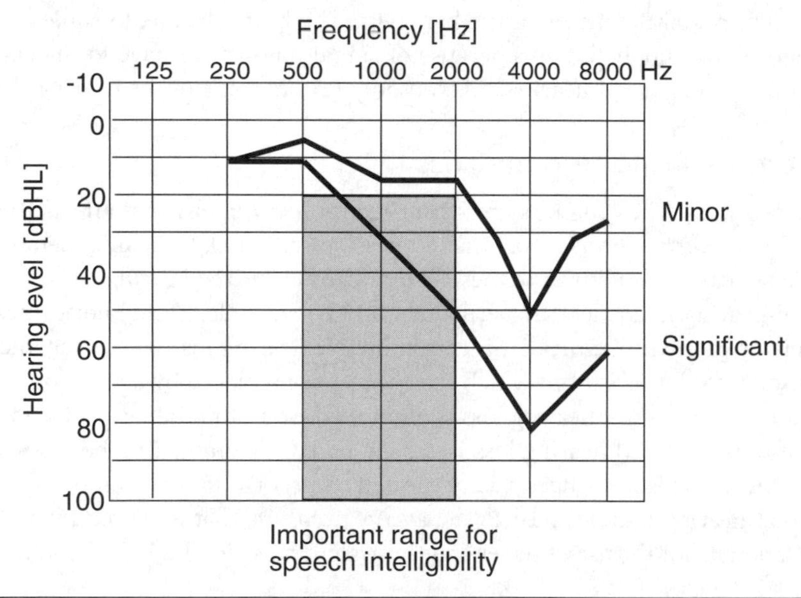

Figure 3.17 Tone audiogram showing examples of minor and significant hearing loss. Significant hearing loss is said to have occurred when the hearing loss curve reaches the shaded area. Compare to Figure 3.30 to see important frequency ranges for voiced and unvoiced sounds.

This text is an attempt at indicating the problem:

- Vowels only—••e •o•• o• •o••e••a•io• •o• •ea•i•• •o•• i• •i••.
- Consonants only—Th• c•st •f c•mp•ns•t••n f•r h••r•ng l•ss •s h•gh.

An additional type of audiometry is *speech audiometry*, where the difficulty in discriminating speech in the presence of noise is assessed.

Many people suffer from *age-related hearing loss* (ARHL) that usually takes the form of high-frequency hearing loss and is associated with a loss of hair cells, sensitive at high frequencies. This loss in the elderly is often called *presbycusis*. ARHL is a result of general wear-and-tear due to noise exposure in society. Health-related conditions such as high cholesterol levels, and habits, such as smoking, that affect the cardiovascular system, seem to influence ARHL. Some typical curves showing hearing loss due to ARHL are shown in Figure 3.18.

Because of changing working habits, it is likely that there will be a smaller gender difference regarding ARHL in the future. It is also important to realize that there is a difference between *hearing disability* and hearing impairment.

Tinnitus

Tinnitus is when a person perceives sounds, even though there is no sound exposure. Tinnitus is often the result of excessive noise exposure to either continuous or transient noise. Even a single exposure of sufficiently strong impulsive noise may cause hearing damage and associated tinnitus. Tinnitus may cause sound impressions that can be sporadic or steady, noisy or tonal. Sometimes the perceived sound can be so strong as to cause annoyance or prevent sleep, among other things. Often, tinnitus is a side result of hearing damage that has resulted in permanent hearing loss.

Tinnitus is often the result of sporadic hearing cell activity. One can think of the tinnitus-induced sound as being caused by hearing amplifying in spite of not having anything to amplify (because of the hearing loss). As in other regenerative systems, excess amplification can lead to instability and oscillation. Usually the tinnitus phenomenon starts as noise that after some time converts into a steady tonal oscillation. Once tinnitus has occurred, rest, absence of stress, and blocking sounds, such as low-level music, can help minimize the subjective discomfort.

Criteria for Excessive Noise

In 1990, the European Community instituted a noise legislation that sets a limit so that action is taken when the daily personal exposure of a worker to noise is likely to exceed 85 dBA. (The EC Directive on Noise 2003/10/EG is the most recent version at the time of writing.) One then has to take appropriate action as specified in the directive:

- Measure noise levels
- Provide ear protectors and train workers in their use
- Inform workers of the risk
- Make hearing checks available

In case the sound level exceeds 90 dBA, further action should be taken.

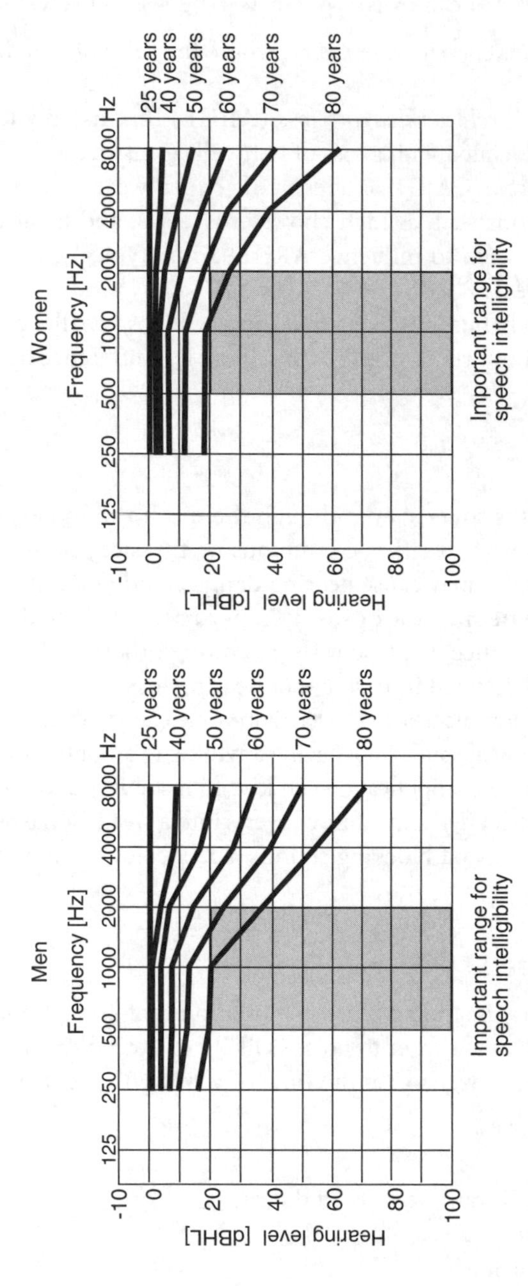

Figure 3.18 Hearing threshold shift due to age-related hearing loss for various age groups (for men and women respectively). (After Ref. 3.6)

The limiting value of 85 dBA does not imply that hearing loss cannot occur at sound levels below this value. For negligible risk, the sound levels have to be below 70 to 75 dBA in conjunction with hearing rest outside working hours.

In many other countries there is an additional national standard for the determination of risk for hearing impairment and hearing damage by excessive sound.

Impulsive Noise

Impulsive noise is particularly troublesome as a cause for noise-induced hearing loss. The reason is that hearing does need a certain amount of time to perceive sounds. Since much of the damage that can be caused by excessive noise is due to mechanical overload in the cochlea, the damage can be fairly instantaneous. Gun noise and other explosion-type noise can result in single exposure damage. The Stapedius reflex needs some time to react, which is not possible in the case of isolated transients.

Even repetitive tone bursts are not perceived as loudly as the steady-state tone having the same amplitude, shown by the data in Figure 3.18. The curves show the length that a burst of sound needs to be perceived as loudly as the steady-state tone. One sees that a duration of approximately 0.2 s is needed for this to occur. Figure 3.19 illustrates the sound level meter characteristics, *IEC impulse*, *IEC fast*, and *IEC slow*. One notes that even the impulse setting does not allow measurement of the peak values of short, transient sounds. Repetitive impulsive noise often takes on the properties of continuous noise.

Ear Defenders

The two most common types of ear defenders are circumaural defenders and earplugs: earmuffs that cover the external ear and insert earplugs that seal the ear from sound, respectively. Some examples of *real ear attenuation* for some protectors are shown in Figure 3.20.

The efficiency of earmuffs and earplugs in providing protection from noise depends on how the devices fit the person. Earplugs and earmuffs can be equally good protectors. The fact that ear muffs are larger does not mean that they necessarily protect better. Earmuffs often suffer from poor protection at low frequencies

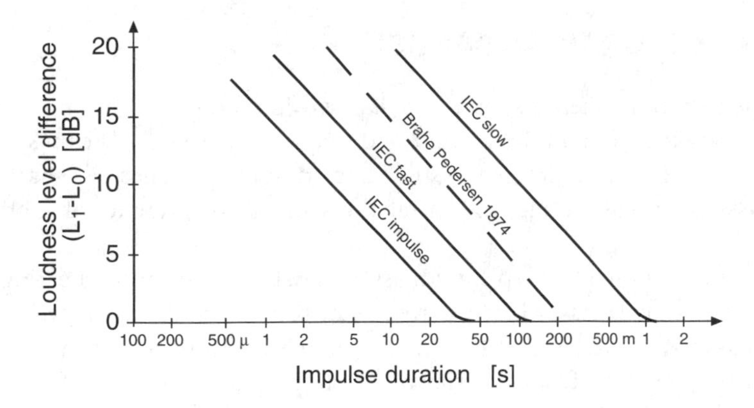

Figure 3.19 Measured loudness level differences between 1 second exposure and tone burst exposure for different burst lengths (at 1 kHz). The curves marked IEC slow, IEC fast, and IEC impulse correspond to the integration characteristics of sound level meters. (After Ref. 3.7)

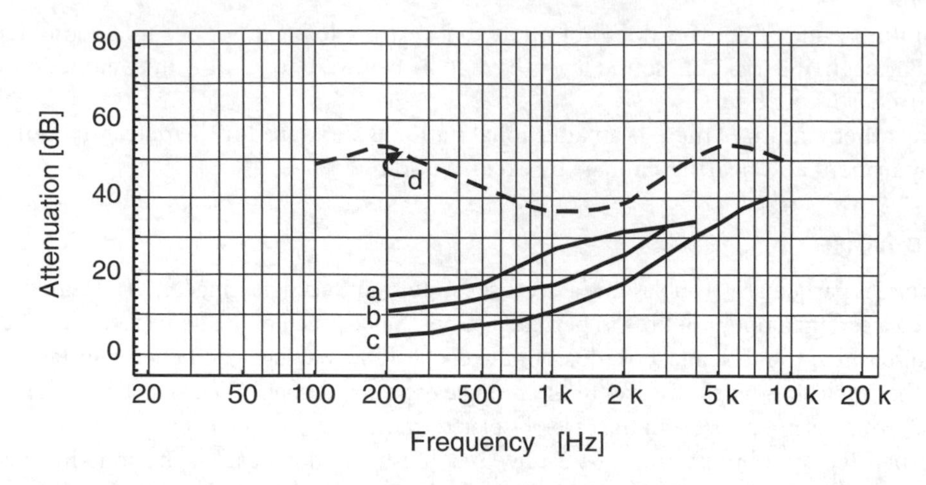

Figure 3.20 Measurement results for the insertion loss by some ear defenders: (a) neoprene insert earplug, (b) earmuff with negligible leakage, (c) earmuff with normal leakage, and (d) bone conduction limit.

because of the resonance due to earmuff mass and head-band compliance. People have different preferences. Many people prefer earplugs since they are easier to carry, lightweight, disposable, and do not show. Earplugs are often preferred in humid climates.

There can be significant differences in attenuation as measured using a manikin or simulator versus the attenuation measured using the response from people. The reason is that it is difficult to take *leakage*, *skin compliance*, and *bone conduction* into account when measuring ear defender insertion loss using a manikin. The skull is set into vibration by surrounding sound. This vibration reaches the cochlea by bone conduction and limits the maximum protection of ear defenders to around 40 dB.

The availability of ear protectors using *active noise cancellation* means that it is, in principle, possible to tailor the insert loss of ear protectors to the spectrum of the unwanted sound.

3.5 MASKING AND CRITICAL BANDS

Masking is the phenomenon when a sound, the *masker* blocks the perception of another sound, the *maskee*. Masking can take place both in the time and frequency domains. The *masking level* is the level by which one sound needs to be increased to still be heard in the presence of a masker. Simplified, one can regard the masking level as the shift of hearing threshold in the presence of another sound, such as a tone or a noise.

Figure 3.21 shows an example of spectral masking. It shows the shifted hearing thresholds of the maskee measured with a tone masker having a frequency of 1 kHz at different levels L_M. The masking is largest when the masker and maskee are close to one another in frequency. The masker's effect is mostly active upwards in frequency so from the viewpoint of speech intelligibility, low frequency masking sounds will also be disturbing, since they will mask speech.

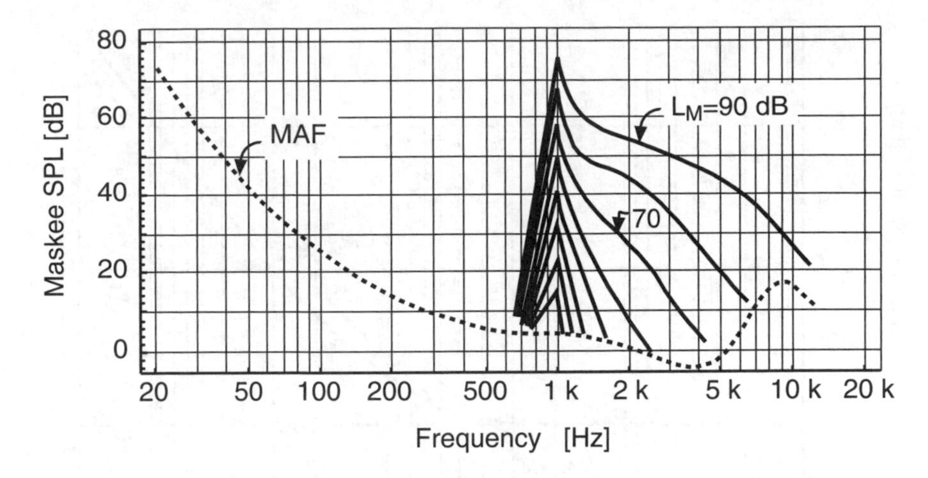

Figure 3.21 Hearing threshold shift of a test tone (maskee) due to masking. The masker is a 1 kHz tone at different sound pressure levels L_M. Solid curves show the resulting hearing threshold of the maskee in the presence of the masker. The dashed line shows the threshold of hearing for tones (MAF). (After Ref. 3.3)

Hearing is characterized by frequency analysis some of which is accomplished already by the basilar membrane. The stimulus is analyzed equivalent to an adaptive filter bank with a large number of parallel frequency selective filters. These are similar to those of bandpass filters used in signal processing by electronic equipment and are called critical band filters. The audio frequency range can be subdivided into 32 critical bands, each having its own "critical band-width".

The critical bandwidth depends on the frequency as shown in Figure 3.22. Up to about 0.5 kHz the critical bandwidth is fairly constant, but at higher frequencies the bandwidth is about that of a third octave band wide bandpass filter. At a frequency of 1 kHz the critical bandwidth is about 160 Hz. There are several ways of measuring the critical bandwidth, these give slightly different results. A common way to measure the critical bandwidth is to use noise having a constant spectral density, increasing the bandwidth of the noise only and to investigate the masking of an added pure tone. Reaching a certain bandwidth, the masking no longer increases as the bandwidth is increased. This bandwidth is the critical bandwidth.

Psychoacoustic experiments have shown that our ability to hear a sound, for example a tone, in the presence of noise centered on the frequency of the tone depends on the bandwidth of the noise. In order for a tone to be heard its power, approximately, has to exceed that of the noise within the critical bandwidth.

In most practical situations, the masker is a noise-like signal with some bandwidth. Wide-band noise will have energy over a bandwidth much wider than the critical bandwidth. It is then necessary to take the masking of several bands into account, since masking is asymmetric and primarily acts upward

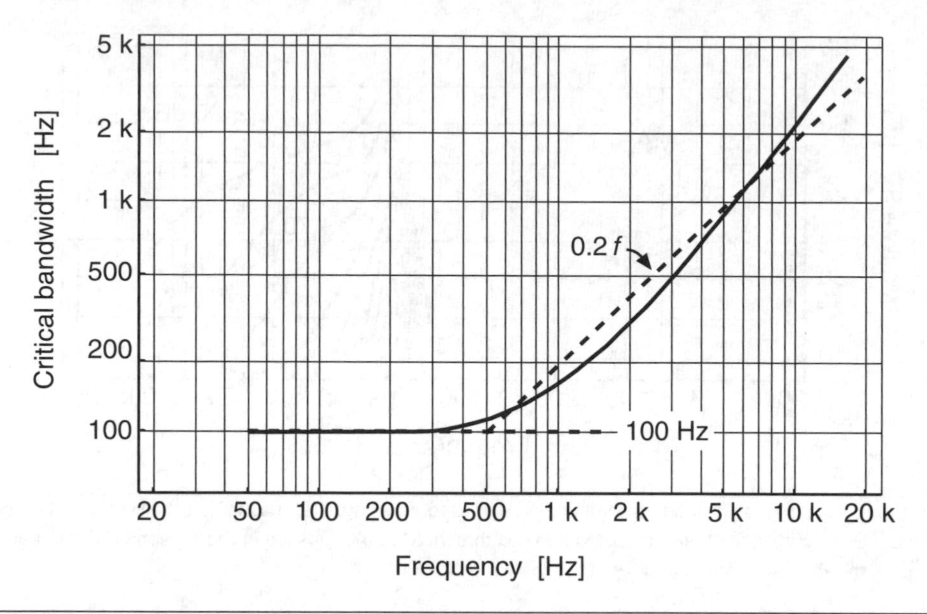

Figure 3.22 The solid curve shows the critical bandwidth as a function of frequency. The broken line marked 0.2*f* shows the third octave bandwidth as a function of center frequency. (After Ref. 3.3)

in frequency. This is done in the determination of loudness and loudness level according to the ISO 532 standard.

3.6 DISTORTION, LINEARITY, AND HARMONICS

Changes to a signal's properties are called *distortion. Linear distortion* is characterized by simple frequency-dependent amplitude changes to various parts of the spectrum. *Nonlinear distortion* is characterized by the generation of new sounds at frequencies that are multiples of the frequency of the original sounds, *harmonics*, and is called *harmonic distortion*. This type of the distortion is called *intermodulation distortion* (see Chapter 17). Some nonlinear processes create *subharmonics*. When the signal contains many frequencies, the distortion will take on a noise-like quality; this is of importance in digital signal processing. Sometimes nonlinear resonant systems may show *chaotic* behavior.

Hearing has its own nonlinear distortion; one can easily note this distortion when listening *monaurally* or *binaurally* to the reproduction of two tones or other sounds. One will note a difference between when the sum of the sounds is presented to both ears and when the two sounds are presented to each ear separately. The nonlinearities that can be detected in this way are a property of the cochlea.

Subjectively, low pass filtered sound reproduction may be superior to full frequency range sound reproduction, if the sound reproduction system is characterized by nonlinearities. Digital signal processing may be used to generate tones that are not present originally in the signal but which hearing expects. This technique is sometimes used to extend the frequency range synthetically, at both low and high frequencies.

Consonance is an example of a nonlinearity that takes place after the cochlea. Consonance is characterized by the property that tones having frequency relationships of the forms 2/1, 3/2, 4/3, and so on are perceived as pleasing.

3.7 BINAURAL HEARING

Binaural hearing is listening to sound using both ears. Binaural hearing leads to a clarity and subjective perception of auditory space that cannot be had when listening monaurally, such as when using only one ear. The reasons are three phenomena: (1) localization, (2) noise suppression, and (3) decreased masking.

One can better discriminate between sounds coming from different directions because of the better localization ability that is the result of the ability of hearing to perform cross-correlation analysis of the signals from the two cochleae. This is sometimes called the cocktail party effect because it allows better discrimination when listening to many sounds coming from various angles to the median plane. The cross-correlation analysis is successful because of the level and time differences between the signals from the cochleae when sounds are incident away from the median plane. The differences are a result of the reflection and shadow action of the head.

Localization Using Interaural Time and Level Differences

In studying binaural hearing, it is convenient to model the head as a sphere, having a radius r of 8 cm, with the ears positioned diametrically, as shown in Figure 3.23. The measured interaural time delays suggest that the equivalent diameter of the average head is approximately 16 cm.

The additional travel distance difference for signals reaching the far ear is $r(\theta) + \sin(\theta)$. For a sound arriving from an angle of $\theta = 0.96$ radians, the difference in arrival time will be approximately 0.45 ms. In the case of a tone having a frequency of 0.1 kHz, this corresponds to a phase difference of 0.045 periods, at 1 kHz a difference of 0.45 periods, and at 10 kHz a difference of 4.5 periods.

Experiments show that at frequencies below approximately 1 kHz hearing primarily uses phase differences for localization of sounds. At higher frequencies, it is primarily the time differences in the signal envelope that are important for localization.

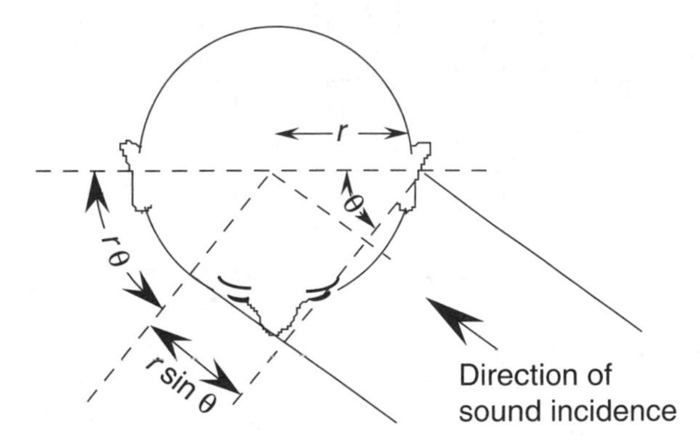

Figure 3.23 A simplified model of the human head for the calculation of the interaural path length difference.

Hearing uses these *inter-ear time differences* (ITDs) in the arrival of sound to the ears for determination of the angle of incidence relative to the median plane. The head causes *inter-ear level differences* (ILDs) of as much as 5 dB at frequencies as low as 1 kHz; at frequencies above 3–4 kHz, the ILD is sufficient for localization.

In case of localization difficulty, small head movements are necessary to decide on the arrival direction. Figure 3.24 shows the localization error using binaural hearing for sounds incident in the horizontal plane (*lateralization error*).

The noise suppression is due to reduced masking. Some undesired sounds, such as noise or excessive reverberation, often arrive simultaneously from many directions, being diffusely distributed in the room. The desired sound can then often be heard more clearly by turning one's head so that one ear faces the sound source. The sound reflecting properties of the head and the correlation ability help us hear the desired sound. The cross-correlation process improves the clarity of the direct sound over the noise and the reverberation.

These properties are of great use to us in daily listening. Binaural sound reproduction uses these advantages. In this technique, sound is picked up by a manikin head and torso, using two microphones mounted in the conchae of pinna replicas. An example of a commercial binaural recording manikin (dummy head) is shown in Figure 5.24. The left and right microphone signals are recorded and then played back to the listener binaurally, using two headphones or using loudspeakers and a crosstalk cancellation filter.

Localization Using Only Spectral and Level Changes

For sounds arriving in the median plane, both ear signals are essentially the same since human ears are fairly symmetric. In this case, the localization ability is poor, since median plane localization to some extent will depend on memory cues.

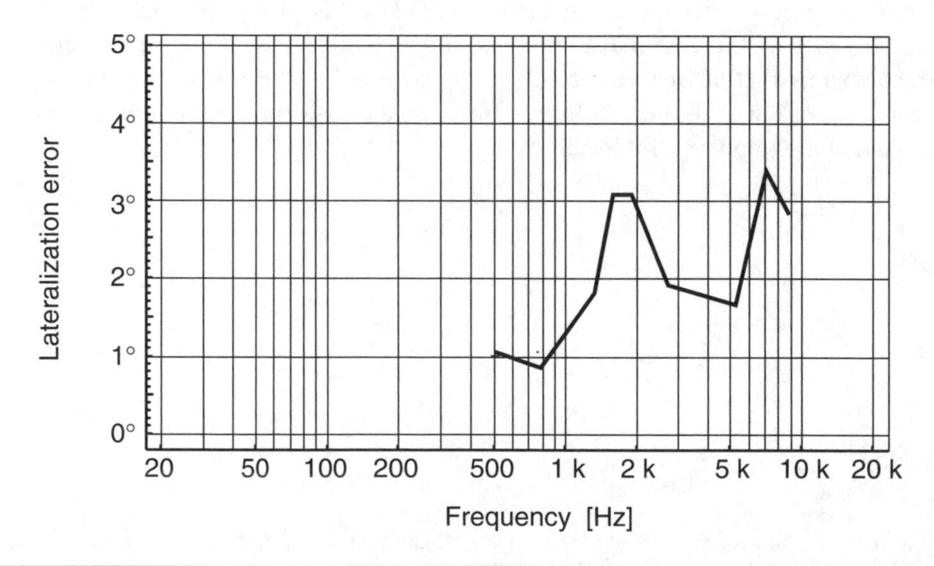

Figure 3.24 Localization uncertainty for tonal sounds incident in the horizontal plane for frontal incidence. (After Ref. 3.3)

Some examples of the spectral differences as a function of angle of arrival in the median plane are shown in Figure 3.25. Tests have shown that one can change the subjectively perceived angle of sound incidence in the median plane by spectral changes to the sound. The result of such an experiment is shown in Figure 3.26. The sound, in this case, was always coming from the same place, but the perceived location in the median plane of a narrowband noise changed, as the center frequency of the noise changed. The localization was in the direction where the HRTFs amplified the respective frequencies.

The poor localization in the median and horizontal plane when using monaural hearing are related; in both cases, the direction finding ability is limited to what is provided by memorized sound cues. Figures 3.27a and b show results from such experiments.

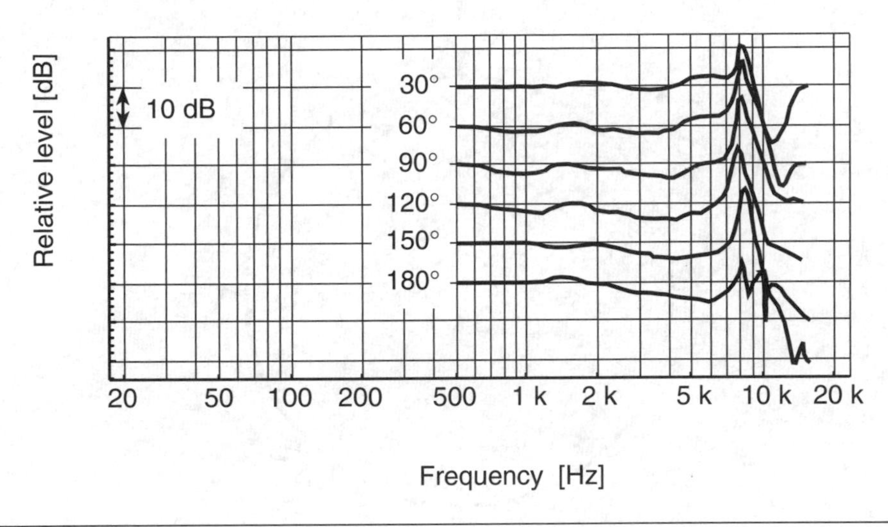

Figure 3.25 Relative frequency response curves for 6 different angles of sound incidence in the median plane (φ = 0). Each curve is normalized to the response at 0.5 kHz for the respective angle. The median plane angle is δ in Figure 3.2. (After Ref. 3.8)

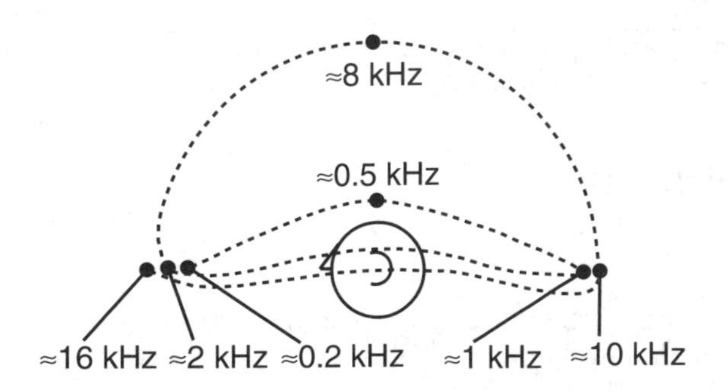

Figure 3.26 Result of experiment showing virtual median plane source positions for various narrow-band noise signals. (After Ref. 3.1)

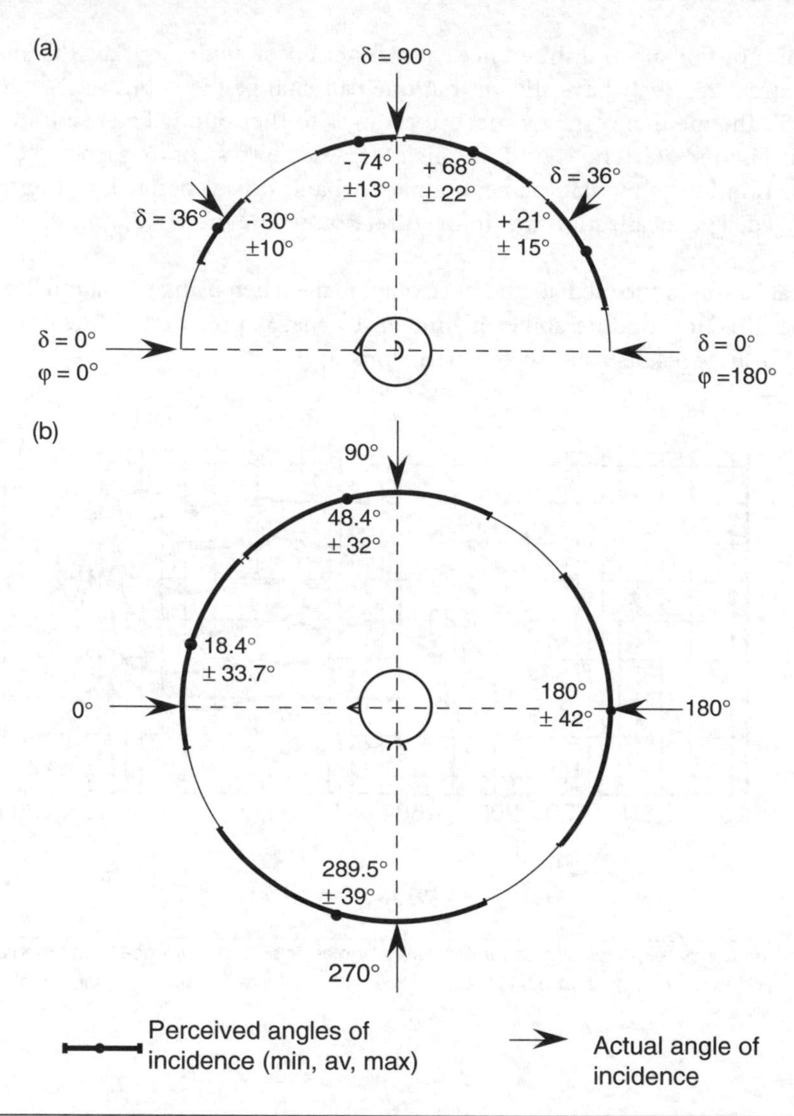

Figure 3.27 Localization and localization uncertainty in the median plane (a). Localization and localization uncertainty in the horizontal plane using one ear (b). (After Ref. 3.1)

3.8 VOICE AND SPEECH

The Voice Mechanism

The fundamental components of the voice mechanism are the lungs, the *vocal cords*, the tongue, the mouth, and the nose, as shown in Figure 3.28. The voice mechanism is powered by air that is exhaled from the lungs.

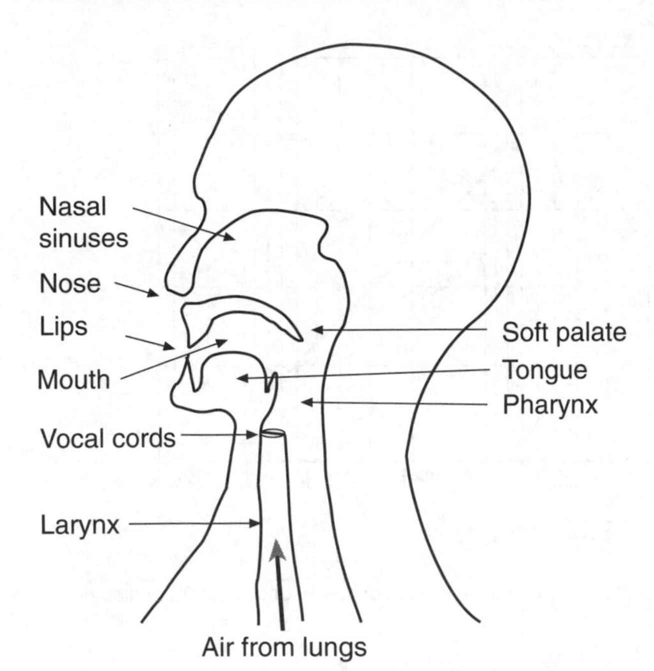

Figure 3.28 Important components in the voice generation system.

There are two basic voice sound components: *voiced* and *unvoiced* sounds. Voiced sounds are a result of the modulation of the air stream by the oscillation of the vocal cords in the larynx. The length of the vocal cord opening and the tension of the vocal cords determine the fundamental frequency of voiced sounds, which differs in men and women. The modulation of the air stream is sawtooth-shaped and is characterized by many harmonics. The acoustical filter, created by the volumes of the throat, the mouth, and the nose, has many resonances that amplify and attenuate the various frequency components in the air stream.

Unvoiced sounds, fricatives such as f and s, and stop consonants such as p and t, are produced by modulation of the air stream using the tongue and lips. The unvoiced sounds have a wide frequency range because of their noise like or transient character.

Spectral Characteristics of Speech

Speech has a wide frequency range, with frequency components from approximately 80 Hz up to 10 kHz. An example of the longtime-averaged speech pressure spectrum level, averaged for men and women, is shown in Figure 3.29; its sound pressure level is approximately 60 dB at a distance of 1 m in front of the talker. The sound power level of the voice is approximately 70 dB. A trained speaker can change the power output of the voice over approximately a 20 dB range. A whisper, entirely unvoiced speech, has an approximate sound power level of 30 dB.

By changing the shape of the volumes—for example, by moving the tongue—one can change the filter parameters and by that the spectra of vocal sound. The peaks in the spectra of vocal sounds are called formants. Figure 3.30 illustrates approximate spectral positions for some speech sounds.

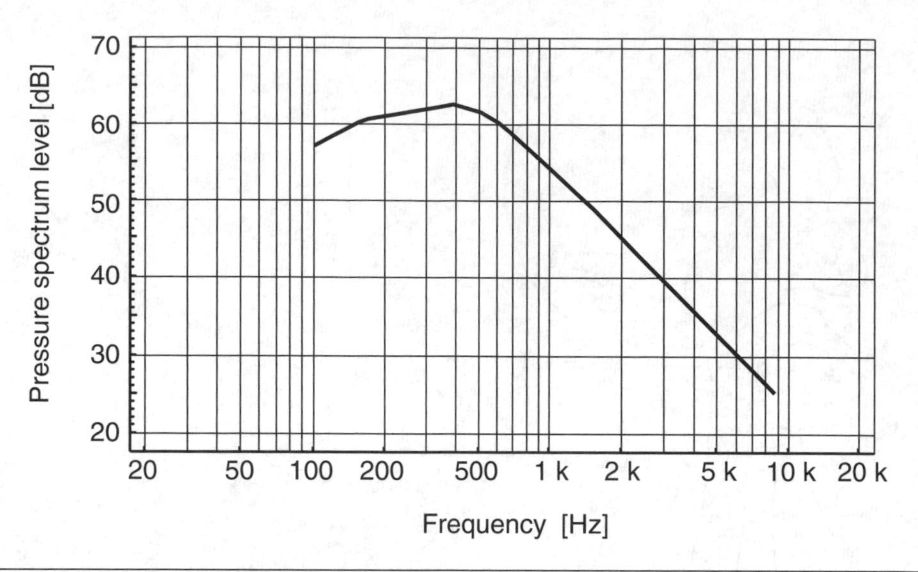

Figure 3.29 An example of longtime averaged speech pressure spectrum level in frontal direction at 0.05 m distance from mouth. (After Ref. 3.9)

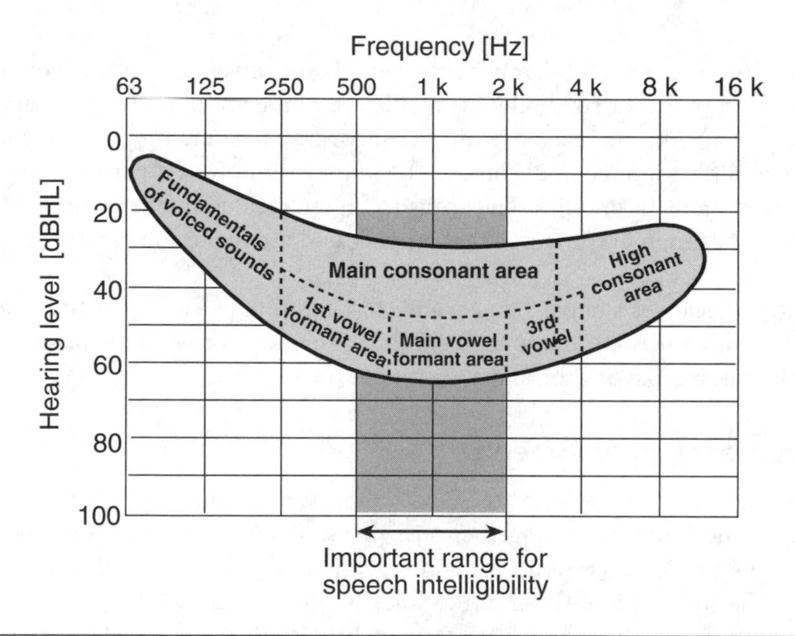

Figure 3.30 Sound pressure levels for voiced and unvoiced speech sound components drawn using the audiogram approach, that is, relative to the threshold of hearing (MAF) shown in Figure 3.12. (After Ref. 3.9)

Figure 3.29 shows further that most of the energy is concentrated at relatively low frequencies due to the voiced sounds. The unvoiced sounds have predominantly high frequency spectral content and are relatively more important for speech intelligibility.

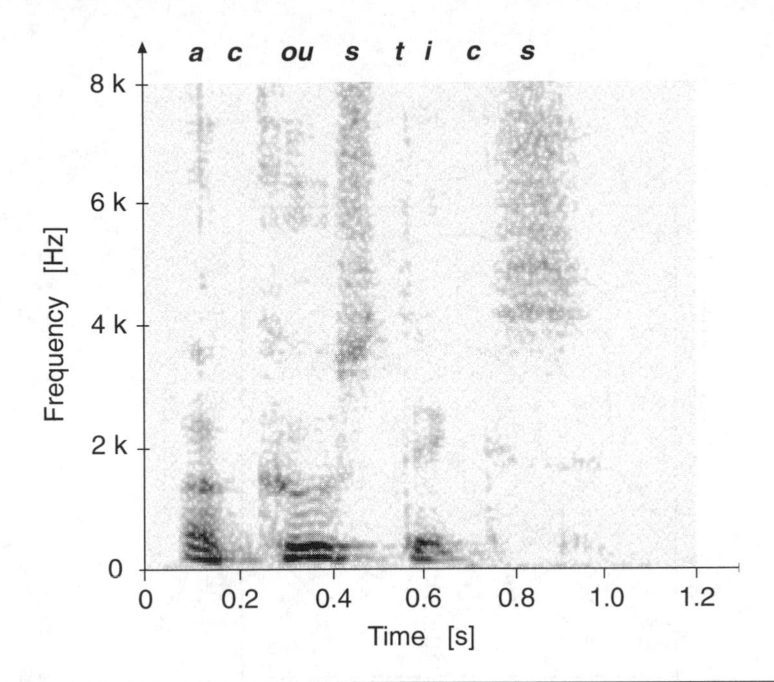

Figure 3.31 A time-frequency analysis for the spoken word acoustics.

It is convenient to show the spectral content of speech using *spectrograms* (time-frequency plots of spectral content). An example of such a spectrogram is given in Figure 3.31. One clearly sees how speech is modulated in various frequency bands. The fundamental modulation frequency of the speech *envelope* is approximately 7 Hz.

Voice Directivity

Speech is primarily radiated by the mouth and nose openings; only at the lowest frequencies is there radiation by the chest. The sound shadow caused by the head will cause some directionality though, as shown by the curves in Figure 3.32.

The directivity is quite substantial at frequencies over 1 kHz; there is a more than 5 dB sound pressure level difference between front and back radiation. Since most of the information in speech is at high frequencies, this means that one should strive to speak in the direction of the listener.

It is clear from this data that any testing of room acoustic quality for speech or song that does not include this directivity will be subject to a systematic error. Fortunately, it is possible to simulate reasonably well the directivity of the human voice by using a glass fiber baffle with a small rectangular hole in front of the loudspeaker driver of a small loudspeaker box as illustrated in Figure 3.33.

Speech Intelligibility

It is of great interest to understand how the relative attenuation of various speech spectrum components affects the *perception of speech*. Many situations critically involve correctly understood speech, such as

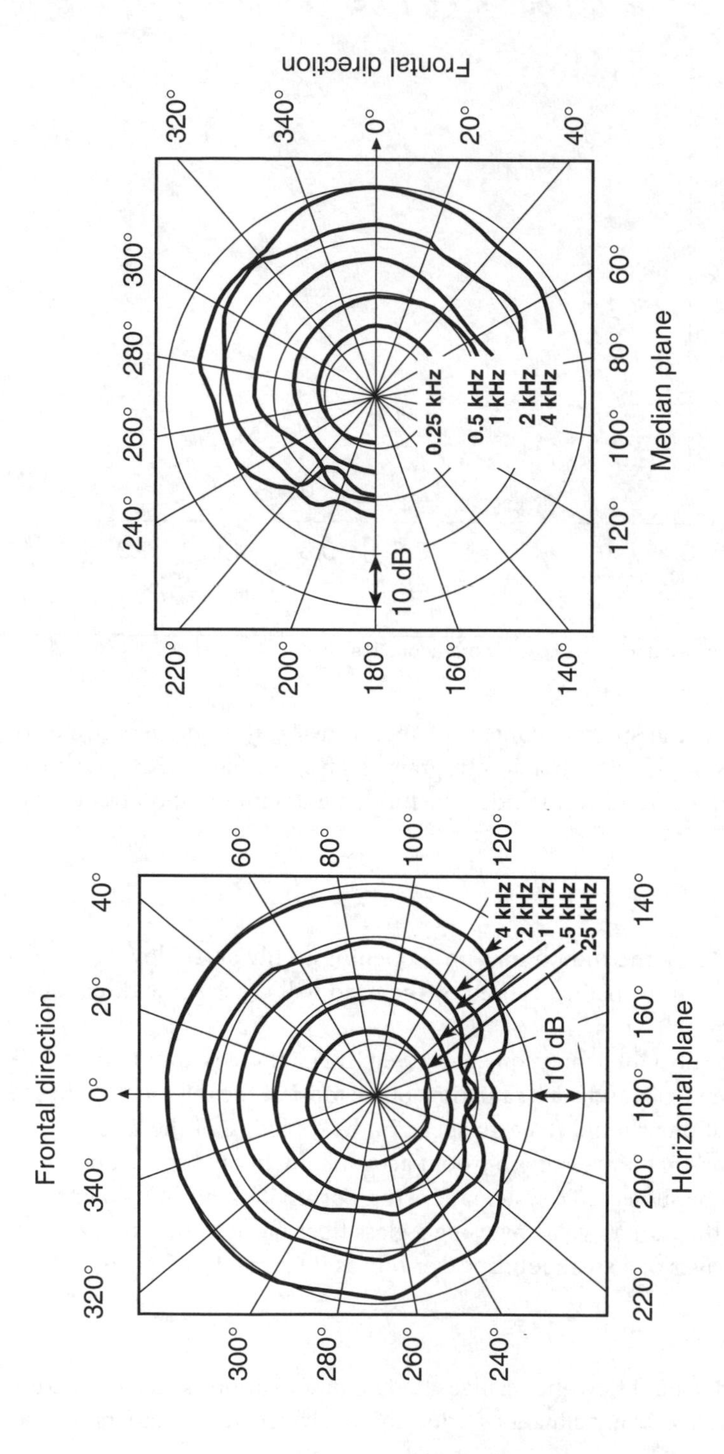

Figure 3.32 Polar plots for voice directivity in the horizontal and median planes. (After Ref. 3.10)

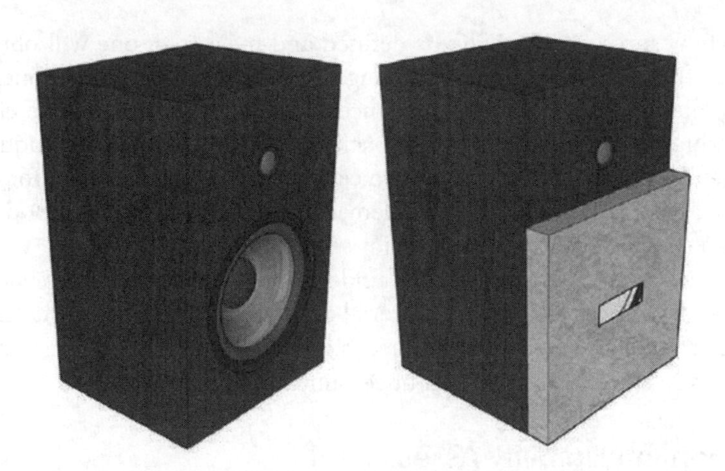

Figure 3.33 Using a commercial 2–way loudspeaker having similar dimensions as the human head (left), it was possible to achieve suitable directivity by attaching a baffle with a 30 by 10 mm hole in front of the low frequency unit (right). The baffle was made of 10 mm thick hard-pressed glass wool. The directivity that resulted was similar to that of the human voice.

classroom teaching, warning systems, or flight control systems, among others. The curves in Figure 3.34 show the relative influence of the various spectral components of speech on speech perception. One sees that most of the speech information is contained in the frequency range between 0.4 kHz and 6 kHz. Many transmission systems decrease this bandwidth even further; for example, the GSM system for mobile telephony uses a 4 kHz upper frequency limit.

It must be noted that speech intelligibility is not the same as speech (sound) quality. Speech quality is related to the result of linear and nonlinear distortion and the way the speech sounds.

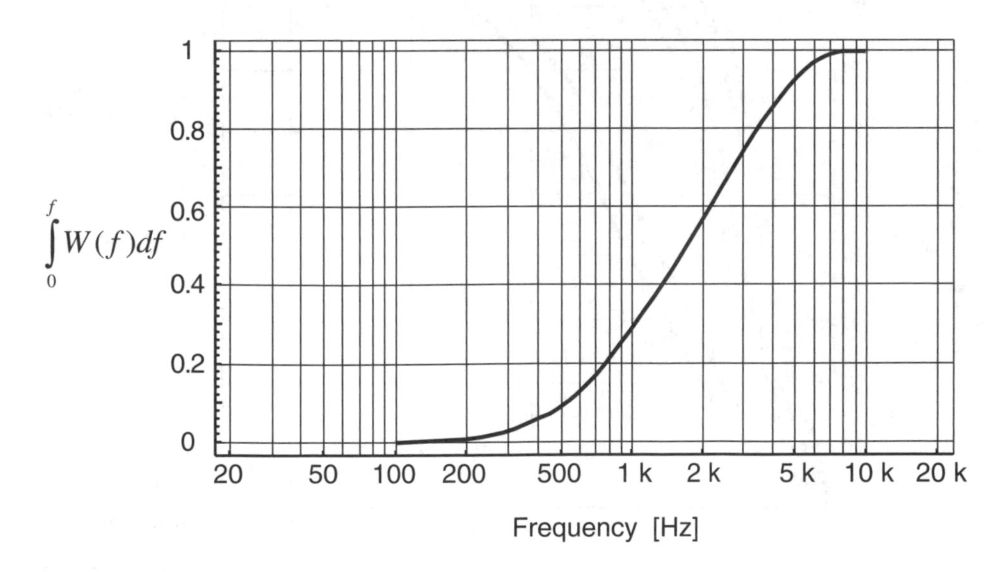

Figure 3.34 The integrated weight function for estimation of the importance of various parts of the spectrum on speech intelligibility. The region of largest slope is the most important for intelligibility. (After Ref. 3.11)

Depending on how *speech intelligibility* is defined and measured, one will obtain varying results. The intelligibility can be estimated by the percentage of correctly perceived phonemes, single syllable nonsense words, single syllable phonetically balanced words, or sentences. The choice of word lists, listener ability (hearing threshold, intellectual capacity), and presentation techniques all influence the intelligibility test results. In addition, the talker's voice spectrum, pronunciation, for example, will influence the test results. Figure 3.35 shows how the percentage of correctly understood syllables, words, or sentences varies with signal-to-noise ratio (S/N) for some different test items.

The percentage at which items are correctly understood is often called *articulation*, and it is usually necessary to specify which type of word material or sounds were used in the test—for example, by specifying sentence articulation. To assess the speech intelligibility it is necessary to use listeners, consequently the results will show large statistical uncertainty.

Objective Speech Intelligibility Assessment

Because of the difficulties, uncertainties, inconveniences, and costs involved in subjective speech intelligibility assessment, many metrics have been devised to estimate the influence of various changes to the speech spectrum and to the modulation of components of the speech spectrum. It is important to remember in using such metrics that they are only approximations and that the results obtained are usually specific for a certain test situation. Most of the metrics also do not even take the directionality of speech into account. One should always be cautious when using speech intelligibility metrics, particu-

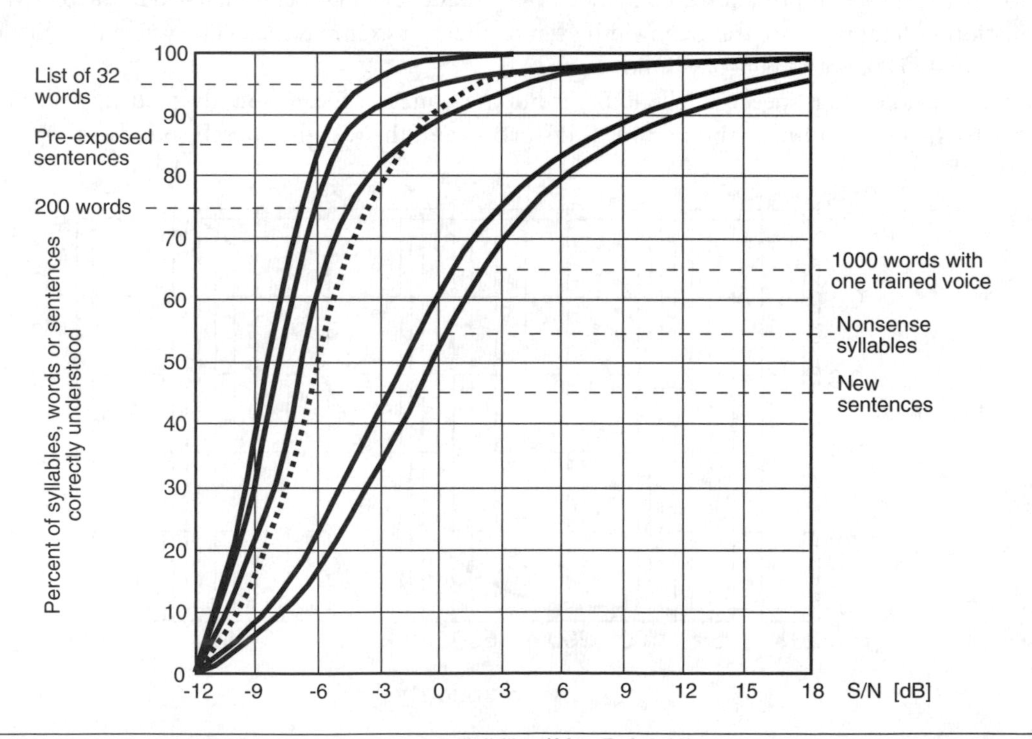

Figure 3.35 The influence of S/N ratio on speech intelligibility. (After Ref. 3.12)

larly within the field of room acoustics. Many metrics, however, function quite well for single channel, analog telephone communication situations.

The *articulation index* (AI) is a metric designed in the 1940s to estimate the quality of radio telephony and was intended to be used for single channel voice communication in the presence of noise. It can be estimated through listening tests, using the number of correctly understood voiced and unvoiced sounds according to a special formula. One can also use a table or nomogram by which the AI can be estimated from the S/N ratio. There is now a more modern and similar index, the *speech intelligibility index* (SII), defined by an ANSI standard. The SII is a simple way of roughly estimating the influence of noise on speech intelligibility.

Articulation loss of consonants (%ALcons) is still a popular metric among designers of speech reinforcement systems for stadiums and other sports arenas. It is calculated from knowledge of the level relationships between direct and reverberant sound and reverberation time.

Modulation damping (MD) is a simple, little used, but good metric based on the modulation characteristics of speech. The test signal consists of white noise that has been frequency weighted using an A filter and then chopped into *noise pulses*. The noise pulses consist of 70 ms of noise and 140 ms of silence. Because of this property, it can be easily generated and used to estimate the influence of both reverberation and noise on the quality of sound transmission. One simply measures the modulation index of the noise before transmission and after reception, then subtracts. The results using the modulation index show quite good agreement with those obtained by subjective speech intelligibility determination.

Speech transmission index (STI) uses the same principle but a more advanced approach than that of MD. STI is based on the transmission of a noise having a speech-like spectrum and a wide modulation spectrum. Essentially the modulation is checked for *98 carrier and modulation frequency combinations* as shown in Table 3.1. The result is then obtained by a weighed summation of the differences incurred by the transmission channel or room. As with MD, STI can be used to investigate the influence of both speech and reverberation. Because of the extensive subjective testing that accompanied the development of STI, it must be regarded as the best metric yet derived for speech intelligibility. Using a computer, the STI value can be determined from measurement of the impulse response of a system and the knowledge of the background noise spectrum. RASTI is a simplified metric similar to STI.

The relationships between STI and various types of speech intelligibility tests are shown in Figure 3.36. One problem with STI as a metric for the acoustical quality of auditoria, is that it (along with the other metrics discussed in this section) does not include the influence of some properties of binaural hearing. Comparing listening to natural speech in auditoria to STI measurement results, one quickly finds that very different speech quality situations can have approximately the same STI values. STI does not measure the sound quality aspects of speech, such as distortion, timbre, and such.

Finally, it is important to note that the metrics described have been developed primarily for testing of linear systems. Transmission systems using compression, perceptual coders (codecs), and the like, should also be tested with other appropriate test systems (see Chapter 16).

Table 3.1 The 96 combinations of carrier and modulation frequencies used in the calculation of STI. The gray squares show the combinations used for RASTI. Each combination is weighted using an empirically found coefficient.

Octave band carrier frequency [Hz]	Modulation frequency [Hz]													
	0.63	0.80	1.00	1.25	1.60	2.00	2.50	3.15	4.00	5.00	6.30	8.00	10.0	12.5
125														
250														
500														
1														
2														
4														
8														

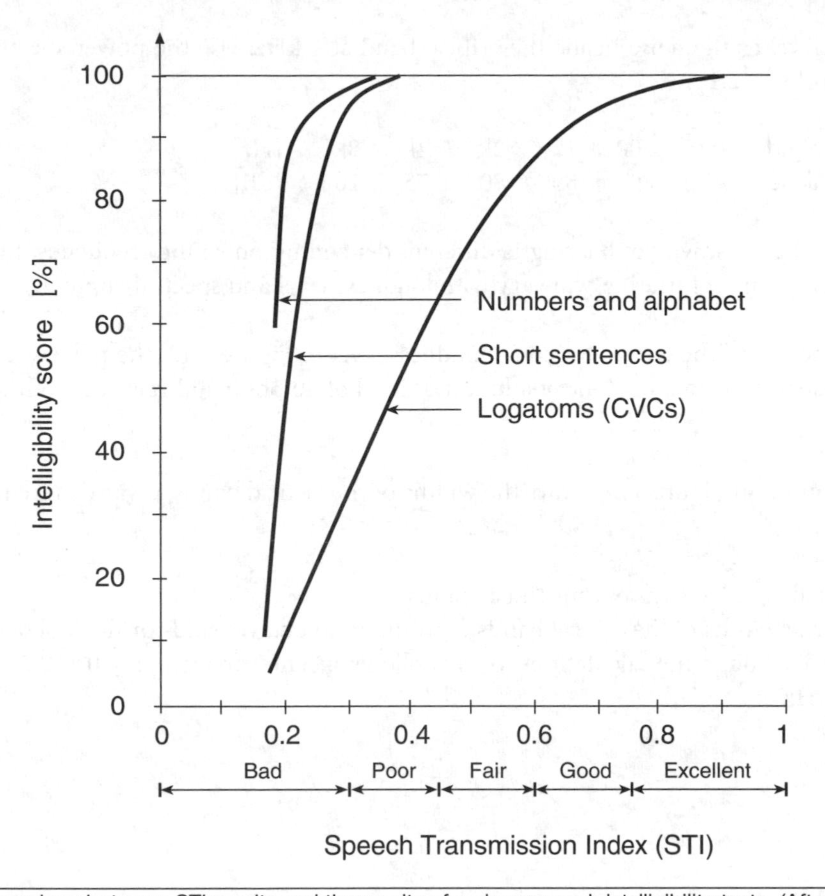

Figure 3.36 Comparison between STI results and the results of various speech intelligibility tests. (After Ref. 3.13)

3.9 PROBLEMS

3.1 You are listening to a 1 kHz Morse code sine signal that is contaminated by white noise.

Task: Will the use of a 100–Hz-wide bandpass filter centered on 1 kHz improve the audibility of the tone?

3.2 Figure 3.20 shows the hearing threshold shift due to masking.

Task: Determine, using Figure 3.20, how much the audibility of a 2 kHz tone is reduced when an additional 1 kHz masker tone having a sound pressure level L_p = 80 dB is added.

3.3

Task: Determine the sound pressure level that a 1 kHz tone must have to be audible in background noise having octave band levels given in the table. Assume that audibility requires that the tone have at least the same level as the noise inside the critical band at 1 kHz. Use the power spectrum level, PSL, discussed in Problem 2.10.

Octave band	250	500	1k	2k	4k	8k	[Hz]
SPL of noise	75	80	85	80	75	70	[dB]

3.4 Because the sensitivity of hearing is different depending on sound frequency, the effective frequency response of hearing will vary with loudness level and spectral content.

Task: Determine, using the curves for equal loudness level in Figure 3.12, the perceived change in frequency response if we listen to a tone at a loudness level of 40 phon and raise the sound pressure level by 30 dB.

3.5 The diagram in Figure 3.21 shows the widths of the critical bands at various frequencies.

Tasks:

a. Calculate the widths of the critical bands.
b. Do the widths of the critical bands correspond to octave bands or third octave bands?
c. Carry through the calculations for the following center frequencies; 0.1, 0.2, 0.4, 1, 2, 4, and 10 kHz.

Basic Room Acoustics 4

4.1 INTRODUCTION

Room acoustics involves a number of issues. Room acoustics is related to architecture, art, music, physics, engineering, and psychology, for example. It is common to subdivide room acoustics into two branches: (1) *technical room acoustics* and (2) *psychological room acoustics*.

Technical room acoustics can be further subdivided into *geometrical*, *statistical*, and *wave theory* room acoustics, which are three different ways of analyzing wave propagation in a confined space. Finally, there is always the possibility of using numerical approaches, such as those offered by the *finite element method* and various *finite time difference methods*.

Geometrical acoustics is similar to ray optics and is consequently a high-frequency approximation using concepts, such as mirrors and rays. For the approach used in geometrical acoustics to be applicable, surfaces should be much larger than the wavelengths of sound. Geometrical acoustics is the basis for many software models of sound propagation in rooms as well as outdoors.

Using *wave theory based room acoustics*, one can obtain exact mathematical solutions to problems in room acoustics, provided one can describe the room and its boundary conditions in a mathematically relevant way and that the problem is simple enough for analytical solutions to be feasible. This is seldom the case when dealing with practical problems in room acoustics. Therefore, the wave theory approach is primarily used to study the fundamental principles of behavior of sound waves in the low-frequency region of room acoustics for which the behavior of individual modes can be studied. It is used for studies of simple idealized room shapes, such as rectangular, cylindrical, and spherical shapes that approximate those of the real room and can be used as a point of departure for a further analysis.

Statistical room acoustics is based on the analysis of how the energy carried by the sound field is distributed in or between rooms and how the energy is absorbed by the sound-absorbing surfaces and objects of the room. Statistical room acoustics can only be used to study the average, approximately stationary behavior of sound under static or quasi-static conditions.

Because of these limitations, the solutions approach offered by statistical room acoustics is particularly useful to study sound due to continuous, wide-frequency range noise sources, such as ventilation, air conditioning, engines, and machinery. For quasistationary sound types, such as speech and music, statistical room acoustics must be used with great care to avoid drawing unfounded conclusions. It is also important to understand the limitations of the statistical approach and that the sound fields in real rooms are usually not ideal from the viewpoint of statistical room acoustics. Statistical room acoustics is a branch of *statistical energy analysis*, and it creates a possibility to study the interaction between sound in panels and rooms.

The purpose of *psychological room acoustics* is to document and explain how humans hear, perceive, and judge sound fields, both simple sound fields in the laboratory and the complex sound fields found in common rooms for speech and music. One major part of psychological room acoustics deals with the mapping of the perceived and judged room acoustics on physically measurable room properties and associated metrics. Psychological room acoustics is the topic of Chapter 5.

All of these approaches need to be understood to be able to successfully design and optimize the acoustic conditions in rooms for speech and music, ranging from small studios to large auditoria. The principles of adjusting and optimizing the room acoustic conditions are sometimes called *applied room acoustics*. In applied room acoustics *physical modeling* and numerical methods, such as the *ray tracing*, *mirror image*, and *finite element* methods are used to obtain better approximate solutions to room acoustic problems. Excellent references are 4.1 and 4.2. Modeling and numerical methods are discussed in Chapter 6.

4.2 GEOMETRICAL ROOM ACOUSTICS

In geometrical room acoustics, we use the same modeling principles as in geometrical optics—the ray tracing and the mirror images methods—to study the propagation of sound. These methods are then often extended by further acoustically relevant approximations. Geometrical room acoustics is useful to study the behavior of sound both indoors and outdoors.

The following starting approximations and assumptions are made in geometrical acoustics:

- Sources and receivers are infinitesimal and omnidirectional
- Sound propagation velocity is constant in the volume
- Surfaces have dimensions much larger than the wavelengths of sound considered
- Surfaces are smooth and nonscattering
- Surfaces have locally reacting reflection properties
- Surfaces have no resonant absorbers (real wall impedance)
- Sources radiate insensitive to surroundings
- Source signal is assumed to be a delta pulse
- Interference effects are neglected
- The phase part of the reflection coefficients is neglected
- Power summation of mirror image or ray contributions is used
- If diffraction and scattering are included in the model they are handled by angularly dependent reflection coefficients.

The Wavelength Criterion

A major problem in dealing with room acoustics is the large wavelength range. The lowest audible frequencies at 20 Hz have wavelengths of about 17 meters, while the highest at 20 kHz have wavelengths of approximately 0.017 meters.

This means there are few surfaces in buildings that can be considered perfectly reflecting in the sense of geometrical acoustics, at least in the high-frequency range. At low frequencies, the room dimensions may simply be too small for the room surfaces to be large compared to wavelength, whereas at high frequencies, the irregularities caused by poor workmanship and unavoidable lack of precision are so large as to make the surfaces more scattering than mirroring.

In practice, it is customary to consider surfaces with random peak irregularities smaller than $\frac{1}{16}$ of a wavelength as flat and thus acting as ideal mirrors. It is also reasonable to consider surfaces having dimensions at least three wavelengths long as large compared to wavelength.

The essential frequency range for many speech applications is 0.5 to 4 kHz—that is, a wavelength range of 0.7 to 0.1 m. In dealing with music, one has to be concerned with a greater frequency range, and then one normally has to do a geometrical acoustics analysis separately for various frequency bands in the frequency range of interest, for example, in octave-wide frequency bands.

Even though geometrical acoustics is incorrect in many respects, it is still a useful approximation as it applies an intuitive time/space approach, yielding results that are easily understood and the consequences of which can be easily demonstrated to nonprofessionals. With experience, it is usually possible to know what kind of errors to expect and to consider the errors in design.

Fresnel Zones

Huygen's principle states that each point on a wavefront may be regarded as a source point in calculating the next wavefront. Studying the results of the experiment shown in Figure 4.1, one finds that this is indeed the case. In this experiment, the source is omnidirectional over the angle subtended by the circular, concentric *zone* discs. The microphone senses the sound being reflected by the discs. Varying

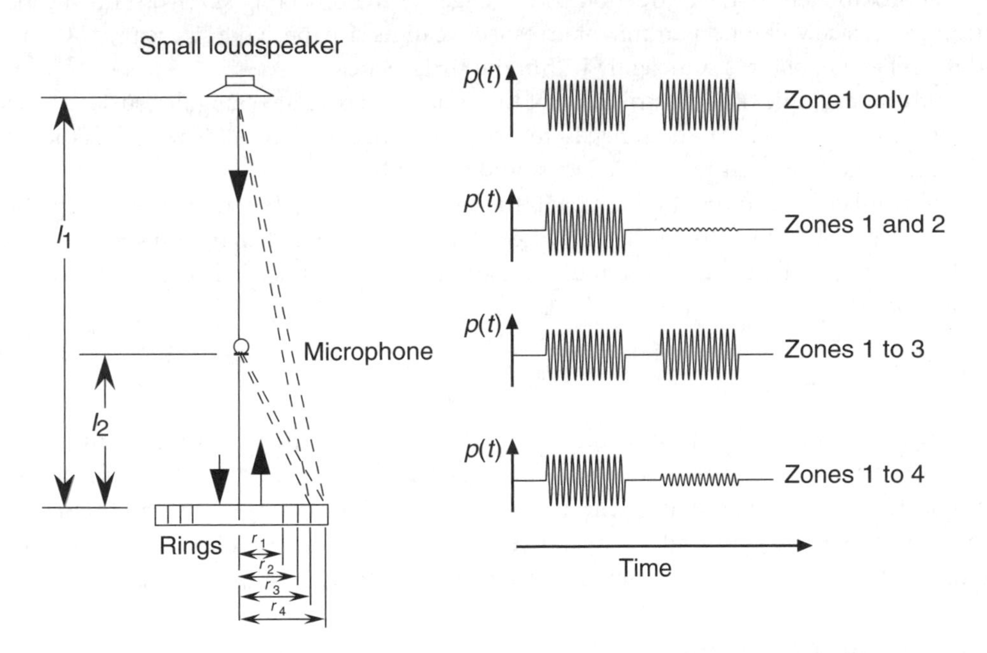

Figure 4.1 Demonstration of the influence of Fresnel zones on reflection by circular disks. The experimental setup consisting of a loudspeaker driver, a microphone, and a Fresnel zone plate using concentric reflecting rings (left). Response from the microphone for a tone burst for various numbers of reflecting Fresnel zones (right). (After Ref. 4.1)

the diameter of the disc assembly as r_1, r_2, and so on, one finds by measurement that the tone bursts will behave as shown in this figure.

The resulting sound pressure at the microphone is due to constructive and destructive interference between the sound pressure contributions by the various discs. The interference can be analyzed by studying the phase and amplitude of the reflected sound contributions by phasor analysis. It is easy to analyze the conditions for out-of-phase and in-phase reflection; that is, when the distance differences between the loudspeaker driver and the microphone—via points (or rather circles) on the disc—are multiples n of one half of the wavelength λ:

$$\begin{aligned} \Delta l &= \sqrt{l_1^2 + r_n^2} + \sqrt{l_2^2 + r_n^2} - l_1 - l_2 = n\frac{\lambda}{2} \\ &(n = 1, 2, 3, \ldots.) \end{aligned} \tag{4.1}$$

If one assumes $l_1, l_2 \gg r_n$, the radii r_n are obtained as:

$$r_n^2 \approx \frac{n\lambda}{\frac{1}{l_1} + \frac{1}{l_2}} \tag{4.2}$$

The zones created between the circles of these radii are called Fresnel zones. As the disc grows larger, the interference between the outermost zones becomes weaker. A useful approach is consequently to only consider the reflection by the innermost zone. The contributions by the outer rim sources can be considered approximately as the diffraction by the edges of the reflecting surface. This means that by this principle a surface will reflect, or mirror, irrespective of its size, once the size is more than some set limit and the *reflection point* is sufficiently within the surface area.

If we study an example of the distribution of the zone areas over a rectangular surface, such as that shown in Figure 4.2, we find that the zones 1b to 6 quickly cancel each other. The higher the frequency, the smaller the center zone has to be to reflect sound efficiently.

We can formulate a condition for how small the surface can be to function as an acceptable reflector of a plane-wave of sound (that is, l_1 very large) for a certain wavelength and receiver distance. It is reasonable to assume that the surface must contain at least the inner half zone. This results in the approximate condition:

$$l_{\text{limit}} \approx \frac{d^2}{2\lambda} \tag{4.3}$$

where d is the smallest surface dimension and l_{limit} is the distance to the observation point.

This condition is sometimes used when one needs to design sound reflectors for auditoria. Of course, a reflector also has to have sufficient surface mass (mass per unit area) so that it remains in place and is not moved by the sound field. A small object can act as a diffuser if it is small compared to the wavelength; the Fresnel zone criterion implies that a small object cannot mirror sound.

Ray Tracing and Mirroring

Assume a monopole sound source emitting sound energy in the form of rays. (Other related possibilities are emission of sound particles, beams, and cones.) Assign a certain fraction of the power radiated by the sound source to each of the rays, inversely proportional to the number of rays.

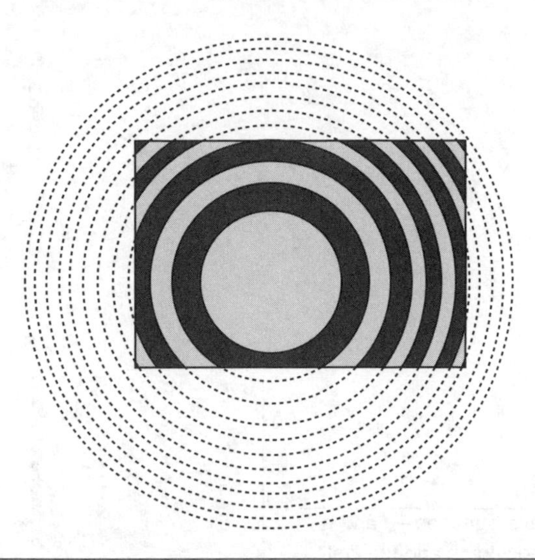

Figure 4.2 Illustration of the principle of Fresnel zones on a rectangular plate when the geometrical reflection point is offset from the center. The dark and light patches reflect out-of-phase with one another. When reflection and incidence occurs at an angle, the circular zone pattern will become elliptical. (After Ref. 4.1)

The path of each ray is determined by *Fermat's principle* that states that rays follow the fastest path, usually a straight line. During its progress, a ray may be reflected as well as bent, depending on the presence of sound-reflecting objects or changes in propagation conditions, such as wind or temperature and impedance gradients as mentioned in Chapter 2. Mirroring implies that the incident and reflecting rays are in the same plane as the normal to the surface and that the angles of incidence and reflection are equal. In the case of a gradual impedance change, the angles of incidence and transmission are related by Snell's law, mentioned in Chapter 2.

To find the rays, we can either send out rays and follow each ray along its path, or we can find the mirror images and then draw rays from the mirror images of the source to the observation point. The case of two opposing mirrors in Figure 4.3 is elementary. If one has several surfaces that are not parallel, one must take more complicated mirroring into account, as shown in Figure 4.4. In this case, we have four different mirror image sources since the ray has been reflected twice (note that two of the image sources have identical locations). The number of times a reflection has been mirrored, or a ray path reflected, is called the mirror image or ray *reflection order*. Figure 4.5 shows an example of the distribution of the mirror images of a two-dimensional rectangular room.

Geometrical acoustics is primarily useful for the study of the behavior of the early sound distribution in rooms. In practical rooms, the unevenness of the walls, the inaccuracies of wall positions, and the sheer multitude of rays or images, make it difficult to analyze the conditions by hand after the first or second order of reflection. Higher reflection orders have to be handled numerically by computerized calculation, which will be further discussed in Chapter 6.

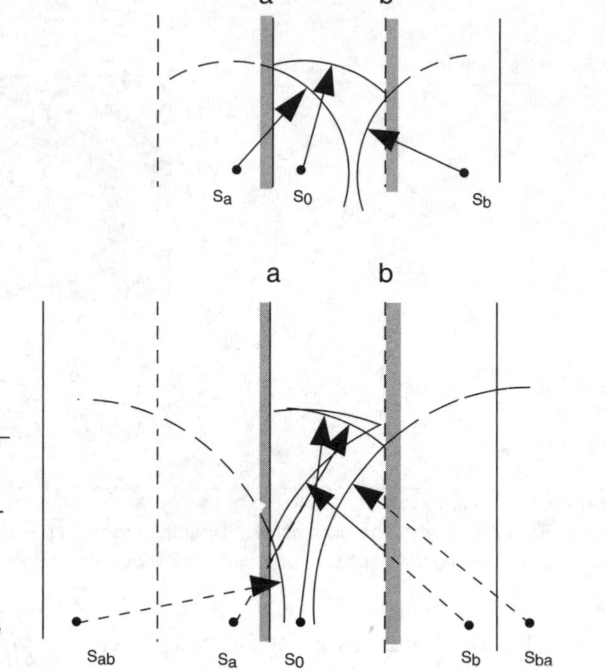

Figure 4.3 Mirroring by two parallel surfaces—a and b. The upper figure shows the sound source and the first order mirror image sources. The lower figure shows conditions after more time and the additional second order mirror images necessary to study the wavefronts. For each time instant, the radius of curvature is determined by the speed of sound and the time from the start of the sound.

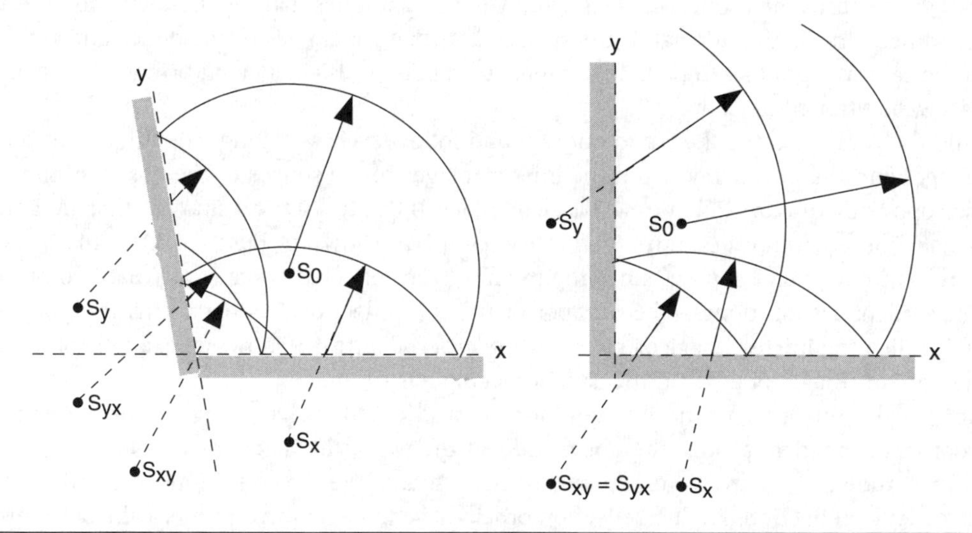

Figure 4.4 Mirroring of sound at the corners of intersecting surfaces, showing first and second order image sources.

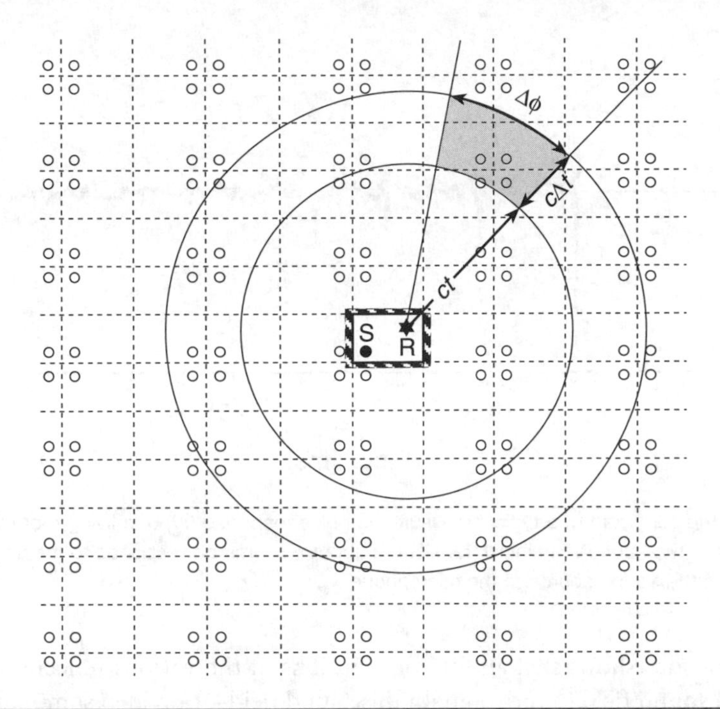

Figure 4.5 Mirroring by the walls of a rectangular room. The sound source is marked by the filled circle at S and the receiver is positioned at the star at R. The figure shows the location of mirror sources of various orders (empty circles). Only the sources in the grey zone contribute to the impulse response from the angle segment $\Delta\phi$ during the time segment Δt.

When surfaces are curved, ray tracing is intuitively simpler to use than mirror imaging. For a curved surface, the surface may be subdivided into smaller surface patches and analyzed as previously, or one can use the mirroring laws of optics.

In practice, room impulse responses are not as simple as one might be led to believe by simple geometrical optics. As shown by the example in Figure 4.6, the measured impulse response of a real room is complicated. This is due to the presence of scattering, diffraction, complex surface impedance, source and receiver impulse responses, for example. The energy behavior, however, can be quite well simulated. This is usually done by calculating the time integral of effective value of the sound pressure.

4.3 STATISTICAL ROOM ACOUSTICS

Diffuse Sound Field

In using statistical room acoustics, we idealize the sound field in a room so that we can limit ourselves to the study of the attack, steady state, and decay of sound energy. A prerequisite for this approach is that the sound has a wide spectrum and that we are not interested in the conditions at a specific frequency but rather in a frequency band wide enough to contain many frequencies.

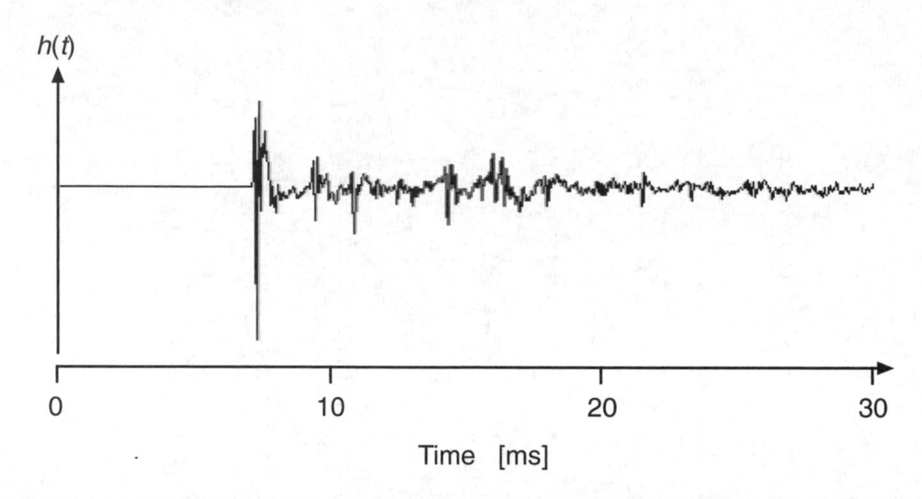

Figure 4.6 Example of the early part of a typical measured impulse response *h*(*t*) of a living room (including the measurement equipment). The time before the initial transient at about t = 7 ms corresponds to the time necessary for the sound to propagate from the sound source to the microphone.

Geometrical acoustics shows that all the contributions of the various reflections, after some time, set up a complicated sound field in the room. In this sound field—provided some simple conditions are met—the sound waves will come from many directions; will have bounced off many walls and objects; and will, on the average, be equally strong from all directions. The sound field will be *diffuse*.

An ideally diffuse sound field is characterized by the following physical properties:

- Energy density is equal at all places.
- All directions of incidence are equally probable.
- Incident sound has random phase.

When the walls of the room under study have low average sound absorption, the sound absorption is well distributed over the different walls, and the room does not have an *extreme* shape, the sound field in the room will have properties similar to those of an ideally *diffuse sound field.*

In reality, there are no ideally diffuse sound fields. When studying the sound fields of rooms, one can always find places where the sound is not diffuse, for example, close to walls where the reflected sound is well correlated to the incident sound. If there are major wall areas that have markedly higher sound absorption and size than others, the conditions are also not met. In practical engineering, however, one can often assume the sound field diffuse at positions being more than one-half wavelength away from wall surfaces.

For steady-state conditions, the sound field is best considered as a superposition of resonant room modes. Another way of formulating the sound field properties necessary for diffusivity is then to say that the sound field under study has to have a sufficiently large frequency range for it to contain many modes, that all modes are approximately equally damped, and that the room geometry is not extreme. We will study this way of expressing the requirements in the section on the wave theoretical approach to room acoustics, later in this chapter.

Attack and Rise Time

When a sound source starts to emit sound in a quiet room, the energy in the room will increase. The process of increase of energy density is known as the *sound attack*. The energy density, which is related to the effective value of sound pressure, is determined by the energy in the room and the room volume.

One can show that the rate of energy increase in the room depends on both the power radiated by the sound source and on the sound that is absorbed—by the surfaces of the room and by the objects present in the room.

The sound absorption of a surface was discussed in Chapter 1 (see Equation 1.58) and is specified by mean sound-absorption coefficient, α, characteristic for the surface at some frequency. The sound-absorption coefficient is related to the incident sound power, W_{inc}, and absorbed power, W_{abs}, as:

$$\alpha = \frac{W_{abs}}{W_{inc}} \tag{4.4}$$

It is common to find the value of α expressed in percentage in various tables of material data. The absorption area of a surface, A, is obtained by multiplying the sound absorption coefficient by the surface area and is expressed in units of m^2S metric sabin. (Note that the unit sabin refers to the imperial system). If there is a sound source in the room that is activated, then, during the attack, the energy in the room increases as:

$$E = E_0\left(1 - e^{-\frac{cA}{4V}t}\right) \tag{4.5}$$

where E_0 is the steady-state energy density that will ultimately be reached.

Later we will note that the attack process is complementary to the decay process. We will also derive an expression for the steady-state energy density as a function of room volume and sound absorption.

Reverberation Time

The rate at which the energy density diminishes after the sound source has been switched off is known as the *rate of decay*. It is often essential to know the rate of decay for sound in a room as a function of the acoustic properties of the room. The rate is sometimes described by a decay constant, but it is more common to express the rate indirectly by specifying the reverberation time of the room.

The *reverberation time*, T (also T_{60} or RT), is defined as the time it takes for the *sound energy density* to drop to one-millionth of its previous value. This corresponds to a sound-pressure level decrease by 60 dB. By using knowledge regarding the room volume, the sound-absorption coefficients of the room surfaces, and their respective sizes, it is possible to calculate the reverberation time.

By using geometrical acoustics, it is possible to trace the ray paths or calculate the mirror image distribution and study the decay process at any position in the room more in detail and also calculate the *local* reverberation time (within the limits of the method). Since the sound field is never ideally diffuse, the reverberation time will differ from one location in the room to the next. Even quite small differences in reverberation time are perceptible. One usually tries to calculate reverberation time to an accuracy of 0.05 s in room acoustics design. Note that it is difficult to measure the reverberation time over a 60 dB span (see Figure 5.16) in practice, particularly at low frequencies.

Examples of the determination of T are shown in Figure 4.7. Generally, one is limited to estimating the decay over a 30 to 40 dB span from which the reverberation time is then extrapolated. Because of this, it is common to give the span information by providing the start and stop levels, for example, for a span extending from −5 dB to −35 dB below start level as $T_{-5,-35}$ or $RT_{-5,-35}$.

In some cases the reverberation curves may be *dual-slope* or *multiple-slope*; that is, show two or more nearly linear slope regions. This behavior can be the result of different groups of modes having different decay constants or the sound field at the observation point being composed of sound coming from different volumes. When listening to reproduced sound, the effective reverberation curves will usually be dual-slope because of the presence of listening room reverberation along with the recorded reverberation.

Typically, such multislope reverberation decay curves can be found for rooms that have one large surface that is more highly absorptive than the others. Dual-slope reverberation processes can also be found when two rooms are coupled to one another through an opening. An example of where such conditions can be met is in a proscenium theatre where the stage and the auditorium are acoustically coupled through the stage opening.

Sabine's Formula

In this section, we will derive a simple expression for the decay rate expressed by reverberation time, using the room volume and the sound-absorption area as variables. Assume that the sound-absorption properties of the walls are the same for all walls, given by a mean sound-absorption coefficient α. Let a wave *packet* radiate at time $t = 0$. Each time the wave is reflected by a wall surface, it will lose a part α of its energy as heat or by transmission to some other acoustic system, and only a part $(1 - \alpha)$ will

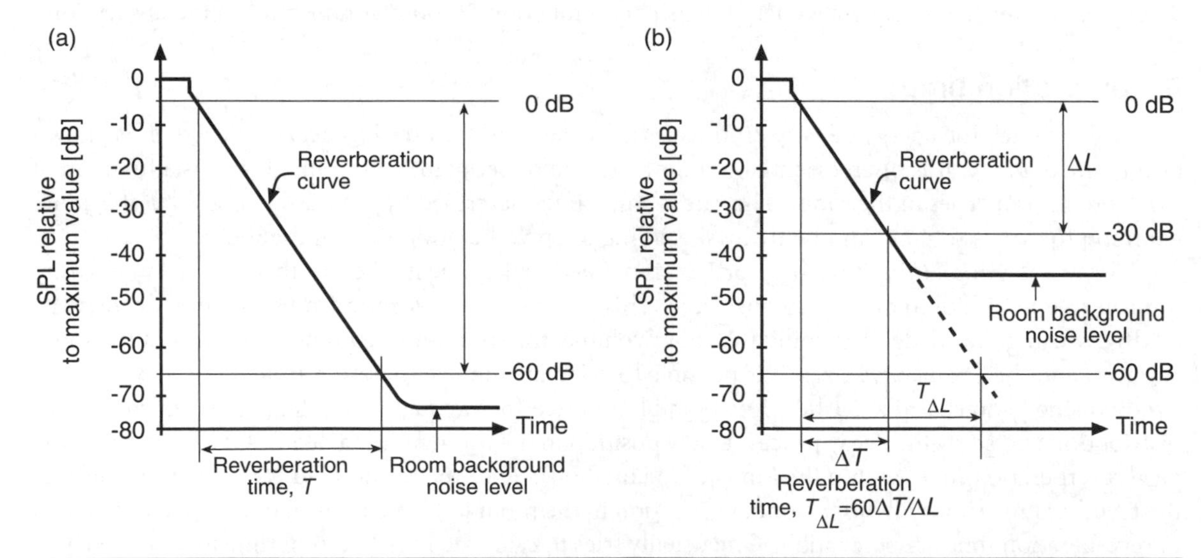

Figure 4.7 Examples of the determination of reverberation time (see also Figure 5.16). To ensure that the reverberant sound is diffuse the start of the measurement is usually 5 dB below the steady-state value (a). Determination of T when room background noise level is high (b). Real reverberation curves will not be characterized by smooth, constant slope, particularly in narrow, low-frequency octave and narrower bands.

remain. Another power loss mechanism is the attenuation of sound by propagation losses in air, which was studied in Chapter 1.

The length of the path between two successive ray-wall collisions can be represented by a mean length, usually called the *mean-free-path length* l_m, which depends on the properties of the room, such as its geometry, the diffusitivity of the walls, and the presence of sound diffusing objects. One can show that the mean-free-path length l_m in a volume V, having a total surface area S, is on the average given by:

$$l_m \approx \frac{4V}{S} \tag{4.6}$$

The distribution of path lengths will depend on the room geometry as shown in Figure 4.8. An even distribution of path lengths is typical of rooms that do not have an *extreme* shape or distribution of sound-absorptive materials over their surfaces.

It is convenient to characterize the decay rate of the energy by a real-valued decay constant, δ, which we can associate with the reverberation time T. The energy decay follows $E(t) = E_0 e^{-2\delta t}$. Since the wave loses a part $(1 - \alpha)$ of its energy each time it is reflected, one can write the remaining energy at each time instant as:

$$E_{\text{wave}} \propto (1-\alpha)^{\frac{ct}{l_m}} e^{-mct} = e^{-2\delta t} \tag{4.7}$$

where the previously mentioned propagation losses in air have been included using the constant m (see Figure 4.9 or Figure 1.3).

Taking the logarithm of both sides in Equation 4.7 one finds:

$$2\delta = mc - \frac{c}{l_m}\ln(1-\alpha) \tag{4.8}$$

The definition of reverberation time gives the relationship between T and δ as:

$$T = \frac{3\ln(10)}{\delta} \approx \frac{6.91}{\delta} \tag{4.9}$$

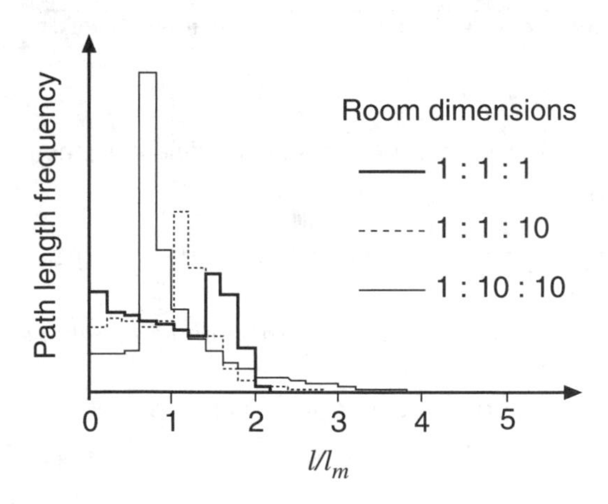

Figure 4.8 Some distributions of the path lengths l in rectangular rooms of different proportions and the respective mean-free-path lengths l_m. Note that the long paths are infrequent. (After Ref. 4.2)

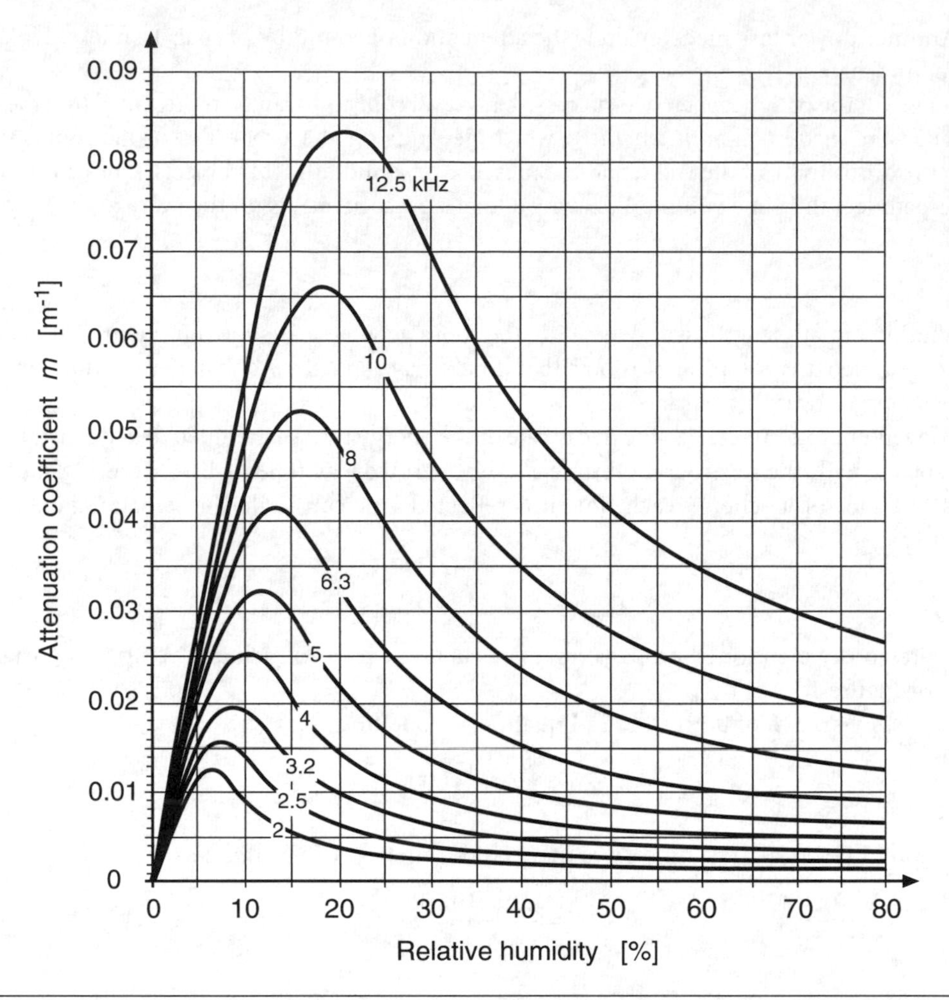

Figure 4.9 Values of the attenuation coefficient *m* for air at a temperature of 20°C, for various values of relative humidity and frequency (in kHz). (After Ref. 4.3)

Dependence of reverberation time on room volume and the sound-absorption coefficient is then obtained as:

$$T = \frac{24\ln(10)}{c_0}\frac{V}{4mV - S\ln(1-\alpha)} \tag{4.10}$$

It is now appropriate to introduce the concepts of *absorption area, equivalent absorption area, Sabine absorption,* and *room absorption.*

The *absorption area* is simply the product of absorption coefficient α and physical area *S*:

$$A = \alpha S \tag{4.11}$$

Note that this expression assumes that a surface absorbs sound in the same way no matter what size and placement.

One can show, by both theory and experiment, that this is not the case (see Chapter 6). Not only are certain dimensions relative to wavelength characterized by higher sound absorption, but splitting a large sound-absorbing area into small patches usually increases the diffusitivity of the sound field. In this way, it increases the effective sound absorption. This is the result of the mean-free-path length becoming shorter due to the more frequent reflections.

The *equivalent absorption area* is a concept used to characterize the sound absorption of irregularly shaped objects for which it is difficult to specify the area or the absorption coefficient. Examples of such objects are humans and chairs. The equivalent absorption area is defined as the absorption area that would absorb the same sound power as the object in a room with a diffuse sound field, such as a *reverberation chamber* (see Chapter 17).

The *room absorption* is the sum of the absorption by all room surfaces and objects and the sound absorption of the air (see Figure 4.9):

$$A_S = 4mV - S\ln(1-\alpha) + \sum_j A_j \approx 4mV + A \tag{4.12}$$

The room absorption A_S represents the total sound absorption of the room and is expressed in units of m²S. The unit metric sabin is sometimes used.

The total surface area of the room S_t, including wall surfaces and the surface areas of other large objects, such as sound-absorptive screens and audience area, is obtained as:

$$S_t = \sum_i S_i \tag{4.13}$$

It is practical to use a mean sound-absorption coefficient α_m defined as:

$$\alpha_m = \frac{\sum_i \alpha_i S_i + \sum_j A_j}{\sum_i S_i} \tag{4.14}$$

Using this definition, we can the calculate the reverberation time using *Sabine's formula*:

$$T = \frac{24\ln(10)}{c_0}\frac{V}{4mV + \alpha_m S_t} \tag{4.15}$$

Equation 4.12 is named after the Harvard scientist W. C. Sabine who was the first to formulate it. The slightly more exact formula in Equation 4.10 is called *Eyring's formula*. The Sabine absorption is the absorption area given by Sabine's formula, by means of the measurement of the reverberation time of the room.

In the case of small rooms and low frequencies, the sound absorption by air can usually be neglected. In large spaces, such as concert halls and factory spaces, having volumes more than 10,000 m³, the sound absorption by air is usually the dominant sound-absorption mechanism at high frequencies.

Assuming that thc losses incurred by air sound absorption are much smaller than the losses at the walls, $m \ll \alpha/l_m$, one obtains a common approximation to the decay constant as:

$$\delta \approx \frac{cS\alpha}{8V} \tag{4.16}$$

Sabine's formula is commonly expressed in an approximate way, using SI units, as:

$$T \approx 0.161 \frac{V}{A_S} \tag{4.17}$$

One can usually use Sabine's formula under the following approximate conditions for diffuse sound fields:

- The room averaged sound-absorption coefficient α is less than 0.3.
- The sound-absorbing surfaces are approximately evenly distributed over the room.
- The shape of the room is not extreme.

Reverberation Curves

The most common metric for characterizing the suitability or quality of a room for a certain speech or music purpose is the reverberation time.

Central to the determination of the reverberation time is the reverberation curve, as discussed earlier. For practical reasons, the reverberation curve is usually studied as the decay of sound pressure level versus time, as shown in Figure 4.7. In the case of a single, exponentially damped oscillation, the curve would be a straight line, the lower part of which would be obscured by noise.

If the decaying signal is noise or otherwise contains a number of frequency components, the reverberation curve becomes wiggly simply because of the summation of the different tones at each instant. In practice, the different frequencies excite different room resonances that typically have unequal damping and reverberation times. These effects make the reverberation curve appear as a *noisy* line, but the latter condition leads to the reverberation curve having double or more slopes. In the low-modal density and low-frequency region, the reverberation curves may show effects of beats because of interference between strongly excited modes close in frequency.

A better way to measure the reverberation time than by direct approximation to the decay curve to evaluate the reverberation process using a method called *backward integration* or Schroeder's method. This method makes the determination of reverberation time slightly more precise. An example of such a curve is shown in Figure 5.14.

In any case, the determination of reverberation time is prone to vary depending on the method used. The ISO 3382 standard gives details on how the reverberation time of an auditorium should be measured. The ISO 354 standard details the measurement of sound absorption in a reverberation chamber. The measurement of sound power in a reverberation chamber s detailed in the ISO 3740 standard.

Sound Pressure Level and Sound Power Level

The sound field in a room acts as a store of acoustic energy. Measurement of the sound power radiated by a sound source is commonly made by measuring the sound pressure level and the room absorption.

As discussed previously, the energy in a room will asymptotically reach a steady-state value, if the sound source is allowed to act indefinitely. We have already noted that the sound pressure level in the ideal diffuse field is constant in a room. Close to the sound source and walls, the sound field is not diffuse, and the sound pressure level will be higher.

We can consider the resulting sound pressure at a point in a room with a noise sound source to be composed of two uncorrelated parts, one due to the direct field and one due to the reverberant, diffuse field. If the two parts are sufficiently uncorrelated (sufficiently wide-band noise) the effective values of each can be added to give the total effective value.

If we assume a monopole sound source in free field, the effective value of sound pressure due to the sound source as a function of distance r must be:

$$\tilde{p}_{\text{dir}}^2 = Z_0 \frac{W_+}{4\pi r^2} \tag{4.18}$$

If the sound source has been active a long time, at least as long as the reverberation time, quasi-static conditions will have been reached. Under static conditions, the power radiated by the sound source will increase the energy in the room until it has reached a level where there will be an equilibrium between the power loss due to room absorption W_-, and the power supplied by the sound source, W_+.

When the sound source is switched off, the energy in the sound field will drop as a result of the sound absorption by the surfaces and objects in the room. Assume the sound source to be switched off at time $t = 0$. The energy of the sound field will then diminish as:

$$E(t) = E_0 e^{-2\delta t} \tag{4.19}$$

It will lose power W_- according to:

$$W_- = -\frac{dE}{dt} = 2\delta E_0 e^{-2\delta t} \tag{4.20}$$

Using our previously defined decay constant, we find that:

$$2\delta = \frac{cA_S}{4V} \tag{4.21}$$

where we have used Equations 4.6 and 4.8 and assumed $\alpha \ll 1$.

Before the start of the decay, at time $t = 0$, there must have been a power balance; that is:

$$W_+ = W_- \tag{4.22}$$

This gives the static energy E_0 as a function of the input power as:

$$W_+ = 2\delta E_0 = \frac{cA_S}{4V} E_0 = \frac{cA_S}{4} \frac{E_0}{V} \tag{4.23}$$

The relationship between energy density P and intensity I in a sound field can be shown to be:

$$I = Pc \tag{4.24}$$

where the energy density in this case simply is the stored acoustic energy divided by room volume:

$$P = \frac{E}{V} \tag{4.25}$$

The relationship between the intensity and the mean square value of the sound pressure in a plane wave is given by Equation 1.37, and the corresponding energy density by Equation 1.38. The energy density relationship, derived for the plane wave, also holds in a diffuse sound field, since we at any point can assume the sound field to be composed of plane-waves with random angles of incidence. The relationship between the mean square sound pressure and the energy density is:

$$\frac{<\tilde{p}_{\text{diff}}^2>}{Z_0} = \frac{E_0}{V}c \tag{4.26}$$

The brackets < > indicate that we mean the *spatial average of the mean square sound pressure,* that is, averaged over room volume.

Using Equation 4.23, we then find:

$$W_+ = \frac{c_0 A_S}{4V}\frac{V<\tilde{p}_{\text{diff}}^2>}{\rho_0 c_0^2} \tag{4.27}$$

Finally, we obtain the effective value of the static sound pressure in the diffuse field as a function of sound source input power as:

$$<\tilde{p}_{\text{diff}}^2> = Z_0\frac{W_+}{(A_S/4)} \tag{4.28}$$

Adding the two effective values for the direct field and the diffuse field, we obtain the total effective value as:

$$\tilde{p}_{\text{tot}}^2 = W_+ Z_0\left(\frac{1}{4\pi r^2}+\frac{4}{A_S}\right) \tag{4.29}$$

As usual, it is more convenient to express the relationship by levels. Using the expressions 2.24 and 2.26 we obtain the desired and useful formula:

$$L_p = L_W + 10\log\left(\frac{1}{4\pi r^2}+\frac{4}{A_S}\right) \tag{4.30}$$

Equation 4.30 is the common formula to calculate the sound pressure level due to a sound source in the reverberant field in a room. This expression is depicted graphically in Figure 4.10 that shows how the sound pressure level drops by 6 dB per distance doubling close to the sound source. Far from the source, in the diffuse field, the sound pressure level will theoretically be constant. The distance at which the direct field and the diffuse field are equally strong is called the *reverberation radius.*

When dealing with a directional sound source, we need to take the directivity of the sound source into account in describing the sound pressure level distribution in the room. The *directivity factor* Γ is related to the directivity and the directivity function as:

$$\Gamma(\theta,\varphi) = D\cdot F^2(\theta,\varphi) \tag{4.31}$$

The directivity factor tells us how strong the radiated intensity compared to that of an omnidirectional source radiating the same acoustic power. In the main lobe of a directional sound source, such as a violin or a large loudspeaker, the intensity (and sound pressure level) will be much higher than in other directions. In the direction of high directivity, the sound from the loudspeaker will reach farther, which is useful in rooms with long reverberation times or in noisy spaces.

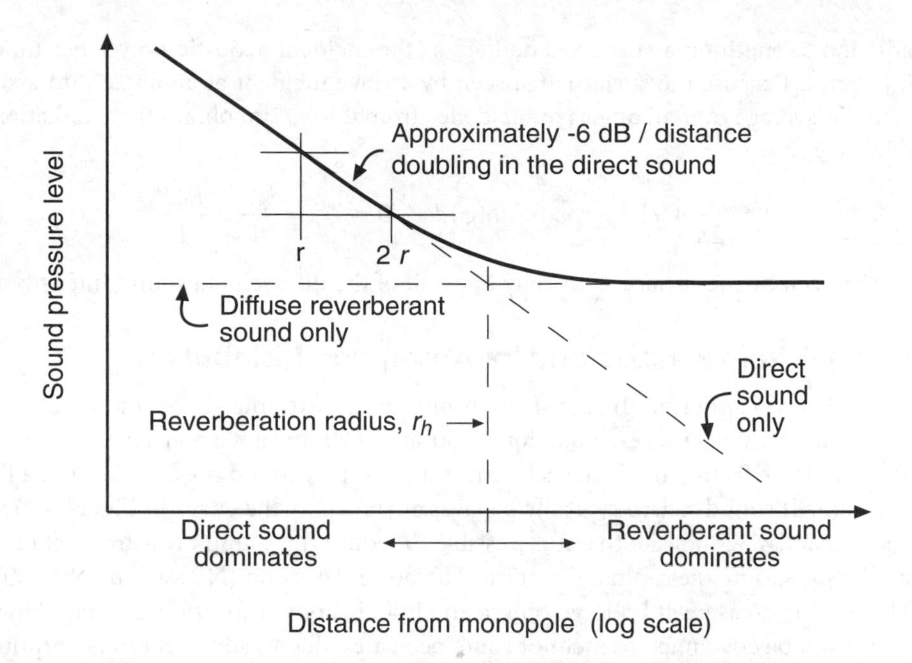

Figure 4.10 The theoretical behavior of the steady-state sound pressure level as a function of distance from a monopole source in a room with a broadband diffuse sound field, according to expression 4.24. The contributions of the direct field and the reverberant field will be equally strong at the reverberation radius r_h .Compare to the discussion and Figure 17.16 in Chapter 17.

One can show that an improved expression for the relationship between sound level as a function of angle in a reverberant space, $L_p(\theta, \varphi)$, and the sound power level of the source L_W, taking the directivity of the sound source and the diffusitivity of the sound field into account is:

$$L_p(\theta,\varphi) = L_W + 10\log\left(\frac{\Gamma(\theta,\varphi)}{4\pi r^2} + \frac{4(1-\alpha_m)}{A_S}\right) \tag{4.32}$$

The quantity RC = $A_S/(1-\alpha_m)$ is called the *room constant.*

As mentioned previously, there are no ideally diffuse sound fields. Close to walls and other sound-reflecting surfaces and objects, the sound pressure level can increase by up to 6 dB due to interference by the phase coherent incident and reflected waves. In corners, the increase can be even higher.

Power Incident on a Surface

The sound field is never perfectly diffuse. In a diffuse sound field, at any given time, one-half of the energy is moving in the opposite direction of the other half. Obviously, this cannot be the case close to a sound-absorbing surface.

The irradiation strength on a surface is defined as the incident acoustic power per unit area and is denoted B [W/m^2]. Because the surface area seen by a wave incident at an angle θ to a surface S is $S\cos(\theta)$, we need to average over all angles of incidence from 0 to $\pi/2$ to obtain the irradiation by a diffuse sound field. We find:

$$B = \frac{1}{2\pi}\int_0^{\pi/2}\int_0^{2\pi} I_{inc}\cos(\theta)\sin(\varphi)\,d\varphi\,d\theta = \frac{I_{inc}}{2} = \frac{I_{diff}}{4} = \frac{Pc_0}{4} \tag{4.33}$$

since the intensity seen by the surface I_{inc} is only one half of the diffuse field sound intensity I_{diff}.

Rooms Having Extreme Shape and/or Absorption Distribution

Because the sound field cannot be diffuse if the room has an extreme shape and/or uneven sound-absorption distribution over surfaces, Equations 4.30 and 4.32 are not applicable to rooms, such as open plan offices or large factory halls. In such spaces, the drop of sound pressure level as a function of distance from a (small) sound source typically behaves as shown by the curve in Figure 4.11.

Equations 4.30 and 4.32 indicate that it is possible to reduce the sound pressure level in the diffuse field of a room by increasing the room absorption. The noise reduction (NR) will be NR = 10 log(RC_2/RC_1) (dB). The level decrease will be large only if the initial room absorption is small. However, one often finds that the subjective improvement of room acoustics, due to added room absorption, is large even if the sound pressure level decrease is small. This subjective impression is due to the reduced reverberation time and the longer reverberation radius.

A room having little room absorption and correspondingly long reverberation time is often called a *live* or *hard* room. A test room for sound-absorption measurement is characterized by very rigid, glossy surfaces and is called a *reverberation chamber*.

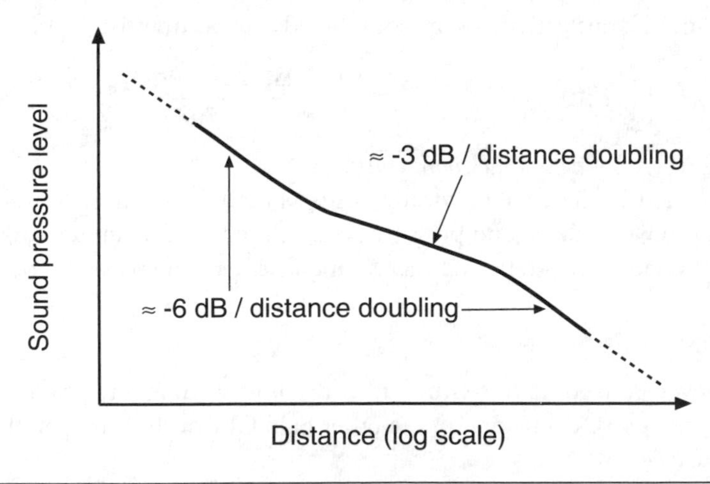

Figure 4.11 The reduction of sound pressure level by distance in a non-Sabine room can follow different distance laws. The type of sound level behavior shown here is typical of that from a small sound source in an open plan office or large factory space having low ceiling height.

A room that is characterized by short reverberation time is said to be *dead* or *dry*. For example, a test chamber to be used for electroacoustic measurements needs its walls to have a reflection coefficient $|\underline{r}| \leq 0.1$, an absorption coefficient $\alpha \geq 0.99$. Such a room is usually called an *anechoic chamber*.

Special purpose test chambers for sound power measurement of machinery may need a rigid floor surface. Such chambers that have only a half space of sound absorption are called *hemi-anechoic*.

4.4 WAVE THEORETICAL APPROACH

Using wave theory, we can find the distribution of the sound pressure in a room even at frequencies so low that the sound field is not diffuse. As an example of this approach, we will study the sound pressure distribution in an idealized rectangular room, but other geometries can also be analyzed using the same approach.

The room under study is shown, with the chosen coordinate system, in Figure 4.12a. It is assumed to have hard, immovable walls. The air in the room is assumed not to absorb any sound power, and the sound source is assumed to be a point source.

The wave equation in Cartesian coordinates is:

$$\frac{\partial^2 \underline{p}}{\partial x^2}+\frac{\partial^2 \underline{p}}{\partial y^2}+\frac{\partial^2 \underline{p}}{\partial z^2}+k^2\underline{p}=0 \tag{4.34}$$

The spatial distribution of the sound pressure over the room is a function of the room coordinates. The spatial distribution of particle velocity is linked to the pressure distribution. Since the wall surfaces are hard and immovable, the particle velocity component perpendicular to the respective walls must be zero:

$$\underline{u}_x\Big|_{x=0,l_x}=0 \quad ; \quad \underline{u}_y\Big|_{y=0,l_y}=0 \quad ; \quad \underline{u}_z\Big|_{z=0,l_z}=0 \tag{4.35}$$

As usual, when dealing with partial differential equations, we guess a reasonable form of solution. We then test the solution's validity by checking if the solution satisfies the differential equation. We expect

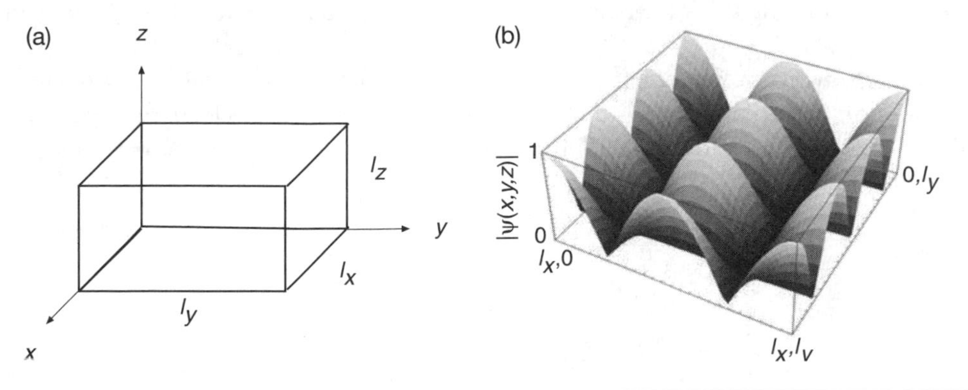

Figure 4.12 The rectangular room under study and its coordinate system (a). The normalized magnitude of the eigenfunction $|\Psi(x, y, z)|$ for the mode q_x: q_y: q_z = 3: 2: 0 (b).

the sound pressure to be high at the walls in analogy with the standing wave in front of a hard wall. In this case, a reasonable guess then is a solution of the form:

$$\underline{p}(x,y,z) = \hat{p}\,\Psi(x,y,z)\,e^{j\omega t} \tag{4.36}$$

where the spatial distribution of sound pressure is given by the function $\psi(x,y,z)$:

$$\Psi(x,y,z) = \cos(k_x x)\cos(k_y y)\cos(k_z z) \tag{4.37}$$

The second order derivative of p with respect to x is:

$$\frac{\partial^2 p}{\partial x^2} = -k_x^2\,\hat{p}\cos(k_x x)\cos(k_y y)\cos(k_z z)\,e^{j\omega t} = -k_x^2\,\underline{p} \tag{4.38}$$

We find the second order derivatives with respect to the y and z directions in the same way. By inserting the second order derivatives with respect to x, y, and z into the differential equation, we obtain the condition:

$$k_x^2 + k_y^2 + k_z^2 = k^2 \tag{4.39}$$

where k is the wave number, $k = \omega/c$, which depends on the frequency of the sound. For the solution (which we guessed) to be a valid one, the condition 4.39 must be met.

The wave number k has the components k_x, k_y, and k_z. These components can be regarded as the wave number components of waves moving around inside the rectangular room, bouncing off the walls. The waves will interfere and lead to standing wave patterns in the same way as in the one-dimensional case of Kundt's tube.

The room will be resonant for the combinations of k_x, k_y, and k_z that fulfill the condition 4.39. At resonance, the energy in the room will build up infinitely if there are no losses. This is called an oscillation mode of the room, an *eigenmode,* or simply *mode.*

The function ψ, which describes the spatial variation of the sound pressure at resonance frequency of a particular eigenmode, is called an *eigenfunction.* Figure 4.12b shows an example of the normalized magnitude of the eigenfunction for such an eigenmode.

To determine the eigenfunctions, we need to find values for the wave number components k_x, k_y, and k_z. We use the condition that the particle velocity perpendicularly to the walls, $\underline{u}_x$, $\underline{u}_y$, and $\underline{u}_z$, must be zero.

We start by studying the conditions in the x direction—that is, at the walls at $x = 0$ and $x = l_x$. The equation of motion, Equation 2.7, gives us an expression for the particle velocity $\underline{u}_x(x,y,z)$ as a function of $\underline{p}(x,y,z)$. Insertion of the solution 4.36 for the differential equation into this expression gives us the particle velocity $\underline{u}_x(x,y,z)$:

$$\underline{u}_x = -\frac{1}{\rho_0}\int\frac{\partial \underline{p}}{\partial x}dt = -\frac{1}{\rho_0}\frac{1}{j\omega}\frac{\partial \underline{p}}{\partial x} \tag{4.40}$$

$$\underline{u}_x = \hat{p}\frac{k_x}{\rho_0}\frac{1}{j\omega}\sin(k_x x)\cos(k_y y)\cos(k_z z) \tag{4.41}$$

The conditions for the particle velocity at the wall at $x = 0$ is obviously fulfilled. Analogously, we find that the solution 4.28 also fulfills the conditions for the particle velocity at the walls at $y = 0$ and $z = 0$.

According to the expression given in Equation 4.40, for zero particle velocity at right angles to the wall at $x = l_x$, the following condition must be met:

$$\sin(k_x l_x) = 0 \tag{4.42}$$

The sine will only be zero when the wave number component k_x is:

$$k_x = \frac{q_x \pi}{l_x} \tag{4.43}$$

where q_x is a natural number 0, 1, 2, . . . In the same way, we find that requirements for the wave number components k_y and k_z are:

$$k_y = \frac{q_y \pi}{l_y} \tag{4.44}$$

$$k_z = \frac{q_z \pi}{l_z} \tag{4.45}$$

Inserting the values for the wave number components into Equation 4.39, we find that the condition for the wave equation to be satisfied (with the specified geometry and boundaries) is:

$$k^2 = \frac{\omega^2}{c_0^2} = \left(\frac{q_x \pi}{l_x}\right)^2 + \left(\frac{q_y \pi}{l_y}\right)^2 + \left(\frac{q_z \pi}{l_z}\right)^2 \tag{4.46}$$

Obviously, there are an infinite number of combinations of q_x, q_y, and q_z. Each of these combinations will identify a possible eigenmode. The *eigenfrequencies* of these modes are determined by the speed of sound and the room dimensions. The combination $(q_x: q_y: q_z)$ is often called the *mode index.*

Eigenfrequencies and Eigenmodes

If we rewrite Equation 4.46, we find that the k-values correspond to frequencies according to the formula:

$$f_{q_x, q_y, q_z} = \frac{c_0}{2} \sqrt{\left(\frac{q_x}{l_x}\right)^2 + \left(\frac{q_y}{l_y}\right)^2 + \left(\frac{q_z}{l_z}\right)^2} \tag{4.47}$$

These frequencies are often called the eigenfrequencies or *resonance frequencies of the room.* For each eigenfrequency, there is at least one eigenmode. For certain room dimensions, many eigenmodes may have identical eigenfrequencies. In this discussion, we have limited ourselves to the case of hard and immovable walls. If the walls can move, have a finite impedance, the resonance frequencies will change.

Each mode in the rectangular room is characterized by a spatial distribution of sound pressure of the form:

$$\underline{p}_{q_x, q_y, q_z}(x, y, z) = \hat{p}_{q_x, q_y, q_z} \cos\left(\frac{q_x \pi x}{l_x}\right) \cos\left(\frac{q_y \pi y}{l_y}\right) \cos\left(\frac{q_z \pi z}{l_z}\right) \tag{4.48}$$

The sound field in a practical room with losses is built up by a superposition of pressure contributions from an infinite number of modes since the losses are accompanied by the resonances having finite bandwidths.

One can show that for a small sound source, having a volume velocity $\underline{U}$, placed at location, r_0, the expression for the total sound pressure at another point, r, due to all the modes, is given by an infinite sum of the pressure contributions of the various modes according to:

$$\underline{p}_{tot}(\omega, \mathrm{r}, \mathrm{r}_0) = \frac{j\omega\rho_0 c_0^2}{V}\underline{U}\sum_{q_x}^{\infty}\sum_{q_y}^{\infty}\sum_{q_z}^{\infty}\frac{\Psi_{q_x,q_y,q_z}(\mathrm{r})\Psi_{q_x,q_y,q_z}(\mathrm{r}_0)}{\Lambda_{q_x,q_y,q_z}\left[\left(\omega_{q_x,q_y,q_z}^2 - \omega^2\right) + 2j\omega_{q_x,q_y,q_z}\delta_{q_x,q_y,q_z}\right]} \tag{4.49}$$

In this expression, δ is the decay constant discussed previously, and the expression is only valid for small values of δ. The expression is also not valid close to the monopole at r_0. The normalizing factor Λ for a mode depends on if the mode is created by waves moving in one ($\Lambda = 1/2$), two ($\Lambda = 1/4$), or three ($\Lambda = 1/8$) dimensions.

An interesting property of Equation 4.49 is that the source and observation points r and r_0 can be exchanged. This property is an example of *reciprocity*.

The steady state frequency is determined by the sound source. How *strongly* any particular mode will be excited depends on the following:

- How close its resonance frequency is to the excitation frequency
- How damped the particular mode is
- Where the source is located in the room

We note from Equation 4.37, that all eigenfunctions always have a large magnitude at the corners of the room in the simple case studied here. We also note that the ratio between the sound pressure and the particle velocity is highest at the corners; that is, the sound field impedance is highest at the corners. This means that a constant volume velocity-type source will excite the room modes best when placed in the corner of a room. Many technical sound sources, such as machines and loudspeakers, are constant volume velocity-type sources.

Using a loudspeaker and a sine-wave generator, we can study the properties of the room modes. The summed sound pressure of several excited room modes is going to depend on how the sound pressure contributions of each mode add up with respect to magnitude and phase. This in turn depends on the particular modes that contribute at a particular location as shown by Equation 4.49. The sound pressure must therefore vary from one position to the next. At high frequencies where many modes leak into one another, the sound pressure level will vary considerably from one position to the next as shown by the experimental curve in Figure 4.13. The sharp minima and maxima are due to interference between many modes.

When the sound source is switched off, the modes will start to oscillate at their natural resonance frequencies (given by Equation 4.47) if the modes are weakly damped and uncoupled. Since the modes having resonance frequencies closest to the excitation frequency are likely to have been excited most strongly, they will tend to dominate the decay process.

Modal Density

If we study Equation 4.47 for the determination of the mode frequencies, we note that it describes the length of the diagonal in *frequency space*, extending from the origin out to a point determined by q_x, q_y, and q_z as shown in Figure 4.14. Note that all modes having lower resonance frequencies than f will have shorter diagonals—in other words, they will be contained within a radius of f Hz.

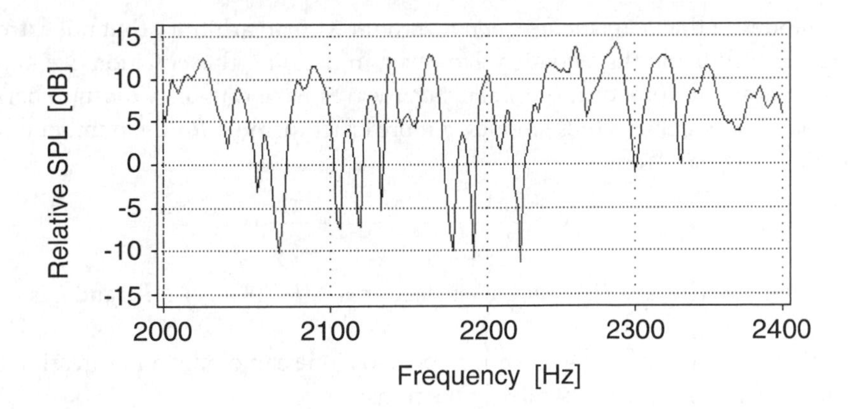

Figure 4.13 An example of the variation of the frequency response in the diffuse field of a living room as a function of frequency obtained from the room impulse response shown in Figure 4.6.

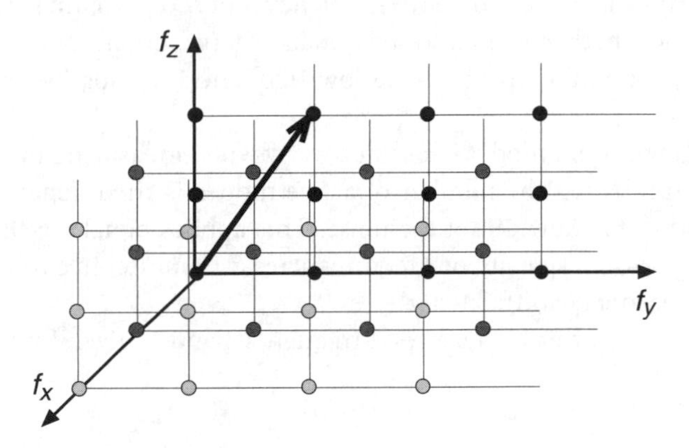

Figure 4.14 Resonance frequency lattice in the *frequency space* for a rectangular room. The arrow marks the lattice point corresponding to a mode index (q_x,q_y,q_z) = (0,1,2). The lattice points are assigned according to Equation 4.47.

The total number of modes below the frequency described by q_x, q_y, and q_z, and l_x, l_y, and l_z can then be calculated by dividing the volume of 1/8 of the sphere by the volume of each frequency box. The reason we do not use the full sphere volume is that the numbers q_x, q_y, and q_z have to be natural numbers—that is, 0, 1, 2, 3, . . . Negative numbers would result in counting each frequency more than one time. Each frequency box has a volume, which is:

$$\Delta V = \frac{c_0^2}{8l_x l_y l_z} \tag{4.50}$$

This now gives us the total number of modes N below some frequency f in a volume V as:

$$N \approx \frac{4\pi V}{3c_0^3} f^3 \tag{4.51}$$

The number of modes in the *dense mode region* in a room with an arbitrary (but not extreme) shape can be estimated rather well using this equation. However, in deriving this equation, we have not included waves that travel parallel to one or more walls. Such waves have one or two q-numbers that are zero. One can show that if one includes these modes, a more exact formula for the number of modes below a frequency f in a rectangular room is:

$$N \approx \frac{4\pi V}{3c_0^3} f^3 + \frac{\pi S}{4c_0^2} f^2 + \frac{\pi L}{8c_0} f \tag{4.52}$$

where V is room volume $(l_x l_y l_z)$, S is total wall surface area $2\,(l_x l_y + l_y l_z + l_x l_z)$, and L is total edge length $4\,(l_x + l_y + l_z)$.

By differentiating N in Equation 4.51 with respect to f, one can obtain an expression for the number of modes ΔN in a narrow frequency band Δf. One finds:

$$\Delta N \approx \frac{4\pi V}{c_0^3} f^2 \Delta f \tag{4.53}$$

The *modal density* $\Delta N/\Delta f$ is dependent on both frequency and room volume. This is one of the reasons that we can, by ear, decide whether a room sounds small or large. Small rooms are unsuitable for music reproduction since they cannot support all the low-frequency sounds by appropriate (and suitably damped) room modes.

The presence of enough room modes is also necessary when measuring the spatially averaged effective value of the sound pressure. The number of active modes is then dependent on both the modal density of the room and the bandwidth of the noise. This applies equally well to the measurement of reverberation time using noise. Typically, one requires at least 10 modes in a frequency band in the room under study to assume reasonable diffusitivity.

The number of modes in a third octave-wide frequency band is given by:

$$\Delta N \approx 0.23 V \frac{f^3}{c^3} \tag{4.54}$$

We note that there will only be a small number of modes strongly excited at low frequencies in small rooms, leading to poor averaging and measurement results. This corresponds to the condition of poor sound field diffusitivity, as discussed in the section on statistical room acoustics.

Complex Eigenfrequencies

Acoustical power losses are always present in real rooms. To simplify the discussion in this section, we assume all losses to be a part of the propagation through air so that the losses are continuous.

The energy of each eigenmode must decay exponentially in a linear system. The sound pressure of a mode can therefore be written:

$$\underline{p}_q = p_q(x,y,z) e^{j\omega_q t} e^{-\delta_q t} \tag{4.55}$$

where $q = [q_x, q_y, q_z]$.

Here δ_q is the mode-specific, real-valued decay constant introduced previously. We can thus write the exponential decay as:

$$e^{j\omega_q t} e^{-\delta_q t} = e^{j\underline{\omega}_q t} \tag{4.56}$$

where

$$\underline{\omega}_q = \omega_q + j\delta_q \tag{4.57}$$

is the complex eigenfrequency.

The resonance curves for the effective sound pressure of a particular mode (as a function of frequency) will have a certain width because of the presence of δ_q, as indicated by Equation 4.49. The resonance curves for the various modes will overlap considerably even at quite low frequencies.

The Q-Value

A metric to describe the *peakiness* of a resonance curve is the *Q-value*. The Q-value is calculated from the effective value of the sound pressure as a function of frequency as shown in Figure 4.15. The Q-value is defined as the resonance frequency divided by frequency width of the resonance curve at the –3dB points and is consequently a dimensionless number.

Low and high Q-values correspond to strongly and weakly damped resonances respectively. The Q-value concept is used in many other applications when dealing with resonances, for example, in conjunction with vibration isolation. Note that a response curve, such as the one shown in Figure 4.13, is the result of the addition of a large number of resonances and that the Q-value of a peak, in this case, is not the same as the Q-value of an individual resonance.

Loss Factor and Reverberation Time

We have already studied the relationship between the decay constant and the reverberation time. One can derive a similar expression for the relationship between the loss factor of an oscillating system and its reverberation time.

The spatial average of the effective value of the sound pressure is proportional to the acoustic energy in the room:

$$E \propto \langle \tilde{p}^2 \rangle \tag{4.58}$$

The decay of the mode will be a function of time:

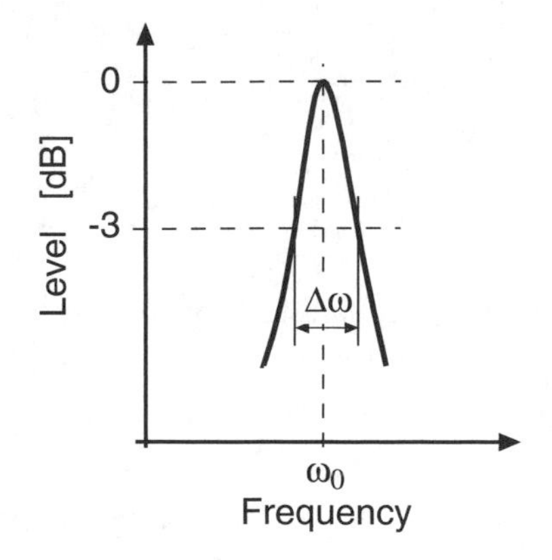

Figure 4.15 The response of a simple first order resonant system. The Q-value is defined as $Q = (\omega_0/\Delta\omega)$ where $\Delta\omega$ is the frequency at which the resonance curve is 3 dB below the peak value at the resonance frequency ω_0.

$$E \propto e^{-2\delta_q t} \tag{4.59}$$

It is often practical to express the decay of the room modes by a *loss factor*, η. The loss factor is usually defined as the relative loss of energy per radian:

$$\eta = \frac{W_+}{E_0}\frac{1}{\omega} \tag{4.60}$$

Using the previously derived expressions for the decay constant one can also write the loss factor as:

$$\eta_q = \frac{2\delta_q}{\omega_q} \tag{4.61}$$

We note also that the loss factor is a dimensionless number.

It is useful to know that the Q-value is the inverse of the loss factor:

$$Q_q = \frac{1}{\eta_q} \tag{4.62}$$

Assume that a room is excited by a sound source that is switched off at time $t = 0$. The power radiated by the source is W_+. The steady-state value of the energy in the room is then determined by the fact that the power loss W_- is equal to the power injected by the sound source W_+. If we use Equations 4.59 and 4.61, we find:

$$W_- = -\frac{dE}{dt} = \omega\eta E_0 e^{-\omega\eta t} \tag{4.63}$$

At time $t = 0$, the energy in the room is given by:

$$W_+ = \omega\eta E_0 \tag{4.64}$$

Using Equations 4.9, 4.61, and 4.64, we obtain the relationships between reverberation time, decay constant, and loss factor as:

$$T = \frac{3\ln(10)}{\delta} = \frac{6\ln(10)}{\omega\eta} \approx \frac{2.2}{f\eta} \tag{4.65}$$

and

$$T = \frac{3\ln(10)}{\delta} = \frac{6\ln(10)Q}{\omega} \approx \frac{2.2Q}{f} \tag{4.66}$$

Wave Theory and Diffuse Sound Fields

The wave theory approach to room acoustics tends to be complicated when we want to solve practical problems characterized by:

- High- and wide-frequency ranges
- Non-rectangular and irregularly shaped rooms
- Rooms that have uneven distribution of sound absorption

It is more practical then to use the approach offered by statistical room acoustics. The relationship between the wave theory for high frequencies and the statistical theory is fairly straightforward.

If one has a large number of modes excited by a wide bandwidth noise, the room does not have an extreme shape, and the modes have approximately the same damping, one can regard the resulting sound field as diffuse. The reason is that each mode is associated with a certain wave vector and consequently sound propagation direction in the room.

If there are many modes, there will be several directions of incidence. If, in addition, the modes are equally damped, the intensity of the sound from each direction will be approximately the same. The requirement for random phase will be satisfied if we excite the room by a wide bandwidth noise signal.

When the damping of the modes is so high that we have wide resonance curves and large *mode overlap*, it is difficult to keep track of the individual modes. An upper-frequency limit, for the application of wave theory, is commonly taken as the frequency at which the Q-value of the resonance curves is less than 3. This can be shown to correspond to mode overlaps of more than 3; that is, at least 3 modes within the half-value widths of the resonance curves of the effective value of the sound pressure.

This limit is the *Schroeder frequency*, f_S, and is approximately given by the expression:

$$f_S \approx 2000\sqrt{\frac{T}{V}} \tag{4.67}$$

Here T is the reverberation time and V the volume of the room. In practice, the calculation of the Schroeder frequency is somewhat iterative.

4.5 PROBLEMS

4.1

Task: Which eigenmodes belong to the octave and the third octave bands centered on 50 Hz, in a rectangular room having the dimensions $L \times W \times H = 10 \times 8 \times 5$ m³?

4.2 The number of eigenmodes per frequency band is of importance both for the sound quality, when the room is used for performance/listening, and for the choice of acoustic measurement methods.

Task: Calculate the number of eigenmodes in each of the third octave bands from 100 to 500 Hz for a room having the dimensions $L \times W \times H = 4.5 \times 4.0 \times 2.6$ m³.

4.3 In a room having the dimensions $L \times W \times H = 10 \times 7 \times 3.5$ m³, the reverberation time, at frequencies high enough for diffuse sound field conditions, was measured as 1.5 s.

Tasks:

a. Determine the absorption area A_S and the mean sound-absorption coefficient α_m using Sabine's formula. Neglect the sound absorption due to air. Is $\alpha_m < 0.3$?

b. Repeat *a* but include sound absorption due to air by using the attenuation coefficient m = 0.002 m^{-1}.
c. Determine the mean sound-absorption coefficient using Eyring's formula. Neglect the sound absorption due to air.

4.4 The reverberation times in the third octave bands in the table were measured in a room having the dimensions $L \times W \times H = 10 \times 7 \times 3.5$ m^3.

f	50	63	80	100	125	160	[Hz]
T	2.5	2.5	2.0	1.7	1.4	1.4	[s]
f	200	250	315	400	500	630	[Hz]
T	1.3	1.3	1.3	1.3	1.3	1.2	[s]

Task: Determine from which frequency one can use the concept of statistical room acoustics using the Schroeder frequency as the criterion.

4.5 Assume an omnidirectional source emitting a wide-frequency range signal with the sound power in a room of 0.01 W is being used. The room dimensions are $L \times W \times H = 10 \times 15 \times 5.5$ m^3, and the absorption area is 50 m^2S. The frequency range considered is above the Schroeder frequency.

Tasks:

a. Calculate the rms sound pressure as a function of distance to the sound source for distances 0.5, 1, 2, 4, and 8 m.
b. Calculate the corresponding sound pressure levels.

4.6 Consider the following three rooms:
a. A living room of volume V = 80 m^3 and having a reverberation time T = 0.5 s
b. A theatre of volume V = 4,000 m^3 and having a reverberation time T = 1.1 s
c. A concert hall of volume V = 15,000 m^3 and having a reverberation time T = 1.7 s

Task: Assuming an omnidirectional sound source, calculate the reverberation radii for the sound fields in the respective rooms.

4.7 A general purpose studio is to be adjusted for recording of speech, which requires low reverberation times in the octave bands 0.5, 1, and 2 kHz. The table lists the present and the desired reverberation times. The volume of the studio is 60 m^3, and its total surface area is 94 m^2.

Octave band	0.5	1	2	[kHz]
$T_{current}$	0.43	0.38	0.38	[s]
$T_{desired}$	0.3 ± 0.02	0.3 ± 0.02	0.3 ± 0.02	[s]

Tasks:

a. Calculate the acceptable interval for the absorption area for each octave band. Sound-absorbing panels having the area 1.2 × 1.2 m^2 are available. The sound-absorption coefficient for these panels are:

Octave band	0.5	1	2	[kHz]
α	87	94	90	[%]

b. Calculate how many panels are required to achieve the desired reverberation times. Assume that the panels are mounted on top of surfaces having the current mean sound-absorption coefficient. Use Sabine's formula and neglect sound absorption due to the medium.

After the adjustments have been done, one finds that the mean sound-absorption coefficients are so high that Eyring's formula should have been used instead of Sabine's formula. In the Eyring formula, the expression $A_S = 4mV - S\ln(1 - \alpha)$ is used to calculate the total absorption of the room A_S.

c. Calculate the resulting reverberation times using Eyring's formula instead of Sabine's formula.

4.8 A small loudspeaker is radiating sound into a rectangular room where the dimensions are $L \times W \times H = 7 \times 5 \times 3$ m^3. The loudspeaker is positioned at the coordinate $(x_0, y_0, z_0) = (0, 0, 0)$. The following signal types are used:

1. A sinusoidal tone having a frequency of 66 Hz
2. Band-limited noise having a center frequency of 2 kHz and a bandwidth of 300 Hz

In both cases a sound pressure level of 75 dB was obtained at the coordinate $(x_1, y_1, z_1) = (2, 3, 1)$.

Task: Calculate the sound pressure level at the coordinate $(x_2, y_2, z_2) = (4, 1, 2)$ for the two cases using an appropriate theory.

4.9 A classroom having the dimensions $L \times W \times H = 12 \times 7 \times 3.5$ m^3 is to be acoustically adjusted so that the reverberation times are 0.8 ± 0.1 s in all octave bands in the range 125–4000 Hz.

The floor is covered by a soft carpet on top of concrete. One wall has 5 windows with a total area of 18 m^2 and has 10 m^2 covered by drapes. The rest of the walls, as well as the ceiling, are made of concrete covered by plaster.

The sound-absorption coefficients of the various surfaces are:

Octave band	125	250	500	1k	2k	4k	[Hz]
Plaster on concrete	2	2	3	2	5	5	[%]
Windows	40	30	20	17	15	10	[%]
Drapes	12	20	42	53	64	62	[%]
Carpet on concrete	9	8	21	26	27	37	[%]
Alternative (a) above	63	42	35	12	8	8	[%]
Alternative (b) above	57	97	68	63	59	40	[%]

Neglect the sound absorption due to doors, persons, and chairs.

Two different sound-absorbing treatments are available:

a. Using 0.006 m thick particle board on 0.05 m high battens on top of the concrete

and/or

b. Using 0.004 m thick perforated particle board on 0.15 m high battens on top of the concrete but with 0.03 m thick glass wool in the air space

Task: Determine the required area of each treatment to achieve the desired design goal. Motivate your choice.

4.10 A person living in an old apartment has a bedroom immediately above the boiler room. The noise from the boiler is disturbing, and its fundamental at 30 Hz corresponds to the lowest eigenfrequency of the bedroom. The eigenmode at this frequency completely dominates the sound field in the bedroom.

The bedroom pillow is located at coordinate (x_1, y_1, z_1)= (0.3, 0.3, 0.3). When the local health inspector measures the sound level this is done using a microphone approximately at the center of the room, in this case at the coordinate (x_2, y_2, z_2) = (2.5, 1.5, 1.5). The sound level L_{pA} at this point was measured as 35 dB, which is considered acceptable.

Task: Calculate the actual sound level at the point of the pillow, if the mode theory for an undamped rectangular room can be applied to the problem.

5 Spatial Sound Perception

5.1 INTRODUCTION

The listener's auditory spatial perception is influenced by many sound field properties. Some of those properties are energy-related, such as reverberation time, the relationships between the levels of direct and reverberant sound, and the spectra of the incoming sound components. Some are spatially-related, for example, the spatial distribution of the incoming sound field components. Still others are dependent on arrival time, such as echoes and other strong sound reflections.

The subjective perception of spatial characteristics often does not map well onto the physically measured dimensions of the spatial sound field properties. As an example, one can mention that the perceived reverberation times in auditoria often do not correspond to those obtained by using instrumentation.

It is therefore necessary to be quite careful when using results from classical physical measurement, such as reverberation time, in judging the quality of an auditorium, such as a theater or concert hall. A careful study of the fine structure of the sound field is usually necessary for expressing oneself knowledgeably about the room acoustic quality. Unfortunately, considerable knowledge about human spatial sound perception seems to be lacking at this stage, even though great progress has been made in recent years.

To study the influence of various sound field parameters in a well controlled and detailed way, it is necessary to move from the auditorium to the laboratory. In *psychoacoustic tests* in the laboratory, it is convenient to use a technique known as *sound field synthesis.* This technique involves the variation of different sound field components, sound source radiation patterns, reflections, noise, and so forth, thus studying the parameters that are believed to influence spatial sound perception.

To use sound field synthesis, one needs to use either binaural sound reproduction or sound reproduction over loudspeakers in an *anechoic chamber*. This is necessary for complete control of the sound field to which the test person is exposed. In a regular room, the test person would be subject to uncontrolled sound field components due to sound reflections from walls, for example. Another possible problem would be background noise.

An anechoic chamber is characterized by low background noise levels, L_{pA} less than 15–20 dB, and by walls that are nearly nonreflective over some limiting frequency, for example 100 Hz. Usually the walls have a sound-absorption coefficient larger than 0.99. A classic system for sound field synthesis in an anechoic chamber uses a large loudspeaker array, such as that shown in Figure 5.1 (see Reference 5.15).

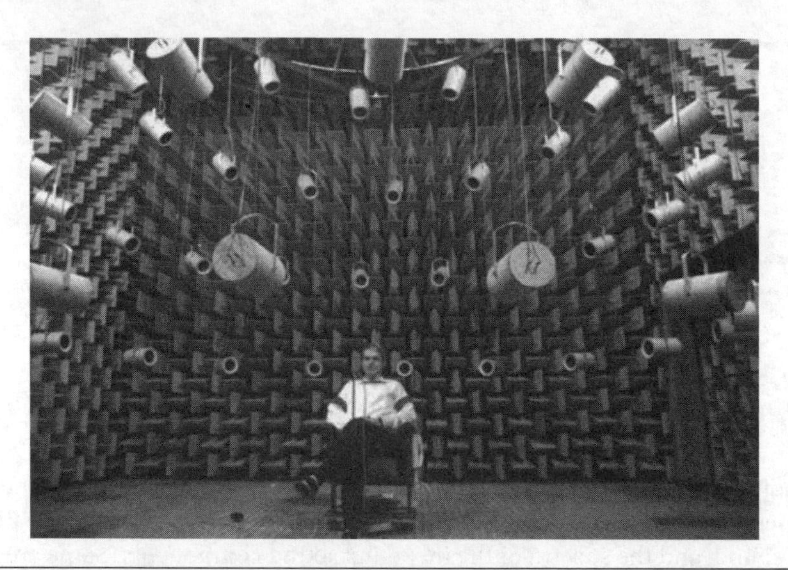

Figure 5.1 An anechoic chamber equipped for experiments in psychoacoustics and sound field synthesis using a large hemispheric loudspeaker array. (Photo by Jakub Kirszenstein).

In addition to the loudspeakers, the equipment includes *reverberation units, delay units,* among others, so that one can create an *illusion of presence*, similar to that of being in the sound field of an actual room.

More modern types of sound field synthesis use loudspeaker equivalent *binaural sound reproduction,* using only two loudspeakers and a technique known as *crosstalk cancellation*. In using crosstalk cancellation, the signals through the loudspeakers are tailored by digital filtering so that they, at the listener, correspond to those signals that would be obtained by conventional binaural listening using headphones.

Of course, one can use direct binaural sound synthesis to achieve simulation of the room sound field components.

The various techniques for achieving audible room sound field simulation are often called *auralization*, in analogy with visualization by computer aided techniques as used in architectural design (see Reference 5.16 and the discussion in Chapter 6).

5.2 SUBJECTIVE IMPRESSION OF SOUND FIELD COMPONENTS

When listening to sound under normal circumstances, the auditory perception is determined not only by what one hears but also by what one sees. In trying to find the fundamental properties of hearing it is necessary, in some way, to remove or neutralize other sensory information so that the listener is not affected, at least by visual sensory input (although visual information related in some way to the test in question may be added for experimental purposes). From this perspective as well, using an anechoic room may have both positive and negative influence on the *test*, the *test person*, and the *test results*.

Since 1950, many psychoacoustic tests have been made to study the behavior of hearing when exposed to sound fields of both small and large complexity (see Reference 5.14).

The following sections relate some of the important results of this work. One must remember that many of these tests were done with a fairly small and select number of test persons, that they were seldom made with the same audio test material, and that the listening conditions varied between laboratories. Critical reading is advised.

Simple Sound Fields

The real or simulated sound field properties are determined by many factors, such as:

- Sound pressure level L_p of the direct sound
- Levels, angles, and arrival times of delayed sound components relative to the direct sound
- Spectra of delayed sound components

There are primarily four phenomena that are observed: (1) *masking*, (2) *localization*, (3) *echo*, and (4) *coloration*.

In the psychoacoustic tests, the test person listens to sounds having various angles of incidence in the horizontal and median planes. In addition, the delay time between the various signals Δt is varied. Such tests allow the researcher to investigate the importance of both time delay between direct sound and (simulated) sound reflections by room walls. Unless otherwise noted, all angles used in this chapter refer to the head-related coordinate system shown in Figure 3.2. Also, all direct sound field components arrive at an angle of (0°, 0°) unless otherwise indicated.

Figures 5.2 and 5.3 show the results of some tests using simple sound fields where recorded speech was used as audio material. Figure 5.2 shows the sensitivity of hearing to a single additional sound field component (that is, besides the direct sound field component), incident at (0°, 0°), as a function of its relative delay time Δt.

In a corresponding way, Figure 5.3 shows how the sensitivity varies as a function of angle of incidence (θ°, 0°) for a delay of Δt of 50 ms. It is noteworthy that the sensitivity increases as delay time is

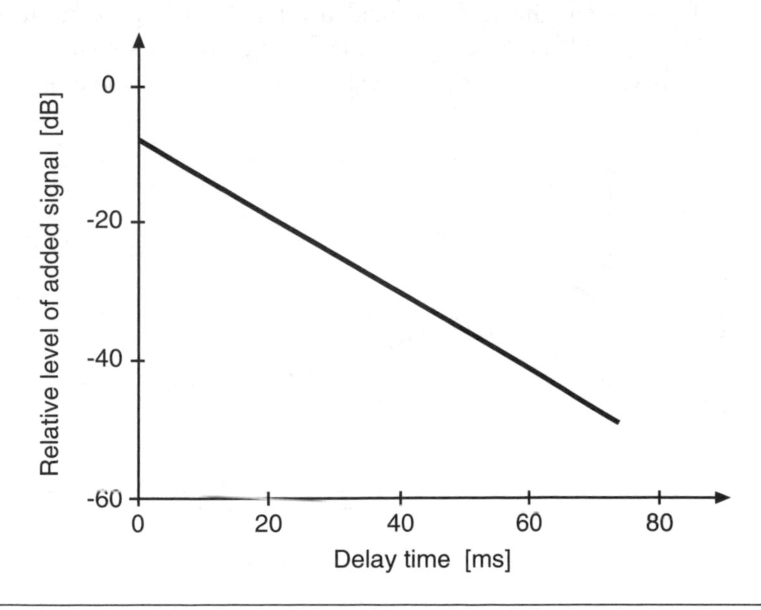

Figure 5.2 Regression line for the relative threshold for noticing a delayed signal added to a direct signal in the case of frontal sound incidence (speech at L_p = 70 dB). (After Ref. 5.1)

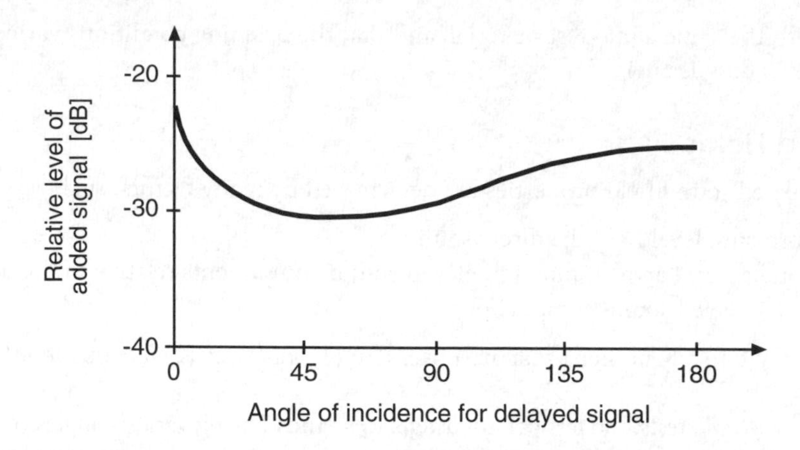

Figure 5.3 Relative threshold for noticing a signal delayed by 50 ms relative to a direct speech signal as a function of horizontal angle for the delayed signal. Both sound components incident in the horizontal plane, direct sound has frontal incidence. (After Ref. 5.2)

increased, indicating reduced masking in Figure 5.2, and that the sensitivity seems to be at a maximum at an incidence angle of (45°, 0°), as shown by the curve in Figure 5.3.

Tests using music as a test signal give somewhat different results, although it is seldom possible to do direct comparisons between tests. Music has a wider range of modulation frequencies than speech and often contains more low frequency modulation. Music is also characteristic in its wide spectrum, which makes it easier to hear interference effects, often as *coloration*—that is, usually unwanted changes in timbre. Another important difference is that music usually comes from many source locations, so it is less relevant to talk about one direction of incidence for the direct sound. It is also worth noting that it is quite difficult to find anechoically recorded music, usually stereo or mono recordings of various anechoic quality have been used for tests. Figures 5.4–5.6 show some important results from

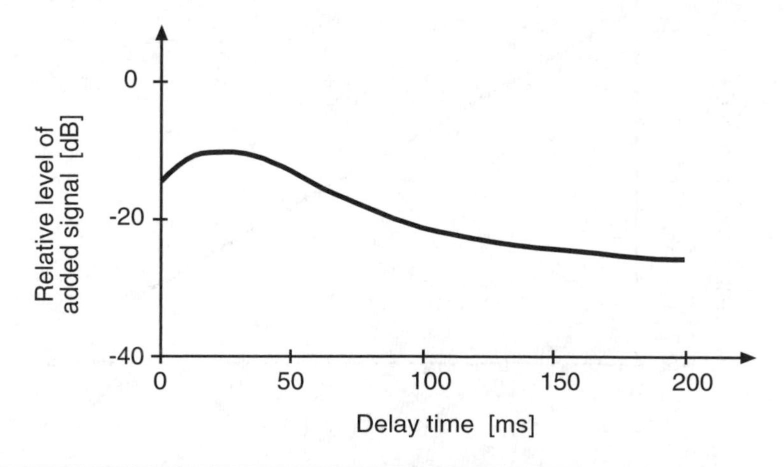

Figure 5.4 Relative for noticing a delayed signal incident at $\varphi = 90°$ added to direct sound incident at $\varphi = 0°$ (mean for six musical works). (After Ref. 5.3)

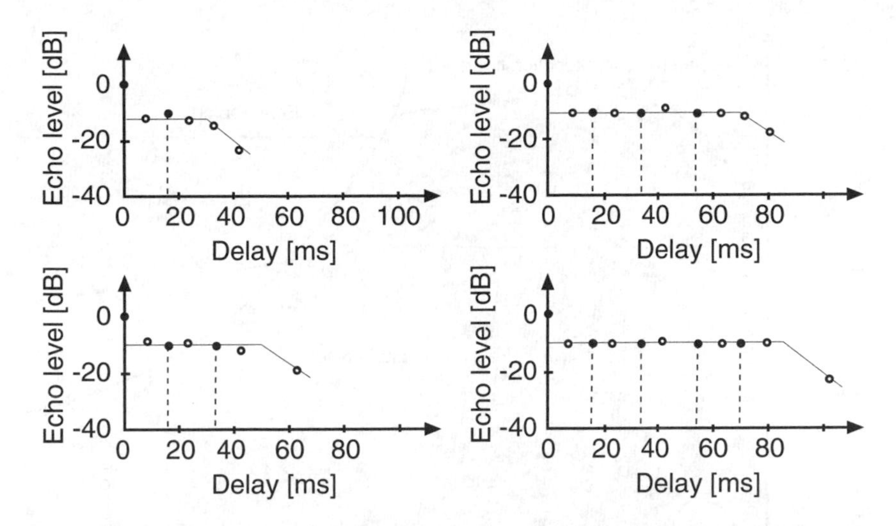

Figure 5.5 Relative threshold for noticing a delayed signal (open circles), added to a sound field of increasing complexity obtained by adding 1, 2, 3, and finally 4 sound field components with varying time delay and level (solid circles at 0 and −10 dB respectively) (speech, $\varphi = 0°$). (After Ref. 5.2)

psychoacoustic tests using recorded music, usually classical music. Figure 5.4 shows that the increased frequency range of music make us more sensitive to early reflections, as compared to the results shown in Figure 5.1 that were obtained for speech.

Usually there are no single sound field reflections in a room; reflections arrive in greater density as time goes. Results of simple tests using multiple sound reflections are shown in Figure 5.5. The graph shows the sensitivity of the test persons to various numbers of added sound field components all having arrival angles of (0°, 0°) and a level of −10 dB relative to the direct sound. Although this sound field structure is not really possible in a normal room, the results show that in the case of several sound field components, the components can mask one another. The results also show that the sensitivity increases only after some 20 ms of time have passed after the arrival of the last component.

Once the additional sound field components have risen above the threshold of perceptibility, they will influence the way the sound field is perceived in many ways; phenomena such as echo and coloration, for example, will be clearly noticeable.

Figure 5.6 shows how various levels and time delays of only one single additional sound field component (90°, 0°) will sound. It is often desired to have good spaciousness; low coloration and absence of directional shift are undesired. These problems, created by early reflected sound, are well known in dealing with studio acoustics.

In summary, it can be noted that hearing is extremely sensitive to added sound components, particularly if they are not spatially displaced and if the time delays involved are of the order of 5 ms or smaller.

The Precedence Effect

This effect, also called the *law of the first wave front,* implies that it is the arrival direction of the first of several coherent sound contributions (temporally and spatially displaced) that determines the subjectively

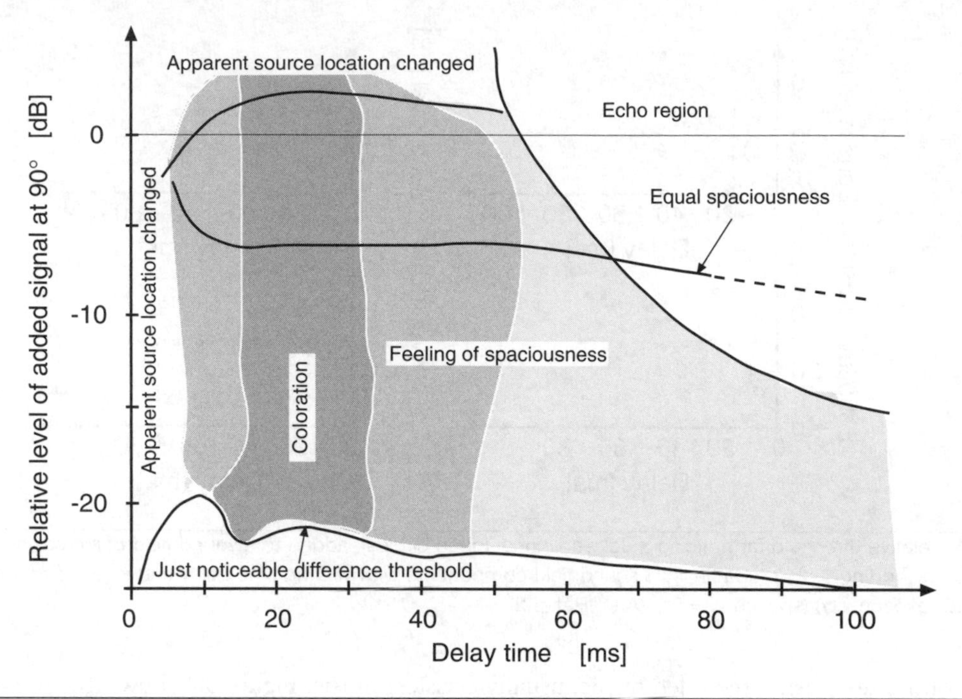

Figure 5.6 Various subjective effects of an added sound field component at 90°, for various delays and levels relative to the direct sound which has frontal incidence (classical music). (After Ref. 5.4)

perceived direction of the sound source. The law of the first wave front usually works quite well in rooms or auditoria, for example, but in some cases, such as when the sound source is obscured or when there are focusing surfaces, the directional impression can shift in an undesirable way. The limits of the principle are therefore of interest.

Psychoacoustic experiments, both regarding stereo sound reproduction and the usefulness of the principle, have given interesting results.

In one experiment, the test subjects were subject to direct sound arriving from the direction (0°, 0°) and the same sound, but variably delayed, arriving from (90°, 0°). The results showed that for the perceived sound source location to be shifted to (45°, 0°), that is, halfway away from the direction of the first wave front, the component arriving at (90°, 0°) had to be approximately 10 dB stronger than the direct sound component, as shown in Figure 5.7.

These results verify the observations discussed initially. The law of the first wave front can be put to good use in many sound reinforcement applications where amplification of sound, without directional shift, is desired. By introducing a small time delay into the amplified signal, the directional impression of the sound source will remain unchanged, while the level of the direct signal is perceived to have increased.

Delay versus Level Differences

In stereo sound reproduction, two loudspeakers and sound reproduction channels are used. The line between the loudspeakers is called the baseline. The listener is situated at a right angle to the baseline

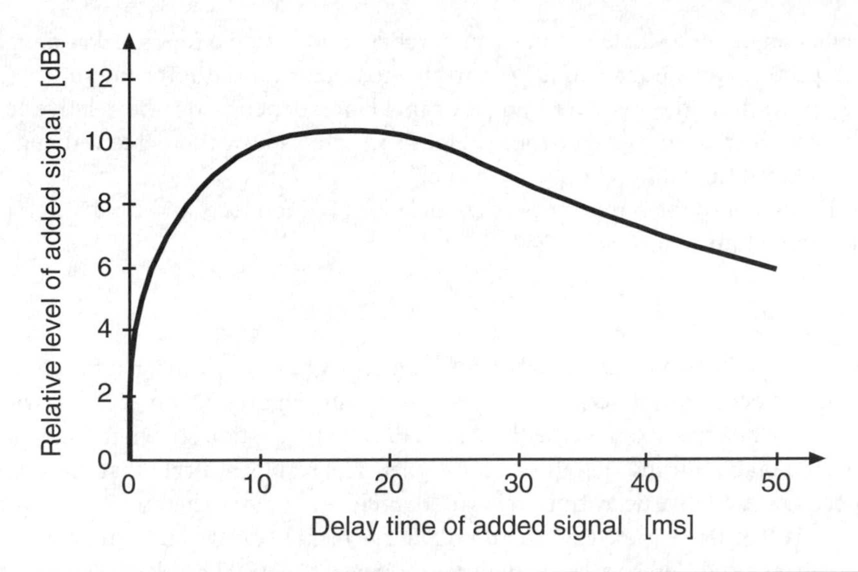

Figure 5.7 The level difference between the added delayed signal and the direct signal that gives both signals the same loudness, as a function of delay time (speech). (After Ref. 5.5)

symmetrically with respect to the loudspeakers. An equilateral triangle setup is often used. Stereophonic sound reproduction is based on the perception of virtual sound sources, sometimes called *phantom sources* (see Reference 5.6).

To render a phantom source on the baseline of the stereo loudspeaker pair, it is necessary that the signals radiated by the two loudspeakers be suitably correlated. Using magnitude and arrival time differences between the sounds from the two loudspeakers, one can move the phantom sources along the baseline between the two loudspeakers, as indicated by Figures 5.8 and 5.9. These figures show the perceived

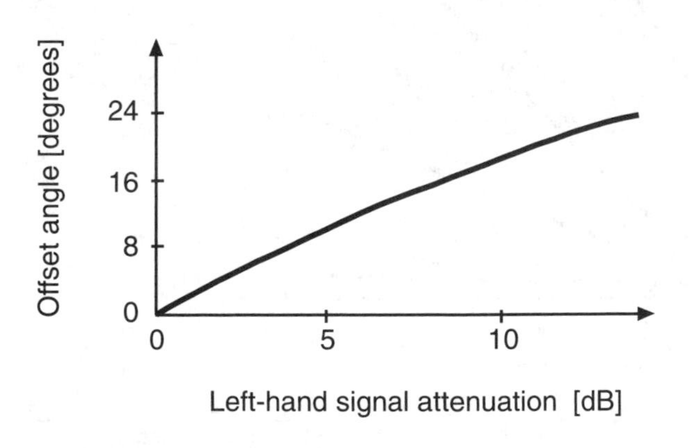

Figure 5.8 Lateral angle position of the phantom source for a mono sound source reproduced by two loudspeakers at $\varphi = \pm 30°$ as function of the reduced level of the left-hand loudspeaker. (After Ref. 5.6)

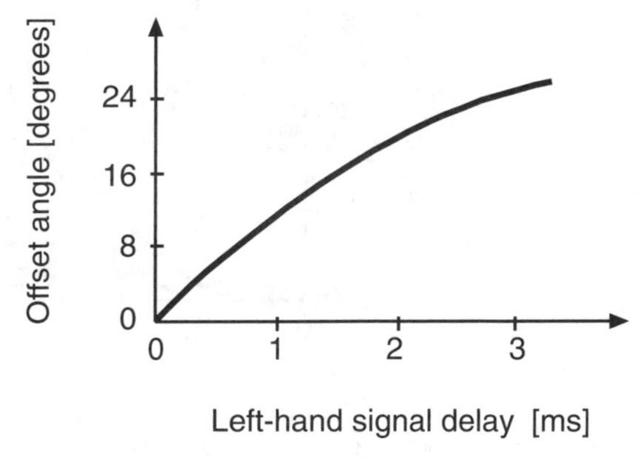

Figure 5.9 Lateral angle position of the phantom source for a mono sound source reproduced by two loudspeakers at $\varphi = \pm 30°$ as function of the extra time delay of the left-hand loudspeaker. (After Ref. 5.6)

location (receiving angle) of a single channel sound radiated by a stereo loudspeaker pair, but where the signals to the respective loudspeakers have been treated to obtain various level and time delays.

Figure 5.8 shows how the perceived angle of incidence depends on the relative level difference between the loudspeaker sounds. Analogously, Figure 5.9 shows how the perceived angle of incidence depends on the relative time difference.

A further discussion of the importance of the balance between level and time differences for stereo sound systems can be found in Chapter 17.

Echo

If the time delay between two coherent signals is long enough, one will hear *echo*. By using the same approach as in the experiments described in conjunction with Figures 5.8 and 5.9 but with both signals arriving from (0°, 0°), one has obtained the thresholds of echo perception shown in Figures 5.10 and 5.11. These results were obtained using speech as a test signal. The results indicate that the sensitivity to echo depends on speech rate, relative delay time, relative sound level, and reverberation time. The most important result shown is that there is a clear *echo threshold* for relative delays larger than around 50 ms that correspond to path length differences larger than approximately 17 m. The risk of echo makes it important to avoid concave, focusing surfaces in auditoria.

The results shown in Figure 5.10 also indicate that if the delay time is short, less than 20 ms, it will be possible to increase the intensity of the delayed signal relative to the direct signal without echo. This result is a further reminder of the temporal masking properties of hearing discussed earlier.

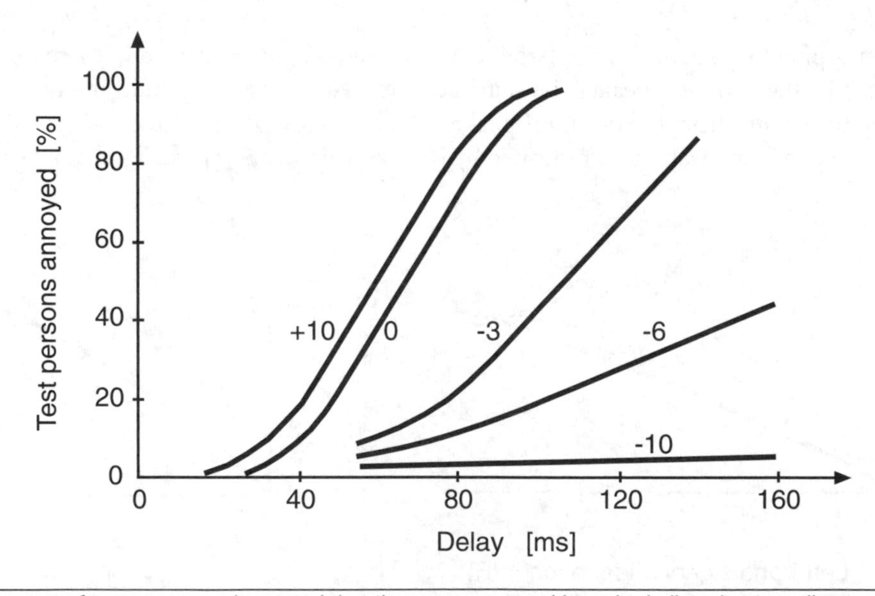

Figure 5.10 Percent of test persons who stated that they were annoyed by echo in listening to a direct sound in combination with a delayed sound (φ = 0°) with the relative level of the echo to the direct sound as parameter. (After Ref. 5.7)

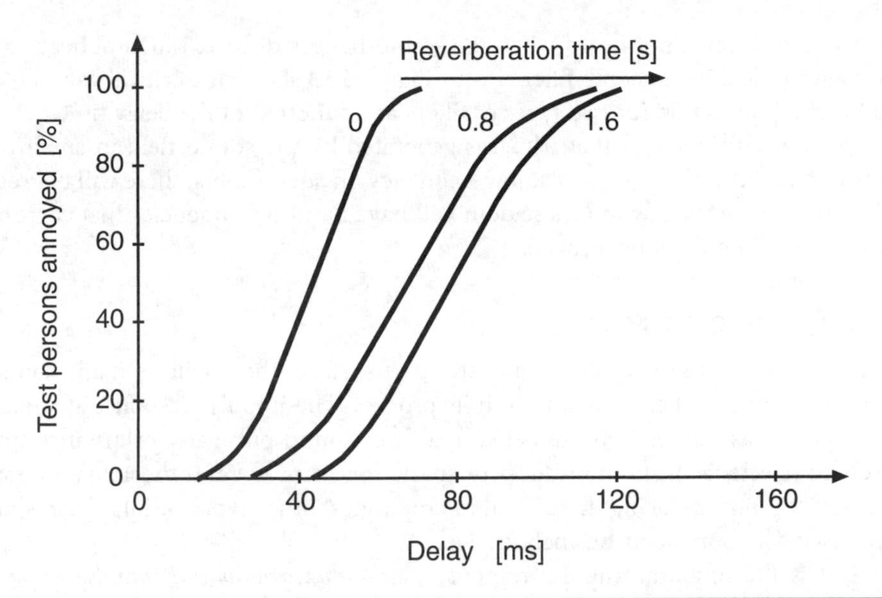

Figure 5.11 Percent of test persons who stated that they were annoyed by echo in listening to a direct sound and reverberation in combination with a delayed sound (φ = 0°), with reverberation time *T* as parameter. (After Ref. 5.7)

One will quickly notice that echoes are more annoying at high frequencies than at low frequencies. At high frequencies, there is simply more time available (more periods) to form an echo perception even with short time delays.

Coloration

When listening to music, *coloration* (timbre change) is usually more of a problem than echo. This is probably due to the typically slow modulation of classical music, although there are musical instruments that, when played in concert, are used to generate short sounds. At one extreme, one has slow organ tones, at the other, short trumpet passages.

Coloration particularly is a problem when binaural hearing cannot be used to separate the directions of incoming sounds, for example, when one is listening to a monaural reproduction of a stereo recording (particularly in AB-type recordings, see Chapter 17), or in auditoria where there are strong sound reflections from the ceiling or from the audience area (see Reference 5.8).

Figure 5.12 shows how the repetitive frequency response variations are introduced by various types of repetitive finite- and infinite-impulse response filters. For example, finite-impulse responses are found in the case of sound field contributions that are correlated, somewhat time delayed, and coming from the same direction. Repetitive infinite-impulse responses are obtained when one has two highly reflective parallel walls dominating the sound field in a room. The acoustic effect is, in this case, called a *flutter echo* or *comb filter effect*.

Comb filter coloration is not heard when the sound field is diffuse. Binaural hearing may resolve monaural ambiguities leading to comb filter effects. Figure 5.13 shows the critical values of the gain factor g for audibility of coloration for two types of filters as a function of the delay time t_0.

Another form of coloration is that which is generated by the sound field in small rooms having sparse, weakly damped room resonances at low frequency. In such rooms, there will be frequency shifts in the reverberation since the few modes seldom will have natural frequencies that correspond exactly to those of the audio from the loudspeakers.

Sequences of Reflected Sound

In a real room, having more than one sound-reflecting surface, there will be many sound field components—natural room reverberation is an infinite process. The impulse response at some point in the room can be regarded as the relationship between a short sound pulse at 1 m distance from an omnidirectional source (electroacoustic transducer or spark, for example) and the actual pressure response at the measurement point. By using directional microphones, other types of impulse responses may be measured and used for room acoustic analysis.

Of course, it is the binaural impulse response pair—*binaural spatial impulse response* (BSIR)—rather than the omnidirectional response, that is of real interest. However, at this time there is little knowledge regarding how such a measured BSIR can be analyzed, so currently analysis is usually limited to the measured omnidirectional *spatial impulse response* (SIR) (sometimes called the *room impulse response*).

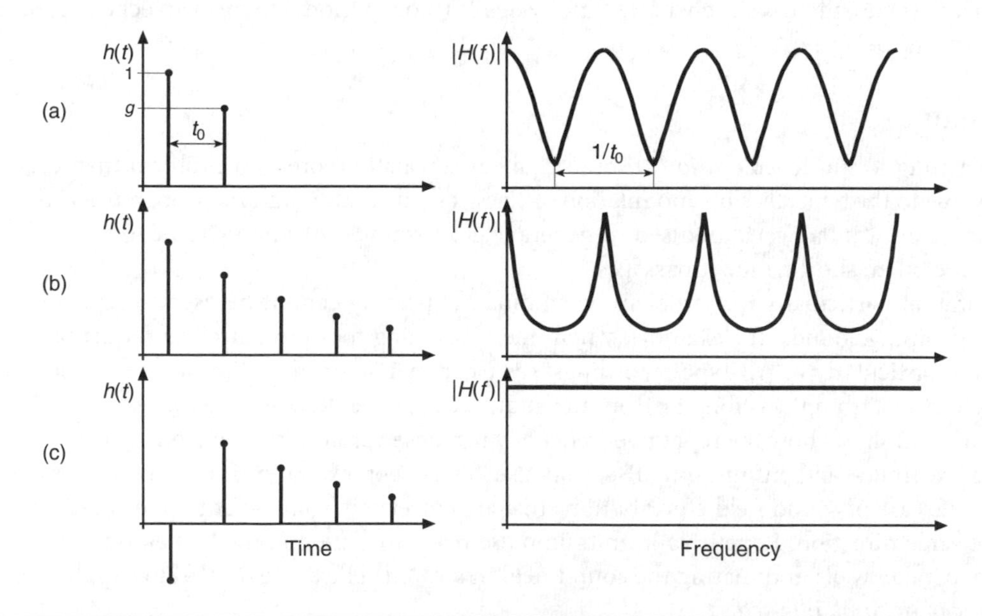

Figure 5.12 Examples of impulse responses and their associated transfer function magnitudes for some comb filters. Delay between impulses is t_0 in all cases. Gain setting is g = 0.7: (a) simple delay, (b) infinite impulse response, repetition with gain g^n, or (c) similar to b but with inverse start transient for all-pass frequency response.

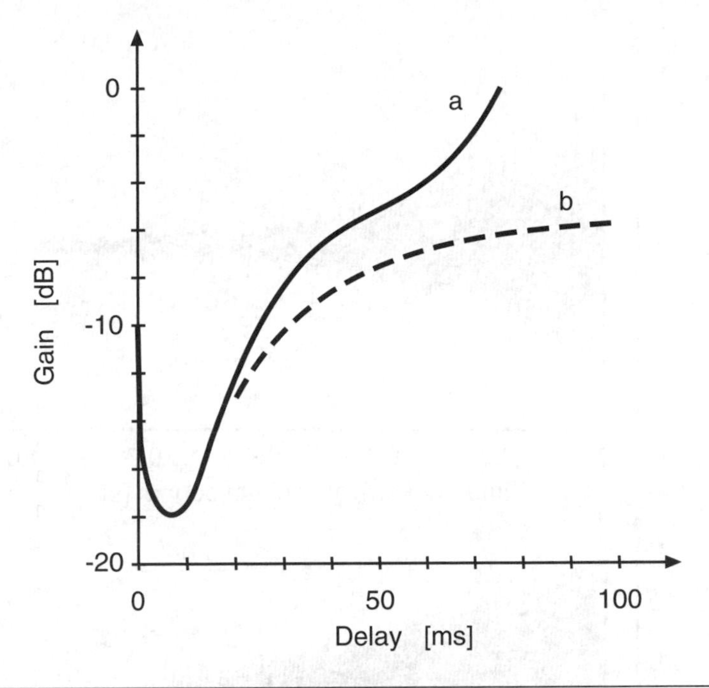

Figure 5.13 Gain factors *g* which give just perceptible coloration of white noise: (a) impulse response $h(t) = \delta(t) - g\delta(t\text{-}t_o)$; (b) impulse response $h(t) = \Sigma g^n\, \delta(t - nt_o)$ (n = 0, 1, 2, 3, 4, . . .). (After Ref. 5.8)

One can show that it is possible to obtain the true reverberation curve by *backward integration* of the SIR. True, here, means the reverberation curve that would be obtained by averaging an infinite number of (squared) impulse response measurement results for the room. The backward integration process is currently the common way to measure reverberation characteristics (see Reference 5.17).

An example of a measured impulse response for a large room is shown in Figure 5.14. The expected discrete reflections are not found in the graph. This is due to the multitude of reflections, the limited resolution of the graph, and primarily because the loudspeaker and sound-reflecting surfaces have a frequency-dependent phase and amplitude properties. Since the SIR behavior is so complex, it is current practice to use various types of energy-based metrics to describe the relevant properties of the room (see Section 5.3).

If one removes or alters sections of the SIR or BSIR and convolves the resulting impulse responses with audio, it is possible to listen to changes in the properties of the room. This is an attractive technique to study the importance of various parts of the impulse responses.

Reverberation Time and Subjective Preference

The reverberation time is the most useful of the room acoustic metrics to judge the acoustical properties of a room and its suitability for a certain use. Experience has shown that the reverberation time that is suitable for various uses and sizes of rooms can be approximated as shown in the graph in Figure 6.9.

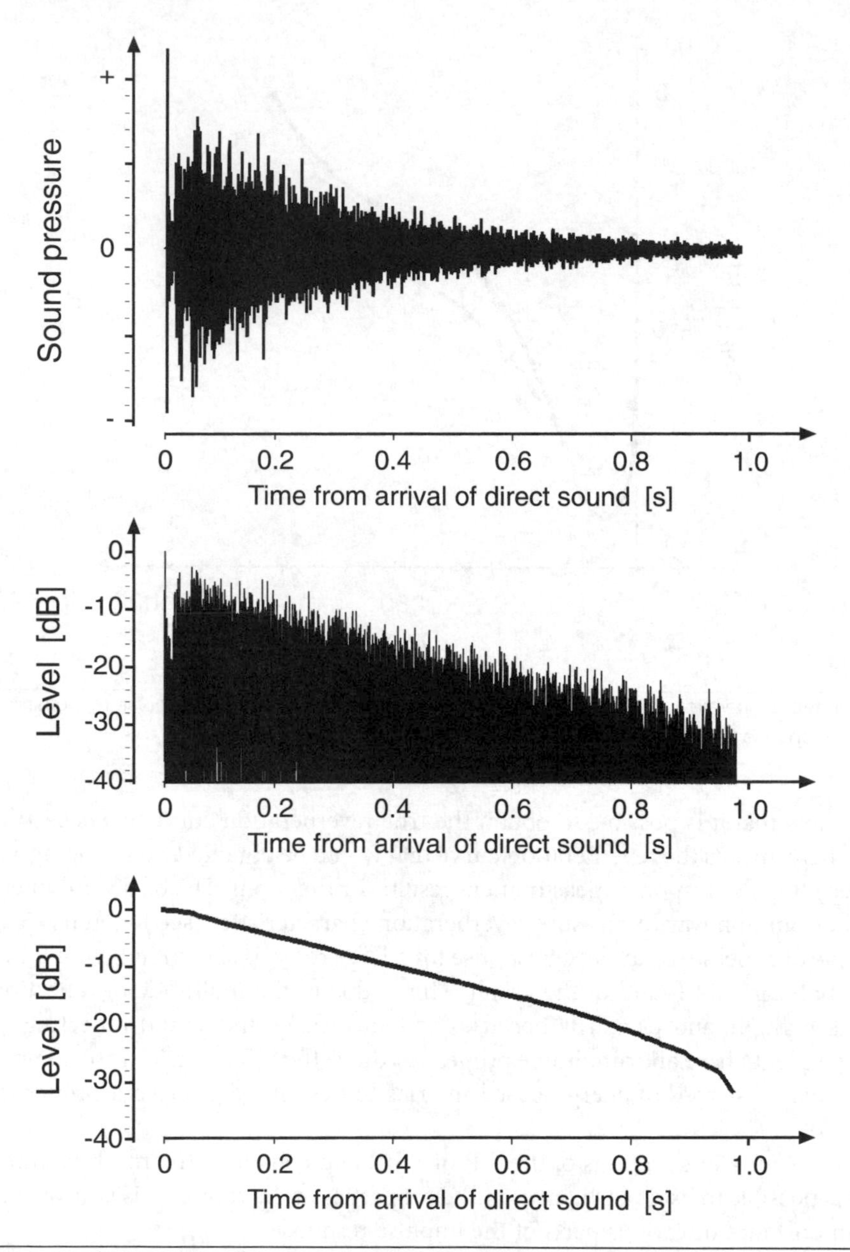

Figure 5.14 Three ways of representing the same measured sound decay characteristics. The top curve shows the sound pressure drop (impulse response); the middle curve shows the SPL drop (echogram); and the bottom curve shows the reverberation curve obtained by reverse summation of energy in the top curve.

Human ability to discriminate between different reverberation times is shown by the curve in Figure 5.15, which shows the average ability to sense a certain change in reverberation time ΔT for various reverberation times T. One can see that for most of the range, a reverberation time change of less than 5% can be heard.

In practice, one finds that people prefer particular ranges of reverberation times for various audio. Speech usually needs short reverberation times, whereas music can need reverberation times from nearly zero to several seconds, depending on its type. Since the reverberation of a room contributes to the loudness of sound, some reverberation is usually required for an optimum speech intelligibility. The reverberation times in rooms for speech should generally not be lower than approximately 0.6 s at the center frequencies; if the reverberation time is shorter, the rooms will sound dry. Without reverberation, the sound quality of speech and music would be dependent on the directivity of the sound source.

The ability of hearing to perceive changes in reverberation time is surprisingly good. A reverberation time change of less than 5% can be noticed (around 1–3 kHz) (see Reference 5.9). Usually one strives to achieve control of the reverberation time to an interval of ± 0.1 s at midrange frequencies.

The reverberation time in rooms for classical music (without amplification) should increase somewhat at low frequencies from 250 Hz and downward; the reverberation time in the 63 Hz octave band may be around 50% higher than in the 1 kHz octave band.

Rooms for speech and rooms used for amplified sound (such as speech or music) should generally not have any reverberation time increase at low frequencies. For these types of rooms, one should strive

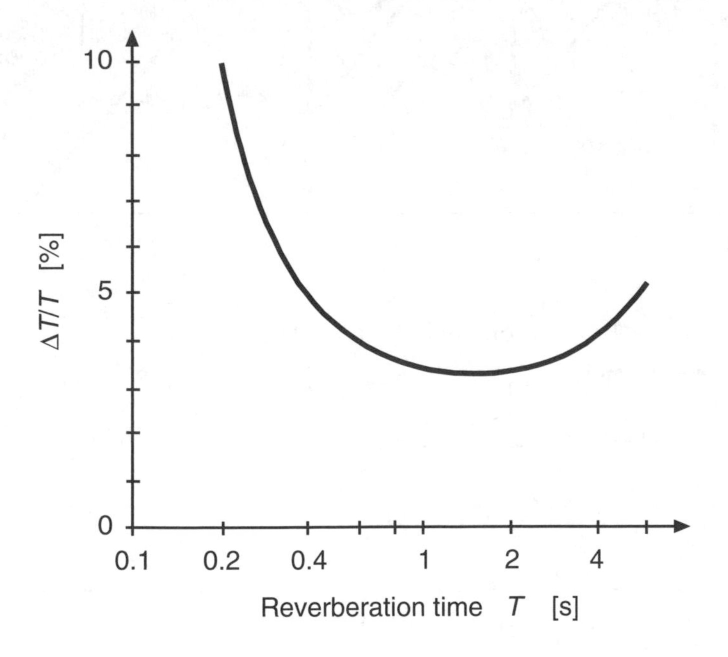

Figure 5.15 Barely noticeable difference in percentage for various reverberation times. (After Ref. 5.9)

for a flat, frequency independent, reverberation time curve. These rooms otherwise have a tendency to sound *boomy*.

One must also be aware of the difficulties inherent in trying to measure reverberation time with audience and performers in a room. The measurements take some time and the audience expects to be entertained rather than be part of an experiment. Considerable experience is required to expertly judge the difference in reverberation characteristics between when a room is occupied and when it is unoccupied.

Early and Late Reverberation

Reverberation time is defined as the time it takes for the energy density in a room to reach a level of −60 dB below that of the start of a sound decay, after a switching of the sound source. Since it is difficult to measure the energy density, the reverberation curve is measured using the sound pressure level decay characteristic, and the results obtained at various locations are averaged.

Psychoacoustic experiments indicate that hearing is primarily focused on approximately the first 0.15 s of the reverberation process in *running* music or speech. This can be understood from the importance of masking (see Figure 5.16). It is only at the end of a piece of music that one will hear the entire reverberation and how the reverberation recedes into the background noise of the room. Usually, at least

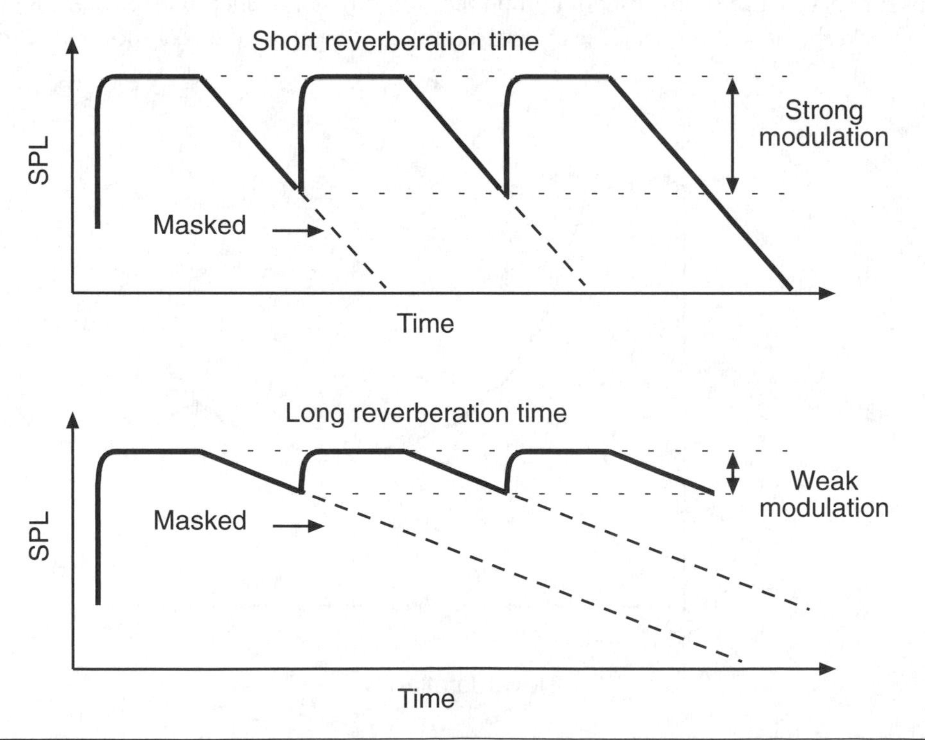

Figure 5.16 Reverberation by previous sound masks later sound. The modulation depth decreases as a result and less detail is heard.

for classical music, this is an important part of the music, so both the *early* and the *late* reverberation characteristics are important when designing high-quality auditoria.

Figures 5.17 and 5.18 show the results of experiments regarding the judgment of perceived reverberation time versus reverberation time measured in various ways. One sees that the early reverberation time is a reasonably good measure of the perceived reverberation time, particularly if judged over the interval of the first 0.16 s of the reverberation process instead of over the first 15 dB of the process.

These results underline the importance of a suitable acoustic design of the room geometry, since it will be the room geometry and the acoustic properties of the walls by which the sound is initially reflected that are important to the early part reverberation process.

Electroacoustic equipment can be used to change the properties of the reverberation process. This is, in effect, what is done when using sound reinforcement systems (see References 5.18 and 5.19).

Dual-slope Reverberation Curves

When doing reverberation time measurements, one quickly notices that the sound level decay curve is not a linear function of time, particularly during the time in which sound propagates the first few mean-free-path lengths. Quite often, one also finds *bends* in the slope of the reverberation curve, even when the sound in the room has become quite diffuse (see Figure 4.7).

Such reverberation curves were originally referred to as *broken* but now the term multi-slope is used. In the case of only one bend, the term is *dual-slope curve*. The reason for the dual-slope is usually that the characteristics that affect such things as the decay, absorption, diffusion, mean-free-path length, and volume are different in various subspaces of the volume. This is particularly the case for rooms in which several smaller subspaces make up one larger space. Churches and cathedrals often have such coupled subspaces.

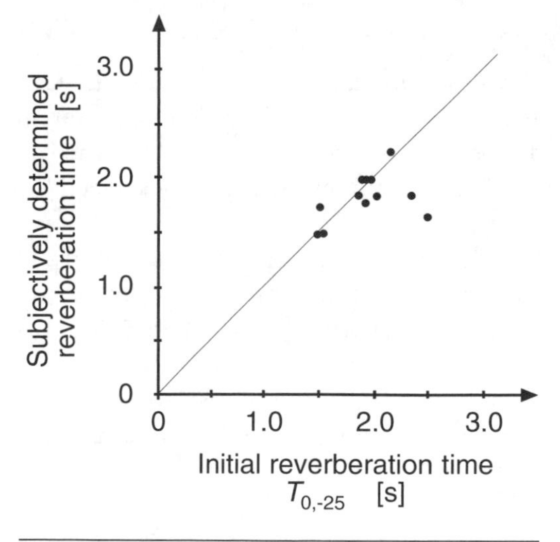

Figure 5.17 Measurement results for the subjectively perceived reverberation time as a function of the reverberation time of the first 25 dB of the reverberation process. (After Ref. 5.10)

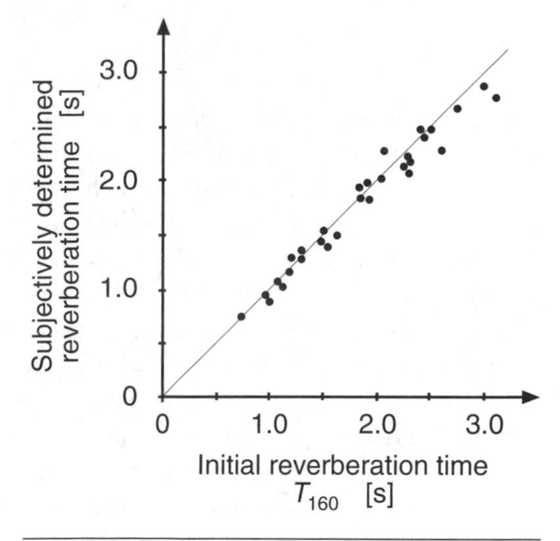

Figure 5.18 Measurement results for the subjectively perceived reverberation time as a function of the reverberation time of the first 0.16 s of the reverberation process. (After Ref. 5.10)

Many people have come to like the acoustics characterized by dual reverberation slopes. The reason for this is probably that when listening to music in the home, over stereo or multiloudspeaker systems, the reverberation in the recording is mixed with the reverberation of the listening space. Since these are seldom the same, the resulting reverberation curves will be characterized by many slopes. Dual-slope decays also make it possible to create listening situations that have extremely good clarity coupled with long reverberation times and much ambience. Some modern concert halls are expressly designed to have acoustically coupled subspaces that have much longer reverberation times than the main auditorium. One prerequisite for the audibility of such long, final slope, reverberation in auditoria, such as concert halls, is that the auditorium features low background noise.

Subjectively Diffuse Sound Fields

It is important to realize that sound fields may appear subjectively diffuse without being diffuse in the physical sense (discussed in Chapter 4). Even a sound field composed of only two components can be perceived as diffuse, for example, with coherent noise from two diametrically placed loudspeakers, at angles (90°, 0°) and (270°, 0°), when loudspeaker two is fed loudspeaker one's signal slightly delayed (approximately 1 ms or more). It is therefore best to be careful in the use of the term diffuse and make it clear whether it is meant in the physical or subjective sense.

The degree of *coherence* between the signals at a person's ears is important for the perception of the subjective diffuseness of a sound field. Of course, a sound field in which there is much frontal direct sound will have higher coherence than a sound field primarily consisting of reverberant sound.

Experiments have shown that the degree of spaciousness experienced when listening to a sound field is related both to the coherence of the signals at the ears and to the *level difference between direct sound and reverberant sound* (German: Hallabstand), H, as defined by:

$$H = L_{p,\text{direct}} - L_{p,\text{reverberation}} \tag{5.1}$$

Figure 5.19 shows some experimental results concerning how the *interaural cross-correlation* (IACC) influences the subjectively perceived spatial extension, as the reported *solid angle* distribution of sound in the upper half space when listening to a band-limited noise signal (see Reference 5.11). One can notice that the spatial extension, which here can be regarded as a measure of the degree of subjectively perceived diffuseness, increases as the value of IACC decreases.

Figure 5.20 shows the result of experiments in which the level difference between direct sound and reverberant sound was changed and how this affects the spaciousness of the sound field. One unit of change on the spaciousness scale corresponds to the *just noticeable difference*—that is, where 50% of the time the test persons were able to discern a difference.

The importance of the early part of the reverberation in relationship to H is also shown by the curves in Figure 5.21. The curves show the subjectively judged reverberation time T_{subj} as a function of H, with reverberation time T as parameter. It confirms the possibility of reducing the subjectively perceived reverberation time by addition of strong early reflections.

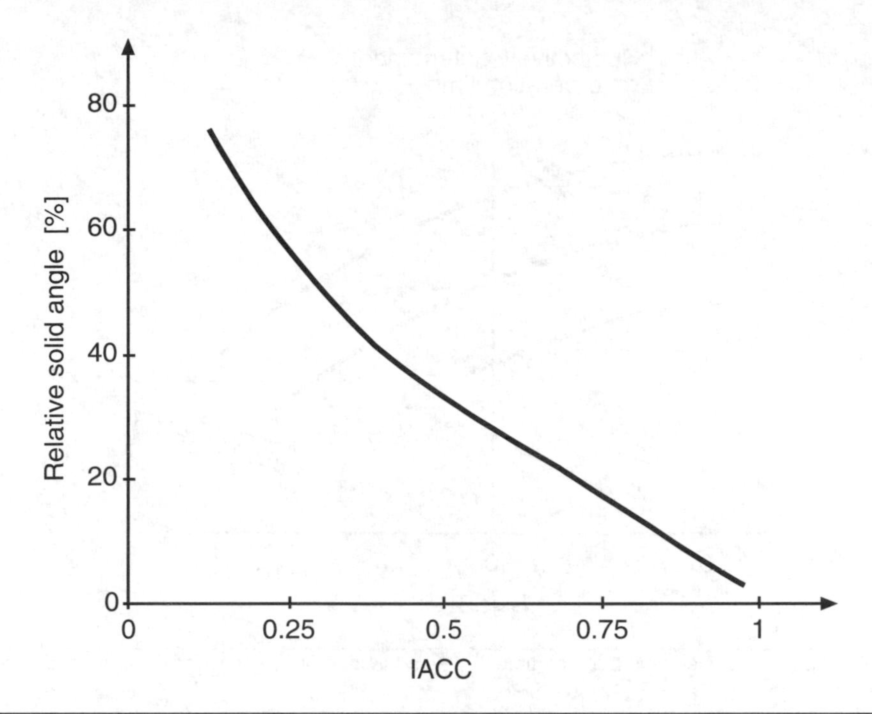

Figure 5.19 Perceived solid angle of the sound field as a function of the interaural cross-correlation. (After Ref. 5.11)

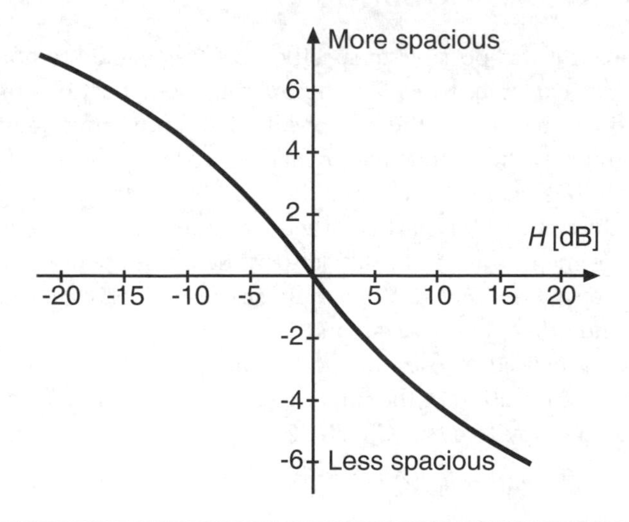

Figure 5.20 Subjectively perceived steps of spaciousness as a function of the level difference H between direct and reverberant sound. Note that high spaciousness is obtained for large negative values of H. (After Ref. 5.12)

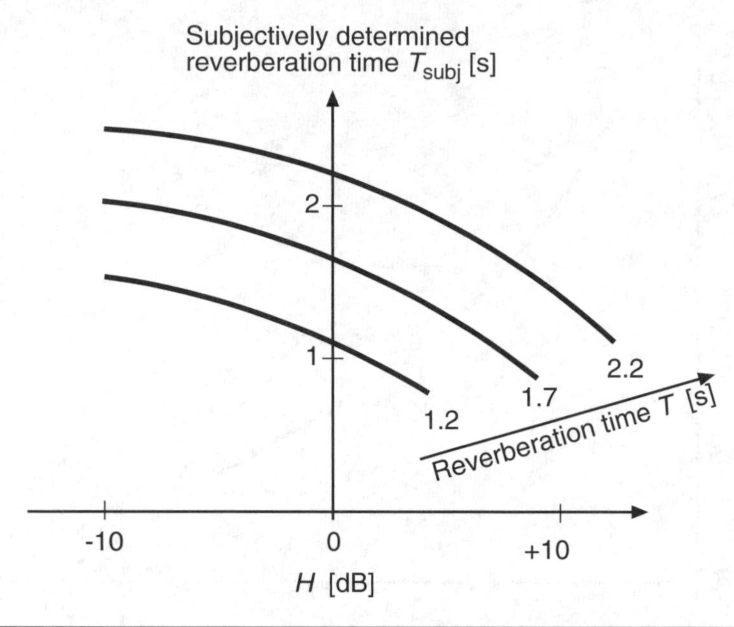

Figure 5.21 Subjectively perceived reverberation time T_{subj} as a function of H with reverberation time T as parameter. (After Ref. 5.12)

5.3 METRICS FOR ROOM ACOUSTICS

It would be advantageous if it was possible to specify the acoustic quality of a room by some simple numbers or indicators, for example, when writing building specifications or for quality control. It turns out that this is difficult because of the lack of sufficient knowledge about the spatial properties of human hearing. A room may show large numerical differences between the current quality metrics measured at various positions in the room while the subjective experience of listening in the room does not bear out these differences. It is particularly the fine structure of the reverberation that is difficult to measure. In experiments done in the laboratory, the influence of visual cues can be removed, but the sound field must be approximated. When doing experiments in real auditoria, it is difficult to remove the visual cues and other cognitive factors.

Nearly all of the currently used metrics for room acoustics are based on the measurement of SIR. The measured SIR, h, is a convolution of the impulse responses of the loudspeaker, microphone, and other parts of the measurement system (see Chapter 2).

$$h(t) = h_{lsp}(t) * h_{room}(t) * h_{mic}(t) * h_{other}(t) \tag{5.2}$$

where h_{lsp}, h_{room}, h_{mic}, and h_{other} are the impulse responses of the loudspeaker, room, microphone, and other equipment (such as electronic filters and sound cards) involved in the measurement process.

The influence of these other, undesired impulse responses is generally disregarded when considering the room acoustics metrics. For simplicity, the room impulse response is often set as the raw measured impulse response or filtered using a suitable bandpass filter, such as an octave band filter. It is important to note, that such filters will have group delay that must be compensated. Using a process called deconvolution, it is possible to eliminate some of the problems due to the electroacoustic equipment.

Along with the music room acoustics metrics listed here, there are also various speech-related metrics discussed in Chapter 3 as well as various metrics related to stage acoustics. References 5.13 and 5.14 are useful for their discussion of the various metrics mentioned in upcoming text.

Strength Index

The most important factor in determining acoustic quality is loudness. A metric designed to estimate the relative loudness of sound in a room is the *strength index* G, sometimes called the room gain. The strength index is based on the measured SIR $h(t)$ and is defined in the following way:

$$G = 10\log\left(\frac{\int_{0\text{ms}}^{\infty} h^2(t)\,dt}{\int_{0\text{ms}}^{5\text{ms}} h_A^2(t)\,dt}\right) \quad [\text{dB}] \tag{5.3}$$

where $h_A(t)$ is the measured impulse response for the direct sound component only, at 10 m distance in anechoic space. Since it is difficult to find such large anechoic spaces, the measurement is usually done at a shorter, convenient distance and then scaled to compensate for the geometric strength difference due to the distance difference, assuming that the level drops by −6dB per distance doubling.

Reverberation Time

From the viewpoint of subjective importance, after loudness, reverberation time is generally the property of a room that is the most important for the subjective impression of room acoustics.

According to the ISO 3382 standard the reverberation time is obtained by extrapolation to 60 dB of a line fitted to the reverberation curve from −5 dB to, for example, −35 dB of decay. The reverberation time is then often written as $T_{-5,-35}$. It is difficult to obtain a good dynamic range when measuring reverberation time in large venues. Unless otherwise noted, the room impulse responses are to be measured using an omnidirectional microphone. The standard also requires the measurements to be done using at least two sending positions and five receiving positions, spaced in a specified way on the stage and in the audience seating area.

Definition

Definition (D) (German: Deutlichkeit) is a metric used when judging the suitability of rooms for speech. The Definition metric is based on the assumption that the sound arriving during the first 50 ms of the reverberation process is useful (beneficial to speech intelligibility), whereas the rest of the sound is detrimental. Definition has been shown to correlate reasonably well to speech intelligibility. Definition is defined as:

$$D = 100\frac{\int_{0\text{ms}}^{50\text{ms}} h^2(t)\,dt}{\int_{0\text{ms}}^{\infty} h^2(t)\,dt} \quad [\%] \tag{5.4}$$

There are many other similar metrics, such as Clarity. Since Definition does not take the signal properties of sound into account, it is not as good a metric as modulation damping and speech transmission index. It also does not take the influence of the background noise of the room into account.

Clarity

Clarity C_{80} is a metric similar to Definition but adapted to evaluation of rooms for music. It is defined as:

$$C_{80} = 10\log\left(\frac{\int_{0\text{ms}}^{80\text{ms}} h^2(t)\,dt}{\int_{80\text{ms}}^{\infty} h^2(t)\,dt}\right) \quad [\text{dB}] \tag{5.5}$$

The index *80* marks the split between the early and late parts of the reverberation process, in this case taken as 80 ms.

Sometimes the split is taken at 50 ms and Clarity C_{50}, is calculated, and used in place of Definition. The two main differences between C_{50} and Definition are the use of different integration times and the use, for C_{50}, of the log of the ratio of energies. The reason for using C_{50} instead of C_{80} is agreement with the time constant used for *D*.

Interaural Cross-Correlation

The IACC is a metric describing the dissimilarity between the left and right ear parts of the binaural room impulse response. One is usually interested in the reverberation sounding diffuse. Since reverberation diffuseness is closely connected to correlation, it is reasonable to assume that it is advantageous that the late reverberation part of the SIR has as low an IACC as possible.

The IACC is defined as:

$$\text{IACC} = \max_{|t|<1\text{ms}} \frac{\int_{0\text{ms}}^{\infty} h_{LE}(t)h_{RE}(t+\tau)\,dt}{\sqrt{\int_{0\text{ms}}^{\infty} h_{LE}^2(t)\,dt \int_{0\text{ms}}^{\infty} h_{RE}^2(t)\,dt}} \tag{5.6}$$

where h_{LE} (*t*) and $h_{RE}(t)$ are the left ear and right ear measured room impulse responses respectively. These are measured using microphones mounted in the ears of a manikin (see Figure 5.22 and 17.6).

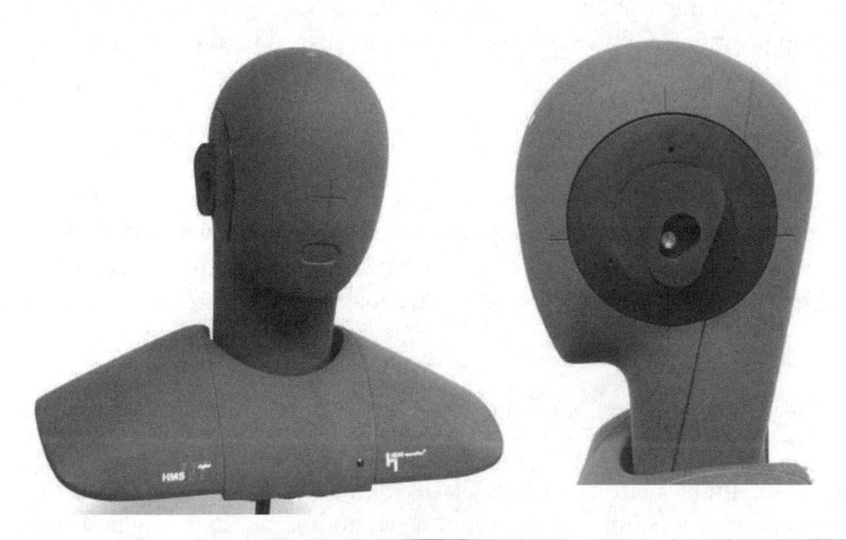

Figure 5.22 A manikin for binaural recording and measurement, and its ear replica, showing the microphone at the bottom of the imitated concha cavity. (Photo by Mendel Kleiner.)

It is important to note that IACC can be used with many different types of test signals and with bandpass-filtered impulse responses. Care is advised when using IACC data and specifying and measuring IACC. Also note that sometimes the IACC is measured over parts of the SIR, for example 0–80 ms and 80–∞ ms.

Originally, IACC was intended for use with binaural recording of running music so that the metric described what was happening in an actual listening situation.

Recently the use of the quantity 1–IACC (in octave bands) has become common (see Reference 5.13).

Lateral Energy Fraction

As discussed previously, it is much easier for hearing to separate between signals if they differ in angle of incidence relative to the median plane. A metric designed to separate between sound arriving in and out of the median plane is the *Lateral Energy Fraction* (L_f) defined as follows:

$$L_f = \frac{\int_{5\text{ms}}^{80\text{ms}} h_L^2(t)\,dt}{\int_{0\text{ms}}^{80\text{ms}} h_T^2(t)\,dt} \tag{5.7}$$

where $h_L(t)$ is the room impulse response measured using a bidirectional microphone and $h_T(t)$ is the room impulse response measured using an omnidirectional microphone.

The exact angular weighting factor for the lateral energy fraction is specified as $h^2 \propto \sin(\varphi)$, where φ is defined as shown in Figure 3.2. In practice, the weighting function $h^2 \propto \sin^2(\varphi)$ is usually used since it is difficult to find suitable microphones otherwise.

Center-of-Gravity Time

A problem common to both definition and clarity is sensitivity to abrupt SIR changes at around 50 and 80 ms respectively. The metric *center-of-gravity time*, t_s, is designed to circumvent this problem by not using any fixed *transition time*. The definition is:

$$t_S = 1000 \frac{\int_{0\text{ms}}^{\infty} t \cdot h^2(t)\,dt}{\int_{0\text{ms}}^{\infty} h^2(t)\,dt} \quad [\text{ms}] \tag{5.8}$$

The center-of-gravity time (German: Schwerpunktzeit) has not had any success in replacing definition and clarity but is used alongside the two older metrics.

Early Decay Time

The importance of the early part of the reverberation process has been stressed several times. An alternate definition of what constitutes the early decay is used for the metric *early decay time* (EDT) which is based on the extrapolation to 60 dB of the reverberation time during the first 10 dB of reverberation decay. The results reported in Figure 5.19 speak against the use of EDT as a metric for evaluation of reverberation time. The EDT values for some octave bands are used in the definition of the *brilliance* metric.

Warmth

Warmth is an important requirement for halls for classical music. It is dependent on the relative energy between reverberant sound at low and medium frequencies respectively. Since the relative energy is in turn dependent on the relative sound absorption, it will also depend on the reverberation time. *Bass ratio* (BR) is a way of expressing this relationship.

$$BR = \frac{T_{125ob} + T_{250ob}}{T_{500ob} + T_{1000ob}} \quad (5.9)$$

Here T_{125ob} is the reverberation time T in the 125 Hz octave band, and so on.

Typical expected BR values are from 1.1 to 1.25 for rooms with $T > 2.2$ s and from 1.1 to 1.45 for rooms with $T < 1.8$ s.

Brilliance

A lack of *brilliance* can be observed in rooms where the high frequency early decay times are much shorter than the midfrequency early decay times. Two metrics have been suggested for brilliance, B_{2kHz} and B_{4kHz} for the 2 kHz and 4 kHz frequency bands respectively (see Reference 5.13).

$$B_{2kHz} = \frac{2EDT_{2000ob}}{EDT_{500ob} + EDT_{1000ob}} \quad (5.10)$$

$$B_{4kHz} = \frac{2EDT_{4000ob}}{EDT_{500ob} + EDT_{1000ob}} \quad (5.11)$$

Here the EDT indices refer to the octave band frequencies for which they were measured. The recommended minimum values for B_{2kHz} and B_{4kHz} are 0.9 and 0.8 respectively for concert halls.

5.4 Problems

5.1 The function $h(t) = Ae^{-\delta t}$ is sometimes used to approximate the envelope of an ideal SIR.

Task: Express the damping constant δ as a function of reverberation time T using the definition of reverberation time.

5.2

Task: Using the function $h(t) = Ae^{-\delta t}$ as the approximation of an ideal SIR envelope, determine how clarity and definition can be expressed as functions of reverberation time T.

5.3 A recording studio uses an omnidirectional microphone for recording and then uses a signal processing mixing console to position the virtual sound source at the desired location between the listener's loudspeakers. This is done using time and level differences between the right and left channels. The phantom sound source can be placed between the loudspeakers in the sketch below, using different amplification factors A_R and A_L for the two channels, according to Figure 5.9.

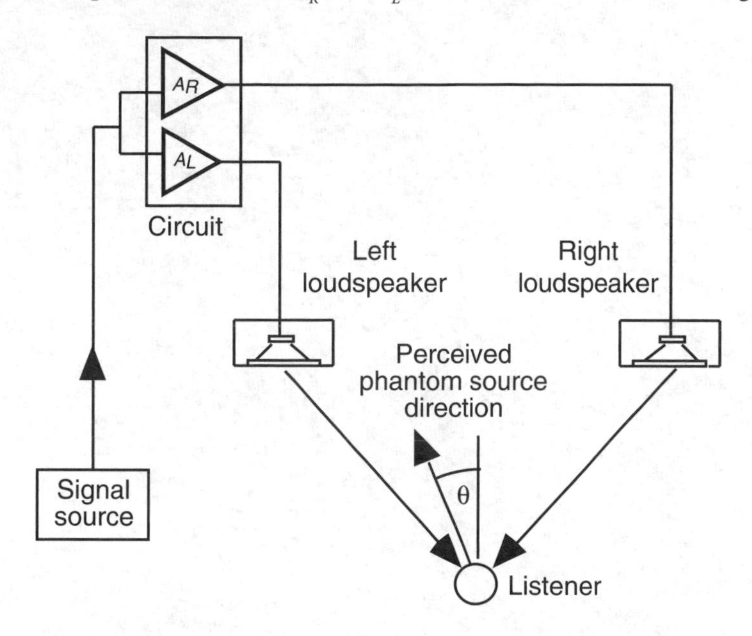

There are two restrictions to the circuit:

1. The circuit is only allowed to give angular change—no amplitude change is allowed.
2. For $\theta = 0°$ the amplification factors must be unity.

Hint: Use the fact that the loudspeaker signals are correlated and have the same time-of-flight.

Task: Calculate the correct amplification factors A_R and A_L, to obtain directions θ from 0 to 24 degrees in 4 degree increments (to the phantom source).

Room Acoustics Planning and Design

6

6.1 INTRODUCTION

The purpose of this chapter is to give examples of the kind of deliberations and concerns that are involved in the planning and design of rooms. Many of the physical and psychoacoustic aspects of room acoustics have already been presented and discussed in Chapters 4 and 5.

The purpose of room acoustics planning and design is usually to provide rooms that allow good communication between people—for example, by providing good acoustics for speech and music. This requires not only specific acoustic properties regarding the way the room guides the sound from the speaker or performer to the audience and other orchestra members, but also that competing sounds, such as noise are suitably controlled.

The requirements for the way in which sound is distributed in the room are not only expressed by sound level and sound level distributions but are also expressed with timbre and temporal characteristics. The temporal characteristics are a result of the room size and shape, as well as the way the audience and other sound-absorbing and scattering areas are distributed over the room's boundaries. It is important to note that there are substantial interindividual preference differences between people, and that the psychoacoustics data presented in Chapter 5 is not exact. Much room acoustic data still needs to be collected and compared to personal preference for improving the mapping between the physical domain of room design and the subjective domain of acoustics appreciation.

Performers and audiences may have conflicting requirements regarding suitable acoustics. Solo performers, such as speakers and singers will usually want to sense the acoustic response of the auditorium, whereas musicians playing in groups or orchestras will usually put more emphasis on the way the room allows them to interact.

Background noise levels must also be sufficiently low to allow good communication. In rooms for music, low levels of background noise are needed not only to keep the noise from masking the *running* music, but also so that people will be able to hear the full dynamic range and spatial qualities of the reverberation processes. Background noise can further influence and predispose people's reactions to the acoustic quality of the room. The sound quality of the noise will be different depending on whether the noise is due to heating, ventilation, and air conditioning noise (usually called HVAC noise), to other activities in the building, or to traffic, neighboring industry, and other exterior noise. This leads to different noise criteria (NC) being necessary depending on the noise and the use of the room.

For many types of rooms, the room acoustical conditions are so important that acoustic planning and design must be introduced in the early sketching stage. Cooperation between architect and acoustician is necessary for excellent acoustics. The possibilities of changing the general room acoustics of a room are limited once the room exists physically, without building a new interior shell or implementing other drastic changes.

Electronic architecture—that is, active sound field control, using microphones and loudspeakers in the room in conjunction with electronic signal processing—may sometimes be a way to improve acoustical conditions so that the room may be useful for some purpose other than the ones allowed by *natural* or *passive* acoustics. *Sound reinforcement systems* are often used to provide better conditions for listening to speech in acoustically difficult environments, such as those characterized by excessive distance or reverberation, or inadequate sound reflections.

It is important to stress that the quality of room acoustics is difficult to measure. Chapter 5 discussed some metrics for room acoustics quality. These metrics often require considerable experience and expertise to be applied correctly. Often, listening and subjective judgment by experts is the only way to assess the room acoustics quality. Hearing is an important measurement device.

6.2 BASIC REQUIREMENTS FOR GOOD ROOM ACOUSTICS

Quiet: Acceptable Noise Levels and Noise Sound Quality

The requirements of noise that does not damage hearing were discussed in Chapter 3. In designing the acoustics of rooms for homes, offices, industry, and entertainment, additional NC must be met. What is appropriate will depend on the use of the room and on the requirement for visual and auditory impressions to be matching. We expect a concert hall to be quiet, whereas we will not be surprised to find an industrial workplace somewhat noisy. Sometimes the noise is added intentionally as *acoustic perfume*, such as Muzak-type background music, broadband noise having an unobtrusive spectral character, or even intentional HVAC noise.

Typically, noisy environments may be found in schools, offices, industry, and restaurants. From the viewpoint of the acoustician, these environments require special consideration: limiting noise so that it does not cause hearing loss and is acceptable for the use of the room and designing sound reinforcement and other sound amplification equipment for maximum speech intelligibility and other program enjoyment.

Noise level requirements will vary between rooms and room uses. Different noise criteria typically must be applied to office equipment for use in small offices and in office landscapes. The maximum noise is often specified by the maximum allowable A-weighted sound pressure level, without considering the spatial characteristics of the sound field relative to the person being exposed to the sound. The sound pressure at the entrance of the ear canal will typically vary by at least ± 10 dB at frequencies above 2 kHz due to the angle of sound incidence.

Not only does a 10 dB sound pressure level increase in the midfrequency range typically correspond to more than a doubling of loudness, the change may also considerably increase the risk of noise-induced hearing loss and influence *task performance*. It is also important to note that an increase in A-weighted sound pressure level does not necessarily increase *annoyance*. The temporal and other characteristics of the noise are important in this respect. One often uses various metrics of *Product Sound Quality* to further describe the properties of noise, see Chapter 17. The noise may also be described using special terminology usually specific to a certain engineering field; for example, in describing the noise of air handling equipment, words such as hiss, rumble, and roar are used.

It is important to stress the need for low background noise. Most noise criteria have been set with persons having normal hearing in mind. In societies with much sound-generating equipment (mechanical as well as electroacoustical), people are subject to extra noise-induced hearing impairments. Persons having hearing impairments or hearing loss generally have difficulty in understanding speech buried in noise and are often more annoyed by noise than persons having normal hearing.

Noisy Environments

Much of the negative influence of heating, ventilation, and air conditioning noise on speech is typically due to masking. Frequency domain masking is primarily active upward in frequency as discussed in Chapter 3, where metrics such as signal-to-noise (S/N) ratio, articulation index, speech transmission index, among others, were introduced. As indicated by Figure 3.34, it is reasonable to expect that the S/N ratio (assuming typical speech and HVAC noise spectra) must be at least 15 to 25 dB for speech intelligibility to be unaffected by noise in the speech frequency range.

The Speech Interference Level (SIL) can be used to estimate the seriousness of noise affecting speech. SIL is defined as the mean of the 0.5, 1, and 2 kHz octave band noise levels (see Reference 6.1). Using the data shown in Figure 6.1, one can roughly determine the interference by a particular noise on

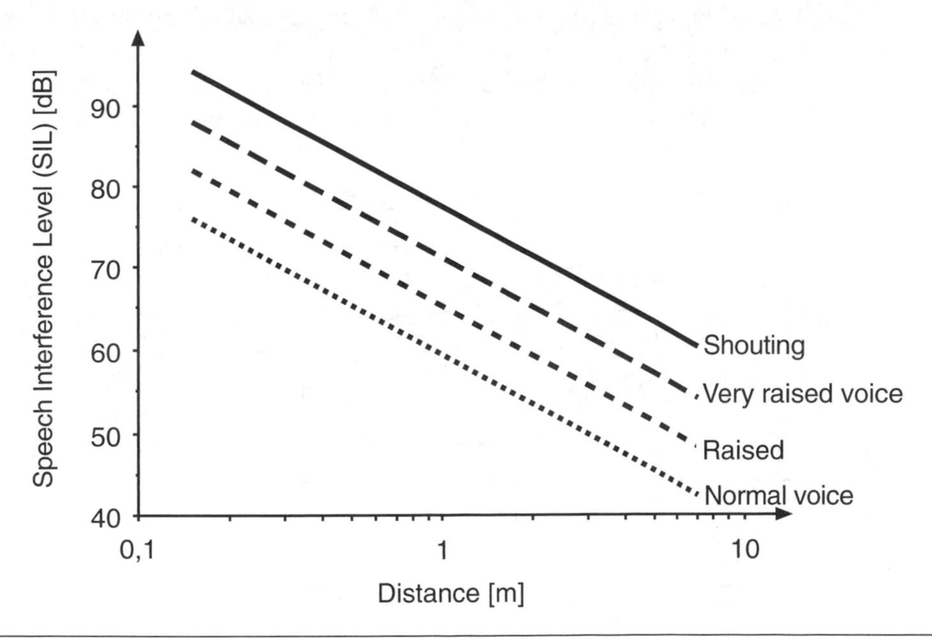

Figure 6.1 The Speech Interference Level (SIL) is defined as the mean of the sound pressure level values for noise spectrum in the 0.5, 1, and 2 kHz octave bands. The graph shows SIL for speech at various distances and for different voice modes (after Ref. 6.1).

speech, not only for offices and other workplaces but also for telephone transmission and similar cases. The data can also be used to determine the necessary sound pressure levels for sound reinforcement equipment.

Quiet Environments

Some environments are characterized by communication activities that are particularly sensitive to noise. This is particularly true for audio and video recording studios, radio and television stations, concert halls, operas, theaters, and similar venues. The background noise limits are usually set during consultation between director, fundraiser, architect, and acoustician.

In rooms for music in particular, the noise levels tolerated are very low. Curves, such as those shown in Figures 6.2 and 6.3, are often used in determining whether a noise spectrum is acceptable, or not. Such curves, specifying maximum octave band sound pressure levels from 31 Hz to 16 kHz, are useful because they, to some degree, take the character of noise into account.

In Europe, it is common to specify the maximum allowable noise level with dBA values. This is inferior compared to the U.S. way of specifying the maximum allowable noise as an NC or RC value. The noise criterion (NC) curves, shown in Figure 6.2, and the A filter curve are based on auditory threshold data. Noise shaped along the NC curves has a tendency to sound both *rumbly* and *hissy*. The NC curves were not shaped to have the best spectrum shape but rather to permit satisfactory speech communication without the noise being annoying (see Reference 6.2).

The room criterion (RC) curves shown in Figure 6.3, on the other hand, have been shaped to be perceptually neutral—they are straight lines with a slope of −5 dB/octave. The RC method involves the determination of both an RC rating and a *spectrum quality descriptor* that determines if the spectrum is rumbly or hissy (see Reference 6.2).

In some cases, one has to be content with a value for sound level L_{pA}. Concert halls usually require sound levels well below 20 dBA. The noise of microphones and HVAC equipment typically have very

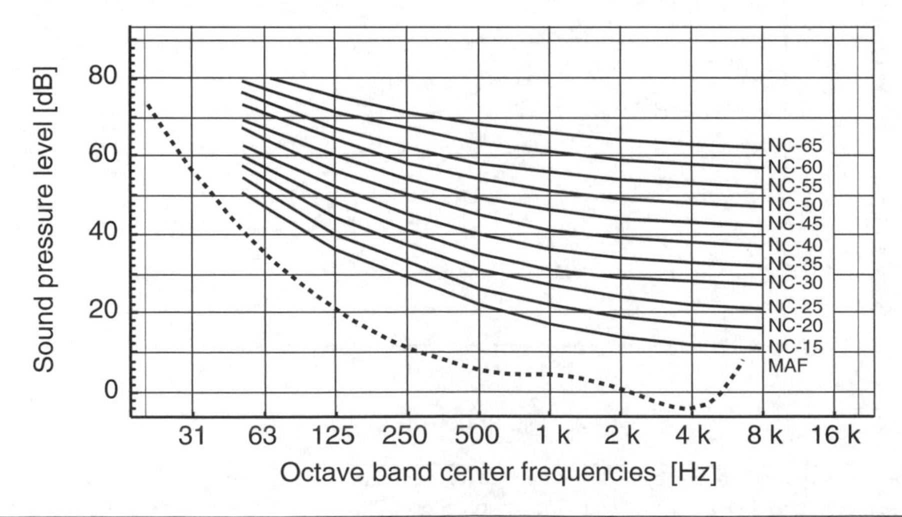

Figure 6.2 These NC curves apply to octave band-filtered noise levels and are often used in room acoustics (after Ref. 6.2).

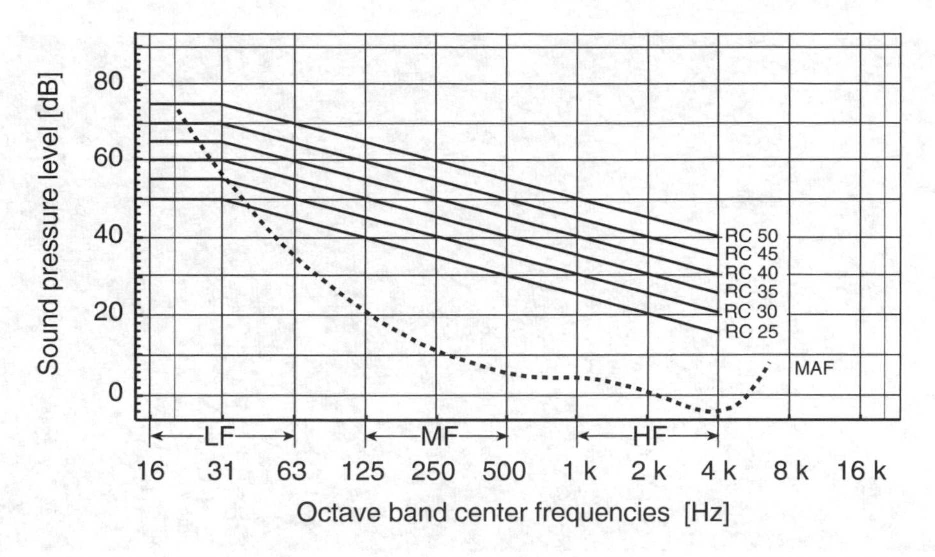

Figure 6.3 The RC curves also apply to octave band-filtered noise levels. Note that there are several editions of these curves—the ones shown are the Mark II curves (after Ref. 6.2).

different spectra. Comparing A-weighted sound levels for different spectra shapes is not relevant from the viewpoint of sound quality, as in the following example.

The acoustic background noise in a sound recording studio and the equivalent acoustic-electronic-background noise of microphones usually have different spectral shapes. Although one can say that the A-weighted sound level of microphone noise should be below the acoustic noise level in the recording room, the fact is that the noises will be perceived differently due to their different sound quality characteristics. The acoustic noise (for example, HVAC and traffic) will generally have spectra dominated by low-frequency noise, whereas microphone noise (due to its electronic generation mechanism) tends to have a *white spectrum*—that is, the octave band levels increase by 3 dB per octave. This gives the microphone noise subjectively a hissy character. Therefore, a microphone having an equivalent 20 dBA sound level internal noise level will probably sound hissy to the recording engineer when used for recording in a concert hall having an acoustic background sound level of 20 dBA. This may be a considerable problem when recording music in large concert halls, such as the one shown in Figure 6.4.

The recognizable properties of noise are usually important in the environments under discussion here. It may sometimes be easier to tolerate a noise having a random character than a noise that is transient, has a specific time pattern, or carries information content. The spatial distribution of the noise is also important. A diffuse sounding noise is often less disturbing to some people than a noise that can be localized in space (although in the latter case, one can perhaps more often find it possible to eliminate the noise source).

It is also important to avoid audible pure tones in the noise spectrum—the audibility of such tones was discussed in Chapter 3. It is reasonable to require the tones to be inaudible, which requires their sound pressure level to be below the critical band level of the other noise in the band. In practice, this results in the need for such tones to be at least 5 dB below the octave band level of the rest of the noise in the band.

Figure 6.4 In a modern concert hall, such as Symphony Hall, Birmingham, United Kingdom, the background noise sound level L_{pA} may be as low as 11 dB. (Photo by Mendel Kleiner.)

6.3 FUNDAMENTALS OF ROOM ACOUSTIC PLANNING

Rooms for Speech

The speech intelligibility properties of rooms intended for verbal communication are important, but it is essential to remember that good speech intelligibility is not the same as naturally sounding speech. One can increase the intelligibility of amplified speech, particularly in noisy surroundings, by various types of signal processing, such as spectral shaping and compression.

For our purpose here one can regard speech as a modulated signal with a complex tonal or noisy spectrum, primarily covering the frequency range 0.25 to 4 kHz that is modulated by frequencies in the 0.2 to 8 Hz range.

Reverberation and noise will decrease the *modulation strength* (also known as the *modulation depth*). A reduction of modulation depth, corresponding to the attenuation of modulation strength (sometimes called the *modulation damping*), results in a reduction of speech intelligibility.

For speech modulation at the listener to not be affected by reverberation, the reverberation time must be short, and the ratio between direct and early reflected sound to late reverberant sound must be high.

There is, however, a practical limit to shortening the reverberation time. Besides the expense of adding sound-absorbent treatment to the room, there is also another aspect. People expect rooms to have a certain reverberation time and level, determined by tradition and visual aspects (e.g., room purpose, shape, and volume).

Speech intelligibility can be shown to increase with increasing reverberation time for very short reverberation times in the range of up to 0.5 s. The reason for this speech intelligibility dependency on reverberation is that reduced sound absorption in the room leads to both an increased reverberation time and an increased sound pressure level. In addition, early sound reflections by more reflective walls help eliminate the influence of speaker sound radiation directivity on the sound pressure. The directivity index of the human voice is approximately 6 to 10 dB in the frequency range that is most important for speech intelligibility.

These are some good rules for auditoria used primarily for the spoken word:

- Keep the travel paths of direct sound and important early reflections short. In practice, it is difficult for an untrained speaker to reach audiences with sufficient speech intelligibility at distances over 20 m without the use of amplification.
- There must be enough early reflections to provide sufficient sound level while at the same time not feeding sound energy into the late part of the reverberant sound field. It is advantageous to strive for early sound reflections to arrive close to the horizontal plane, rather than from above, to have good speech sound quality. Sometimes for practical reasons, however, it is necessary to make use of overhead sound reflectors as shown in Figure 6.5.
- Mirror-like (specular) reflections must not exceed the level of the reverberant sound if they arrive more than 30 ms after the arrival of the direct sound, such reflections may be perceived as echoes. This means that sound being focused by concave surfaces must have short propagation. The focusing effect may be remedied using alternate focal distances and sound-absorptive or scattering treatments in the travel paths as shown in Figure 6.6.

Figure 6.5 A modern lecture hall equipped with sound-reflecting panels over the sending end of the room in Chalmers University of Technology, Gothenburg, Sweden. (Photo by Mendel Kleiner.)

Figure 6.6 When the sound absorption of the walls is high, a room having high concave ceilings is susceptible to echoes. The ceiling in this church has been made extremely irregular to scatter sound and avoid echo. (Photo by Mendel Kleiner.)

- Take the directivity of the human voice into account. In small rooms, such as classrooms and small auditoria having relatively hard walls, the speaker will usually be quite close to some sound-reflecting surfaces. This leads to strong early reflections that will add to the direct sound in such a way as to increase the speech intelligibility. In large auditoria, such as theaters or assembly and lecture halls, it is important to place the seating area within an approximately 120 degree arc in front of the speaker. The effective speech intelligibility will increase when the listeners can see the mouth movements of the talker.
- The reverberation time should be in the range of 0.6 to 1.0 seconds, increasing with the size of the room, see Figure 6.9. The reverberation time should also be constant to within 10% over the 125 Hz to 8 kHz octave bands, for small- and medium-sized rooms (see Figure 5.17). It is advantageous if added sound-absorbent material is placed so that it does not interfere with the propagation of important early sound reflections. This means that one should leave the central part of the ceiling nonabsorptive in auditoria, classrooms, and other verbal communication rooms.
- For small studios and control rooms, such as those used at radio stations (see Figure 6.7), having volumes smaller than 25 m^3, it is common to reduce the reverberation time to the

Figure 6.7 Control rooms usually need to have short reverberation times. It is often necessary to remove undesirable reflections by sound absorption or sound scattering. School of Music, Gothenburg, Sweden. (Photo by Mendel Kleiner.)

range 0.2 to 0.4 s. It is also usually necessary to control the resonant modes at low frequencies using various types of porous or resonant sound absorbers.

Rooms for Music

Over the ages, the acoustic properties of the performance space have influenced the way music is written and performed. Composers of music have adapted their works to the reverberation characteristics of the spaces that were familiar to them as likely places for the performance of the particular music. Classical secular music from the baroque period was generally intended for performance in small spaces, having short reflection paths and highly reflective surfaces. Much of the classical music written during the nineteenth century, however, was written for the much larger rooms built in that era that were associated with considerably more reverberant characteristics and nearly twice as long reverberation times (see Reference 6.3).

Today it is common for musical performance spaces to be built acoustically flexible, so that many types of music can be performed successfully—one such flexible venue is shown in Figure 6.8.

Much of the popular music of today does not require a dedicated acoustically reflective environment. Using loudspeakers on and off stage, one can simulate or create the desired acoustical characteristics for the space. *Sound reinforcement systems* and *electronic reverberation enhancements systems*—particularly those outdoors—can be used effectively to simulate many types of spaces.

Some general recommendations for rooms intended for the performance and enjoyment of classical music are (see Reference 6.4):

- The *background noise levels* must be low.
- The *sound level* must be optimized—it should be neither too loud nor too low. It is defined by the *room gain*.

Figure 6.8 In this computer-controlled auditorium at IRCAM, Paris, the room acoustic conditions could be changed drastically using rotating wall sections. The sections were designed to be used as reflectors, scatterers, or absorbers (photo by Leif Rydén).

- There must be suitable *clarity*, and the *spatial* and *temporal distribution of the early reflections* must be good.
- The *reverberation time* must be appropriate for the music expected to be played.

The sound level in a real room will generally not behave exactly in the way described by the simple room acoustic theory presented in Chapter 4. The reason for this is the nondiffuseness of the sound field, sound absorption is not evenly distributed over the room surfaces. In the quasidiffuse sound field in a real room, the level will stay reasonably constant but will drop off with increasing distance. This is particularly noticeable in the audience area of, for instance, a concert hall. Far back on the main floor the sound level will also drop off due to negative interference from the sound reflected off of the seating in front, as discussed in Chapter 1.

The optimal *reverberation time* depends on the size of the room and on the type of performance that is expected to dominate the use of the room (see Figure 6.9). It is common to provide for several auditoria within one venue, which allows for the design of rooms that can complement one another regarding acoustical characteristics.

If it turns out to be necessary to build only one main auditorium, the auditorium can possibly be designed to provide for *variable acoustics* by passive or by active means. *Passive acoustic variation* can be achieved using variable absorption or volume. *Active acoustic variation* can be achieved by electronic sound field simulation or reverberation enhancement using microphones, digital signal processing, amplifiers, and loudspeakers (see Figure 6.10).

The reverberation process can often be considered as subdivided into several parts from the viewpoint of hearing, at least regarding large rooms having reverberation times between 1 and 3 seconds. Usually one considers *direct sound, early,* and *late* reverberation.

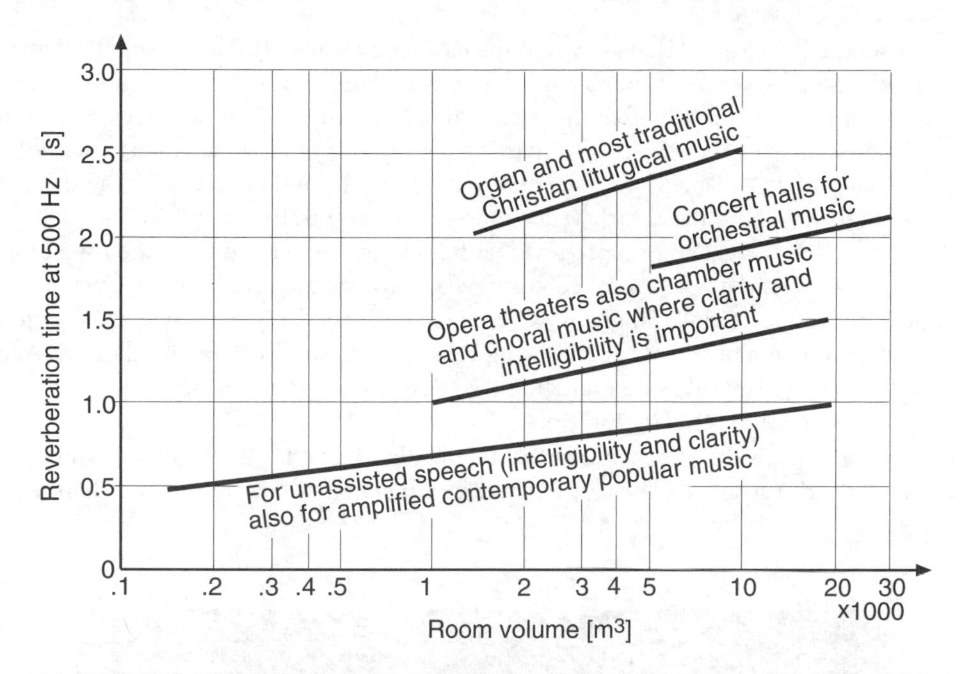

Figure 6.9 Typical recommended 0.5 kHz reverberation times for some common types of auditoria.

Figure 6.10 Using an electronic reverberation enhancement system one can achieve high flexibility in room acoustic conditions. The photo shows an active system for control of acoustics on a stage using loudspeakers integrated into a stage shell to achieve a proper blend of early reflections and local reverberation. Installation in Motala Community Center, Sweden. (Photo by Mendel Kleiner.)

The direct sound is the sound that arrives to the listener by way of the shortest path. The path may, at times, be interrupted by corners and balcony fronts, for example. It is also common to regard some reflections that follow immediately after the direct sound (having a time delay relative to the direct sound of less than approximately 5 ms) as a part of the direct sound. This is particularly the case if they are in the same vertical plane of the listener, such as from the audience area or the stage. From the listener's perceptual viewpoint, these early reflections are *fused* with the direct sound.

The early part of the reverberation process that follows the direct sound in the time range of 5 to 80 ms (5 to 50 ms for speech) is often called *early reflections*. Figure 6.11 shows reflectors suspended over the stage at the Concert Hall, Singletary Center for the Arts, University of Kentucky to enhance the early reflections. The hall was designed using results from sound field simulation, "Auditorium Synthesis", and by using physical scale modeling (see Reference 6.17). The acoustics of the hall combines high clarity with long reverberation time.

The *late reverberation* is the reverberation that follows the early reflections—that is, more than 80 ms after the arrival of the direct sound. The *very late reverberation* that follows more than about 160 ms after

Figure 6.11 This photo shows the Concert Hall, Singletary Center for the Arts, University of Kentucky, Lexington, KY. Using sound reflectors suspended over the stage, one can add early reflections at the listener. In wide auditoria, overhead sound reflectors are usually necessary to provide both good stage acoustics and sufficient early reflected sound for the audience. It is important that the reflectors are large, have suitable scattering properties, and have enough mass so that they will provide strong and even sound coverage. The shape, number, placement, and adjustment of the reflectors are all factors that will influence the results. "Continental seating" is used in this hall. Note the sound reinforcement system's large central cluster loudspeaker array that is used to amplify speech (photo by Dwight Newton, courtesy of University of Kentucky College of Fine Arts).

the direct sound is usually heard only at the end of a musical piece. It could be argued that this very late reverberation is extremely important since it is the last part of any piece of music heard in a concert hall.

When measuring the reverberation time we usually obtain a decay curve that shows both the early and late parts of the reverberation process. A physical reverberation process, at least when one looks at a wide frequency range response, is not characterized by the exponential decay of the theoretical process discussed in Chapter 4. The real wide bandwidth decay in the frequency range above the Schroeder frequency has many irregularities caused by the approximately discrete reflection sequence of sound in the room. The process will usually approximate that of one or more exponential processes. The part of the reverberation process that is from the start of the process out to a time limit of 0.16 s after the start is particularly important for the subjective determination of reverberation time with running music.

Note that the reverberation process will depend on the location of the sound source and the microphone position in any given room. The ISO 3382 standard gives advice on good practice for the measurement of reverberation time of performance venues.

Some target values for the reverberation time T for some different types of rooms can be found in Figure 6.9. Large rooms for classical music will typically require a volume of around 12 m^3 per square meter of audience area to obtain a sufficiently long reverberation time at 1 kHz where the sound absorption of the audience and the chairs is the highest. One often strives for a frequency-independent reverberation time within ± 5% in the octave bands from 250 Hz to 4 kHz.

In the case of large rooms, such as concert halls (volumes more than approximately 5,000 m^3), the reverberation time will drop off in the 4 kHz octave band and above due to the sound absorption of air, as discussed in Chapter 1. It is common practice to leave surfaces for final adjustment of the reverberation characteristics of a room after completion.

Generally, the sound absorption by common building constructions and surfaces is fairly low in the octave bands at 125 and below. It is common to allow an increase of reverberation time down to the 63 Hz octave band of approximately 30% over the value at 0.5 kHz. This practice results in good bass response and a warm sounding reverberation. Sometimes surfaces, such as the stage floor, may act as secondary sound radiators at low frequencies for cellos and double basses that further enhance the *warmth*.

A room having a short reverberation time is usually considered *dry* and acoustically uninteresting. The desire for high clarity however makes reverberation times in the range 0.2 to 0.4 s desirable for control rooms, such as the one shown in Figure 6.7.

Another main factor influencing room acoustic quality is the spatial and temporal distribution of the early reflections.

The *time gaps* between the direct sound and major early reflections, are important and should be irregular and not arrive later than approximately 30 ms. The early reflections should be wide-band (not have a limited frequency range). One should avoid strong overhead reflections appearing in the same vertical plane as the direct sound. Since hearing cannot differentiate between signals having the same interaural delay time, such overhead reflections will cause comb filter effects unless masked by other reflections coming from the sides. A diffuse or curved ceiling will redirect these reflections to the walls.

The spatial distribution of the early reflected sound is important, particularly for creating the impression of *auditory source width*. One should strive for sound reflection angles close to the horizontal plane to maximize the advantages of our binaural hearing. It is advantageous if the directional distribution of the early reflections are such that the sound field appears to be *symmetrical* when listening. One should also strive for an even angular distribution of the reverberant sound to enhance the feeling of *envelopment* and diffuseness in the sound field.

It is important to realize that subjective diffusitivity and physical diffusitivity are not well correlated. One can have good subjective sound field diffusitivity without the sound field being diffuse in a physical sense. Good physical diffusitivity does not guarantee good listening conditions in a room.

The side walls should be somewhat scattering; for example, by using shelves, balconies, unevenly placed sound-absorbing patches or objects having sizes in the range of 0.5 to 0.05 m. Figures 6.12–6.14 show various ways of achieving high scattering by wall and ceiling surfaces. *Flutter echo* is the term given to repetitive short path sound reflections, usually appearing when reflections between two close and parallel walls dominate the sound field. It can be removed by splaying the walls, absorption or scattering.

One phenomenon of particular importance in long concert halls is that of cancellation on the main floor of direct sound by the first reflection of sound off the audience in the frequency range between 125 Hz and 1 kHz. This reflection will be at a small angle leading to phase reversal of the sound in the frequency range mentioned (that is, a reflection coefficient of $R = -1$). Since the sounds cannot be separated because of the short time difference between the direct and reflected sounds (typically less than 5 ms), there will be cancellation. The cancellation will result in *coloration*—timbral change in the sound quality of the *direct* sound.

According to the *law of the first wave front*, the direct sound determines our perception of the timbre of the sound source. Consequently, it is important to design the seating area in such a way that the coloration is minimized. This requires that:

- The chairs are designed appropriately.
- The seating area has a sufficient slope.
- The stage is sufficiently high.
- There are good sight lines to all seats.

A room having good properties regarding background noise, early reflections, reverberation time, and scattered sound will be close to ideal from the audience's acoustical point of view.

In addition, it is important that the acoustical conditions on stage give musicians good possibilities for hearing one another and give support for their own playing. The stage and its surrounding walls and ceiling must be designed so that the sound is redirected appropriately to all parts of the orchestra as well as to the audience. The stage and the auditorium must, in addition, be free of severe acoustical defects, such as echo and nonlinear distortion (for instance, a rattle).

Recording Studios

The acoustical requirements for music recording rooms are similar to those that apply to rooms for listening. Because of the different ways in which a listener interprets the sound by direct listening and by listening to sound picked up by microphones, there will be both difficulties (absence of binaural signal

Figure 6.12 One can scatter the reflections from surfaces using diffusors to soften the early reflections. This photo shows testing of diffusors to create good ensemble characteristics for a concert hall stage in Baltimore's Joseph Meyerhoff Symphony Hall. (Photo by Mendel Kleiner.)

Figure 6.13 Formerly, high sound field diffusitivity was achieved by uneven surfaces, niches, and rich ornamentation, as in the Wiener Musikverein Concert Hall, Vienna shown here. (Photo by Mendel Kleiner.)

Figure 6.14 Currently, the surface unevenness is achieved by simple means and is primarily introduced to provide desired diffuse sound reflection as in the San Francisco Symphony, San Francisco, CA. (Photo by Mendel Kleiner.)

processing) and advantages (close-up recording, directional microphones, and various forms of signal enhancement). (See Reference 6.5.)

By using various microphone techniques, one can adapt the recording to the room acoustical conditions. Since the visual impression is not as important, more effort can be put into optimization of the room acoustics.

It is common for studios to have various types of sound reflecting, scattering, and absorbing panels, both freestanding and wall mounted. A recording studio venue will also usually contain many small rooms, or subspaces, for particular instruments. These will have splayed surfaces to avoid flutter echo.

Any recording studio must be quiet, since the listener will likely have to listen to the same recording several times. The background noise sound level in a studio for recording of acoustical instruments should preferably be below 15 dBA, as mentioned previously. It is important to observe that the noises from traffic and HVAC have noise spectra that are different from that of typical microphones.

Television studios generally pose acoustical difficulties, from the viewpoint of both noise and room acoustics. Close-up microphone placement (a wearable broadcasting microphone, for example) is often necessary to allow visually unobtrusive sound pickup (see Reference 6.6).

Rooms for Sound Reproduction

Rooms for sound reproduction are typically much smaller than those for concerts or classical music recording—typically having volumes in the range of 25 to 75 m^3. This leads to problems at both low and high frequencies.

In sound reproduction the acoustical characteristics of the transfer paths between loudspeaker and listener are overlaid on the recorded signals.

It is necessary to differentiate between listening rooms in studios (control rooms of various kinds) and rooms in dwellings (living rooms, for instance). The studio environment is usually subject to a more critical evaluation and listening than the dwelling environment. Also, there is likely to be more interest and money available for creating a good listening environment in a studio than in a home. The freedom to design the room with acoustics in mind will also be larger since fewer compromises have to be made because of other uses.

Both environments require fairly short reverberation times. As a rough approximation, one should strive for a listening room reverberation time no longer than approximately half that of the recording venue if the recording has been made so that the reverberation is clearly audible and dominant (for example, using omnidirectional microphones far beyond the room's reverberation radius). Typically, a reverberation time in the interval between 0.3–0.7 s may be desired over the entire audio frequency range.

An important issue in playback rooms is to avoid early reflected sound incident in the median plane, since such sound components will cause comb filter effects as discussed previously. The components can be eliminated by using absorptive or scattering treatment (or simple redirection) of the surfaces responsible for the undesired reflections. Sometimes it is also useful to diffuse the wall patches that are responsible for the first side wall reflections. A specification of listening room characteristics is made in the IEC 268–13 standard that strives to formulate the acoustical characteristics of a typical listening room. A room that fulfills the standard is shown in Figure 6.16.

Figure 6.15 Wenger's small commercial portable cabin for music rehearsal featuring electronically variable reverberation. (Photo by Mendel Kleiner.)

In one technique often found in control rooms one half of the room (where the loudspeakers are) is treated to be highly sound absorptive, and the other half (where one listens) is "live" and treated to be diffusely reflecting. This classic approach allows a distinct and clear sound around the listener in combination with a feeling of space. More modern approaches are creating a "reflection free" zone around the listener or making walls and the ceiling diffusely reflecting.

Small rooms usually have problems in accurate music sound reproduction because of the low modal density and poor damping of modes at low frequencies as mentioned previously. This makes it virtually impossible to reproduce music accurately at low frequencies unless these low frequency modes are well damped. Using various types of signal correction by means of digital signal processing, it is possible to circumvent this problem to some extent. The use of various types of low-frequency sound absorption is in any case a must. Slit and Helmholtz resonators, as well as membrane-type sound absorbers, are suitable for this task. Figure 6.17 shows a combination of a diffuser and low-frequency sound absorber.

Correct positioning of the loudspeakers, relative to the room modes and the listening position, will ensure the least variation in frequency response. It is practical to use the reciprocity technique for finding the optimal positions.

Figure 6.16 The IEC listening room at the Section of Acoustics at the Aalborg University is a good example of an advanced IEC listening room (photo courtesy of AAU, Denmark).

Figure 6.17 An example of a combination of slit absorber and diffuser for broadband control of the acoustical conditions in a recording studio in the Gothenburg Opera, Sweden. (Photo by Mendel Kleiner.)

6.4 TOOLS FOR PREDICTION OF ROOM ACOUSTIC RESPONSE

The acoustic properties of a room can be predicted in many different ways using *computer aided design* (CAD) with varying results.

For *low frequencies*, calculations using software based on the *finite element method, boundary element method*, or *finite-difference time-domain method* are, in principle, quite exact, but comparatively slow. For the modeling, one needs the impedance data for the boundaries—without which the results may be quite erroneous, particularly regarding the location of resonance frequencies, mode shapes, and damping. The necessary impedance data however is seldom available. The generation of a suitable mesh is also a problem since few current mesh generators are suited to quickly and effectively meshing the geometries common in room acoustics tasks (see References 6.11 to 6.17).

For problems involving *medium and high frequencies*, geometrical acoustics is usually used. Initial design considerations usually include manual ray tracing and reverberation time calculation using Sabine's or *Fitzroy's* formula (see Reference 4.1). Sabine's formula essentially requires the sound field in the room to be diffuse, which is seldom the case. Fitzroy's formula is particularly useful in predicting the reverberation time for those rectangular rooms where sound absorption is unevenly distributed between the walls. For building acoustics and some uses in architectural acoustics software based on statistical energy analysis

(SEA), modeling is also used (see Chapter 10). This approach is based on the study of energy flow between resonant systems.

Ray tracing and mirror image method software is now common in most acoustics consultants' offices. Using CAD, it is possible to predict not only reverberation time, clarity, lateral fraction, and other room acoustics metrics but also to predict the impulse response. Many other modeling methods are, however, also possible (see Reference 6.8).

The following methods are (or have been) used in room acoustics modeling:

- Optical ray tracing using laser
- Optical intensity modeling using lamps
- Ultrasonic scale modeling by the impulse method
- Ultrasonic scale modeling by the intensity method
- Ultrasonic scale modeling by the transmission method
- Physical acoustics modeling using Finite Element (FEM) or Finite Difference Time Domain (FDTD) methods software for low frequency, small room modeling not discussed in this book
- Geometrical acoustics modeling using ray tracing or mirror image modeling software
- SEA modeling using specialized software

The optical methods must be considered obsolete except for some special uses. Optical ray tracing is based on the use of physical scale models (typical scale values are 1:10 to 1:100). The sound source is simulated using a laser or a small light source surrounded by a grill so that light is emitted in sectors. By using a scale model where reflecting surfaces are modeled by mirrors, it is possible to study the distribution of early reflected sound—sound that has been reflected only once or twice against surfaces. The advantage of the method is its simplicity and low cost. It can often be used directly in the architect's scale model. A major disadvantage of the method is that it does not take sound diffraction by edges or scattering by surface irregularities into account.

Optical intensity modeling is based on the same type of physical scale models as those used in optical ray tracing, but in this case, the static light distribution is studied as an approximation to the diffuse field sound intensity.

Ultrasonic scale modeling typically uses models in scales from 1:4 to 1:16, although some researchers have shown useful results even with scales as small as 1:50. (Outdoor sound propagation is typically modeled ultrasonically using a scale of 1:100.) Using spark source excitation, one can study the impulse response of the room. Such a scale model is shown in Figure 6.18.

The static sound distribution can be calculated from the impulse response measurements or directly studied using, for example, jet noise sources.

The following rules apply to ultrasonic scale modeling at 1:m scale:

- The frequency scale is transposed by a factor of m:1.
- The time scale is transposed by a factor of 1:m.
- The sound attenuation by air is reduced by a factor of m:1.
- The acoustic impedances of the materials used in the scale model must be similar to those in the real room (after appropriate frequency scaling).

One of the major problems in ultrasonic scale modeling is the correct scaling of the attenuation of sound by air. Another major problem can be found in the difficulties in designing sound sources and

Figure 6.18 Model of a church built for ultrasonic scale modeling. Note scale model of manikin for binaural sound pickup next to the hand. (Photo by Mendel Kleiner.)

microphones that can cover the frequency range of interest (typically 200 Hz to 200 kHz) with adequate directivity and sensitivity. One can approximate the correct scaling of sound attenuation by air, using dry air (typically having a relative humidity of less than 2%) or using a pure gas, such as nitrogen in the scale model. These methods have been shown to give good results not only regarding the simulation of reverberation time and levels but also regarding impulse response. Research has shown that it is possible to digitally compensate impulse response measurements made using scale models without these precautions, and still obtain quite good simulations.

A large advantage of physical ultrasonic scale modeling is that it is easy to simulate the acoustic conditions in irregularly shaped rooms, and in rooms with uneven sound absorption. In addition, ultrasonic scale modeling allows for inclusion of diffraction and scattering as well as the frequency-dependent sound absorption by various surfaces.

Using computer software, it is possible to predict the acoustic properties of a room in much more detail than by manual methods. Once the geometrical and acoustical data for the room have been entered into the computational model, one can calculate the way sound from one or several sources is incident at various listening positions. Usually one is interested in the spatial impulse response expressed as $p(t)$ relative to the sound pressure at 1 m distance from an omnidirectional source. If one is interested in auralization of the room, it is necessary to know the angles of incidence (φ, δ) for the various sound components contributing to the room impulse response. Using the room impulse response, it is possible to calculate the various room acoustic quality indices, as previously mentioned.

Most software for the prediction of room acoustics is based on sound propagation modeling using geometrical acoustics and, therefore, can only take diffraction, scattering, and real acoustic surface impedance into account in an approximate way. Some software will approximate the influence of these acoustic phenomena on the propagation of sound and thus yield improved modeling. Most such software is based on ray tracing or mirror image modeling or possibly a hybrid between the two.

Modeling using the *ray tracing method* typically assumes a large number of rays (1000 to 1,000,000) being sent out from a sound source. The sound source can be assumed to be omnidirectional or directional. In the latter case, one can either assign different intensities to the rays or adjust the ray density to simulate the source directivity. One then follows each ray on its path and studies how the ray is reflected by the surfaces of the room and the objects it contains. Curved surfaces are generally simulated by piecewise flat surfaces. By studying the number of rays passing through a test surface or test volume (such as a sphere) at the listening position of interest, one can calculate the room impulse response at that position.

In most ray tracing models, the sound absorption is modeled by adjusting the ray intensity appropriately after calculation of the room impulse response. Scattering can be simulated by assigning various randomization rules to the angle of reflection. *Pyramid tracing* and *cone tracing* are variations on simple ray tracing.

Modeling of sound propagation using the *mirror image method* assumes that sound reflected by a flat, hard surface originates from a mirror source behind the surface. One needs to calculate the various mirror images created by the (flat) surfaces of the room and its objects. For a room having a complex

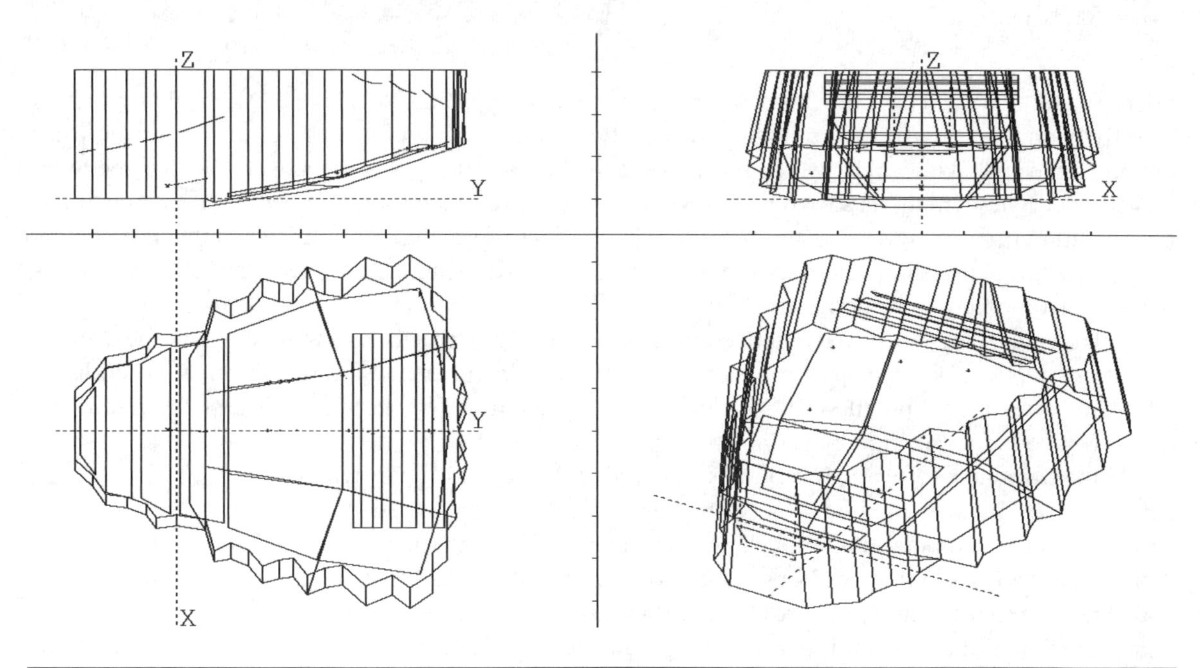

Figure 6.19 CAD using ray tracing or mirror image method-based software can be used to predict the acoustics (see also Figure 6.20). This model of the room shown in Figure 6.11 was done by Julie Byrne, MSc.

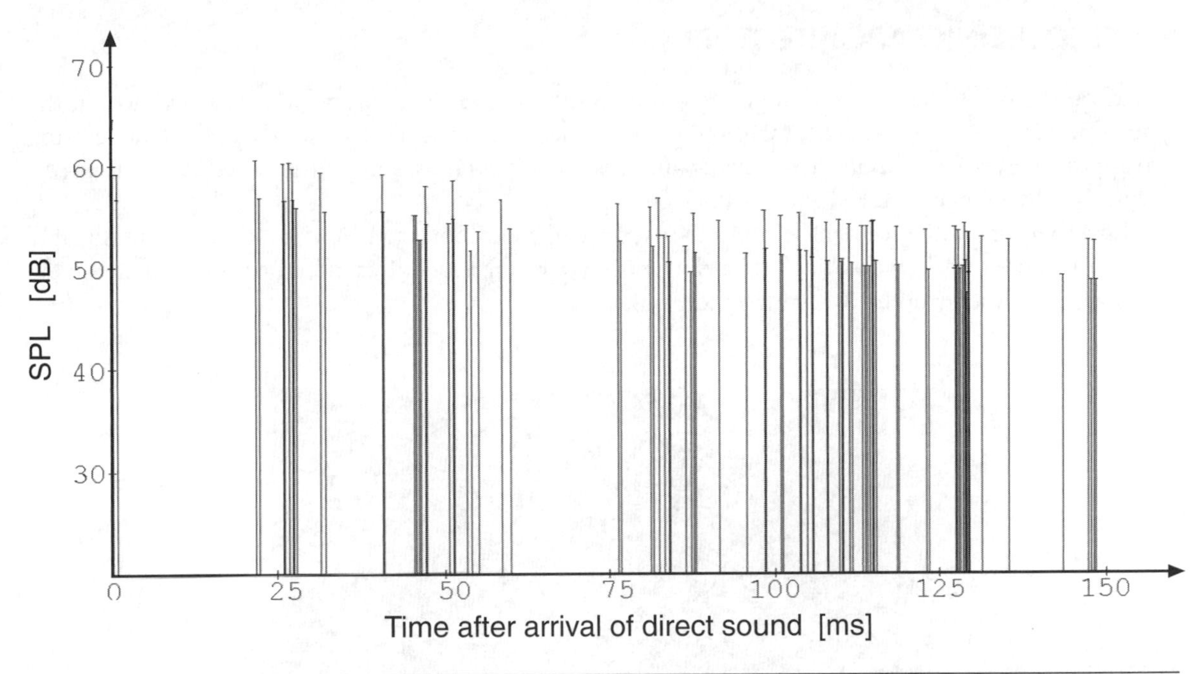

Figure 6.20 A calculated echogram for mid-stage to last row on center in the room model shown in Figure 6.19 (arbitrary reference level).

geometry, it may be necessary to calculate many millions of mirror sources, as sound is reflected multiple times by the various surfaces. The relevant sources must then be sorted out using rules for visibility. A disadvantage of the method is that it does not allow for easy inclusion of scattering and diffraction. Figure 6.20 shows the calculated echogram for the room shown in Figure 6.19 (set for not taking frequency-dependent sound absorption, scattering, or diffraction into account). One can see that the response consists of *discrete* reflections.

Hybrid models are typically based on the use of both ray tracing and mirror image modeling. Ray tracing is used initially to find the possible sound paths and then mirror image modeling is applied to find the exact sound paths. As with ray tracing, a disadvantage is the need for modeling the propagation of a large number of rays.

Both ray tracing and mirror image modeling are similarly computationally intensive. Neither method allows for exact room acoustic modeling since geometrical acoustics is used. Good modeling of absorption, diffraction, and scattering remains a problem. In any case the prediction will be limited by the experience and skill of the user. The late part of the room impulse response corresponding to the late part of the reverberation is usually too computationally intensive to be handled by a direct approach, and various approximations are generally used. Some of these may be based on the use of various statistical assumptions and random generated numbers. Such methods may lead to unevenness in the calculated full room impulse responses.

6.5 ELECTRONIC ARCHITECTURE

It is advantageous to be able to change the room acoustic properties of performance spaces to suit the type of performance, for example, various types of orchestras, music, and voice. This can be done using two approaches, *reverberation enhancement* and *sound field synthesis,* sometimes called *electronic architecture* (see References 6.4, 6.9, and 6.10).

Reverberation enhancement is typically used when the room acoustic conditions are reasonably good, but where reverberation time or level may need to be increased. It is often possible to increase the reverberation time by up to 50% with good results.

Figure 6.21 Sound reinforcement systems may use conventional direct-radiating or horn-coupled loudspeaker array elements to make up an array-type loudspeaker column to achieve the desired directivity (image courtesy JBL/Harman).

Sound field synthesis may include reverberation enhancement, but will usually include also the generation of specific early sound field components corresponding to early reflections by (imaginary) surfaces. Sound field synthesis is sometimes used to reduce the subjectively perceived reverberation time T_{subj} by reducing the relative level of the late reverberant sound. The conventional *public address system* is an example of primitive use of *sound reinforcement* to provide this effect, for example, in churches and other public meeting places (see Figure 6.21).

6.6 AURALIZATION

Audible sound field simulation is usually called *auralization*. Auralization of an environment (e.g., for speech and music) can be considered as analogous to visualization of that environment. The ambition is to make it possible to listen to the sounds in the environment without being in the actual environment (see Reference 6.8).

Computer auralization of an environment will start by entering data for the geometry of the room and the objects in the room into a computer database to be used by a mathematical model based on ray tracing or mirror image modeling. Once the software has calculated the response, it will generate a file containing a list of information about which sound components are incident at the desired listening positions:

- Levels relative to the direct sound for reflections
- Time delays relative to the direct sound for reflections
- Propagation path lengths

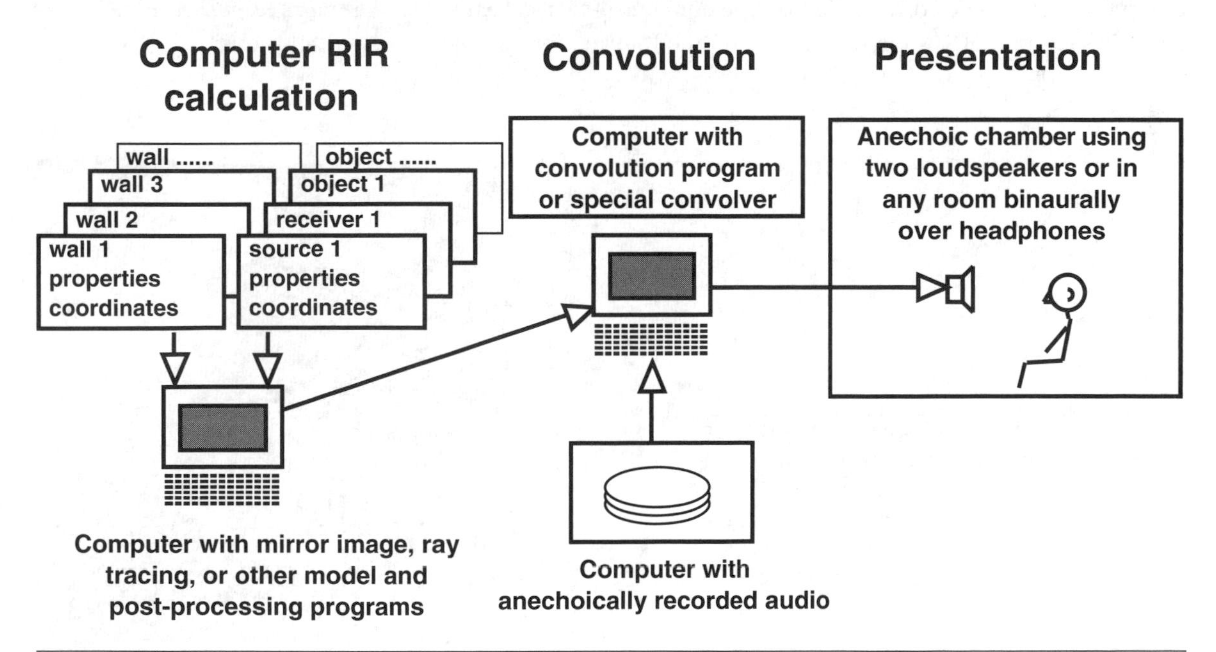

Figure 6.22 Functional diagram for fully computed auralization (After Ref. 6.8).

- Angles of radiation
- Angles of incidence

Using knowledge about the sound reflecting and scattering properties of the various surfaces, one can then add this information to the initially generated *generic* room impulse response (see Figure 6.22).

There are two main presentation methods used for auralization—binaural playback and playback by surround sound systems, such as *Ambisonics*. Binaural playback can be done either directly using headphones or by a stereo sound system using *crosstalk cancellation* (XTC). Surround sound systems need at least four loudspeakers—seven to ten loudspeakers are typically used.

In practice, XTC requires a quiet room in which the reverberation time is short, such as a semi-anechoic room. The aim is to provide the listener or listeners with sound signals at the ears corresponding to those that would be obtained when listening directly using headphones. This requires that the leakage of sound between the ears due to the loudspeaker presentation is cancelled using cancellation signals generated using knowledge about the transfer paths between the loudspeakers and the ears. Presentation using crosstalk cancellation generally gives better frontal localization than ordinary headphone presentation.

For auralization to be presented binaurally, information about the way the head influences the incoming sound waves must be added. Because of the way sound is reflected off the human body, the sound pressures at the ears are going to depend on the angles of incidence and on the frequency. At low frequencies, below 100 Hz, the effects will be negligible but already at approximately 250 Hz, the torso will contribute to the sound pressure. At frequencies above 5 kHz, there will be major effects as shown in Chapter 3 because of the reflection and diffraction of sound by the head. The frequency responses of the sound pressure at the left and right ears relative to the pressure in absence of the body are called the *head-related transfer functions* (HRTFs). Once the head-related impulse responses (HRIR) have been obtained, they can be converted into time domain signals as *head-related impulse responses*. Free libraries of HRTFs and HRIRs may be found on the Internet for various angular resolutions.

Figure 6.23 The Bose Auditioner® system is a commercial system for auralization using CAD. The binaural room impulse responses are convolved with music or speech so that it is possible to listen to a simulation of what it would be like to listen in the actual room being simulated. Crosstalk cancellation is used to render the binaural signals in this setup (photo courtesy of Bose Corporation).

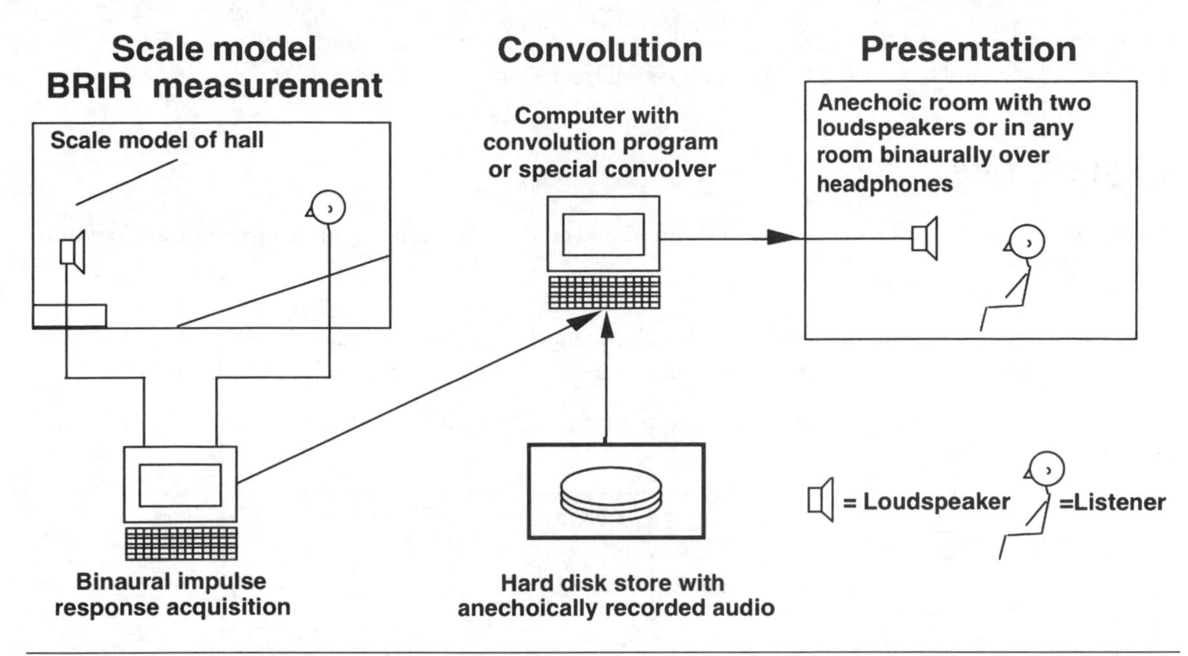

Figure 6.24 Functional diagram for auralization using the indirect ultrasonic scale modeling method (After Ref. 6.8).

Once the *generic spatial impulse responses* for the positions of the source and receiver in the room have been found, these can be appropriately *convolved* with the HRIRs and with the impulse responses of the sound-reflecting surfaces—the impulse responses of the various reflection coefficients described in Chapter 2. This yields the final binaural room impulse responses.

The binaural room impulse responses can then be convolved with *anechoically recorded* speech and music so that one can listen to what the sound of the room would be. The convolution can be done using software in a general purpose computer or by a dedicated *hardware convolver* (digital filter).

The convolved binaural signals can then be played back over headphones or further processed by a crosstalk cancellation network to allow for loudspeaker presentation. (See Figure 6.23.)

Auralization can also be done using a scale modeling approach. Direct modeling where the audio signals are played back as well as recorded in the model is quite problematic and an indirect approach where the binaural room impulse response is measured digitally and is subject to suitable digital signal processing is favored. The principle of this approach is shown in Figure 6.24. The main reasons for this is the possibility for a better S/N ratio and lower nonlinear distortion than using the direct method. In the indirect approach, the main difficulties lie in obtaining a suitably scaled manikin with torso, head, and ears, all correctly modeled to scale. Once the binaural impulse responses have been obtained, these are then convolved with the desired anechoic audio signal as in the case of fully computerized auralization described initially.

The advantages of auralization are many. By listening to auralization of rooms, costly mistakes in planning and building can be avoided, and different wall treatments and seating arrangements can be investigated. Auralization can also be used to study the properties of sound reinforcement systems, to reduce tuning time at building completion. Along with its use in architecture, auralization is used in various forms of virtual reality rendering, games, information systems, among others.

6.7 PROBLEMS

6.1 Two loudspeakers in an auditorium are connected to the same signal source but are radiating different sound power.

Task: Determine, using the figure on page 162 and Figure 5.11, if the sound at listening position A will be considered as good or poor, since one of the sounds will arrive delayed relative to the other.

Hint: Only consider the direct sound components from the loudspeakers and assume that the loudspeakers are omnidirectional.

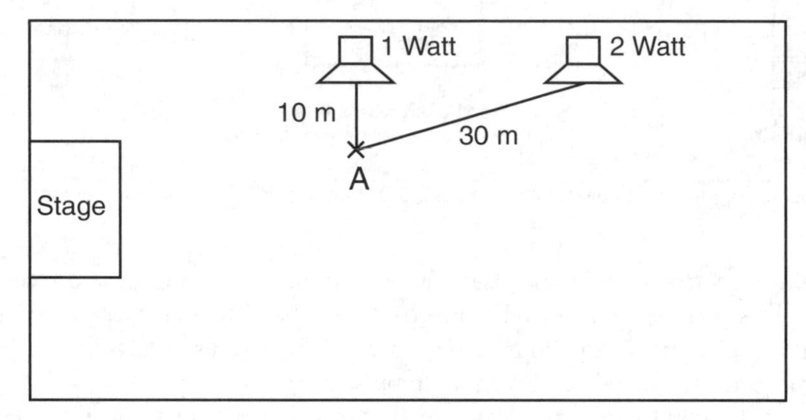

6.2 One wants to use a loudspeaker to improve the acoustical conditions in an outdoor theatre. Assume that the sound source is omnidirectional and all surfaces are sound absorptive. Use the figure to determine the required loudspeaker playback levels. The distances from loudspeaker H and actor A to the respective listening positions are:

	Pos. 1	Pos. 2	Pos. 3	Pos. 4	
Distance to actor	24	14	16	33	[m]
Distance to loudspeaker	15	14	9	16	[m]

The distance between the markers in the figure is 5 m.

Task: Calculate the necessary delay times for the four positions in the figure to avoid echoes.

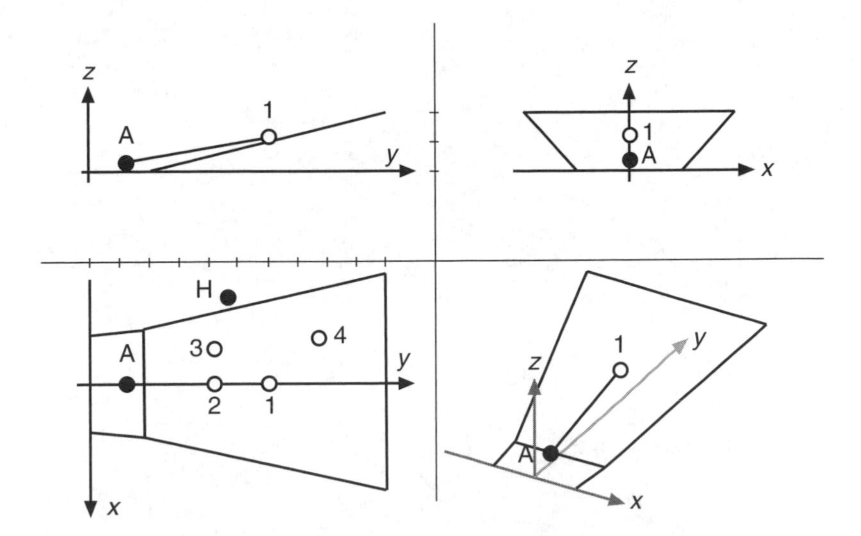

6.3 One wants to use a loudspeaker to improve the listening conditions in a concert hall and mounts a loudspeaker above the stage. The direct sound level of the loudspeaker is adjusted to be 6 dB above that of the speaker in the direction of interest. The volume of the hall is 40,000 m^3 and the reverberation time T is 1.7 s. Assume the sound field in the reverberant field to be diffuse and that sound field components can be added on an intensity basis. Further, assume that the delay between the direct sound components from the speaker and loudspeaker is less than 0.010 s and that the directivity index of the loudspeaker in the direction of interest is 6 dB.

The directivity D is defined as the sound level at a certain distance from a directional sound source relative to the sound level that would be obtained from an omnidirectional sound source radiating the same power. The directivity index (DI) is defined as $DI = 10 \log(D)$.

Task: Calculate the level difference between direct sound and reverberant sound $H = L_{p,\,\text{direct}} - L_{p,\,\text{reverb}}$. Find the approximate subjective reverberation time T_{subj} for various seats at distances of 10, 20, and 30 m respectively from the speaker or loudspeaker with and without the loudspeaker active. Use Figures 5.10, 5.11 and 5.21 in making your decision.

Absorbers, Reflectors, and Diffusers

7

7.1 INTRODUCTION

Sound absorbers are used to reduce the sound levels in noisy rooms and to optimize the acoustic conditions in rooms for voice and music. Absorbers are also used to reduce sound transmission through ducts and silencers. The sound energy is converted into heat by the absorber.

From an architect's viewpoint, absorbers can be grouped into natural absorbers and added absorbers. *Natural absorbers* are the audience and the performers, as well as such sound-absorptive surfaces, such as drapes, upholstered chairs and other furniture, walls, ceilings, carpets, and floors. In most cases, the desired acoustic properties of the room cannot be achieved by the raw surfaces and objects in the room. Materials used for walls, ceilings, and floors seldom have the necessary acoustic properties. *Added absorbers* are used to control reverberation time and other acoustic properties of the room further. Obviously, the division is not clear—an additional absorber can become a "natural" absorber once it is integrated into the building.

A more clear subdivision can be made between resonant and nonresonant absorbers. *Nonresonant absorbers* are usually made of porous materials and often simply called *porous absorbers*. This group includes mineral or glass wool, textiles, and carpets. The *resonant absorbers* are made using resonant acoustic or vibratory systems, such as air columns, volumes, membranes, and plates.

Most materials have quite frequency-dependent sound absorption—porous materials are much more sound absorbing at high frequencies than at low frequencies. By combining porous materials and resonant systems one can provide frequency uniform sound-absorption.

7.2 ABSORPTION COEFFICIENT

The Characterization of Sound-Absorption Properties

Materials are characterized by their sound-absorption coefficient, α. The sound-absorption coefficient α is defined as:

$$\alpha = \frac{W_{abs}}{W_{inc}} \tag{7.1}$$

where W_{inc} is the power of the incident sound and W_{abs} is the power removed from the sound field by the absorber, including power that is transmitted through the absorber, away from the acoustical system under consideration.

In most cases, one can neglect the sound power transmitted through the absorber. There are, however, two special cases that need consideration—open windows and thin sound-transparent drapes, screens, and such. In the first case, one does not have absorption in the sense that sound power is escaping from the room, rather than being converted to heat. At high frequencies, the open window will have an absorption coefficient close to unity. In the second case, if there is an absorber behind the drape or screen, the sound is transported through it (with some losses) and absorbed. If the surface behind the screen or drape is hard, a standing wave field is set up due to the interference with the reflected wave. If the drape or screen is at the particle velocity maxima of this standing wave, the sound absorption may be very high, depending on the flow resistance of the material. Note that the sound-absorption characteristic will be frequency-dependent in this case unless the drapes are in deep folds.

In some cases, it is more relevant to express the absorption as absorption area rather than absorption coefficient. This applies to, for example, resonant absorbers and to absorbers where the geometrical surface can be hard to define. This is typical of audience and performers, as well as some other types of natural absorbers.

The symbol α is used to indicate the absorption coefficient, but it is sometimes necessary to subdivide the use of α into the following categories:

Theoretical values:

α_0 perpendicular incidence (plane wave, infinite surface)
$\alpha_0(\varphi)$ a particular angle φ (plane wave, infinite surface)
α_d diffuse incidence

Measured values:

α_r using the impedance tube method (semiplanar)
α_s using the reverberation chamber method (semidiffuse)

Measurement Using an Impedance Tube

The *impedance tube* (the Kundt's tube or plane wave tube) can be used in two ways, either to measure the standing wave ratio along the tube or to measure the transfer function between two points in the tube (not discussed here). The standing wave method is based on the knowledge of the sound pressure amplitude maxima and minima and their locations along a tube in which a wave is incident on a test sample and was discussed in Chapter 1. The applicable standard is ISO 10534–1. The reflection coefficient measured this way would correspond to that of an ideal plane wave. Because waves can propagate in the tube also in other modes than the plane wave (usually called the 0:th mode), it is necessary to make sure that only plane waves are measured. The cutoff frequency is defined as the frequency at which there are also nonplanar waves that can propagate in the tube. A commercial circular Kundt's tube usually has an internal diameter in the range of 0.02 to 0.1 m. A large tube has a low cutoff frequency, as discussed in Chapter 1.

There are two ways out of this dilemma. The simplest is to limit one's measurements to frequencies below which the modes do not occur. In a rectangular tube, the modes will be similar to those discussed regarding the sound field in the rectangular room. For a rectangular tube having the crosswise dimensions d_1 and d_2 and where $d_1 < d_2$, the measurements should only be done at frequencies lower than:

$$d_2 < 0.5\lambda \tag{7.2}$$

In a circular tube, there will be waves having circular and radial propagation components. For a circular tube having a rigid wall, the corresponding condition for the diametered can be shown to be:

$$d < 0.59\lambda \tag{7.3}$$

A second way out of the measurement dilemma is to use a microphone array over the cross section of the tube. If this array senses the in- and out-of-phase components of the sound field in the tube equally, the summed microphone array signal will be that of the plane wave. These arrays are usually made using probe tubes leading to one microphone and only applied to measurement tubes having a rectangular cross section. Note that close to the loudspeaker, there may be near-field pressure asymmetries; few loudspeakers have planar diaphragms.

From the standing wave ratio γ measured using the tube, the plane wave sound-absorption coefficient is calculated as (see Problem 1.4):

$$\alpha = \frac{4}{2 + \gamma + \dfrac{1}{\gamma}} \tag{7.4}$$

Measurement of *A* Using a Reverberation Chamber

The *reverberation chamber method* is based on the measurement of reverberation time in a chamber having hard walls, with and without the test sample. The conditions and problems that apply to the reverberation chamber method are described in detail in Chapter 4. It should be noted that the reverberation chamber can also be used to measure the scattering properties of objects.

Because the sound field properties of the reverberation chamber will influence the angular distribution of sound incidence on the absorption sample under measurement, different measurement results will be obtained with different chambers. Since high absorption coefficient values are preferred by manufacturers, it is natural to strive for room designs that give such values within the bounds of typical standards documents, such as ISO 354.

Using Absorption Area

In some cases, it is more relevant to characterize the sound absorption by an *absorption area* rather than by an absorption coefficient. This applies to resonant absorbers and to absorbers where the geometrical surface can be hard to define. This is typical of audience and performers, as well as some other types of natural absorbers.

For a large surface, the absorption area is the product of the surface area and the absorption coefficient. Note that the absorption area may be given in either *metric sabins*, or in *sabins* that are given in square ft. (For example, 10 m^2 of a material having $\alpha = 0.9$ has an absorption area of 9 metric sabins.) In this book, we only use metric sabins.

7.3 POROUS ABSORBERS

The conversion of acoustic to thermal energy in porous absorbers is due to the viscous behavior of air flowing in the canals, pores, and air pockets created by the skeleton of the material. Porous materials that do not have open cells or pores cannot function as porous absorbers—an example of such a material is expanded polystyrene. If the flow is blocked by, for example, paint or other surface coverings, the absorption coefficient will usually drop considerably. Because of the elasticity of the material, there may still be some absorption, however. It is important to differ between the sound absorption and the heat insulation properties of porous materials. Porous materials may be good thermal insulators and still have poor sound-absorption properties. Expanded polystyrene and many other porous plastic materials that have closed pores are poor sound absorbers.

One can model the surface of a porous material as having the structure shown in Figure 7.1. The material is regarded as a skeleton that forms narrow air passages. A practical model is to regard the passages as parallel and extremely narrow tubes. The tubes make it unnecessary to be concerned by propagation of sound sideways. Because of the propagation losses in the material, this is a reasonable assumption. Such a material is said to be *locally reacting*—that is, its acoustic behavior is similar to that described in Section 1.7.

Assume that the open area of a tube has a surface area S_h and that each tube requires a surface area S. Since the volume velocity must be the same on both sides of the boundary, the particle velocity outside and inside the tube must have the ratio S/S_h, which results in an impedance transformation having the same ratio. Insertion of the impedance ratio into the expressions for reflection coefficient r and the absorption coefficient α_0 gives:

$$\alpha = \frac{4SS_h}{\left(S + S_h\right)^2} \tag{7.5}$$

One notes that even a small surface ratio of 0.5 yields an absorption coefficient of 0.9. In the model, the surface ratio is also a measure of the *porosity* of the material σ:

$$\sigma = \frac{V_{\text{pore}}}{V_{\text{total}}} \tag{7.6}$$

where V_{pore} is the pore volume and V_{total} the total volume required to hold the material.

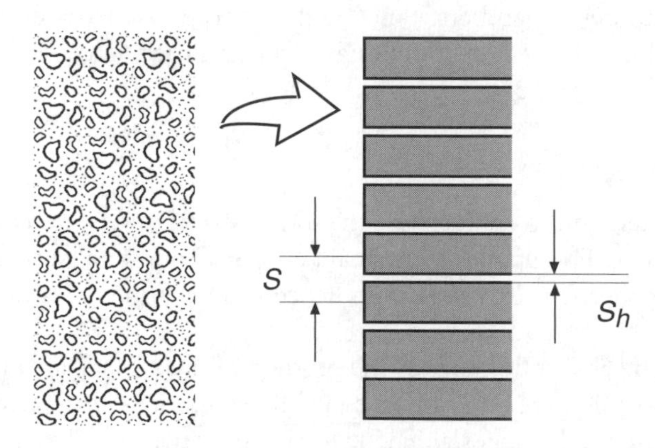

Figure 7.1 In a simplified model of a porous sound absorber one can replace the random pores by thin tubes extending into the material.

This approach is only valid at high frequencies, and the absorption coefficient will approach the value given by Equation 7.1 asymptotically. The porosity is decisive for the absorption properties. Commercial products for sound absorption, such as glass wool, may have a porosity larger than 0.95.

Because of the losses induced by the viscosity of air, the acoustic energy carried in the wave will be converted into heat in the air in the canals. This heat is then transferred to the skeleton of the material. If the canals are narrow, the viscous losses will be considerable. The energy conversion results in isothermal wave propagation and an associated wave propagation speed of down to about 280 m/s. The losses can be modeled by having a complex wave number.

The losses are due to the material's *flow resistance* (the ratio between the pressure needed for the flow, and the flow speed, Ns/m^3). The absorption coefficient of the material α_d will increase as the frequency increases. The absorption coefficient will also vary as a function of properties such as the thickness, extension, placement, and form of the material. The impedance of most materials will be characterized by a small mass component in series with the resistive part.

The most typical placement of sound-absorptive materials is against a hard surface. In this case, the wave propagating inside the material will be reflected as it meets the surface. When the absorber is suspended from a ceiling, of course, both sides will contribute to the sound absorption. The propagation loss of many glass and mineral wool sheets used for sound absorption is about 1 dB per centimeter. Two extreme cases are of special interest:

- *High flow resistance*: In this case, the wave will be rapidly attenuated inside the material, and there will be little reflection at the back side of the material. However, the wave incident on the material will meet a surface having a relatively high impedance and the absorption will be low. This applies to the case of various compressed mineral and glass fiber boards.
- *Low flow resistance*: Since the losses during the wave propagation inside the material will be fairly low in this case, the wave will be quite intense as it is reflected by the hard surface, unless the material has the sufficient thickness. When the wave propagates back toward the outer surface of the material it will be reflected again, and so on. Because of the repeated reflections, one may notice a frequency-dependent sound-absorption coefficient. However, the wave will not be particularly reflected at the outer surface, so the sound-absorption coefficient can be high if the thickness is large.

One can note these types of behaviors in the computed data shown in Figure 7.3, that show how the sound-absorption coefficient varies as a function of the flow resistance of the material. In this and following graphs, unless otherwise indicated, the absorber is assumed to be mounted directly against a hard surface. The computed data shown in Figure 7.2 clearly indicates the influence of material thickness.

The curves in Figure 7.3 show how the absorption at low frequencies increases as the flow resistance is increased, with the penalty of decreased sound absorption at high frequencies. It is clear from these examples that any porous absorber should be optimized with regard to its required sound-absorption properties by proper choice or design of its mounting, thickness, and flow resistance.

Since the losses are associated with flow resistance and material thickness, it is not optimal to position a material characterized by a low flow resistance flush to a hard surface. The particle velocity will be low close to such a surface, because of the interfering, reflected wave. The particle velocity, however, is high in front of the surface at distances of $\lambda/4$ ($+ n\lambda/2$ where $n = 1, 2, 3, \ldots$), so the effective sound

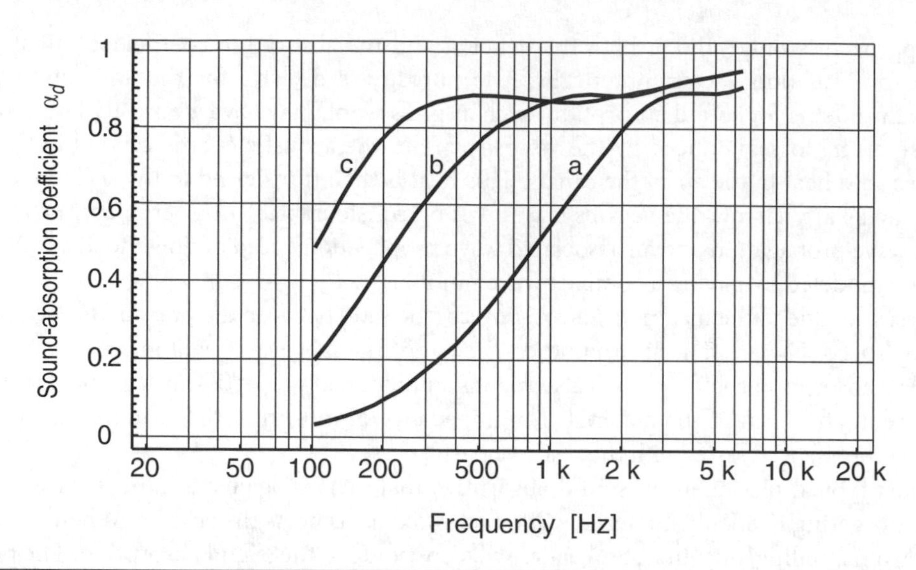

Figure 7.2 Calculated sound-absorption coefficients of a porous sound-absorbing material of different thickness, as a function of frequency. The absorber is assumed to be mounted directly against a hard surface. The specific flow resistance was set to 10^4 kg/m^{3}s and the absorber area was set to 10^2 m^2 to avoid diffraction effects. Curves based on third octave band values: (a) 0.05 m, (b) 0.1 m, and (c) 0.15 m. (After Ref. 7.1)

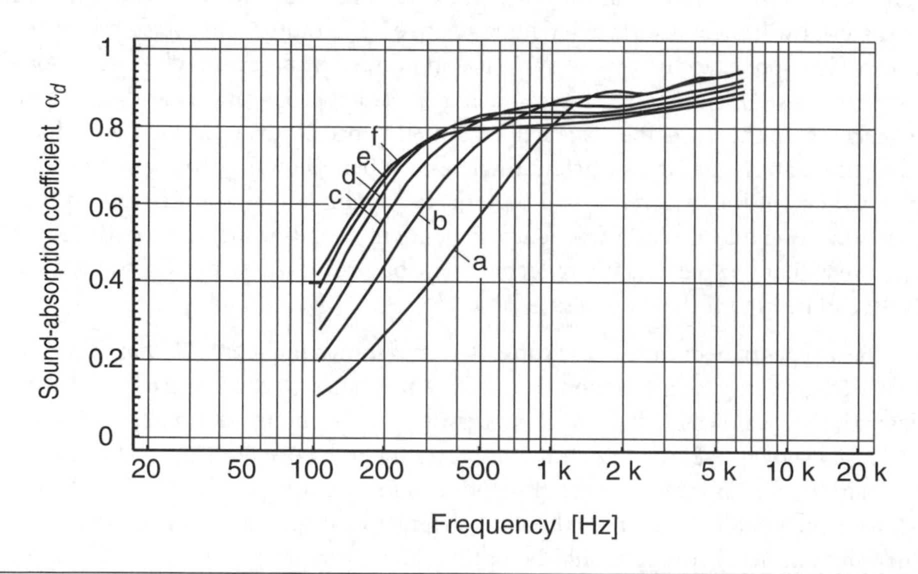

Figure 7.3 Calculated sound-absorption coefficients of porous sound-absorbing materials having different flow resistance as a function of frequency. The absorbers are assumed to be mounted directly against a hard surface. The absorber thickness was set to 0.1 m and the absorber area was set to 10^2 m^2 to avoid diffraction effects. Curves based on third octave band values: (a) $5 \cdot 10^3$ kg/m^{3}s, (b) $1 \cdot 10^4$ kg/m^{3}s, (c) $1.5 \cdot 10^4$ kg/m^{3}s, (d) $2 \cdot 10^4$ kg/m^{3}s, (e) $2.5 \cdot 10^4$ kg/m^{3}s, and (f) $3 \cdot 10^4$ kg/m^{3}s. (After Ref. 7.1)

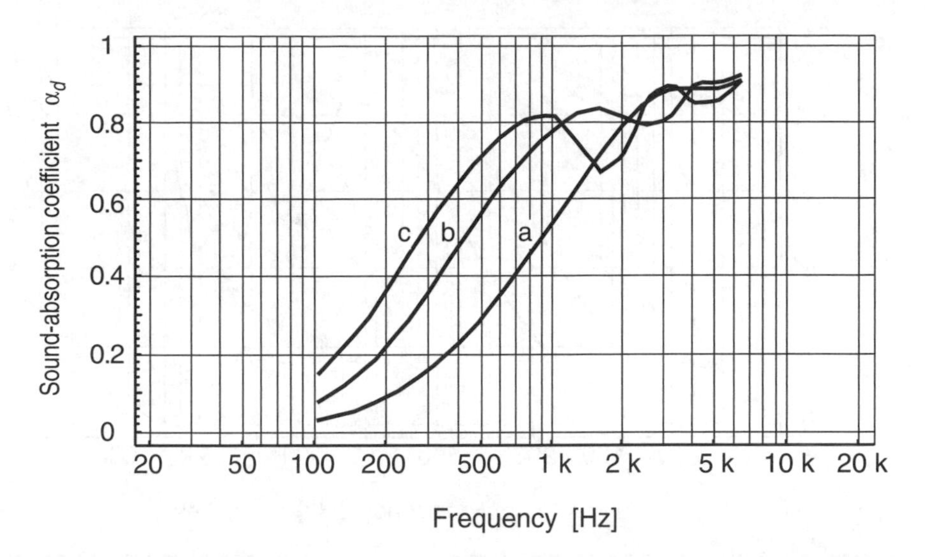

Figure 7.4 Calculated sound-absorption coefficients of a 0.05 m thick porous sound-absorbing material mounted at different distances *d* from a rigid surface as a function of frequency. The specific flow resistance was set to 10^4 kg/m^{3}s and the absorber area was set to 10^2 m^2 to avoid diffraction effects. Curves based on third octave band values: (a) no air space, material mounted directly on surface, (b) d = 0.05 m, and (c) d = 0.15 m. (After Ref. 7.1)

absorption is going to be high if such a material is positioned at these particle velocity maxima. The data in Figure 7.5 also shows this behavior; the absorption peaks of curve b are at frequencies corresponding to air space depths of $\lambda/4$, $3\lambda/4$, etc. This can also be noted in curve c of Figure 7.4, so the phenomenon exists even with materials having considerable thickness.

Since the absorption maxima are so frequency specific, thin, flow resistive materials (such as curtains) have to be draped to achieve reasonable frequency-independent sound absorption.

To increase the sound absorption at low frequencies, one should use a suitable combination of material flow resistance, thickness, and air space distance. The data shown in Figure 7.4 exemplify the beneficial influence of such air space.

The decreased speed of sound in porous sound-absorptive materials (typically about 280 m/s) is due to non-adiabatic sound propagation. Because of the shorter wavelengths a somewhat thinner material can be used to achieve a desired sound absorption than if the wavelength were that of free propagation. A suitably designed porous absorber, mounted against a hard surface and having a thickness of 0.15 m, will have good sound-absorption properties down to a frequency of approximately 400 Hz.

Porous absorbers are easy to use and offer good sound-absorbing properties but have a number of disadvantages and problems. They are:

- Sensitive to mechanical damage
- Dust collectors—particularly close to air exhausts and intakes
- Likely to release fibers hazardous to health
- Difficult to paint since the pore openings will be clogged
- Cannot be washed because pores will be clogged by water due to capillarity
- Prone to condensation because of lack of a diffusion barrier

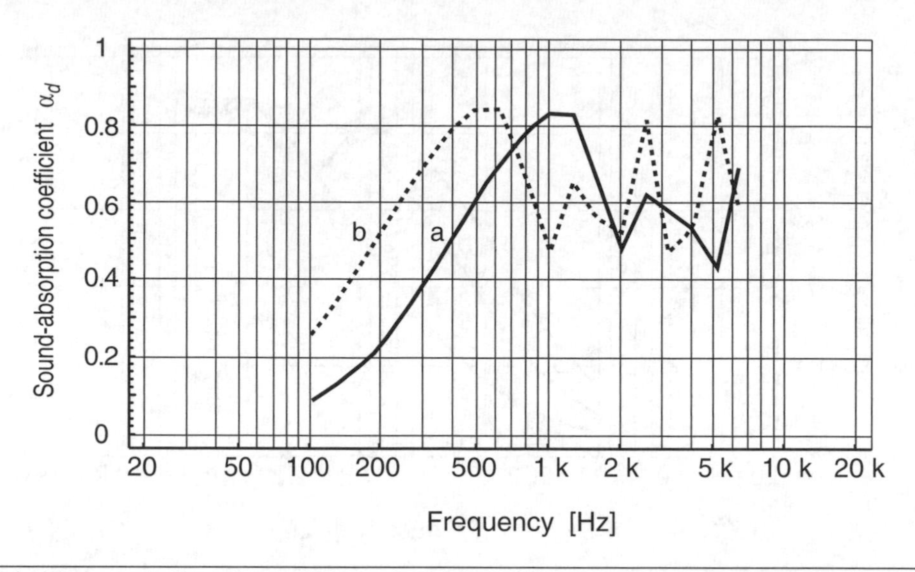

Figure 7.5 Calculated sound-absorption coefficients of a 0.005 m thick porous sound-absorbing material mounted at different distances from a hard surface as a function of frequency. The specific flow resistance was set to 10^5 kg/m^{3}s and the absorber area was set to 10^2 m^2 to avoid diffraction effects. Curves based on third octave band values. (a) 0.1 m and (b) 0.2 m. (After Ref. 7.1)

To avoid some or all of these disadvantages, it is possible to cover the surface of a porous material by various sheets made of glass fiber cloth or plastic and metal foils, as well as various boards made of gypsum or wood, for example. Such covers may have different degrees of limpness, mass, and compliance. Such protective sheets will also help avoid material fibers from contaminating the surrounding air, which may be a health hazard. The apparent surface mass of the limp sheet will increase if the sheet touches the absorber.

As an example, assume a free, limp sheet, characterized by mass only, on which a plane wave impinges perpendicularly, in front of an otherwise perfect sound absorber, having an impedance of Z_0, that is, $\alpha = 1$, as shown in Figure 7.6. We use the impedance analogy concept discussed in Section 1.8 because it is intuitive.

Assume further that the sheet has an area S and a mass M. The mass per unit area then is M/S, and the specific acoustic impedance seen by the incident wave will be:

$$\underline{Z}_2 = \frac{j\omega M}{S} + Z_0 \tag{7.7}$$

Inserting Equation 7.7 into the expression for the reflection coefficient Equation 1.53, we find:

$$\underline{r} = \frac{\dfrac{j\omega M}{S} + Z_0 - Z_0}{\dfrac{j\omega M}{S} + Z_0 + Z_0} = \frac{j\omega M}{j\omega M + 2Z_0 S} \tag{7.8}$$

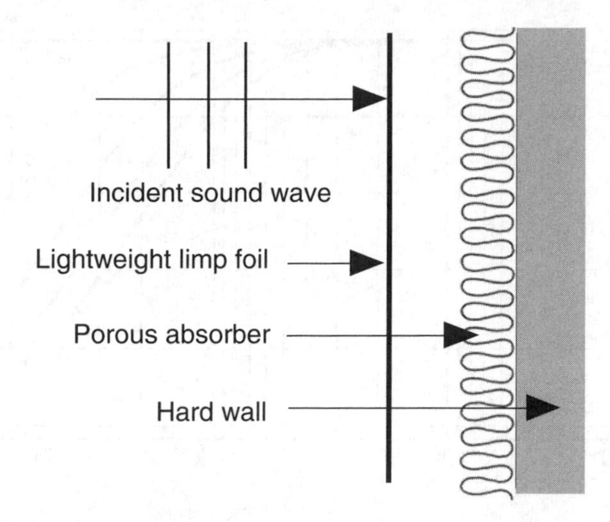

Figure 7.6 A porous sound absorber mounted against a hard wall and covered by a thin, lightweight, limp protective foil.

In practice, it is better to have the sound-absorption coefficient expressed using the concept of *s*urface mass—that is, the *mass per unit area m"* [kg/m²]. Using Equation 7.4, we obtain the absorption coefficient as:

$$\alpha_0 = \frac{1}{1+\left(\frac{\omega m''}{2Z_0}\right)^2} \tag{7.9}$$

One notes that the sound absorption decreases as the frequency increases, as a result of the inertia of the sheet.

If an absorption coefficient larger than 0.8 is considered necessary for good sound absorption, the mass per unit area must be lower than:

$$m'' = \frac{Z_0}{\omega} \approx \frac{65}{f} \tag{7.10}$$

This, in turn, means that any protective sheet must be extremely lightweight, as also indicated by the data shown in Figure 7.7.

For the limp foil to have low effective surface mass, it must also be mounted so that it does not touch the porous absorber, since otherwise its effective mass will increase. Most surface protected commercial sound-absorptive materials have the foil attached to the skeleton of the material. This requires the foil to have a low compliance, so that it can move over the pore entrances in spite of the contact.

A constriction to the air (e.g., a perforated or slit sheet used to cover the absorber) will also behave as an acoustic mass as discussed in Chapter 1. The presence of this acoustic mass in front of the absorber, due to the air in the holes, will result in loss of high frequency sound absorption (assuming a wavelength much larger than the inter-hole distance).

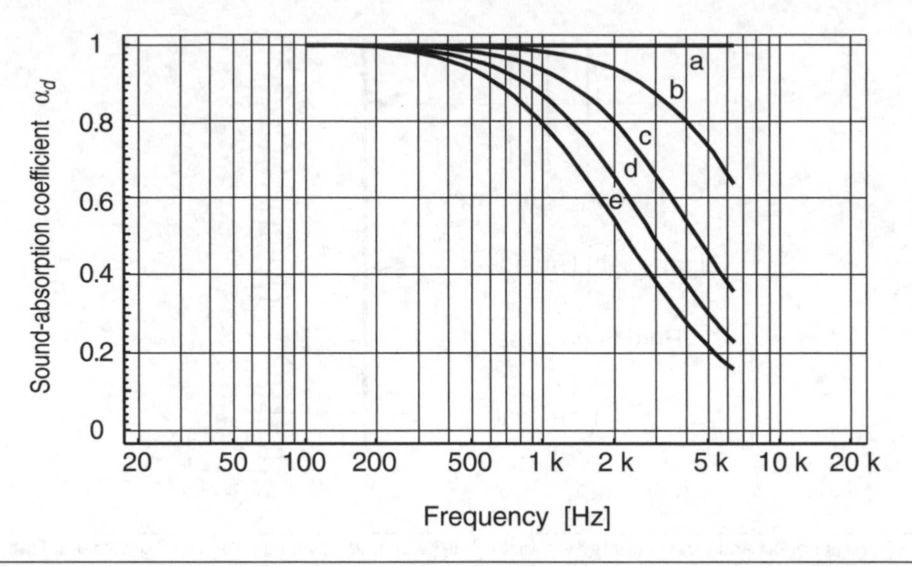

Figure 7.7 Calculated sound-absorption coefficients of an ideal porous sound-absorbing material protected by limp foils having different surface mass as a function of frequency. Curves based on third octave band values: (a) No foil, (b) 0.025 kg/m², (c) 0.05 kg/m², (d) 0.075 kg/m², and (e) 0.1 kg/m². (After Ref. 7.1)

As an example, consider a thin, free, perforated sheet in front of an ideal absorber. The sheet has a quadratic hole pattern, with an inter-hole distance of b, and circular holes of diameter $2a$.

The acoustic mass associated with each such hole in a thin sheet is M_A. The resulting expression for the sound-absorption coefficient can be shown to be:

$$\alpha_0 = \frac{1}{1+\left(\frac{\omega M_A S_{cc}}{2Z_0}\right)^2} \tag{7.11}$$

where

$$S_{cc} = b^2 \tag{7.12}$$

and

$$M_A = \frac{\rho_0}{2a} \tag{7.13}$$

One obtains an expression, corresponding to expression 7.10, for the upper limiting frequency for $\alpha_0 >$ 0.8 as:

$$\omega < \frac{2ac_0}{b^2} \tag{7.14}$$

If the inter-hole distance is of the order of one half wavelength or larger, one will have appreciable sound reflection from the surface between the holes.

For the mass effect by holes of air in a surface to be negligible, the perforation rate—hole area divided by surface area—has to be larger than 0.3. Of course, increasing the thickness of the perforated sheet will lead to more stringent requirements.

Until now, sound incidence has been assumed as perpendicular to the surface of the absorbers. Extending the calculations to the case of nonperpendicular incidence, one finds that a locally reacting surface, having a real impedance of Z_2, has an angle-dependent sound-absorption coefficient $\alpha_0(\varphi)$ that varies by the incidence angle φ as:

$$\alpha_0(\varphi) \approx \frac{1}{\frac{1}{2} + \frac{1}{4}\left(\zeta\cos(\varphi) + \frac{1}{\zeta\cos(\varphi)}\right)} \tag{7.15}$$

where

$$\zeta = \frac{Z_2}{Z_0} \tag{7.16}$$

Figure 7.8 shows how the angle-dependent sound-absorption coefficient behaves as a function of φ.

The sound-absorption coefficient for diffuse sound incidence α_d is usually used for room acoustics since it represents an average of $\alpha_0(\varphi)$ for all angles. In a diffuse sound field, sound waves having oblique incidence carry most of the energy, so usually $\alpha_d > \alpha_0$.

Ideally, one would want to know the sound-absorption coefficient for the particular sound field that one is dealing with. Since the absorber influences the sound field, there is not an immediate solution. Typically, one has to be satisfied by two types of absorption-coefficient data, α_s and α_r, which are measured approximations to α_d and α_0.

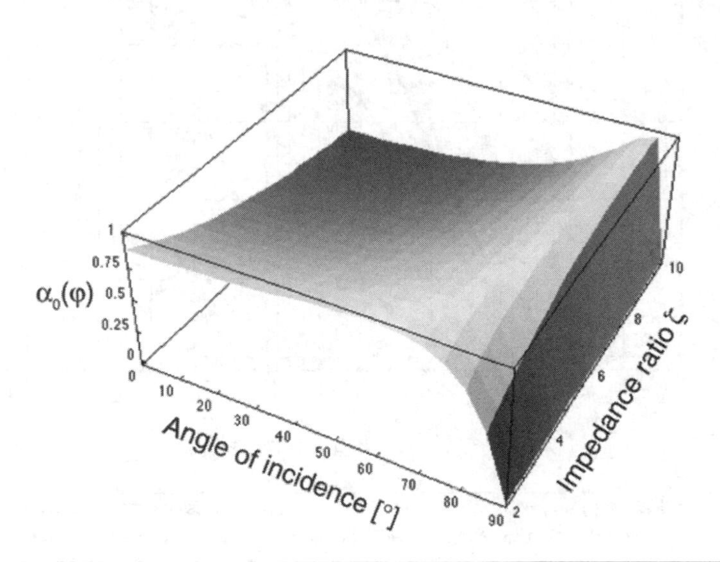

Figure 7.8 The sound-absorption coefficient of a porous sound absorber as a function of the impedance ratio $\zeta = Z_2/Z_0$ and the angle of incidence of the sound wave.

The measurement of α_S is done using reverberation chambers, usually according to the ISO 354 standard, and requires test sample areas of at least 10 m^2. The measured absorption coefficient will vary between laboratories as a result of different laboratory room geometry, diffuser, and sample placement.

When developing new absorption materials and combinations, it is impractical and expensive to use such large samples. One instead uses the standing wave tube method to measure α_r. There is usually considerable difference between the α_r and α_S values so care should be exercised in the use of measurement results.

Until now, the absorber surface has been assumed to be infinite or at least much larger than the wavelength of interest. Because of diffraction, however, one finds that the sound absorption of any surface varies with the global surface geometry. Figure 7.9 shows calculated α_d data for different sizes of a square sound-absorptive panel mounted in a hard surface.

One notes that the absorption coefficient varies with frequency and that maximum absorption is obtained when the length of the square panel is about twice the wavelength. This shows that it may be advantageous to use many smaller panels, mounted at sufficient distance from one another, if one wants to maximize the sound absorption of a certain amount of material. Such use also leads to increased sound field diffusitivity in a room, since the absorbing patches will act as diffusers. For patches that are small compared to the wavelength of sound, the sound absorption will be independent of the direction of sound incidence (see Reference 7.2).

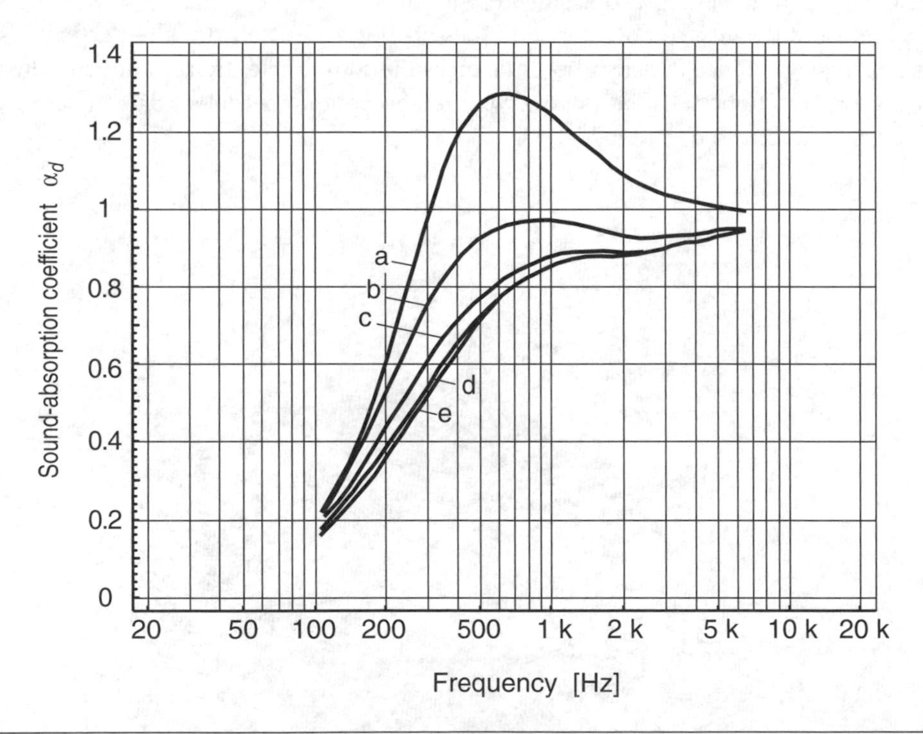

Figure 7.9 Calculated sound-absorption coefficients of a square patch of porous sound-absorbing material of different areas on a rigid surface, as a function of frequency. The specific flow resistance was set to 10^4 kg/m³s and the absorber thickness was set to 0.1 m. Curves based on third octave band values: (a) 1 m², (b) 10 m², (c) 100 m², (d) 1,000 m², and (e) 10,000 m². (After Ref. 7.1)

7.4 RESONANCE ABSORBERS

The sound absorption by resonance absorbers relies on the losses in their acoustical and mechanical constructions that are set in motion by sound. The most common types of resonance absorbers are *Helmholtz, membrane, panel* and *slit absorbers.* Thin panels, walls, and windows can also act as resonant sound absorbers. One can also combine porous and resonant absorber action in a material. Porous materials can be designed to hold resonant elements and vice versa.

Resonance absorbers are primarily used for absorbing sound having frequencies below 400 Hz. In isolated cases, one can design resonance absorbers to cover a wide frequency range, including high frequencies. One example of such a construction, having reasonably smooth, wide range sound absorption, is a wall of *air-bricks* behind which is an airspace with a porous absorber (see Figure 7.13).

A section of a membrane absorber, shown in Figure 7.10, is a mechano-acoustical resonator where the mass of a flexible membrane—for example, impervious cloth or a plastic foil—is mounted so that a closed airspace is formed between the membrane and the underlying wall. This air space acts as a compliance, as explained in Chapter 1. Typical membrane absorbers are windows, wooden floors on joists, and wooden wall paneling on battens.

The losses consist of the mechanical losses in the membrane and the acoustical losses in the air space that can be filled with a porous absorber. Note that the addition of a porous absorber will affect the resonance frequency and the bandwidth of the absorption peak.

Because the resonant elements, the membrane, and the airspace behind it are small compared to wavelength (as it will turn out), it is convenient to use the concept of acoustical impedance to find the condition for resonance.

Using the ideas introduced in Chapter 1, we can write the acoustical impedance for a membrane absorber as:

$$\underline{Z}_{A,\text{tot}} = j\omega M_A + R_A + \frac{1}{j\omega C_A} \tag{7.17}$$

That is, it consists of mass, compliance, and loss. Remembering the relationship between mechanical and acoustical impedance, one can write the components of the acoustic impedance as:

$$M_A = \frac{M}{S^2} = \frac{m''}{S} \tag{7.18}$$

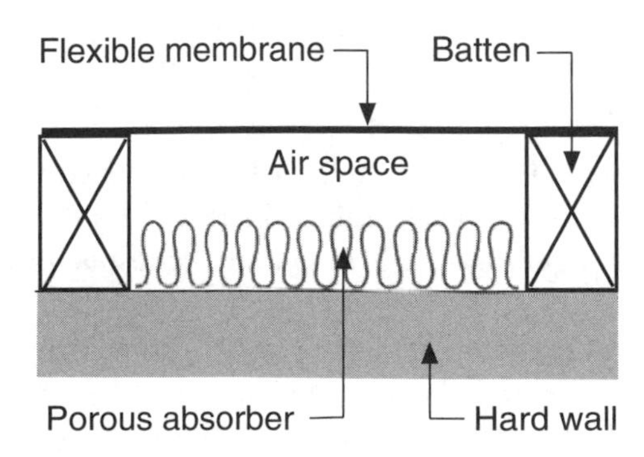

Figure 7.10 The basic design of a membrane absorber. The porous absorber may be placed in different ways in the air space to achieve different resonance and sound-absorption characteristics of the design.

and

$$C_A = \frac{Sd}{Z_0 c_0} \tag{7.19}$$

where m is the mass per unit area of the membrane, S the surface area, and d the depth of the air space.

It is difficult to estimate R_A in advance. The losses are associated with mechanical losses in the membrane, losses at the edges of the membrane, reradiation loss, and losses in the air space. The losses are best found by measurement of the bandwidth of the resonance. The concept of a Q-value to characterize the damping is useful here also. Membrane absorbers made of wood panels typically show low Q-values, on the order of 0.5 to 1.

The resonance frequency f_0 can be obtained by setting the imaginary part of Equation 7.17 to zero that gives (for perpendicular sound incidence):

$$f_0 = \frac{c_0}{2\pi}\sqrt{\frac{\rho_0}{m''d}} \tag{7.20}$$

For diffuse field incidence, the resonance frequency is about 30% higher.

Typical values for f_0 are in the 50 to 400 Hz range when one uses panels having a thickness of about a centimeter and air spaces with a depth of a few centimeters. In these cases, one usually finds that α_S is 0.8 or lower.

If one wants high absorption over a small frequency range, one must use a thin panel and a large airspace distance. Thin panels often exhibit multiple absorption peaks due to air space room resonances. Membrane absorbers and resonator panels are discussed in detail in Reference 7.11.

A Helmholtz resonator is a single resonator where both mass and compliance are due to acoustical components. An open bottle is an example of a simple Helmholtz resonator. A Helmholtz resonator can also be integrated into a wall as shown in Figure 7.11.

The neck constriction is the acoustical mass component and the enclosed airspace the acoustical compliance. Since the Helmholtz resonator has intrinsically low losses, the reradiation of energy to the surroundings is an important factor in design. The resonator can be placed free or inserted in a wall or other surface; the absorption maximum will depend on placement (including placement relative to room corners).

Using the expressions for the acoustic components found in Chapter 2, one finds that the resonance frequency f_0 is determined by:

$$f_0 \approx \frac{c_0}{2\pi}\sqrt{\frac{\pi a^2}{V\left(l_n + \frac{\pi}{2}a\right)}} \tag{7.21}$$

where a is the neck radius, l_n the neck length, and V the enclosed air volume. Note that the end correction for an open tube has been incorporated when writing the effective neck length. Since the end correction will depend on the acoustic environment of the mouth opening, it is best to tune the resonator at its place of operation.

One can adjust the losses of a Helmholtz absorber by inserting a flow resistance in the throat or by adding a porous absorber inside the compliant air volume. The highest absorption will be found when the losses in the acoustic system of the bottle match those of the reradiation. The absorption peak is

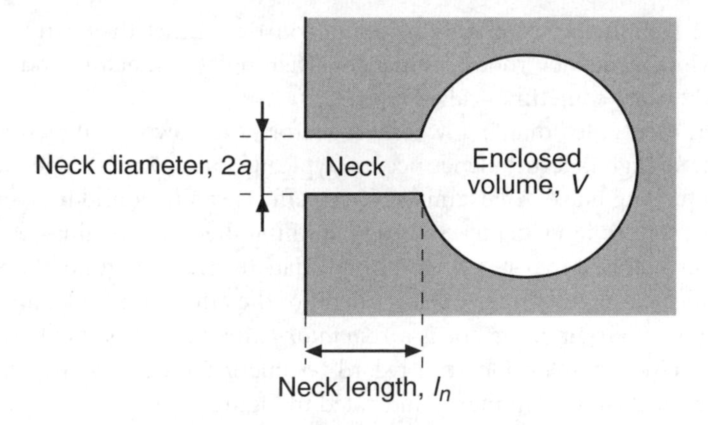

Figure 7.11 Section showing a Helmholtz resonator. The Helmholtz resonator consists of an air volume coupled to the outside air by an open tube. Here the resonator is shown integrated into a wall.

then somewhat narrow. By using a number of resonators having different resonance frequencies, one can achieve more wide band absorption than by using a single resonator.

The absorption of a Helmholtz absorber is best described by the idea of an absorption area, since the absorption area can be much larger than the area of the neck. One can show that the maximum absorption area A_{max} that can be obtained for a Helmholtz absorber inserted in a wall is:

$$A_{max} = \frac{\lambda_0^2}{2\pi} \tag{7.22}$$

where λ_0 is the wavelength of sound at the resonance frequency. The A_{max} absorption area is only obtained at the resonance frequency and the absorption peak is then quite narrow in frequency. Figure 7.12 shows measurement results that illustrate how the absorber draws in sound.

A wide bandwidth requires a large enclosed volume. Note that the shape of the volume is unimportant as long as its airspace has small dimensions compared to the wavelength.

If one listens close to a low loss Helmholtz resonator, it is sometimes possible to hear the ringing sound of the reradiation. Examples of such resonators are some musical instruments, for example guitars and violins.

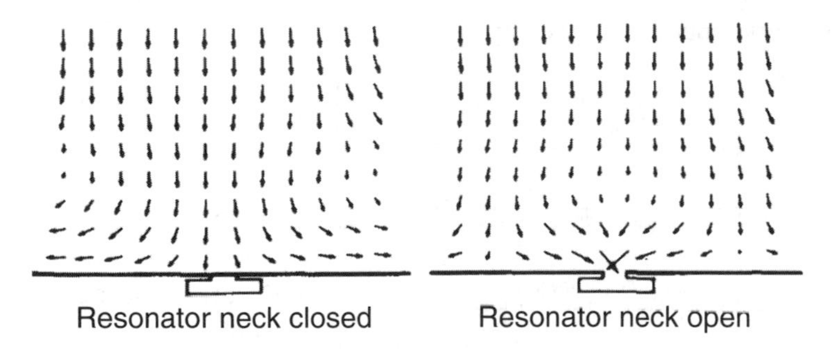

Figure 7.12 Measurement results showing power flow (intensity vectors) over a hard surface with an embedded Helmholtz resonator for normal sound incidence. Note how the open neck resonator draws in energy over a wide area. (From Ref. 7.12)

An advantage of Helmholtz resonators in room acoustics is that they can be used to control the decay of individual low frequency room resonances. Helmholtz resonators and other tuned low frequency sound absorbers are sometimes called *bass traps.*

Finally it must be mentioned that, in low loss resonators, the flow velocity u of the air in the resonator neck may become so high that turbulence occurs (typically for u >7 m/s). The turbulence is a nonlinearity that causes disturbing noise. An example of such effects can be found in bass reflex loudspeakers using open ports, as discussed in Chapter 8. By streamlining the port openings, it is possible to reduce the turbulence. In these cases, it may be difficult to calculate the neck length of the port, and it is usually necessary to use numerical simulation or measurement of the effective port length.

An important type of air-air resonator is a *resonator panel* that consists of panels having perforations, such as slits or circular holes. One can regard resonator panels as a large number of individual Helmholtz resonators next to one another as indicated in Figure 7.13.

Because of symmetry, it is possible to remove the internal side walls of the air volumes. The resonance frequency of such a panel can be determined in the same way as for a single resonator; the enclosed air volume will be:

$$V = S_{cc}d \tag{7.23}$$

where d is the depth of the airspace, $S_{cc} = b^2$, and b is the distance between hole centers, assuming a square hole pattern. If the perforations or holes occupy more than 30% of the surface area, the panel will be quite transparent to sound, and there will be no resonance. Note that the panel will act as a reflector at high frequencies.

As mentioned, the holes do not need to be cylindric—slits and other hole shapes can be used as well, as long as the maximum crosswise dimensions are much smaller than the wavelength. By damping the resonance using flow resistive materials, such as cloth or mineral wool, one can obtain wide bandwidth, typically at least one third octave band. The sound-absorption coefficient at the resonance frequency is typically high; α_S may be close to unity.

It is often easier to manufacture a resonant sound-absorbing panel by using slits rather than circular holes, for example, by using strips with a small airspace between the strips or by using a saw to make slits in a gypsum or particle board sheet. The resonance frequency of such a slit resonator panel having thin slits can be calculated using:

$$f_0 \approx \frac{c_0}{2\pi}\sqrt{\frac{S_{\text{slit}}}{Vl_{\text{eff}}}} \tag{7.24}$$

where the effective neck length l_{eff} is given by:

$$l_{\text{eff}} = l_n + b\left(\frac{1}{2} + \frac{2}{\pi}\ln\left(\frac{c_0}{\pi b f_0}\right)\right) \tag{7.25}$$

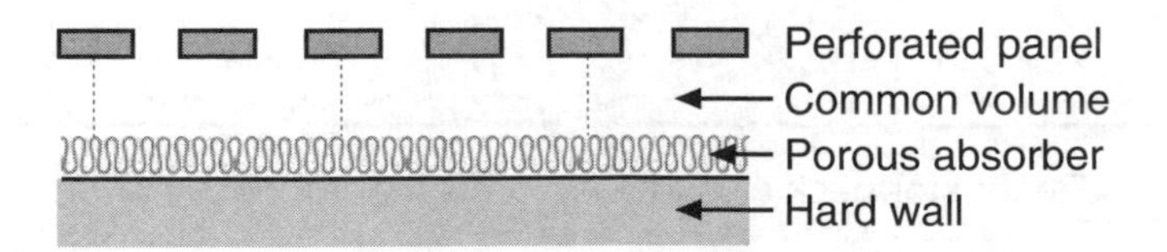

Figure 7.13 The resonator panel uses the acoustic mass of the air in the panel holes to resonate with the acoustic compliance of the trapped air. Porus absorber sheet added to broaden the resonance.

Figure 7.14 The air-brick absorber can be thought of as a panel absorber having a thick panel or as a Helmholtz absorber with a subdivided hole. (Photo by Mendel Kleiner.)

where S_{slit} is the slit area per unit (of panel), l_n the thickness of the panel, b the slit width, and V the enclosed volume per unit area. The expression requires that the slit width be much smaller than the wavelength at the resonance frequency. An iterative approach to the determination of the resonance frequency is needed since Equation 7.25 includes the resonance frequency inside the logarithm. Slit absorbers are often characterized by a fairly high sound absorption over a comparatively wide frequency range, that is they have a low Q-value.

Air-brick sound absorbers with a closed airspace behind the bricks (shown in Figure 7.14) function as Helmholtz absorbers but will have a wider frequency range of useful sound absorption because of additional high frequency sound absorption at frequencies where the length of the holes are a multiple of half-wavelengths.

Translucent Absorbers

Most materials and constructions used for absorbers are opaque and will impede the flow of light. Using the mechanical absorber idea discussed in the previous paragraph, it is possible to make *translucent sound absorbers*, using either glass or plastics. Microperforation can be achieved using lasers or mechanical means. The holes generated will exhibit comparatively high flow resistance. Such microperforated panels and foils may be used with enclosed airspace to design various membrane and resonator panel sound absorbers. One such design using micro-perforated plastic foil sails is shown in Figure 7.15.

Absorbers Using Mechanical Resonance

It must be noted that any surface that can be put into motion by sound waves is likely to act as a sound absorber, because of its transmission or its internal losses due to friction. Some mechanical sound absorbers are mechanically similar loudspeakers (without drive mechanism) and will act as resonance absorbers, as will loudspeakers.

Figure 7.15 Microperforated foils can be successfully employed as translucent sound absorbers. (Microsorber photo courtesy of KAEFER, Germany)

7.5 VARIABLE ABSORBERS

Ideally one wants to be able to adjust the sound absorption of the additional absorbing areas in a room to that required in the specific case, since the natural absorption in a room varies (e.g., as a function of the number of people present). As mentioned previously, various uses require different reverberation times. Using *variable absorbers*, it may be possible to adjust the reverberation time to the desired value irrespective of the number of people in the audience. Some examples of the construction of such variable absorbers are shown in Figure 7.16.

7.6 AUDIENCE ABSORPTION

The natural sound absorption by the audience is often the dominating sound-absorption component at mid frequencies in enclosed performance spaces, such as theaters and concert halls. At high frequencies, usually above 4 kHz for large rooms, the sound absorption induced by lossy sound propagation in the air will dominate as discussed in Chapter 1.

If one wants a certain reverberation time (determined by the air volume of the room), a certain volume per audience area is needed, depending on seating. The audience sound-absorption coefficient depends on the number of persons per unit area and the type of upholstered chairs. The effective audience sound-absorption coefficient also depends on the diffusitivity of the sound field so the design of

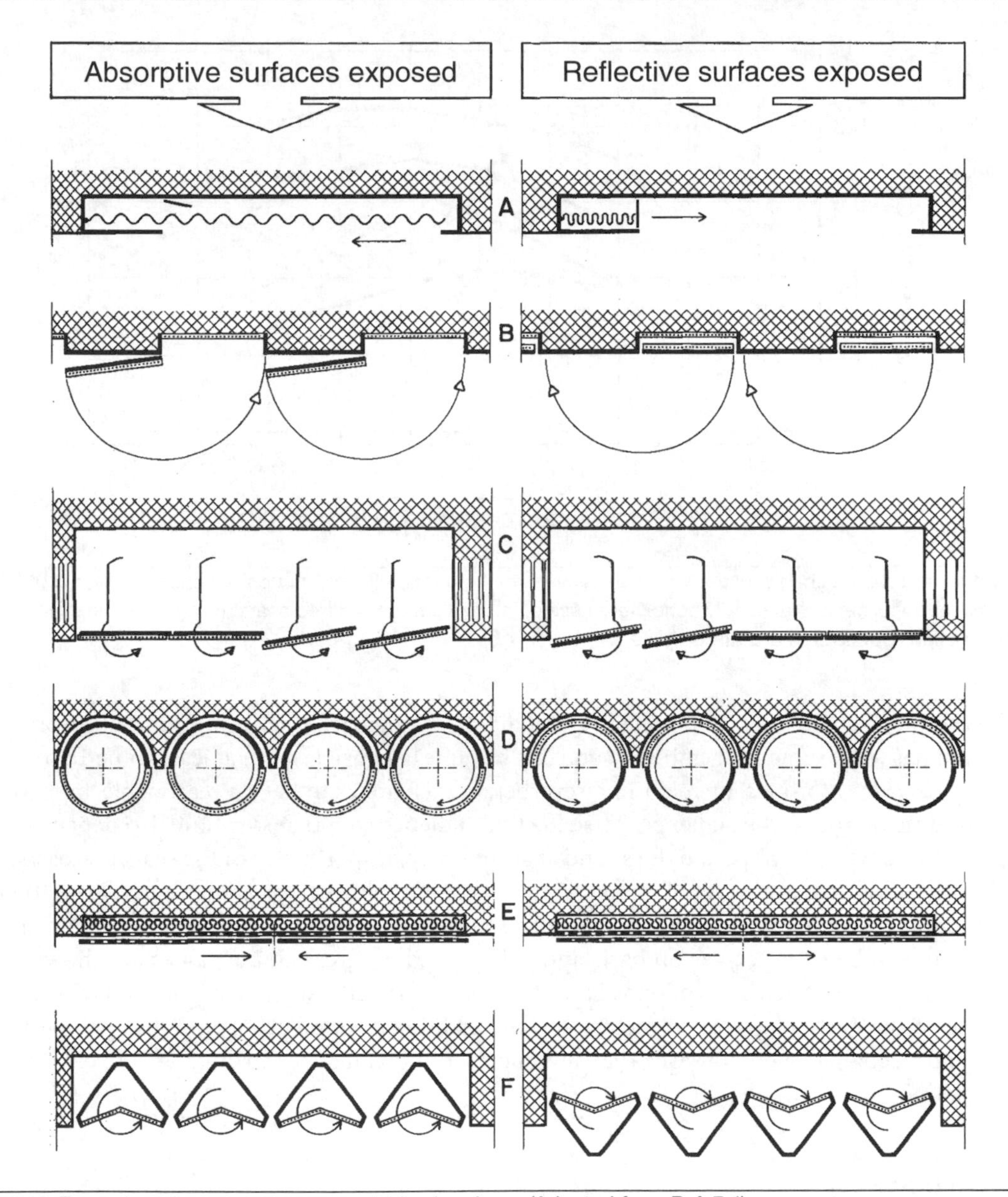

Figure 7.16 Examples of constructions of variable absorbers. (Adapted from Ref. 7.4)

large spaces to obtain a particular, suitable reverberation time may be difficult. Some examples of the sound-absorption coefficient of concert hall seats are shown in Figure 7.17.

Because of the sound field properties at the edges of the seating area, it is necessary to compensate for the sound that enters from the sides. The full *acoustical* audience area is approximately the seating block floor area plus areas of strips 0.5 m wide around the audience seating blocks, except for sides at balcony rails or walls.

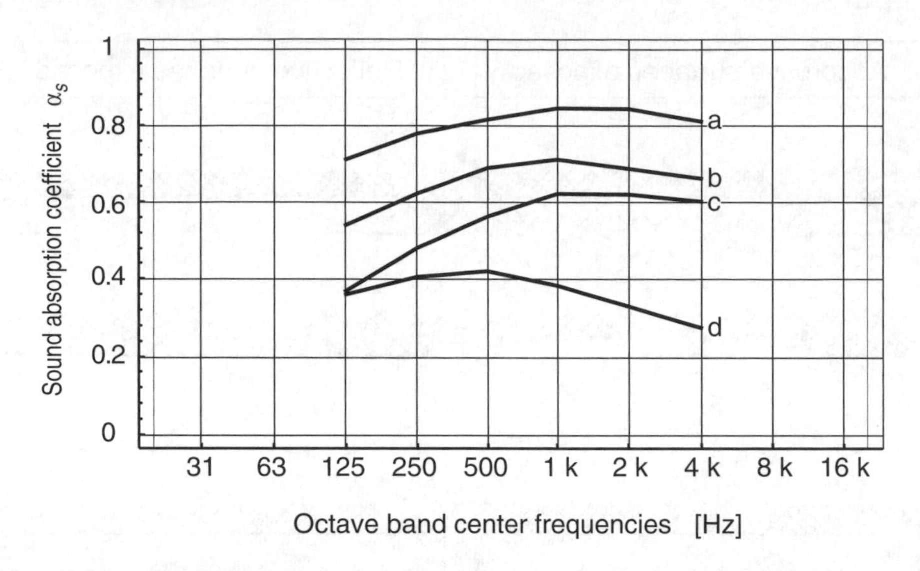

Figure 7.17 Unoccupied chair sound-absorption coefficients from four different groups of concert halls, according to degree of seat upholstering: (a) heavily upholstered seats, (b) medium upholstered seats, (c) lightly upholstered seats, and (d) extra lightly upholstered seats. (Adapted from Ref. 7.5)

The design of chairs present a special problem when they have the same sound-absorption. Such chair design has to be somewhat empirical and can be quite expensive, since at least some 16 to 25 chairs are usually necessary for measurement in a reverberation chamber to have a reasonable approximation to the sound field around the audience. The effective audience sound absorption also depends on edge effects, since one typically must add the sound absorption by the perimeter of the seating areas/sections. Typically, one adds an extra width of about 0.5 m to the sides of the floor area covered by seating.

The sound absorption by individuals (persons standing or sitting far apart, such as a few musicians on stage) will still be on a per-person basis and will depend on dress. Some values are shown in Table 7.2 (see Reference 7.8). Such absorption is hard to measure correctly, since it will depend on the incident sound field in the particular case. In most cases, the additional sound absorption by such persons will be small and negligible. Note however that audience and individuals contribute effectively to the scattering of sound.

7.7 REFLECTORS

Reflectors are usually used to distribute the sound from a speaker or an orchestra to the audience or to improve the acoustical conditions for musicians on stage. The room impulse response can sometimes be improved by careful use of reflectors. Reflectors can be planar or curved. Sometimes curved reflectors are used as diffusers.

For a reflector to be efficient, its surface must have dimensions much larger than the wavelength of sound to be reflected and sufficient mass to prevent it from being set into motion by the incoming sound. The mass per unit area m should be such that $|\underline{r}| > 0.9$ in the frequency range of interest.

Note that the angle of incidence affects the efficiency of the reflector and that groups of reflectors will show different behavior than single reflectors. In addition, the reflection properties of reflectors will depend on their surface structure.

Figure 7.18 shows an example of calculated reflection coefficient magnitude, on the normal, for sound perpendicularly incident on a free, plane, immovable reflecting strip. One notes that the reflection coefficient is small at low frequencies but at high frequencies it will be close to unity (see the discussion of Fresnel zones in Chapter 4).

At low frequencies, and for large angles of incidence, the reflection by strips and other types of planar reflectors will tend to become diffuse. Because of interference, the reflections by groups of such reflectors will tend to be directional. Their reflection properties are best investigated using physical scale modeling.

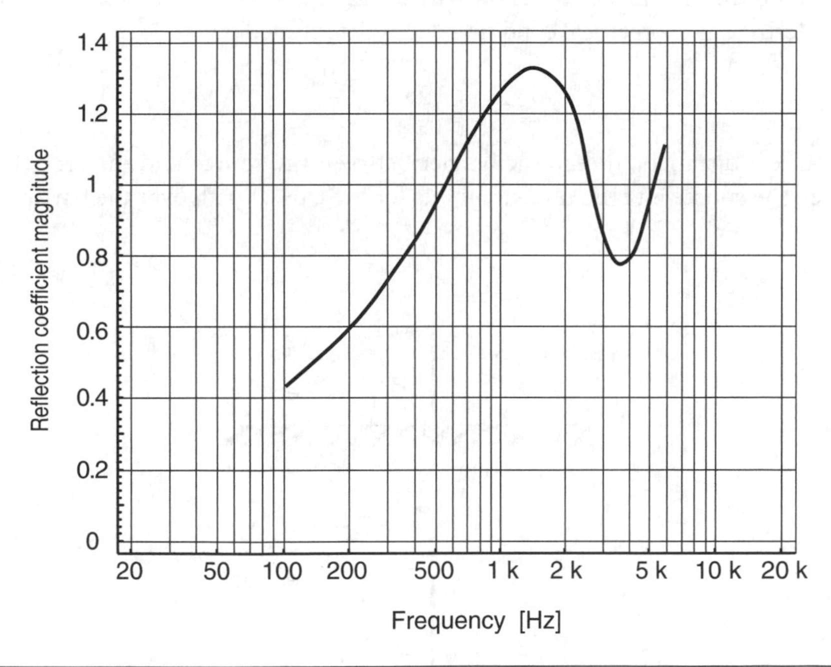

Figure 7.18 Example of the reflection coefficient for a free rectangular panel strip, 2.1 m wide, for perpendicular sound incidence. (Adapted from Ref. 7.6)

7.8 BARRIERS

In many cases, something is needed to reduce the sound level of some activity in a room, without building a wall to separate the noise-generating part of the room from the rest. A partial wall or a *barrier* can then be used to achieve a sound shadow. (Note that there are cases in which items such as pillars may cause an unwanted shadow.)

The effectiveness of a barrier is determined by the following factors:

- Wavelength of sound
- Angle of sound incidence relative to the barrier
- Acoustic and mechanical properties of the barrier

Often the frequency range of interest is 200 Hz to 4 kHz, which corresponds to a wavelength range of 0.08 to 1.7 m. This means that the barrier height is of the order of the longest wavelength. The result is considerable diffraction around the barrier and poor shadow action. Figure 7.19 shows the geometry used in the following discussion.

One can show that the insertion loss by the barrier, ΔL [dB], depends on the barrier geometry and the locations of the source and receiver. The shorter the wavelength and the larger the difference between the path via the edge and the path in absence of the barrier, the larger insertion loss will be. The parameter X_0 is used to characterize this geometrical condition and is determined by:

$$X_0 = \sqrt{2\pi \frac{R_1 - R}{\lambda}} \tag{7.26}$$

where R is the direct distance through the barrier between the source and the receiver points, R_1 the distance between the source and the receiver points for the closest path over the barrier edge, and λ the wavelength.

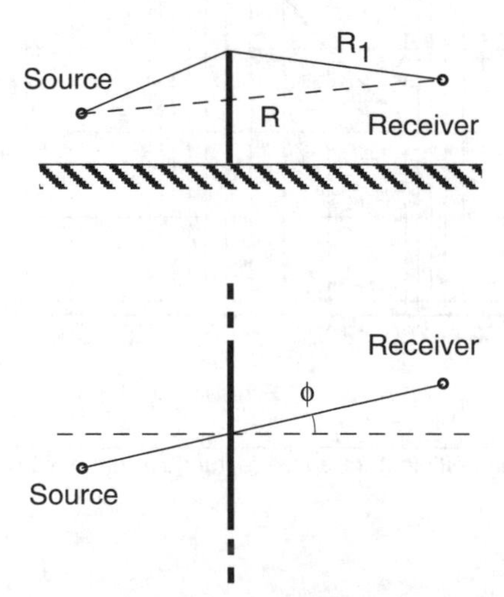

Figure 7.19 Geometry for calculation of the insertion loss of an infinitely long barrier. *R* is the direct distance through the barrier. R_1 is the distance of the closest path over the barrier edge.

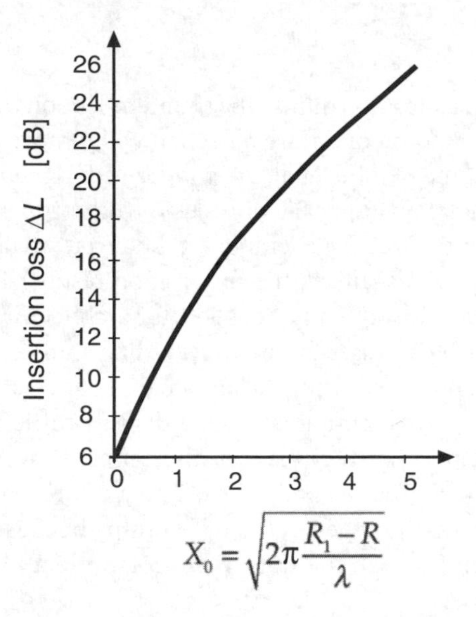

Figure 7.20 Insertion loss of a barrier in a free field as a function of the parameter X. Note the modification of $X(\varphi)$ needed in case $\varphi > 0$. (Adapted from Ref. 7.7)

For a point source near an infinitely wide barrier, made of a totally sound-absorptive material, the insertion loss ΔL achieved by using the barrier is approximately:

$$\Delta L \approx 11 + 20\log(X_0) \tag{7.27}$$

for $X_0 >> 1$.

On the line of sight, the loss is always 6 dB. In the case that X_0 is not much larger than unity, the curve shown in Figure 7.20 may be used to estimate the insertion loss. One notes that the insertion loss increases by 3 dB/octave for small values of X_0.

If the angle φ between the direct propagation path and the barrier is smaller than 90°, the parameter X_0 needs to be replaced by the slightly modified value X_φ:

$$X_\varphi = X_0\sqrt{\cos(\varphi)} \tag{7.28}$$

The difference in insertion loss between a sound-reflective and sound-absorptive barrier is small. A thick barrier will usually result in better insertion loss than a thin barrier, since sound has to propagate over two edges.

It is important that the barrier has sufficient sound insulation, so that sound is not transmitted through the barrier. This typically requires a mass per unit area of more than 10 kg/m^2 for barriers of the type common in open plan offices. It is also necessary to consider the sound reflection from the ceiling above the barrier.

7.9 DIFFUSERS

A common problem in room acoustics is to diffuse the sound reflection by planar surfaces. In most rooms for music and speech, a suitable amount of diffuse reflection will remove unwanted behavior such as flutter echo and improve the smoothness of the audio. A *diffuser* is designed to primarily scatter and diffuse sound so that the reflection is not planar. For the diffuser to work properly, it needs surface irregularities a quarter wavelength or larger at the frequency of interest, but even small surface unevenness is advantageous. Niches 0.3 to 0.4 m deep are usually sufficient for good results. The irregularities should be randomized for optimum effect, but this is often not possible for architectural reasons. Figure 7.21 shows an example of a commercially available diffuser for use in recording studios and home theaters.

Efficient scattering can be obtained using surfaces having specially designed depth profiles that have low spatial autocorrelation. An example of such a depth profile is shown in Figure 7.22. These diffusers operate on the principle that the phase of the reflected wave from the surface should be random for diffuse reflection (see Reference 7.9). Assume a plane wave is incident at an angle ϑ on a plane surface consisting of strips that generate phase jumps because of the distance the incoming wave has to travel down and up through the slots. If all slots reflect identically, one obtains the pressure at far distance:

$$\underline{p}(k,d,\vartheta) \propto \sum_n R_n e^{jkd\sin(\vartheta)} \tag{7.29}$$

Figure 7.21 An example of a commercially available two-dimensional diffuser for use in recording studios and home theaters. (Photo by Mendel Kleiner.)

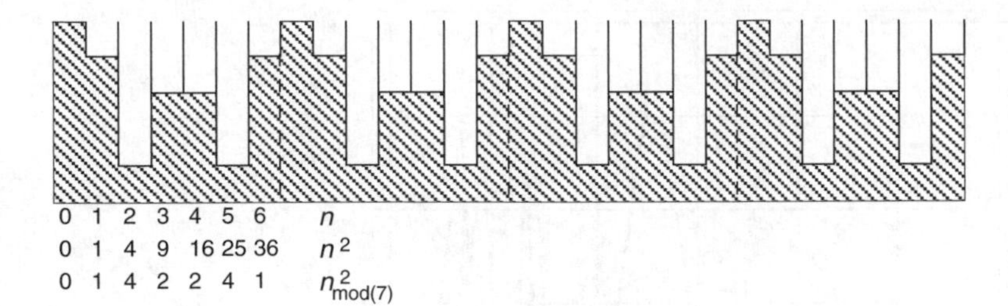

Figure 7.22 An example of the depth profile of a sound diffuser based on the use of a quadratic residue sequence techniques (period = 7 slots). Note that the slit dividers are necessary for proper function.

where n indicates the strip number, R_n the reflection coefficient of the strip, and the slot is d wide. If there are different phase jumps ψ_n at each respective slit, one instead finds:

$$\underline{p}(k,d,\vartheta) \propto \sum_n R_n e^{jkd\sin(\vartheta)+\psi_n} \tag{7.30}$$

Assume that the phase jumps are generated by slits having depths described by a quadratic residue sequence (QRD). A key property of QRD sequences is that they, over the length of the sequence, behave as white noise, that is, as a surface that has random *height*. This means that the phase jumps will be random as the slits reflect the incoming wave. The phase jumps will have the form:

$$h_n \propto \psi_n \tag{7.31}$$

where

$$\psi_n = 2\pi \frac{n^2}{p} \tag{7.32}$$

where $n = 2$ and p is a prime. Since the phase repeats at multiples of 2π, one can replace s^n by the remainder s^n that is obtained from:

$$s_n = n^2_{\text{mod}(p)} \tag{7.33}$$

Using $p = 7$ and $n = 2$, one obtains the sequence $s_n = 0, 1, 4, 2, 2, 4, 1$ after which the sequence repeats periodically.

Since the wave traveling down the slit reflects at the bottom of the slit with a reflection coefficient of 1, the wave returns to the top with a total phase change $2kh_n$. The slit depth turns out to be:

$$h_n = s_n \frac{\lambda_l}{4\pi} = \pi \frac{s_n}{pk_l} \tag{7.34}$$

where k_l is the wave number at the lowest frequency. The lowest nominal working frequency of such a diffuser is the frequency at which the maximum slot depth is approximately $\lambda/4$. The upper operating frequency f_u is given by:

$$f_u = \frac{f_l}{p-1} \tag{7.35}$$

The slots should have widths smaller than $\lambda/2$ at the highest operating frequency, and the slot walls should be thin. It is clear that these diffusers are primarily mid-frequency range devices for practical manufacturing reasons. When sound is incident parallel to the slot walls there will be little scattering; it is only when sound is incident at right angles to the slot that the diffuser works optimally. Also note that the theory presented above assumes incident plane waves.

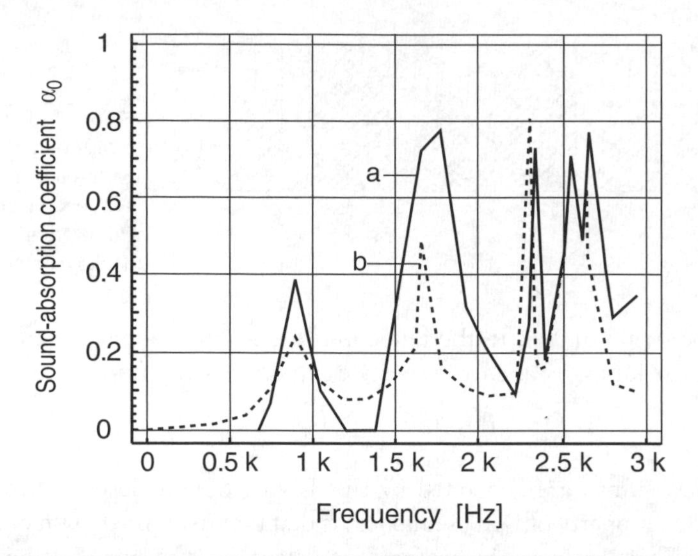

Figure 7.23 Sound-absorption coefficients of a one-dimensional diffuser similar in design to that shown in Figure 7.21. Note the use of a linear frequency axis: (a) measurement and (b) prediction. (Adapted from Ref. 7.10)

While the mathematically defined diffusers discussed so far have had one-dimensional shape variation, they can be calculated for two-dimensional depth shapes as well. Improved properties such as reduced sound absorption can be obtained by low-pass filtering of the surface shape.

Since diffusers of this type are based on acoustic resonance, the slits are similar to quarter-wave resonators. It follows that there is likely to be considerable sound absorption, due to not only the losses at the bottom and along the well walls, but also because of the high particle velocities immediately above the wells, between one well and another. Another problem in the use of these types of diffusers is the tendency of users to set up arrays of diffusers, leading to coloration because of periodicity in the reflected sound. For these reasons, it is much better to use truly randomized diffusers consisting, for example, of an assortment of spherical scatterers of different diameters than to use arrays of identical hemicylindrical or QRD-type diffusers. As the data in Figure 7.23 shows, resonant diffusers similar to the one in Figure 7.22 are characterized by peaky, uneven sound absorption.

Other types of efficient scattering can be obtained using other depth profiles having low spatial auto-correlation (see Reference 7.9). Primitive root based diffusors for example reject specular reflection.

Free objects in the room will act as "volume" diffusers. Their scattering properties will depend on their impedance and their size relative to the wavelength. Orchestra members form effective scatters and barriers on the stage at medium and high frequencies.

7.10 ABSORPTION DATA EXAMPLES

Indicative examples of typical sound-absorption properties of many materials and constructions may be found in Table 7.1. Table 7.2 shows examples of measured data for the absorption area for individuals (see Reference 7.8). References 7.13 and 7.4 contain much data.

Table 7.1 Examples of typical measured sound-absorption coefficients for various materials and constructions

	Octave band center frequencies					
	125 Hz	250 Hz	500 Hz	1 kHz	2 kHz	4 kHz
Walls						
Concrete block, grout-filled, and unpainted	0.01	0.02	0.03	0.04	0.05	0.07
Concrete block, grout-filled, and painted	0.01	0.01	0.02	0.02	0.02	0.03
Poured concrete, unpainted	0.01	0.02	0.04	0.06	0.08	0.10
Normal, unpainted concrete block	0.36	0.44	0.31	0.29	0.34	0.21
Painted concrete block	0.10	0.05	0.07	0.09	0.08	0.05
6 mm heavy plate glass	0.18	0.06	0.04	0.03	0.02	0.02
2.4 mm regular window glass	0.36	0.25	0.18	0.12	0.07	0.04
12 mm plasterboard, wood studs spaced 0.4 m	0.29	0.10	0.05	0.04	0.12	0.09
Plasterboard, 1.6 × 0.8 panels, 62 mm apart, studs 0.4 m (voids filled with glass fiber)	0.55	0.14	0.08	0.04	0.10	0.10
Same, but panels 128 mm apart	0.28	0.12	0.10	0.07	0.10	0.10
Marble and glazed brick	0.01	0.01	0.01	0.01	0.02	0.02
Painted plaster on concrete	0.01	0.01	0.01	0.03	0.04	0.05
Plaster on concrete block or 25 mm on metal studs	0.12	0.09	0.07	0.05	0.05	0.04
Plaster 16 mm on metal studs	0.14	0.10	0.06	0.05	0.04	0.03
6 mm wood over air space	0.42	0.21	0.10	0.08	0.06	0.06
9–10 mm wood over air space	0.28	0.22	0.17	0.09	0.08	0.06
25 mm wood over air space	0.19	0.14	0.09	0.06	0.06	0.05
Venetian blinds, open	0.06	0.05	0.07	0.15	0.13	0.17
Velour drapes, 280 g/m^2	0.03	0.04	0.11	0.17	0.24	0.35
Fabric curtains, 400 /m^2, hung 1/2 length	0.07	0.31	0.49	0.75	0.70	0.60
Same, but 720 g/m^2	0.14	0.68	0.35	0.83	0.49	0.76
Shredded wood panels, 50 mm, on concrete	0.15	0.26	0.62	0.94	0.64	0.92
12 mm wood, holes 5 mm o.c., 11% open, over void	0.37	0.41	0.63	0.85	0.96	0.92
Same with 50 mm glass fiber in void	0.40	0.90	0.80	0.50	0.40	0.30

Continues

Table 7.1 *Continued*

	Octave band center frequencies					
	125 Hz	**250 Hz**	**500 Hz**	**1 kHz**	**2 kHz**	**4 kHz**
Floors						
Terrazo	0.01	0.01	0.02	0.02	0.02	0.02
Glazed marble	0.01	0.01	0.01	0.01	0.02	0.02
Linoleum, vinyl, or neoprene tile on concrete	0.02	0.03	0.03	0.03	0.03	0.02
Wood on joists	0.10	0.07	0.06	0.06	0.06	0.06
Wood on concrete	0.04	0.07	0.06	0.06	0.06	0.05
Carpet on concrete	0.02	0.06	0.14	0.37	0.60	0.65
Carpet on expanded (sponge) neoprene	0.08	0.24	0.57	0.69	0.71	0.73
Carpet on felt under pad	0.08	0.27	0.39	0.34	0.48	0.63
Indoor–outdoor thin carpet	0.10	0.05	0.10	0.20	0.45	0.65
Ceilings (also check floors and walls for similar material)						
Suspended 12 mm plasterboard	0.29	0.10	0.05	0.04	0.07	0.09
Same with steel suspension system frame	0.15	0.10	0.05	0.04	0.07	0.09
Plaster on lath suspended with framing	0.14	0.10	0.06	0.05	0.04	0.03
Glass fiber acoustic tile suspended	0.76	0.93	0.83	0.99	0.99	0.94
Wood fiber acoustic tile suspended	0.59	0.51	0.53	0.71	0.88	0.74
Acoustic tile, 50 mm on drywall, suspended	0.08	0.29	0.75	0.98	0.93	0.76
Same, air space between tile and drywall	0.38	0.60	0.78	0.80	0.70	0.75
Glass fiber, 360 g/m^2, suspended	0.65	0.71	0.82	0.86	0.76	0.62
Same, but 1120 g/m^2	0.38	0.23	0.17	0.15	0.09	0.06
Luminous ceiling, typical with openings	0.07	0.11	0.20	0.32	0.60	0.85
Hanging absorbers, 450 mm, 450 mm on centers	0.07	0.20	0.40	0.52	0.60	0.67
Same, but 159 mm on centers	0.10	0.29	0.62	0.72	0.93	0.98
Seats and people (area basis, not per person)						
Unoccupied seats, light upholstery, 20 mm	0.36	0.47	0.57	0.62	0.62	0.60
Same, occupied	0.51	0.64	0.75	0.80	0.82	0.83

Unoccupied seats, medium upholstery, 50–100 mm	0.54	0.62	0.68	0.70	0.68	0.66
Same, occupied	0.62	0.72	0.8	0.83	0.84	0.85
Unoccupied seats, heavy upholstery, 200 mm+	0.70	0.76	0.81	0.84	0.84	0.81
Same, occupied	0.72	0.80	0.86	0.89	0.90	0.90
Wood benches and pews, occupied	0.57	0.61	0.75	0.86	0.90	0.86
Same, unoccupied	0.15	0.19	0.22	0.39	0.38	0.30
Students in tablet-arm seats	0.30	0.41	0.49	0.84	0.87	0.84
Special material						
Organ at Boston Symphony Hall, total sabins measured by covering chamber opening with heavy, hard, sound-reflecting material and comparing reverberation times	0.11	0.11	0.15	0.11	0.26	0.41
Organ at Tokyo Opera City, total sabins measured by comparing reverberation times before and after free-standing organ installation	0.65	0.44	0.35	0.33	0.32	0.31
Surfaces outdoors						
Lawn grass, 50 mm high	0.11	0.26	0.60	0.69	0.92	0.99
Fresh snow	0.45	0.75	0.90	0.95	0.95	0.95
Plowed soil	0.15	0.25	0.40	0.55	0.60	0.60
Spruce trees 2.7 meters tall	0.03	0.06	0.11	0.17	0.27	0.31
Water in a swimming pool	0.01	0.01	0.01	0.02	0.02	0.03
Openings						
Open entrances off lobbies and corridors: 0.50–1.00						
Air supply and return grilles: 0.15–0.50						
Stage-house proscenium opening: 0.25–0.75						
Residual-absorption coefficient	0.14	0.12	0.10	0.09	0.08	0.07

Table 7.2 Sound absorption of individuals (From Ref. 7.13)

All values are in metric sabins	Octave band center frequencies					
	125 Hz	250 Hz	500 Hz	1 kHz	2 kHz	4 kHz
Individuals						
Man in suit, standing	0.15	0.25	0.60	0.95	1.15	1.15
Man in suit, sitting	0.15	0.25	0.55	0.80	0.90	0.90
Woman in summer dress, standing	0.05	0.10	0. 25	0.40	0.60	0.75
Woman in summer dress, sitting	0.05	0.10	0.10	0.35	0.45	0.60

7.11 PROBLEMS

7.1 A commercial porous absorber (mineral wool) has been measured using the transfer function method to have a normalized sound field impedance $\underline{Z}_2/Z_0$, as shown in the table, for perpendicular sound incidence. Other measurements using the standing wave method have given the α_r values shown in the table.

f [Hz]	$\underline{Z}_2/Z_0$	α_r
100	3.7-j 8.5	0.17
125	2.8-j 7.0	0.20
160	2.9-j 5.5	0.28
200	3.3-j 4.6	0.33
250	2.6-j 3.3	0.44
315	1.7-j 2.6	0.49

Task: Determine the sound-absorption coefficient from normalized sound field impedance and compare these values to the measured α_r values.

7.2 A perfectly absorbing porous absorber is covered by a thin plastic film having a mass per unit area of 0.18 kg/m^2 that is mounted immediately in front of the absorber. The plastic film does not touch the absorber.

Task: Calculate the resulting sound-absorption coefficient at 1 kHz.

Hint: The total impedance of the foil and absorber is obtained as the sum of their respective impedance components (mass and resistance).

7.3 A porous sound-absorbing panel has the absorption coefficient 0.8 at a frequency of 1 kHz. For hygienic reasons, it is desired to cover the panel with a thin plastic foil. This will result in a reduced sound-absorption coefficient due to sound reflection. The transmission loss R_0 of the foil is:

$$R_0 = 20\log\left|1+\frac{j\omega m''}{2Z_0}\right|$$

Task: Calculate the maximum acceptable mass per unit area of the foil that will not result in an absorption coefficient α less than 0.5. Assume perpendicular sound incidence.

Hint: Only consider the direct reflection and the first transmitted reflection from inside the panel.

7.4 A sound absorber having significant absorption at 100 Hz is desired. The available space (depth) is limited, so the maximum thickness of the absorber must be 0.05 m or less. Materials at your disposal are battens of various thicknesses (2, 5, or 10 cm), 0.009 m gypsum sheets having a mass/unit area of 10 kg/m², 0.009 m perforated gypsum sheets with holes of 0.004 m diameter at 0.10 m spacing, and mineral wool sheets having a density of 50 kg/m³ and various thicknesses.

Task: Which types of absorbers can be designed with these materials? Which one is to be preferred? Explain your choice.

7.5 *Task*: Match the sound-absorption curves in the graph with the absorbers listed. Assume the absorbers to be mounted in front of a hard wall. Explain your answer.

a. 0.050 m thick mineral wool board
b. 0.025 m thick mineral wool board
c. 0.025 m thick mineral wool board immediately behind a perforated tile having a relative hole area of 20%.
d. As in Case c but a relative hole area of 2.7%.
e. As in Case d but with a 0.14 m thick air space between the perforated panel and the mineral wool board that is mounted on the wall.

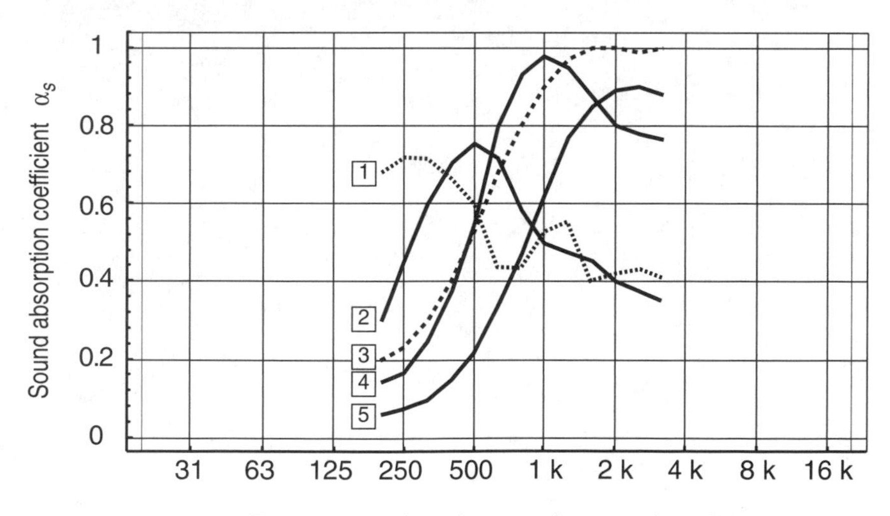

7.6 On measuring the sound-absorption coefficient for a sound absorber made using air-bricks, sound-absorption maxima were obtained at the following frequencies; 315 Hz, 2.5 kHz and 5 kHz. The brick holes were facing the room, with a 0.05 m air space in front of the wall. Each brick is 0.26 m wide, 0.13 m high and 0.065 m thick, and has 78 holes, each having a diameter of 0.01 m.

Task: Try to find the sound-absorption mechanisms in the construction that can explain the measured sound-absorption coefficient values.

Waves in Solids 8

8.1 INTRODUCTION

Many acoustic problems involve wave propagation in solid media and structures. Examples of such problems are resonant vibration in loudspeaker diaphragms, sound transmission through walls and cabinets, and mounting of electronics in cars and other vibrating objects. An example of a resonant high frequency vibrational pattern of a poor loudspeaker diaphragm is shown in Figure 8.1.

In contrast to the case of sound propagation in gases, sound propagation in solids can involve many types of waves. Some of the most important and common wave types will be discussed in this chapter. A characteristic of sound transmission in solid structures is conversion between wave types at boundaries and edges. The wave conversion also distributes energy between different modal systems.

Gases oppose volume changes, resulting in longitudinal waves as a response to a disturbance in a gas volume. Solid media not only exhibit elasticity but also shear. This makes it possible for transverse waves and many other types of waves to exist in solid media. Infinite solid media can carry both longitudinal and transverse waves. Finite solid media can also carry many other wave types, particularly *bending, transverse, torsional, quasilongitudinal,* and various types of *surface waves*. Torsional waves, for example, were often used in spring reverberator units such as the one shown in Figure 8.2.

8.2 WAVE TYPES IN INFINITE MEDIA

Longitudinal Waves

The *longitudinal wave* is the basic form of wave motion in many materials. A longitudinal wave in air is characterized by sound pressure and by particle motion in the propagation direction of the wave. Analogously, due to the symmetry, in the case of solids the longitudinal wave is characterized by tension as well as by particle motion in the direction of the wave. The motional pattern for longitudinal waves is shown in Figure 8.3.

The wave equation for one-dimensional longitudinal waves in solids is:

$$\frac{\partial^2}{\partial x^2}\left(\underline{\sigma}_x, \underline{u}_x\right) + k_L^2\left(\underline{\sigma}_x, \underline{u}_x\right) = 0 \tag{8.1}$$

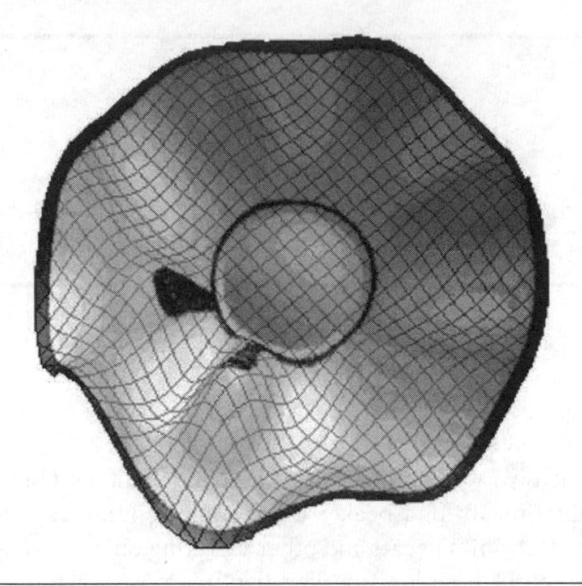

Figure 8.1 Deformation pattern of a conical loudspeaker diaphragm, showing "breakup" at one of the diaphragm's bending wave resonances. Yamaha NS10 loudspeaker paper cone diaphragm at 1050 Hz. (Photo by Mendel Kleiner.)

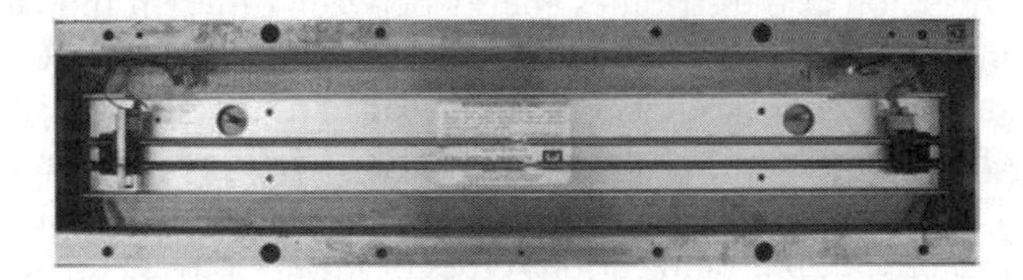

Figure 8.2 Reverberation units such as this one may use the low propagation velocity of longitudinal and torsional waves in metal spring coils to generate the desired delays and reflection patterns needed to synthesize reverberation for guitar amplifier applications. (Photo by Mendel Kleiner.)

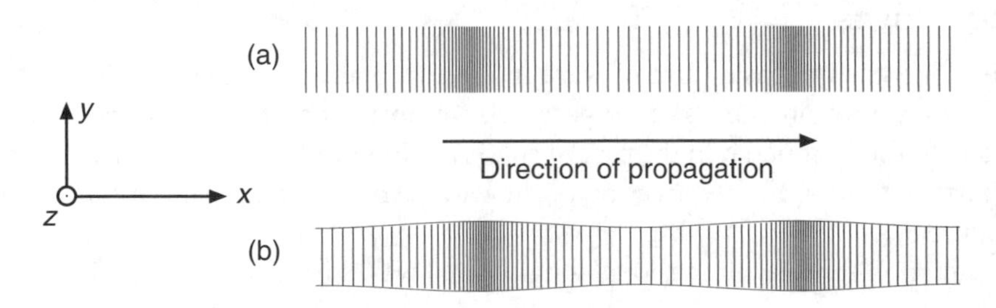

Figure 8.3 An example of the instantaneous deformation pattern (a) of a longitudinal wave, and (b) of a quasilongitudinal wave in a plate.

where $\underline{u}_x$ is the particle velocity in the x-direction, $\underline{\sigma}_x$ the tension in the x-direction, and k_L the wave number ($k_L = \omega/c_L$). The wave propagates with a velocity c_L [m/s], which depends on the elastic constants of the solid:

$$c_L = \sqrt{\frac{D}{\rho}} \tag{8.2}$$

where D is a constant representing the *longitudinal stiffness* of the material (expressed in units of Pascal) and ρ its density.

One notes that the one-dimensional longitudinal wave equation for solids has the same form as the wave equation for gases. This means that all waves propagate with the same speed; such propagation is said to be nondispersive. The longitudinal velocity is the fastest propagation velocity for waves in solids.

Young's modulus, E [Pa], is related to the longitudinal stiffness, D, in the following way:

$$E = D\left(1 - \frac{2\upsilon^2}{1-\upsilon}\right) \tag{8.3}$$

where υ is *Poisson's ratio* for the material. Poisson's ratio is a measure of the cross-contraction of a material as it is elongated. Poisson's ratio is sometimes called the cross-contraction number and must be below 0.5. For many materials, υ is found to be approximately 0.3, resulting in $D \approx 1.35\ E$.

Ideal longitudinal waves can exist only in infinite solids. In reality, because of cross-contraction, one usually finds quasilongitudinal waves, which will be discussed later.

Transverse Waves

The *transverse wave* is characterized by shear forces and particle velocity perpendicular to the propagation direction of the wave. The motional pattern of transverse waves is shown in Figure 8.4.

The wave equation for transverse waves is written in the following way for the case of wave propagation in the x-direction and particle velocity in the y-direction:

$$\frac{\partial^2}{\partial x^2}\left(\underline{\tau}_{xy}, \underline{u}_y\right) + k_T^2\left(\underline{\tau}_{xy}, \underline{u}_y\right) = 0 \tag{8.4}$$

where $\underline{u}_y$ is the particle velocity in the y-direction, $\underline{\tau}_{xy}$ the shear force, and k_T the wave number ($k_T = \omega/c_T$). The wave propagates with a velocity c_T [m/s], which depends on the elastic constants of the solid as:

$$c_T = \sqrt{\frac{G}{\rho}} \tag{8.5}$$

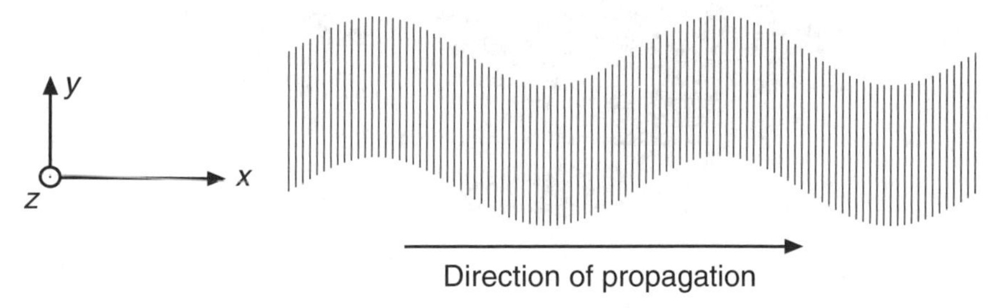

Figure 8.4 An example of the instantaneous deformation pattern of a transverse wave.

where G is the *shear modulus* and ρ the density of the solid. The shear modulus is given by:

$$G = \frac{E}{2(1+\upsilon)} \tag{8.6}$$

Assuming a Poisson's ratio of 0.3, one obtains the ratio between the transverse and longitudinal velocities as $c_T/c_L \approx 0.5$.

Ideal transverse waves can only be found in infinite bodies. In a limited material, such as a large sheet, there is some cross-contraction. This results in quasitransverse waves in the plane of the sheet; such wave motion has limited radiation capability because of the small particle velocity perpendicular to the plane of the sheet.

8.3 WAVE TYPES IN MEDIA OF LIMITED EXTENSION

Quasilongitudinal Waves in Sheets

The cross-contraction phenomenon of solids, characterized by Poisson's ratio, makes it impossible to have ideal longitudinal waves. It is possible though to have nearly longitudinal waves inside a material with an extension that is large compared to the wavelength. Due to the cross-contraction in a thin sheet, the longitudinal wave will be replaced by a quasilongitudinal wave with the motional pattern shown in Figure 8.3.

The quasilongitudinal wave will move at reduced speed because of the cross-contraction. The velocity c_{QL} of a quasilongitudinal wave in a sheet (in the plane of the sheet) is given by an equation similar to Equation 8.1 but with c_L and D replaced by c_{QL} and E_Q where:

$$c_{QL} = \sqrt{\frac{E_Q}{\rho}} \tag{8.7}$$

and

$$E_Q = \frac{E}{1-\upsilon^2} \tag{8.8}$$

Figure 8.5 Loudspeakers using pyramidal, homogeneous plastic foam diaphragms such as this KEF B139 will tend to have their frequency response affected by quasilongitudinal wave resonances in the diaphragm. (Photo by Mendel Kleiner.)

The wave impedance of such a quasilongitudinal wave is given by:

$$Z_{QL} = \sqrt{E_Q \rho} = \rho c_{QL} \tag{8.9}$$

Note also that the quasilongitudinal wave is characterized by poor sound radiation capability since the out-of-plane movement resulting from the cross-contraction is small.

Membrane Vibration

A *membrane* is a thin sheet in which the restoring force due to stiffness is small compared to that by membrane tension. The tension is produced by in-plane forces at the membrane edges. Typical examples of membranes in audio are those of condenser microphones and some electrostatic loudspeakers. Because membranes are often coupled to the air in cavities, the resonance frequencies of their modes will be different from free membranes. For example, this applies to the membrane of condenser microphones (see Reference 8.3). Membrane vibration in electrostatic loudspeakers is usually well damped by the radiation load of air except for the lowest frequency modes, as shown by Figure 14.25.

Membranes are often avoided in loudspeaker engineering as the preferred technology is for non-resonant pistonic motion in transducer radiating surfaces. In microphones, however, the small size, simple manufacturing, and relatively low impedance of membranes are attractive from an engineering viewpoint.

Transverse Waves in Sheets

Two types of *transverse waves* can occur in sheets: *in-plane waves* and *out-of-plane waves*. The sound radiation by in-plane waves is usually not a problem from the viewpoint of noise because the cross-contraction is small.

The out-of-plane waves can be excited using a sound field, vibrator, or other force or pressure generator. Their radiation depends on the mode index. Transverse waves are important in sandwich-type materials where a honeycomb core is bounded by thin foils such as the one shown in Figure 8.6 (see Reference 8.4). The transverse waves in sandwich materials have a velocity that is determined by the sheet materials and the core. The phase velocity of the waves will vary with wave type and frequency as shown in Figure 8.7.

The mode index, sheet geometry, and the speed of the transverse waves will determine the modal pattern on the sheet. If the distance between the antinodes (the points where the amplitude of the standing wave pattern is a maximum) is small compared to wavelength, the radiation from the antinodes will cancel out, as from dipoles and quadrupoles (described in Chapter 2). Each mode will have its own particular radiation pattern and radiation efficiency (see Reference 8.1).

Bending Waves in Sheets

Waves in solids are of particular acoustical interest since, under certain conditions, they may couple well to the surrounding air; that is, the wave motion in the solid can radiate energy easily.

The most important of such waves in sheets is the *bending wave*, in which motion is due to interaction between tension, shear, and mass. Figure 8.8 shows the motional pattern of a bending wave. The bending wave field variables are: rotation, transverse particle velocity, bending moment, and shear forces. The rotation is around the z axis in the figure.

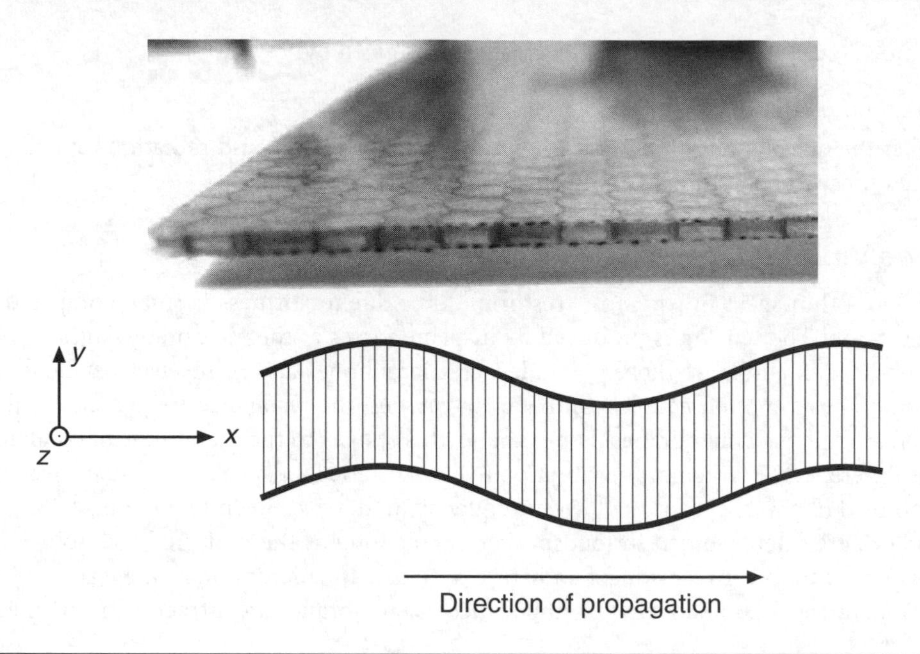

Figure 8.6 An example of a thin sandwich sheet and of an instantaneous deformation pattern of a shear wave in such a sheet. (Photo by Mendel Kleiner.)

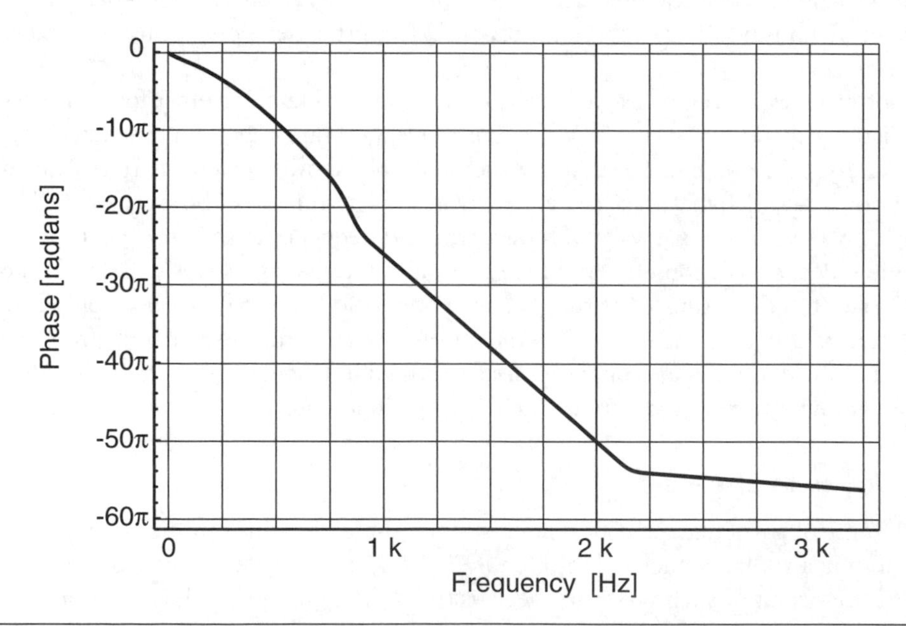

Figure 8.7 Measured phase delay over a 3 m long sandwich beam. Curve shows unfolded phase in radians as a function of frequency (after Ref. 8.5). Three different wave types are present, having different wave propagation speeds.

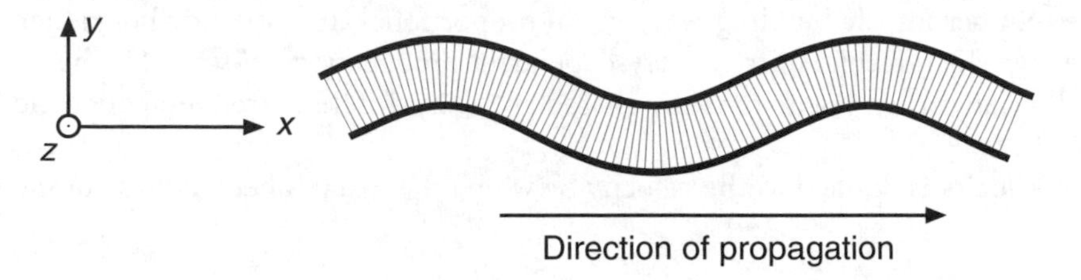

Figure 8.8 An example of an instantaneous deformation pattern of a bending wave in a thin sheet.

Because of the four bending field variables, the bending wave equation will have a somewhat different form from the wave equations studied earlier. Expressed for particle velocity u_y at right angles to the sheet, the equation is written:

$$\frac{\partial^4 \underline{u}_y}{\partial x^4} - \frac{\omega^2 m''}{B'} \underline{u}_y = 0 \tag{8.10}$$

where B' is the *bending stiffness per unit length* [Nm] and m'' the mass per unit area of the sheet. The bending stiffness B' for a thin sheet is given by:

$$B' = \frac{Eh^3}{12\left(1 - \upsilon^2\right)} \tag{8.11}$$

where E is the Young's modulus and h the thickness of the sheet.

The sheet can be considered thin as long as its thickness h is much smaller than the bending wave's wavelength λ_B, expressed by the approximate condition:

$$\lambda_B \geq 6h \tag{8.12}$$

One notes that the wave equation of the bending wave is quite different from previously studied wave equations, in that it is a fourth order equation. The more complex wave motion also results in quite different wave propagation velocity behavior. The solution to a wave equation—in this case expressed for particle velocity at right angles to the sheet—is usually assumed to be of the form:

$$\underline{u}_y = Ae^{\pm jk_B x} \tag{8.13}$$

where k_B is the wave number for the bending wave. Inserting this solution into Equation 8.10 gives the wave number as:

$$k_B = \sqrt[4]{\frac{\omega^2 m''}{B'}} \tag{8.14}$$

However, because of the fourth order wave equation, one finds that solutions of the type:

$$\underline{u}_y = Ae^{\pm k_B x} \tag{8.15}$$

are also valid solutions to the bending wave equation. These solutions are the mathematical formulation of nonpropagating waves. One finds such waves close to discontinuities—such as driving points, edges, and so on. Both solutions give the same expression for the wave number.

The wave number for the bending wave is not proportional to frequency but rather to the square root of frequency, which results in frequency-dependent propagation velocity. The bending wave is an example of a *dispersive wave*. Such waves are characterized by different, frequency-dependent *phase* and *group velocities*.

The phase velocity is defined as the velocity by which the instantaneous phase of the wave motion propagates:

$$c_B = \frac{\omega}{k_B} \tag{8.16}$$

The group velocity, however, is defined as the velocity by which the envelope of the wave, which "carries" the energy of the wave, propagates—and is written c_{gB}. The group velocity is given by:

$$c_{gB} = \frac{1}{dk_B / d\omega} \tag{8.17}$$

Using these definitions, one obtains the group and phase velocities for the bending wave in a sheet as:

$$c_B = \frac{\omega}{k_B} = \sqrt[4]{\frac{\omega^2 B'}{m''}} \tag{8.18}$$

The group velocity is:

$$c_{gB} = \frac{1}{dk_B / d\omega} = 2c_B \tag{8.19}$$

Note that both of these velocities are much smaller than the quasilongitudinal wave speed for the sheet.

One should also note that since the velocities increase proportionally to the square root of ω, one cannot use traditional correlation and impulse measurement methods to determine the propagation paths of bending waves.

It is instructive to draw the relationship described by Equation 8.18 in a log-log diagram, as shown in Figure 8.9, together with the line describing the wave number as a function of frequency for sound in air. The bending wave phase velocity is equal to that of sound waves in the surrounding fluid at the so-called *critical frequency*, f_c. The critical frequency is sometimes called the *coincidence frequency*.

A bending wave having a frequency below f_c is called a slow wave, whereas a bending wave having a frequency above f_c is called a fast wave.

Using the condition $c_B = c$, one finds the expression for the critical frequency as:

$$f_c = \frac{c_0^2}{2\pi}\sqrt{\frac{m''}{B'}} \tag{8.20}$$

The bending stiffness is decisive for f_c—the stiffer the sheet, the lower the critical frequency. When using material data, it is convenient to use the *coincidence number* K_c [m/s], defined as:

$$K_c = hf_c \tag{8.21}$$

Table 8.1 shows the coincidence number for some materials that are commonly needed. Typically, the coincidence number is in the 10 to 40 range. It is noteworthy that, for common construction material thicknesses, the critical frequency of the bending waves falls inside the audio frequency range. This also applies to typical loudspeaker and microphone construction materials, such as various plastics and papers.

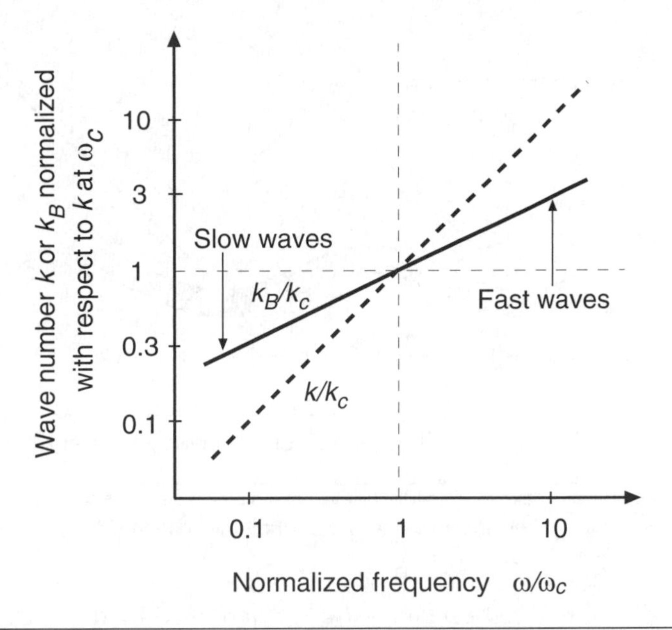

Figure 8.9 Wave number characteristics as functions of frequency. At the critical frequency $\omega = \omega_c$ and the wave numbers are $k = k_B = k_c$. Note: k_B is the wave number for bending waves in a sheet and k is the wave number for sound waves in air.

Bending waves can be excited along a line or at a point on the sheet. Bending waves that propagate at a speed given by the bending wave equation (Equation 8.10) are called *free bending waves*. Free bending waves propagate with the phase velocity, c_B. The wavelength of a free bending wave is consequently determined by the coincidence number of the material and the thickness of the sheet. Such waves are commonly generated by mechanical excitation—for example, on a shop floor. Far away from the point of excitation, circular bending waves will also resemble linear bending waves.

Since the sheet is compliant (because it can bend), sheet motion can be excited by an incident sound wave, as shown in Figure 8.10. This is what commonly happens when a sound wave is incident on a wall in a building. The figure shows the instantaneous extension of a sheet on which a sound wave is incident at an angle φ. The sound wave forces its pressure pattern onto the sheet.

Table 8.1 The coincidence number and the critical frequency for some common construction materials and typical thicknesses.

Material	Coincidence number	Typical thickness	Critical frequency
Concrete	18 m/s	$1.6 \cdot 10^{-1}$ m	0.1 kHz
Lightweight concrete	36 m/s	$7 \cdot 10^{-2}$ m	0.5 kHz
Gypsum	32 m/s	$1 \cdot 10^{-2}$ m	2.5 kHz
Steel	12 m/s	$1 \cdot 10^{-3}$ m	11.7 kHz
Glass	14 m/s	$3 \cdot 10^{-3}$ m	3.5 kHz

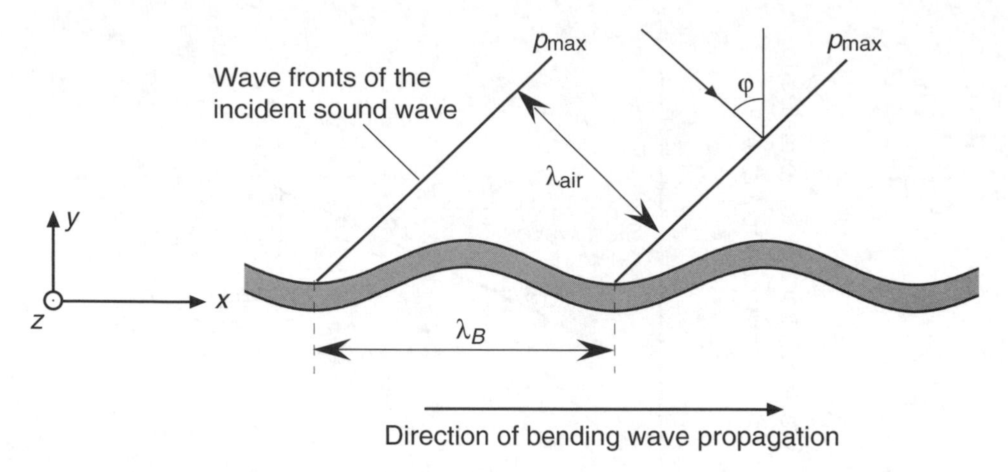

Figure 8.10 The case of sound waves exciting bending waves in a sheet. The wavelength of the bending wave is determined by the condition $\lambda_B \sin(\varphi) = \lambda_{air}$. The sheet radiates waves from the second side (not shown).

The wavelength of such a *forced bending wave* is determined by the angle of incidence and the wavelength of sound in the fluid. When the forced wave impinges on a discontinuity, such as an edge, there will be a wave conversion and free bending waves will be excited together with nonpropagating motion of the type $e^{\pm k_B x}$.

Wave conversion will take place when a wave (bending, quasitransverse, quasilongitudinal wave, or whatever) in a sheet is incident on a discontinuity. A sheet will typically have bending, quasitransverse, and quasilongitudinal waves present at the same time, if there is vibrational energy in the sheet. The vibrational energy is distributed between the modes of the various wave subsystems, which can be regarded as *coupled rooms*. This makes it difficult to measure the quasilongitudinal and quasitransverse waves, since their wave motion (out of the sheet plane) is small compared to that of the bending waves.

Since bending waves are typical for limp sheets, their coupling to waves in air is of considerable interest. The coupling is important from both the viewpoint of reception and radiation. The waves also have a low *mechanical input impedance*. The mechanical input impedance $\underline{Z}_{MB}$, when exciting an infinite sheet at a point, is:

$$\underline{Z}_{MB} = 8\sqrt{m''B'} \tag{8.22}$$

For a finite sheet, the input impedance will oscillate between mass and compliance behavior, just as the acoustic input impedance into a tube or a room. Because of losses, the mechanical input impedance of a sheet will tend toward the value given by expression 8.22 even for a finite sheet, at least for high frequencies. Taking frequency averages over suitably wide bands will further increase this tendency.

The *modal densities* of the wave fields in a limited sheet will also be affected by the different phase velocities of the wave types. The different wave types will each set up their own modal system. These will be coupled due to edge effects, etc. The modal density is low, however, because of the high propagation velocities, since the waves only propagate in a two-dimensional space. The quasilongitudinal and quasitransverse modal systems will have modal densities that are proportional to frequency, because of the two-dimensional vibrational systems.

The bending wave modal density turns out to be constant, irrespective of frequency, because of the frequency dependent phase velocity of the bending waves. The bending wave resonance frequencies of a rectangular sheet, freely suspended at its edges, are given by:

$$f_{q_x,q_y} = \frac{\pi}{2}\sqrt{\frac{B'}{m''}}\left(\left(\frac{q_x}{l_x}\right)^2 + \left(\frac{q_y}{l_y}\right)^2\right) \tag{8.23}$$

where l_x and l_y are the lengths of the sides of the sheet and q_x and q_y are natural numbers (that is, 0, 1, 2, 3, . . .).

At high frequencies, the average bending wave mode density $\Delta N/\Delta f$ in a sheet having a surface area S is given by:

$$\frac{\Delta N}{\Delta f} \approx \frac{S}{2}\sqrt{\frac{m''}{B'}} \tag{8.24}$$

A simple calculation using the data in Table 8.1 shows that a 0.15 m thick and 20 m^2 large concrete slab has a distance between modes of approximately 15 Hz. In comparison, a 0.001 m thick and 2 m^2 large steel sheet—similar to the one used for the reverberator shown in Figure 8.11—has a distance between modes of approximately 6 Hz.

Good averaging over modes in acoustic and vibration measurement requires diffuse sound fields, typically at least 10 modes present in the frequency range of interest. For building acoustics measurements (125 Hz to 4 kHz) that typically are accomplished using octave band frequency resolution, the modal density will usually be satisfactory except in the lowest octave bands.

The modal density also influences the way that the reverberation in a sheet sounds. This gave the reverberation sheets that were used for simulation of room reverberation, their special sound quality. The impulse response of such plates are sometimes implemented in modern digital signal processing units and software.

Figure 8.11 The discontinued EMT-140 plate reverberation unit was an advanced example of the use of the slow bending waves for simulation of room reverberation. The later EMT model 240 used gold foil to increase the mode density (photo courtesy of EMT Studiotechnik GmbH, Germany).

8.4 STRUCTURE-BORNE SOUND IN LOSSY SHEETS

All real sound propagation is affected by losses. In the case of sound propagation in solids, one usually regards the radiation of vibrational energy to the surroundings as energy loss. As in the case of sound waves in air, it is convenient to take the losses into account in calculating the propagation by including a complex wave number. Whereas sound wave losses are introduced *discretely* by sound absorbers at walls—internal losses being small—this is not practical in the case of wave motion in solids.

The approach is similar to that used when handling losses in the case of sound propagation in air in rooms. The wave motion in sheets is damped using some form of continuous damping along the surface or in a layer inside the sheet—similar to the wall sound absorption and the damping by air in the case of sound in rooms.

In the case of sound propagation in sheets, the losses are usually expressed by using a damping coefficient instead of the equivalent of the absorption area used in room acoustics. The loss factor is included by making the mechanical constants complex, for example, the Young's modulus.

The phase difference between extension $\underline{\varepsilon}$ and tension $\underline{\sigma}$ along a sheet is given by Hooke's law:

$$\underline{\sigma} = \underline{E}\underline{\varepsilon} \tag{8.25}$$

where $\underline{E}$ is the complex modulus of elasticity given by:

$$\underline{E} = E_0(1 + j\eta) \tag{8.26}$$

The loss factor η is typically in the range from 10^{-3} to 10^{-2} for many construction materials but can be increased to approximate unity by applying some kind of treatment to the sheet absorbing the energy of the wave fields.

Since the propagation velocities of quasilongitudinal and quasitransverse waves, c_L and c_T, are proportional to the square root of E, one finds:

$$\underline{c}^2 = c^2(1 + j\eta) \tag{8.27}$$

with c either being c_L or c_T. This gives the respective wave number as:

$$\underline{k}^2 = \frac{\omega^2}{\underline{c}^2} = k^2(1 - j\eta) \tag{8.28}$$

Provided that the loss factor η is much smaller than unity, one obtains the complex wave number for quasilongitudinal and quasitransverse waves as:

$$\underline{k} \approx k\left(1 - j\frac{\eta}{2}\right) \tag{8.29}$$

for $\eta \ll 1$. In analogy, since c_B is proportional to $E^{1/4}$, one finds the complex wave number for bending waves in sheets as:

$$\underline{k_B} = k_B(1 + j\eta)^{-1/4} \approx k_B\left(1 - j\frac{\eta}{4}\right) \tag{8.30}$$

The imaginary part of the wave number describes how quickly the wave decays along the propagation path. For quasilongitudinal and quasitransverse waves, one finds that the particle velocity, at right angles to the plane of the sheet, diminishes as:

$$\tilde{u}_y \propto e^{-k\eta x/2} = e^{-\pi\eta x/\lambda} \tag{8.31}$$

It is often practical to express the extra attenuation (on top of the geometrical attenuation) as a level difference per unit length, just as in the case of sound in air.

For quasilongitudinal and quasitransverse waves, the damping is given by:

$$\Delta L_{T,L} \approx \frac{27.2\eta}{\lambda_{T,L}} \quad [\text{dB/m}] \tag{8.32}$$

For free bending waves the damping is given by:

$$\Delta L_B \approx \frac{13.6\eta}{\lambda_B} \quad [\text{dB/m}] \tag{8.33}$$

Since bending waves typically have much shorter wavelengths than quasilongitudinal and quasitransverse waves, their damping is generally much larger per unit length.

Using various forms of damping layers, one can usually increase the losses considerably. Extra losses can also be added by air pumping, dry friction, and magnetic hysteresis. In an otherwise *undamped* sheet, such losses will be the dominant loss mechanisms.

8.5 DAMPING BY VISCOELASTIC LAYERS

As previously mentioned, it is often desirable to damp waves in sheets—metal sheets and other sheets, such as bending waves in building structures as well as the boxes and diaphragms of loudspeakers and other electroacoustic transducers—to reduce sound radiation and vibration propagation.

A simple way to increase the damping of bending waves is to add a layer of a *viscoelastic material* to the sheet, either externally or internally. Another way is to couple the sheet to another structure that has high damping.

A viscoelastic layer exhibits both viscosity and elasticity. Polymers have long molecular chains and allow many combinations of elastic and viscous properties through choice of chain lengths and forks. Chewing gum is a typical example of a viscoelastic material—other examples are most polymers, such as rubber and plastics. Usually the viscous properties are extremely temperature-dependent. The viscosity leads to energy conversion from vibration to heat.

The success of the application of viscoelastic damping depends on several factors: the temperature, the type of excitation, the wave types involved, the coupling to other structures, and the frequency of vibration (since this determines the wavelengths of the different waves). The addition of a viscoelastic layer also changes the mass per unit area and the bending stiffness per unit length. Consequently, it is not sufficient just to study the loss factor.

It is advantageous to use the viscoelastic layers as close to the source of vibration as possible because the application is only needed over a smaller area and because the bending waves usually dominate the wave type spectrum close to the source.

Finally, it must be emphasized that increasing the loss factor primarily affects the resonant vibration. The forced vibrations are usually not affected, except by the addition of mass or stiffness due to the layer.

Viscoelastic Materials

A feature of viscoelastic materials is that the loss factor is much higher, and the shear modulus much lower, than for typical construction materials. Most viscoelastic materials exhibit large temperature and frequency dependence in the Young's modulus, the shear modulus, and loss factor. An example of typical behavior is shown in Figure 8.12.

The shear modulus of a viscoelastic material has a real part $G_{ve:re}$ and an imaginary part $G_{ve:im}$. The parts have different frequency dependencies. The imaginary part of the shear modulus increases quite rapidly in a specific range as temperature is reduced or frequency increased. A rule of thumb is that a change in temperature from 5 to 7°C corresponds to a change in frequency by a factor of 10. The damping is fairly broadband.

Viscoelastic materials are generally not suited as construction materials. Because of their cost, they are typically used in thin layers, either on or in the sheet—possibly together with various distance layers.

Free Layers

By attaching one or more viscoelastic layers to the surface, or surfaces, of a sheet, one obtains both extension and bending in the *free layer*. Generally, one obtains the best use of the material if most of the total elastic energy is stored in the viscoelastic layer.

The loss factor for bending waves in a sheet with a single-sided free, soft viscoelastic layer is given by:

$$\eta_B = \eta_{ve} \frac{h_{ve} h_n^2 \operatorname{Re}[\underline{E}_{ve}]}{B'} + \eta_{pl} \tag{8.34}$$

where η_B is the combined total loss factor, η_{ve} loss factor of the viscoelastic layer, η_{pl} loss factor of the sheet, $\operatorname{Re}[\underline{E}_{ve}]$ real part of the complex Young's modulus of the viscoelastic layer, h_{pl} thickness of the sheet, h_{ve} thickness of the viscoelastic layer, h_n distance between the neutral planes of the two layers. For soft layers $h_n = (h_{ve} + h_{pl})/2$, and B' bending stiffness of the combined construction.

The equation shows that the viscoelastic layer should be thick and have as high a Young's modulus as possible. Often Young's modulus for viscoelastic layers is approximately 100 times smaller than that of the sheet on which it is applied. Free viscoelastic layers are therefore primarily used to damp thin sheets and sheet metal. One also observes that the loss factor of the combination has approximately the same frequency dependence as that of the viscoelastic layer.

Application of a free viscoelastic layer is often used as an after-treatment to resonant sheets. An example of such treatment to a loudspeaker diaphragm is shown in Figure 8.13. While free layers can be used in this way, their use is at a disadvantage because of the mass of the viscoelastic layer that needs to be thick. This limits the use of free viscoelastic layers for loudspeaker diaphragm applications since a loudspeaker's diaphragm mass needs to be small.

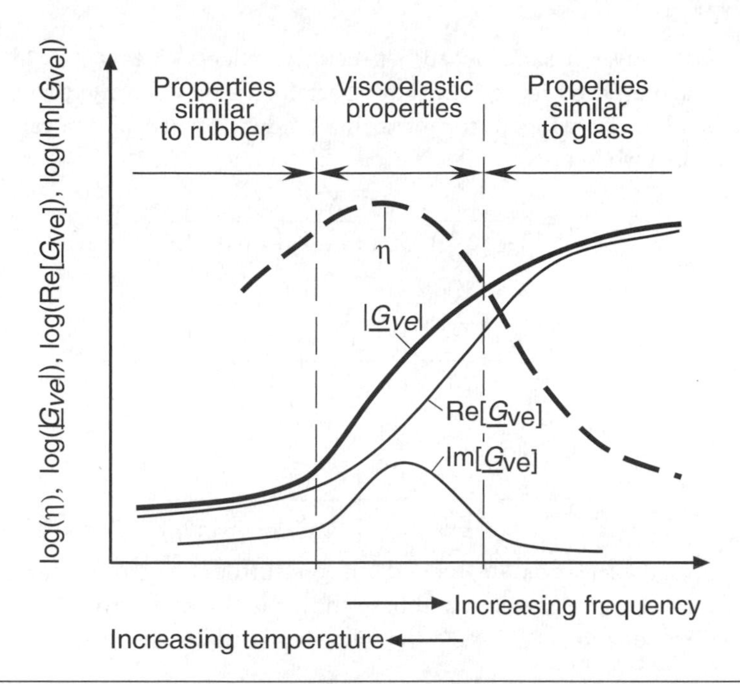

Figure 8.12 A graph of the typical properties of a viscoelastic material as a function of frequency and temperature.

Figure 8.13 Early loudspeakers having a diaphragm made of plastic material, such as this KEF B110, needed to have their diaphragm surfaces treated with viscoelastic compounds to damp the bending wave resonances since the plastics used had small loss factors. (Photo by Mendel Kleiner.)

Is there any advantage to using a free layer on both sides of a sheet? If the layers are thick, it does not matter if the layer is added by one-half to each side. If the layer is thin, it will contribute better to the combined loss factor if it is applied to one side only.

To increase the losses, one can add a flexible structure known as a *spacing layer*. An example is a honeycomb structure between the sheet and the viscoelastic layer. This layer should not store any energy ideally and should be very rigid to shear forces.

Constrained Layers

A *constrained viscoelastic layer* is sandwiched between two sheets so that it is in contact with both sheets. Its damping is achieved by its resistance to shear forces between the two attached sheets (see Reference 8.1). The bending wave loss factor for such a sandwich construction can be estimated using the diagram in Figure 8.13 where:

$$\eta_B = \frac{\eta_{ve} XY}{1+(2+Y)X+(1+Y)\left(1+\eta_{ve}^2\right)X^2} \tag{8.35}$$

with

$$X = \frac{\mathrm{Re}\left[\underline{G}_{ve}\right]}{k_p^2 h_{ve}}\left[\frac{1}{E_{p,1}h_{p,1}}+\frac{1}{E_{p,2}h_{p,2}}\right] \tag{8.36}$$

and

$$\frac{1}{Y} = \frac{E_{p,1}h_{p,1}^3 + E_{p,2}h_{p,2}^3}{12h_n}\left[\frac{1}{E_{p,1}h_{p,1}}+\frac{1}{E_{p,2}h_{p,2}}\right] \tag{8.37}$$

One usually denotes X as the shear parameter and Y as the stiffness or geometric parameter.

If both sheets are of the same material and have the same thickness, the Y parameter will depend only on the geometry. In Equations 8.35–8.37, the following symbols are used:

η_B loss factor of the sandwich construction
η_{ve} loss factor of the viscoelastic layer
$\mathrm{Re}[\underline{G}_{ve}]$ real part of the complex shear modulus of the viscoelastic layer
$E_{p,1}$; $E_{p,2}$ Young's modulus of sheet 1 and 2 respectively
$h_{p,1}$; $h_{p,2}$ thickness of sheet 1 and 2 respectively
h_{ve} thickness of the viscoelastic layer
h_n distance between the neutral planes of the two layers; $h_n = h_n + (h_{p,1} + h_{p,2})/2$
k_p wave number of the sandwich panel at the frequency of interest
λ bending wave wavelength for sandwich panel at the frequency of interest

Equation 8.37 shows that the shear parameter X is proportional to the square of the wavelength of the bending wave. The curve for the loss factor will move upward in frequency when the viscoelastic layer's thickness is reduced and when the real part of the shear modulus $\mathrm{Re}[\underline{G}_{ve}]$ increases, as shown in Figure 8.14.

Note that the loss factor curves show wide bandwidths, and that one can have considerable damping even with a thin viscoelastic layer. These are the main advantages of the constrained layer technique over that of free layers.

If both sheets have the same thickness and are of the same material, the maximum combined loss factor is approximately one-third of the loss factor of the viscoelastic layer. It is not necessary to add the viscoelastic layer to the entire sheet. One can use partial treatment by adding the viscoelastic material in strips or patches. The optimum working frequency will then be adjusted by a factor proportional to the ratio of the untreated and treated surface area. This applies also to free viscoelastic layers. One must note though that it is possible to create new modal systems in this way.

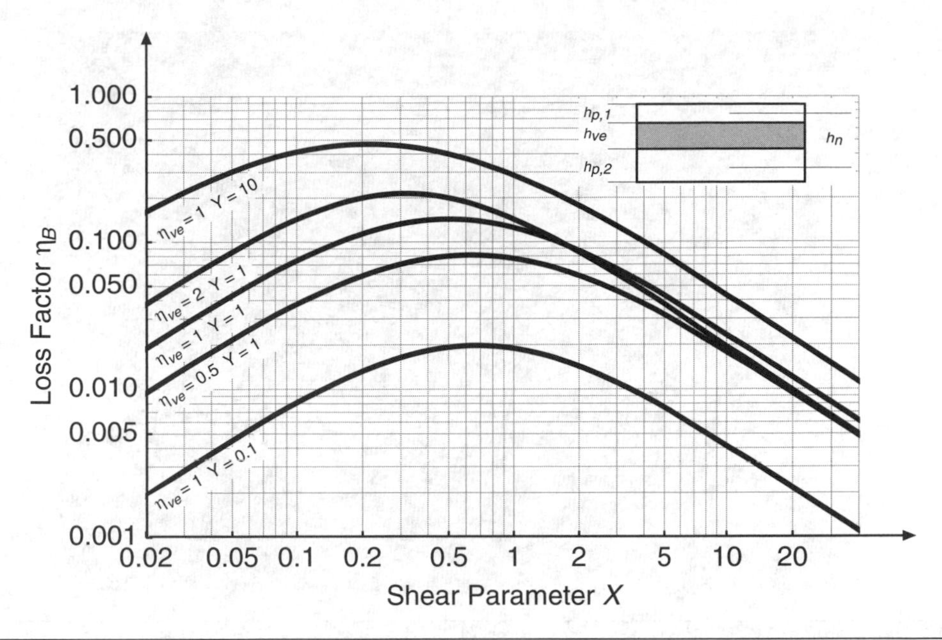

Figure 8.14 The resulting loss factor η_B for a sheet having a constrained viscoelastic layer as a function of the shear parameter X for some values of the geometry/stiffness parameter Y and layer loss factor η_{VE} (adapted from Ref. 8.1).

Figure 8.15 shows a loudspeaker diaphragm that is provided with a system of viscoelastic wedges to absorb radially propagating bending waves in the diaphragm, analogous to the way sound-absorbing wedges are used in anechoic chambers. Constrained layers are ideal for loudspeaker diaphragm applications since the added mass of the viscoelastic layer can be made small.

8.6 DAMPING BY SAND AND OTHER LOSSY MATERIALS

Loudspeaker cabinets and walls can often be well-damped using sand, gravel, and similar materials. Adding damping, by a damping compound such as damping glue, to a sheet does not necessarily lead to less sound radiation by bending wave vibration in the sheet, as discussed in Chapter 9. Because of the vibration generation mechanisms, however, it may often be advantageous to add mass to a structure affected by vibration generating forces. The added mass will increase the input impedance of the structure considerably without any acoustical adverse effects, since the sand has such high damping that the resonances will generally be unimportant. Heavy bracing of the structure, on the other hand, may lead to increased input impedance, but the structure's resonances will remain and primarily only be shifted upward in frequency. Figure 8.16 shows a loudspeaker cabinet where the outside walls are dual-sheet panels resulting in airspace that can be filled with sand. This kind of construction is also used in other applications where high sound insulation over a wide frequency range is desired, such as in concert hall walls.

Figure 8.15 This Manager MSW loudspeaker, uses damped bending waves excited from an electrodynamically excited rod attached to the center of the spherical diaphragm made of a rubber-like material. Note the star-shaped damping arrangement intended to act as an anechoic termination for the bending waves expanding from the center. The intention is to prevent radial bending wave resonances (photo courtesy of Manger Products, Germany).

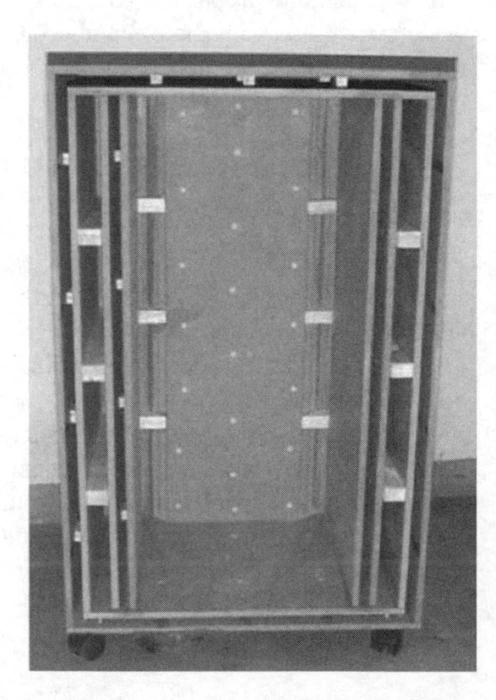

Figure 8.16 A loudspeaker enclosure featuring outer walls with cavities to be filled with sand (photo by Claudio Bonavolta).

8.7 PROBLEMS

8.1 A certain steel sheet has thickness $h = 0.001$ m, area $S = 2$ m^2, modulus of elasticity $E = 2 \cdot 10^{11}$ N/m, density $\rho = 7{,}800$ kg/m^3, and Poisson's ratio $\upsilon = 0.3$.

Task: Calculate the number of modes in the third octave bands 100 Hz and 800 Hz respectively for the freely supported steel sheet and draw the modal density as function of frequency.

8.2 A vibration-sensitive apparatus for the manufacture of integrated circuits is to be mounted on a wall. The wall consists of 0.07 m thick lightweight concrete having density $\rho = 500$ kg/m^3, critical number 38 m/s, and loss factor $\eta = 0.02$. Due to other machines in the room, there is a sound pressure level of 70 dB in the 1 kHz octave band. The apparatus is particularly sensitive to vibration in the 1 kHz octave band, and the maximum acceleration on its supports is 0.1 m/s^2 rms at 1 kHz.

Task: Calculate the resulting acceleration of the wall due to the sound field.

Hint: The sound power W_{abs} absorbed by the wall is:

$$W_{abs} = \frac{S\alpha_S \langle \tilde{p}^2 \rangle}{4Z_0}$$

where S is the surface area of the wall, and $\langle p^2 \rangle$ is the spatially averaged squared rms sound pressure over the wall. The absorption coefficient of the wall can be written:

$$\alpha_S = \frac{2\pi Z_0}{\omega m''} s \frac{f_c}{f}$$

where m'' is mass per unit area, s is the radiation factor (see Chapter 9), and f_c is the critical frequency.

The spatially averaged mean square vibration velocity $\langle \tilde{u}^2 \rangle$ of the wall is obtained from:

$$\langle \tilde{u}^2 \rangle = W_{abs} \frac{1}{\omega \eta S m''}$$

where η is the loss factor of the wall.

8.3 Using a room or a sheet to act as a reverberator are two separate ways of obtaining *artificial reverberation*. In the case of the sheet, a bending wave field is set up. The resulting sound (room) or vibration (sheet) is sensed, amplified, and added to speech or music as artificial reverberation. If the respective modal density is too low, the reverberation is perceived as *colored* and cannot be used. The shaded area in the diagram on page 216 shows in which modal density range the reverberation is perceived as colored.

Task: Determine which one of the possible alternatives—*a*, *b*, or *c*—can be used satisfactorily.

a. A room of volume $3 \times 4 \times 5$ m³.
b. A steel sheet having thickness $h = 0.5$ mm and surface dimensions = 1.2×1.8 m².
c. A gold foil having thickness $h = 20$ μm and surface dimensions = 0.3×0.4 m².

Material	ρ [kg/m³]	*E* [N/m²]	ν
Steel	$7.7 \cdot 10^3$	$20 \cdot 10^{11}$	0.3
Gold	$19.0 \cdot 10^3$	$8 \cdot 10^{10}$	0.3

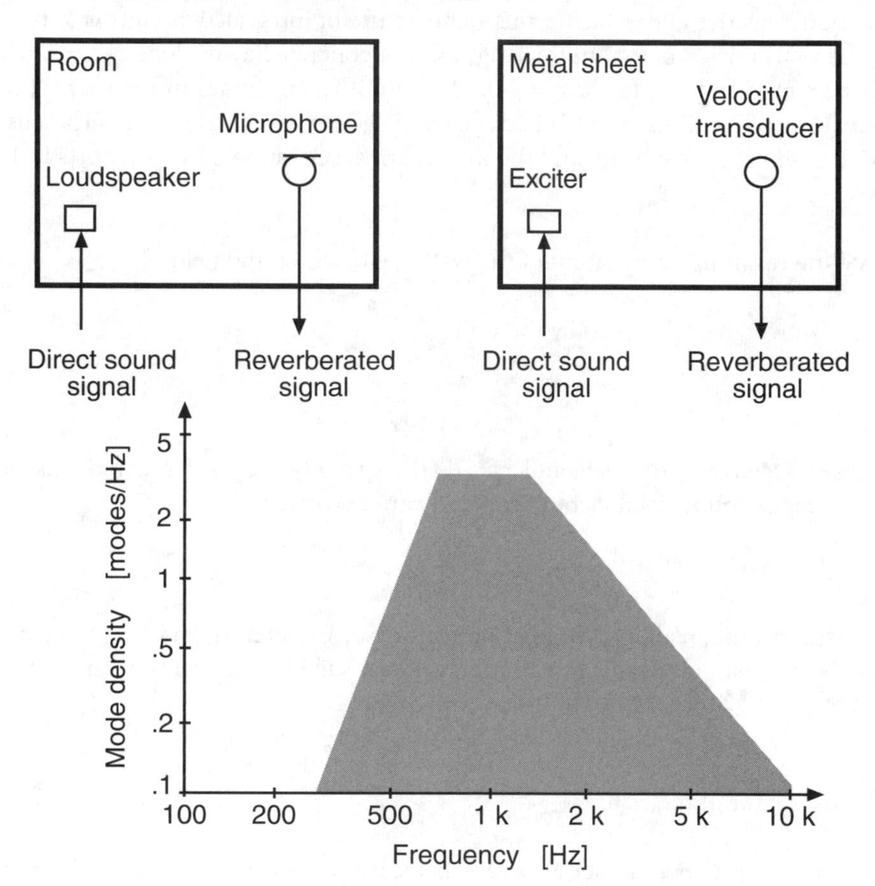

Sound Radiation and Generation 9

9.1 INTRODUCTION

Common to both sound radiation and sound generation is wave motion caused by disturbing the gaseous medium. The disturbance can have many causes and take many forms as shown by Figure 9.1.

The most common type of technical sound source is probably *the vibrating surface*, for example, the vibrating membrane or sheet, such as metal sheets on machine covers or parts, floors, walls, and ceilings in buildings, loudspeaker box walls, and loudspeaker diaphragms. For many types of plates, bending waves are the most important form of vibration from the viewpoint of sound radiation.

Sound can be generated in gases and other fluids by several mechanisms such as *pulsating flow*, *instability*, *turbulence*, and in fluids also by *cavitation*.

Heating of the medium by ionization is yet another type of sound source occurring in *plasma loudspeakers* (see Figure 9.2), *lightning*, *sparks*, and *random electric discharge* (for example, around high-tension power lines).

This chapter focuses on two types of sound generation—sound from surfaces carrying bending waves and sound by turbulence in gases. The important form of sound radiation by piston-like diaphragms, such as ideal loudspeaker diaphragms, is the topic of Chapter 14.

The measurable quantities for sound radiation are the total radiated sound power, intensity, sound pressure, particle vibration, surface vibration velocity, area, and geometry of the radiating surface. Using these quantities, one can define various metrics to describe the *coupling* between the vibration of the radiating surface and the power radiated to the surrounding medium.

9.2 COMMON METRICS FOR SOUND RADIATION

Some common metrics to describe the sound radiation properties of panels and membranes are *radiation resistance, radiation ratio*, and *radiation loss factor*. In analogy with electrical circuits, one can regard the radiation of acoustic power from an acoustic system as a loss in a resistive component. This is often a convenient approach when dealing with electroacoustic systems. In building acoustics, however, the radiation ratio and the radiation loss factor are typically used to calculate the power radiated by vibrating panels.

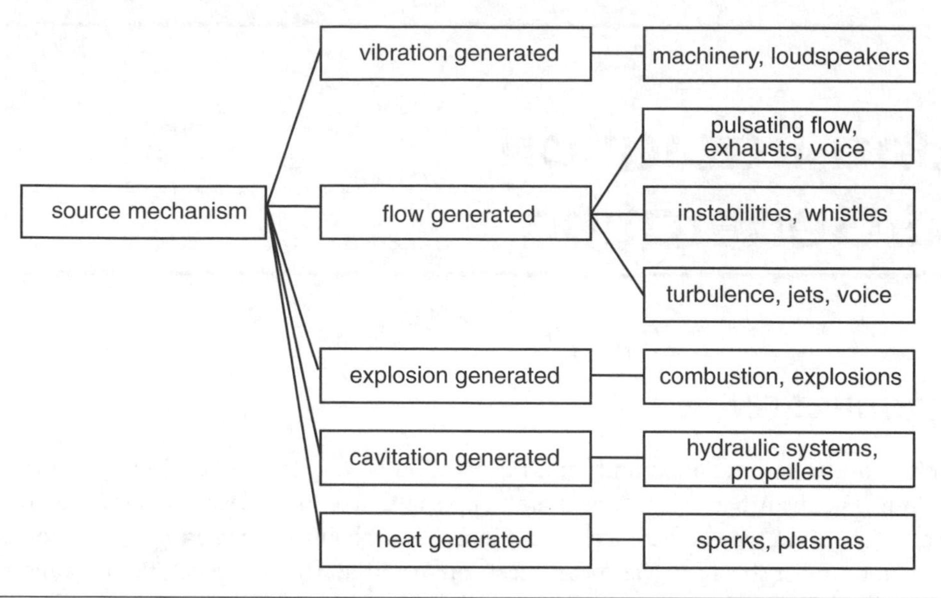

Figure 9.1 A schematic classification of various sound generation mechanisms.

Figure 9.2 A modern plasma loudspeaker using *hot* plasma generated by a high-frequency electric field (photo courtesy of Ulrich Haumann).

Radiation Resistance

The *radiation resistance*—the real part of the radiation impedance—can be defined for both acoustic and mechanic systems. The *mechanical radiation resistance* R_{MR} is related to the acoustical radiated power W_{rad} as:

$$W_{\text{rad}} = \tilde{u}^2 R_{MR} \tag{9.1}$$

where $\tilde{u}$ is the root-mean-square value of the vibration velocity, assumed to be perpendicular to the surface and equal over the entire area.

The complex *mechanical radiation impedance*:

$$\underline{Z}_{MR} = R_{MR} + jX_{MR} \tag{9.2}$$

is often used when working with electroacoustical and electromechanical analogies to characterize the radiation in a way analogous to that used for electrical circuits. The radiation impedance has real and imaginary parts—the real part represents the sound radiation while the imaginary part represents the stored acoustic energy usually in the near field of the radiator.

Radiation impedance is typically used when studying sound radiation by plane surfaces that move with the same velocity and phase over their entire sound radiating area—although, the theory is generally also applied to the various types of diaphragms and membranes used for loudspeakers and microphones that tend to have conical and hemispherical shapes. As long as the depth of the cone or dome is much smaller than the wavelength under study, the theory for planar surfaces holds reasonably well. The radiation impedance is usually considered for each side separately. Horns (such as those shown in Figure 9.3) are often used to increase the radiation impedance seen by the loudspeaker's diaphragm.

Radiation Ratio

The *radiation ratio* was addressed in Chapter 8 and is defined as:

$$s = \frac{W_{\text{rad}}}{Z_0 S \langle \tilde{u}^2 \rangle} \tag{9.3}$$

where W_{rad} is the radiated power by *one side* of the surface, S the radiating surface area, and $\langle \tilde{u}^2 \rangle$ the mean square of the vibration velocity in the normal of the surface, averaged over the surface.

Figure 9.3 Horns are sometimes used to increase the radiation impedance sensed by the loudspeaker diaphragm (photo courtesy of Avantgarde Acoustic GmbH, Germany).

A radiation ratio of unity corresponds to the radiation that will be obtained from one side of an infinitely large plane piston vibrating in the direction of its normal, equally at all points. The radiation ratio is used extensively to characterize the radiation ability of sheets carrying bending waves.

The radiation ratio varies due to a number of factors. If one has some knowledge about the sound radiation characteristics of a panel, for example, by prior measurement data, one can estimate the sound power radiated by a surface by measurement of its vibration. In using this approach, one can determine the sound radiation of various surfaces of a vibrating object—for example, the walls of a room. The radiation ratio, consequently, is an important number. A library of such data is useful to the noise and vibration consultant. Radiation ratio data are usually shown in logarithmic form as $10 \log(s)$.

Radiation Loss Factor

The sound *radiation loss factor*, η_{rad}, is sometimes used to characterize the effect of its sound radiation on the damping of a vibrating surface. This loss factor is defined in analogy to the definition of η in Chapter 2 as:

$$\eta_{rad} = \frac{W_{rad}}{E_k \omega} \tag{9.4}$$

where E_k is the mean of the total kinetic energy carried by the surface; ω, the angular frequency; and W_{rad}, the total power lost by sound radiation.

It is often difficult, by measurement, to separate the losses due to radiation from those caused by edge losses and internal damping.

9.3 SOUND RADIATION BY VIBRATING SURFACES

Sound radiation can be studied using analytical and numerical techniques. While analytical techniques are physically instructive, numerical techniques have the advantage of allowing the sound radiation of nearly any surface structure or vibration pattern and surrounding medium to be analyzed. In this book, we will only use analytical techniques.

Piston radiation and bending wave sound radiation are the two most interesting types of radiation, but other sound radiation—such as that by pistons coupled to horns and radiation by membranes—is also of great practical interest.

9.4 SOUND RADIATION BY VIBRATING PISTONS AND MEMBRANES

We will first study the sound radiation for the case of a monopole situated in free space and on a hard surface. Using Equations 1.24 and 1.25, we find that for a small free sound source having the volume velocity of $\underline{U}_D$, the sound pressure at a distance r will be:

$$\underline{p}(r) = j\omega 4\pi a^2 \underline{u} \rho_0 \frac{e^{-jkr}}{4\pi r} = j\omega \underline{U}_D \rho_0 \frac{e^{-jkr}}{4\pi r} \tag{9.5}$$

When the monopole is on top of a hard surface, the resulting sound at the observation point will be the linear sum of the sound from the monopole and its mirror image, which is also a monopole.

Since the monopoles are assumed infinitely small, they will acoustically be at nearly the same position. The resulting sound pressure will then be twice that of one of the monopoles in free space.

A sound radiating piston or membrane must have a velocity component normal to its surface. To simplify the mathematics and physics, the piston in our case is assumed to be *mounted in an infinite baffle*. The vibrating piston can be considered as a surface covered by small monopoles, where each monopole represents the volume velocity generated by a small patch of the piston surface. When the piston is vibrating, each element on the piston surface area (on both the front and back sides of the piston) can be thought of as an individual, in-phase monopole, generating volume velocity. Because the baffle is assumed, infinite radiation from each side of the piston can be considered independently.

Because of linearity, we can again use the principle of superposition of pressure to determine total sound pressure at the observation point.

Each monopole will contribute to the total sound pressure at a distance as if it has a part of the total volume velocity $d\underline{U}_D = \underline{u}_D dS$, due to the vibrational velocity of the diaphragm $\underline{u}_D$ and the size of the surface element of each surface element dS. The total pressure at the point of observation at a distance r can then be obtained by summing the sound pressure contributions $d\underline{p}$ from each surface element dS of the entire piston surface:

$$\underline{p}(r) = j\omega\rho_0 \int_S \frac{\underline{u}_D e^{-jkr'}}{2\pi r'} dS \tag{9.6}$$

In Chapter 14, we will use this technique to study the sound radiation from circular loudspeaker diaphragms.

If the piston is not moving in phase over its area or if the piston vibration velocity changes over its area, the same technique to find the sound pressure can still be used with appropriate summation of the various sound pressure contributions from the different piston patches.

It must be noted that we have assumed the volume velocity of the diaphragm to be unaffected by the impedance of the surrounding medium. While this is often the case for massive loudspeaker diaphragms, such as of those of electrodynamic loudspeakers, the vibrations of the thin low-mass membranes, for example electrostatic loudspeakers, are considerably damped by the surrounding air.

The radiated acoustic power can be estimated by integration of the sound intensity for the far field of the piston or membrane, where particle velocity and sound pressure are in phase. One prerequisite for far field conditions for a pistonic source is that the sum pressure diminishes with distance as $1/r$.

9.5 SOUND RADIATION BY VIBRATING SHEETS CARRYING BENDING WAVES

When studying the sound radiation of a wall or other large structure, it is best considered as a sheet having a vibration velocity that varies over the sheet. The sound radiation depends, along with the vibratory characteristics of the structure-borne sound, on the acoustic properties of the surrounding medium at the specific location, the sheet geometry, and other factors. For sound radiation, the sheet must have a vibration component perpendicular to its surface.

Bending Waves in an Infinite Sheet

Assume an infinite sheet carrying a single frequency bending wave with a linear wave front and constant amplitude. One of the boundary conditions at the surface of the sheet requires that the phase velocity of the induced sound field along the sheet be the same as that of the vibration of the sheet. Another boundary condition requires that the particle velocity component of the induced sound field at right angles to the sheet be the same as the particle velocity of the sheet at right angles to the sheet.

Using these conditions, one can show that the radiation ratio must be:

$$s = \mathrm{Re}\left[\frac{1}{\sqrt{1-\left(\frac{k_B}{k}\right)^2}}\right] \tag{9.7}$$

The expression shows that there are three different cases:

1. $\lambda_B < \lambda_{air}$ low frequency $f < f_c$
2. $\lambda_B = \lambda_{air}$ critical frequency $f = f_c$
3. $\lambda_B > \lambda_{air}$ high frequency $f > f_c$

At frequencies below the critical frequency there cannot be any radiation of sound power to the far field. This can be understood intuitively by studying Figure 9.4. Since the distance between out of phase areas is smaller than one-half wavelength of sound in the medium surrounding the sheet, destructive interference will prevent sound radiation. One sometimes calls this condition an *aerodynamic short circuit.* We studied this type of cancellation for dipoles and quadrupoles in Chapter 1. In consequence, with the absence of radiated sound power, the radiation ratio must be zero.

Note, however, that close to the surface there will be audible sound pressure and particle velocity. This sound field carries reactive power and is sometimes called a *reactive near-field.* The sound pressure in the near field drops quickly with distance. It is important to remember though that the flow velocities in the near field may be very high.

The bending wave vibration of a panel can be damped by a porous sound absorber placed nearby due to the acoustic losses introduced by the flow resistance of the porous absorber. This property is used

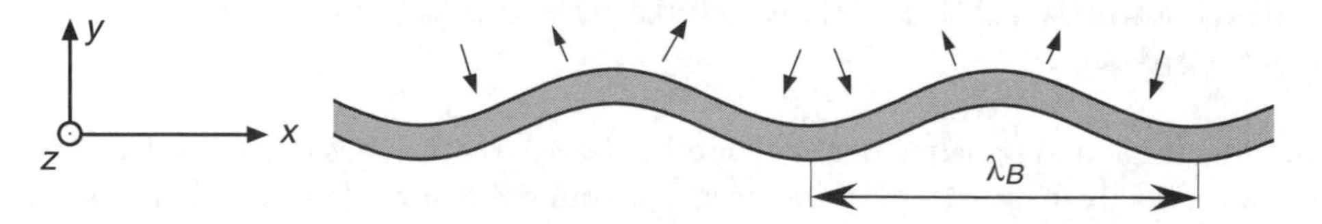

Figure 9.4 Volume flow in the near field at one side of a sheet carrying a bending wave in the x direction when $\lambda_B < \lambda_{air}$, that is at low frequencies $f < f_c$, (only one side shown).

in the design of the EMT reverberation plate shown in Figure 8.10. The reverberation time is adjusted by moving a porous sound absorber various distances from the steel sheet.

At the critical frequency, the sound radiation will be as a wave that is *radiated parallel to the sheet.* At frequencies above the critical frequency—under the assumed ideal conditions—the sheet will radiate plane waves off both its sides. (Remember that the radiation ratio only describes radiation to one side.) The plane waves radiate at an oblique angle φ as shown in Figure 9.5.

The radiation angle φ is obtained from the expression:

$$\lambda_B \sin(\varphi) = \lambda \tag{9.8}$$

The radiation will depend on the matching of the sound field in the sheet to that in the surrounding medium.

Figure 9.6 shows, in summary, the basic frequency dependence of the radiation ratio for an unattenuated plane bending wave in an infinite plane sheet. If the sheet is finite, the wave field in the medium surrounding the sheet will lose some of the symmetry that is necessary for the cancellation of sound radiation at frequencies below the critical frequency. The radiation will be primarily from the areas close to the discontinuity.

Bending Waves in Finite Sheets

The larger the dimensions of the sheet compared to the wavelength of sound in the medium, the smaller the sound radiation since the relative influence of the sound radiation by edges and corners will diminish due to their relatively smaller areas. The radiation ratios for some such cases are shown in Figure 9.7. The data shown in this figure applies to the case of a strip of material, on two edges, carrying an (unattenuated) bending wave that radiates from the ends of the strip (at the edges).

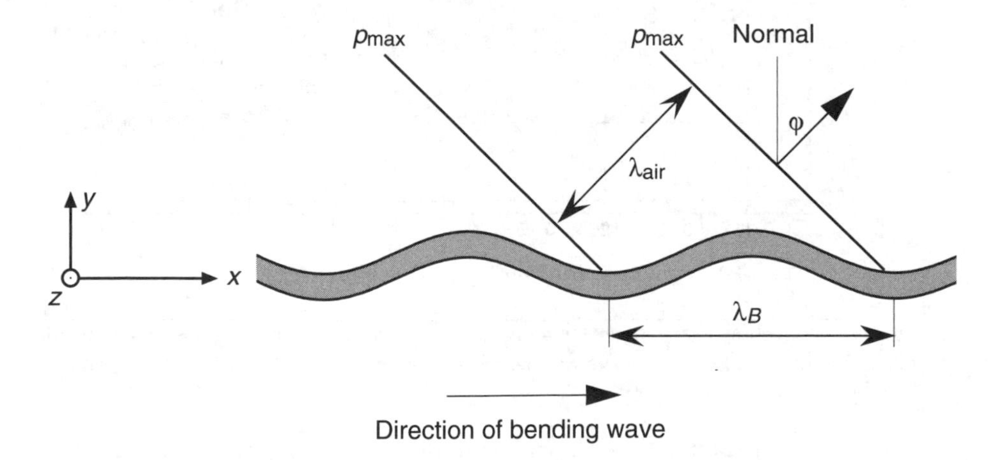

Figure 9.5 Sound radiation by bending waves at one side of a sheet at high frequencies, that is, $f > f_c$ will be at an angle ϕ to the sheet. The wavelength of the bending wave is longer than that of the longitudinal wave in air $\lambda_B > \lambda_{air}$. (only one side shown).

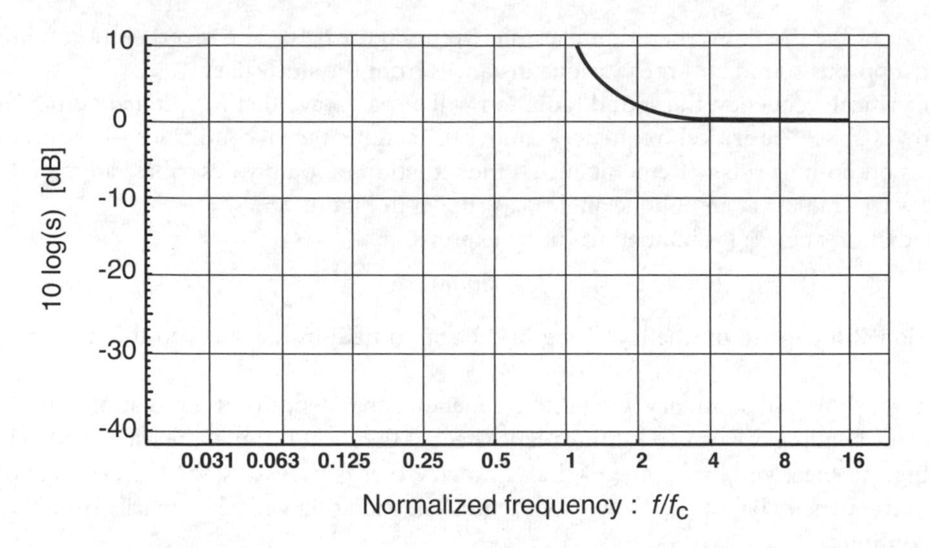

Figure 9.6 The radiation ratio, shown as 10 log(s), as a function of frequency for an undamped plane bending wave in an infinite sheet.

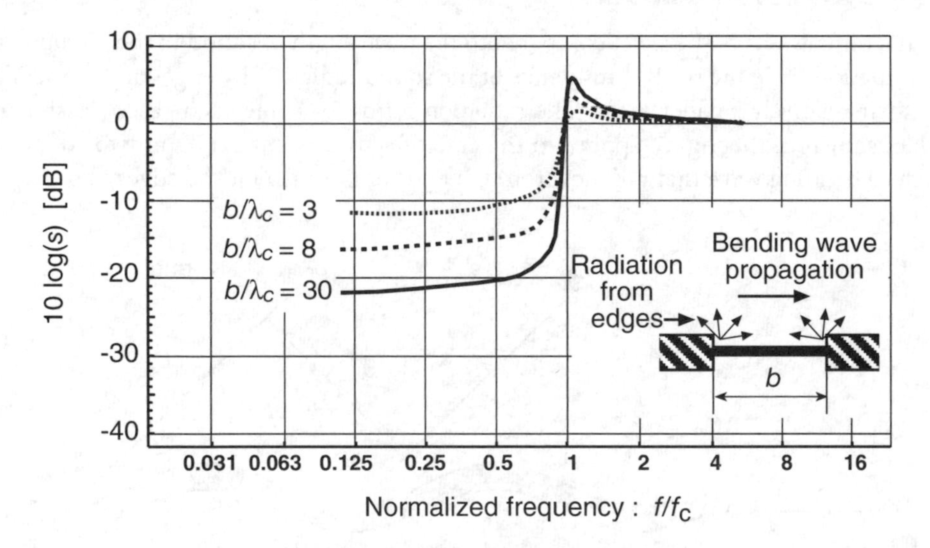

Figure 9.7 The radiation ratio, shown as 10 log(s), as a function of frequency for an undamped plane bending wave in a finite sheet of width b for some values of b/λ_c. The wavelength at the critical frequency of the sheet is λ_c.

Note that the symmetry leading to aerodynamic sound cancellation is also disturbed when a sheet is stiffened—for example, by beams, battens or point attachments. This results in sound radiation at frequencies below the critical frequency. Increasing the stiffness of a sheet does not necessarily lead to lower sound radiation by bending waves. The flat panel loudspeaker shown in Figure 9.8 has a suitable bending stiffness to mass ratio.

Figure 9.8 The NXT distributed mode loudspeaker uses the special properties of a sandwich sheet to achieve a suitable distribution of modes for a flat power radiation response. The sheet is extremely resonant and is primarily damped by edge and radiation losses. Note the two electrodynamic shakers. (Photo by Mendel Kleiner.)

Bending Waves in Damped Sheets

Another case of reduced symmetry is that of a sheet in which the bending wave is damped perhaps by internal losses. Again, the reduced symmetry results in sound radiation below the critical frequency, as shown by the radiation ratio curves shown in Figure 9.9. Increased damping is accompanied by more radiation, thus damping a sheet does not necessarily lead to less sound radiation.

Reduction of Radiation from Bending Waves in Sheets

As Equation 9.2 indicates, there are only a few fundamental ways of limiting the sound radiation by bending waves in sheets.

One can reduce the coupling to the medium by using finely perforated sheets, since this leads to an aerodynamic short circuit between the out-of-phase front and back sides of the sheet. Another way is to use framework constructions instead of sheet constructions. It is usually advisable to strive for high critical frequencies, if possible. Sheets should be attached where the vibration velocity levels are small. It is often advantageous to attach the sheet at a number of points instead of along a line, since the attachment point or line will cause a disruption in the symmetry of the field.

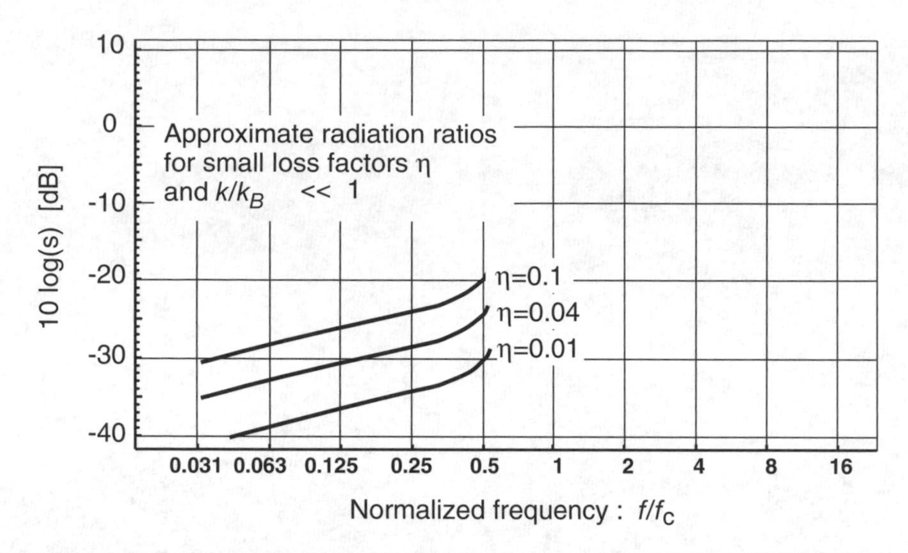

Figure 9.9 The radiation ratio, shown as 10 log(s), as a function of frequency for a plane bending wave in a sheet for some cases of damping at frequencies below the critical frequency.

9.6 SOUND GENERATION BY FLOW

The flow of gas or fluid can also result in sound generation, due to pulsating flow, vortices, turbulence, and cavitation. The elementary sound sources can be of *monopole*, *dipole*, or *quadrupole* types. The term *turbulence* is generally used to describe the property of the flow when the sound generated has a wide noise-like spectrum.

The pulsating flow from an exhaust tube or a siren are examples of *monopole* type aerodynamic sources. When a gas flows past a body, such as a rod, vortices are generated—diametrically in case of a rod—creating a *dipole* type aerodynamic source. An example of this type of sound is the whine from telephone wires on a windy day. Note that the dipole force on the rod can cause large movement if the resonant frequency is the same as the frequency of the dipole sources. *Quadrupole* sources are easily created around the flow from orifices because of viscous stress within the flow. The core flow will be surrounded by vortices, having a quadrupole property. This type of noise source is characteristic of jet aircraft engines and other high-speed gas exhausts.

Noise due to turbulent flow makes it easy to detect the rotating propellers of submarines and other ships. The noise from landing aircraft is dominated by noise due to turbulence around the fuselage. In audio engineering, it is sometimes necessary to design the bass-reflex ports of loudspeaker cabinets so that turbulence is not created.

The noise at close distance to the turbulence or vortices can be disturbing. Typically, the flow in air becomes turbulent at flow speeds above 7 m/s. Turbulent flow outside is one of the main sources of noise in the interior of an aircraft or car. Even when biking, the turbulence generated by the air flow around the head is easily audible.

Turbulence-generated noise due to wind can make it difficult to measure sound outdoors. The turbulence created by the microphone capsule and body is close to the sound sensing membrane of the microphone. This results in high sound pressure at the membrane and associated poor signal-to-noise ratio in the measurement.

One can reduce the problem by using various forms of *wind screens*. These work by removing the region of turbulence from the membrane. Foam plastics or fine mesh cloth on a wire grid may be used for this purpose. In spite of the larger volume of turbulence, the noise sensed by the microphone is diminished since the attenuation of sound by the increase in geometrical distance is so strong. An example of this effect is given in Chapter 12.

Acoustic Efficiency

Air flow at orifices is typically characterized by presence of all the three source types mentioned. The sound power W generated is proportional to the mechanical power in the flow W_{mec}. One writes:

$$W = \eta W_{mec} \tag{9.9}$$

where η is the *acoustic efficiency*.

In the case of a streaming gas jet out of an orifice in a sheet, the mechanical stream power of the jet is given by:

$$W_{mec} = \Delta p u_{flow} S = \frac{\rho_0 u_{flow}^3 S}{2} \tag{9.10}$$

where ρ_0 is the density of the gas stream, u_{flow} the flow speed of the gas at the orifice, and S the area of the orifice. The pressure drop across the orifice Δp determines the speed of the gas flow at the orifice.

The acoustic efficiency is determined by the flow speed of the gas u_{flow} and by the speed of sound in the surrounding medium.

The acoustic efficiency of the noise generation by flow is:

$$\eta = c_m \frac{u_{flow}}{c_0} + c_d \left(\frac{u_{flow}}{c_0} \right)^3 + c_q \left(\frac{u_{flow}}{c_0} \right)^5 \tag{9.11}$$

where c_m, c_d, and c_q are the efficiency coefficients of monopole, dipole, and quadrupole sound generation respectively. Since the mechanical stream power is proportional to the flow speed u_{flow}^3, the contributions to the noise by the three source types will depend on flow speed as u_{flow}^4 for monopole, u_{flow}^6 for dipole, and u_{flow}^8 for quadrupole sources (see Reference 9.1).

The balance between volume flow and flow speed is important. If one, for example, wants to feed air to a room through a vent opening having a grille, dipole sound sources turn out to be the most important at typical flow speeds. For a certain volume flow, one can reduce the flow speed u_{flow} and increase the vent area S, or vice versa. Equations 9.10 and 9.11 show that to minimize flow noise it is advantageous to limit the flow speed by increasing the flow area. This is a general property of flows. Turbulence noise may be disturbing in low frequency sound reproduction by ported loudspeaker boxes, such as the one shown in Figure 9.10.

The spectrum of the turbulence-generated sound will typically be broadband (as shown by the curve in Figure 9.11), unless the associated mechanical structure is resonant. In air, the peak in the spectrum will be at a frequency f_{peak} given by:

$$f_{peak} \approx 0.2 \frac{u_{flow}}{d} \tag{9.12}$$

Figure 9.10 Note the flare of the three bass-reflex port openings of this SVSound™ PB12-Plus subwoofer. Many ported loudspeaker enclosures use a flared port to reduce the *swooshing* noise associated with turbulence generated by turbulence in air at high flow velocities. An application of this technology can be seen in the loudspeaker shown here (photo courtesy of SVSound™ [all rights reserved]).

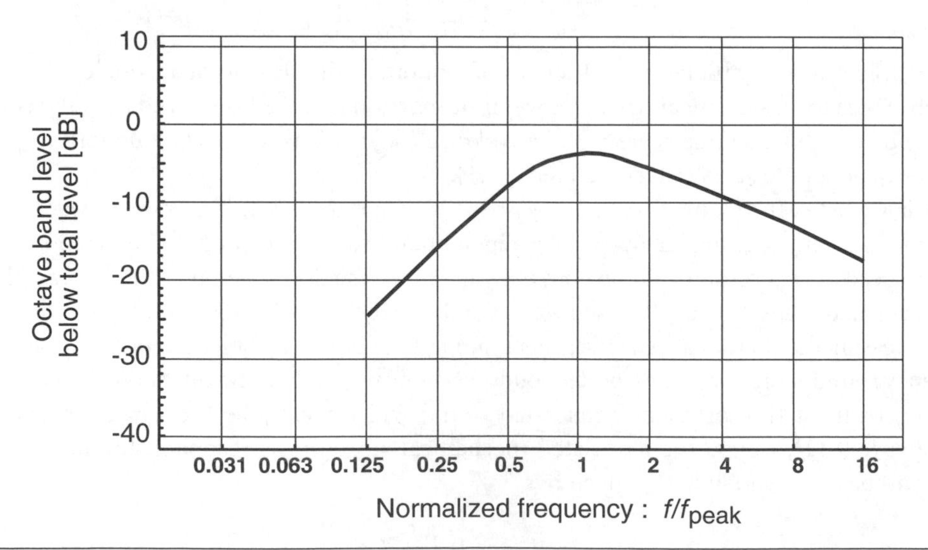

Figure 9.11 Typical noise power spectrum for turbulence generated by air blowing through a grille.

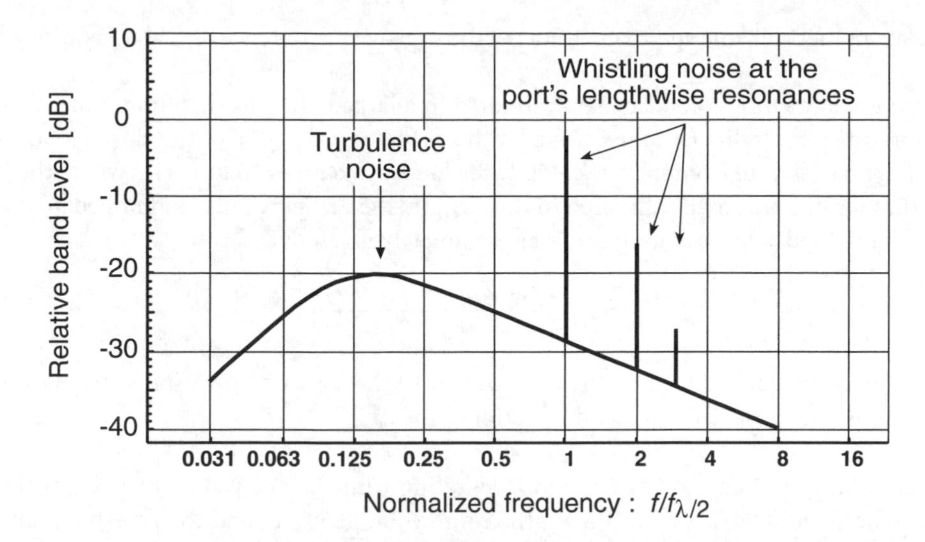

Figure 9.12 Typical noise power spectrum generated by high speed air blowing through a bass-reflex port at the Helmholtz resonance frequency. Whistling noise frequency is normalized to the half-wavelength resonance frequency of the port tube $f_{\lambda/2}$.

where d is the cross-sectional width of the obstruction and u the flow speed. The constant 0.2 is the so-called Strouhal's number for this case (see Reference 9.1).

Flow-generated noise may be a problem in loudspeaker design. In ported loudspeaker enclosures, such as bass-reflex and bandpass loudspeaker boxes and the one shown in Figure 9.10, the particle (flow) velocity may become large in the Helmholtz resonance frequency region. There are two sound effects involved in shaping the spectrum characteristic shown in Figure 9.12.

One is the pulsating tonal sound effect due to vortex excitation of the port lengthwise resonance. The other is a wide band turbulence-noise component generated by vortex shedding immediately in front of the port openings. Whistling noise will be modulated at double the frequency of the excitation, particularly in the frequency region close to the port's Helmholtz resonance frequency, since the flow speed is high at both negative and positive flow speed peaks (see Reference 9.3). These types of noises may also occur in conventional Helmholtz resonators used for sound absorption.

To reduce these noises, it is important to apply the rule of keeping the particle (flow) velocity low by using a wide port and to avoid sharp edges at the port openings.

9.7 PROBLEMS

9.1 Two identical loudspeakers in an outdoor arena are connected to the same signal source and radiating sound to reach the audience. The audience is seated on a plane field that is grass covered and has high sound absorption. Assume the loudspeakers are mounted at a height of 6 m above the audience's head level, and at a distance 10 m apart.

Task: Calculate numerically the approximate loci of the negative interference dips at a frequency of 340 Hz.

9.2 Assume that a circular loudspeaker mounted in a large baffle, its diaphragm has a radius a and is vibrating perpendicular to its surface with a velocity $\underline{u}$. Because of the closed loudspeaker box, there is no radiation from the back side of the loudspeaker. For frequencies where the loudspeaker's dimensions are small compared to wavelength, the real part radiation impedance seen by one side of the loudspeaker diaphragm is approximately given by:

$$R_{MR} = \frac{\pi a^2 Z_0}{2}\left[1 - \frac{J_1(2ka)}{ka}\right]$$

Task: Compare the power radiated by the loudspeaker to that radiated by a spherical, monopole loudspeaker having the same surface area and vibration velocity.

9.3 A piano lid is excited by a shaker and is vibrating with the octave band velocities shown in the following table. The lid's area is 2 m², the room volume 50 m³, and the reverberation time 0.8 s. The diffuse field sound levels in the room are also given by the table.

f	0.5	1	2	4	[kHz]
L_u	135	122	113	112	[dB] re 10^{-9} m/s
L_p	89.9	84.9	80.9	79.9	[dB] re $2 \cdot 10^{-5}$ Pa

Task: Calculate the radiation ratio of the piano lid.

9.4 A small loudspeaker having a radius of 0.15 m is mounted at the top, in the side of a tall rectangular box that is 1.2 m high and has a square cross-section with sides that are 0.4 m. During measurement of the acceleration levels of the loudspeaker cone and the box walls, it was found that the vibration levels of the side walls were on the average 10 dB below that of the cone at a frequency of 90 Hz.

Task: Calculate the approximate level difference between the radiated sound power from the loudspeaker box walls and the cone in an anechoic environment.

9.5 A ported box loudspeaker similar to the one shown in Figure 9.10, but with only one port, has its port resonance tuned to 34 Hz. The circular port area is 0.01 m². During a test of the loudspeaker, it was fed a sine-wave signal at the port resonance frequency. A certain drive level turbulence-generated noise was beginning to be heard from the port opening. Experience has shown that when port air velocities reach about 7 m/s, turbulence becomes noticeable.

Task: Estimate the acoustic power radiated by the port. Estimate the spectrum of the noise, assuming $c_m \approx 10^{-8}$.

Sound Isolation

10

10.1 INTRODUCTION

Building acoustics is the branch of architectural acoustics dealing with the transmission of sound through partitions, enclosures, and openings. Building acoustics is often discussed in terms of sound transmission including airborne, structure-borne, and impact. In most cases, the noise transmitted into a room is due to many sources and comes by many ways.

Sound may be transmitted from room to room as shown in Figure 10.1. The transmission can be: (1) directly through the wall—direct transmission, (2) by an interconnecting structure as vibration—flanking transmission, or (3) through open-air paths.

Airborne sound generally means sound transmission through walls, windows, doors, ceilings and floors, provided that the structure is initially excited by sound in air. It is the sound that is incident on a building partition, moves the partition, and is reradiated in an adjacent space by the partition's vibration. Typical examples of transmitted airborne sounds are voice, sound from loudspeakers, noise from handheld tools, and noise from traffic. The *transmission loss* of a construction is a measure of the sound insulating properties of a wall or other partition.

Impact sound is the sound that is generated in neighboring rooms in a building when such things as floors, walls, and stairs are subject to impacts by footsteps, hammers, and the like. *Impact sound pressure* is a measure of the ability of the building to transmit and radiate sound that is transmitted mechanically to the structure. Note the difference between *airborne* and *structure-borne* sound.

Structure-borne sound generally means sound excited mechanically that travels from the source to the receiver mainly by way of mechanical paths. Structure-borne sound can come from steps in staircases, noise from elevators and their machinery, noise from the closing of doors, noise due to heating, ventilation, and air-conditioning machinery (HVAC), ducts, and outlets.

10.2 INSULATION AGAINST AIRBORNE SOUND

Assume two adjacent rooms separated by a wall with a sound source in one of the rooms that is known as the *sending room*. The walls of the sending room will vibrate due to the sound pressure of the incident sound. Normally, the wall is constructed in such a way that the side of the wall exposed to the *receiving room* will vibrate as well, due to the sound in the sending room.

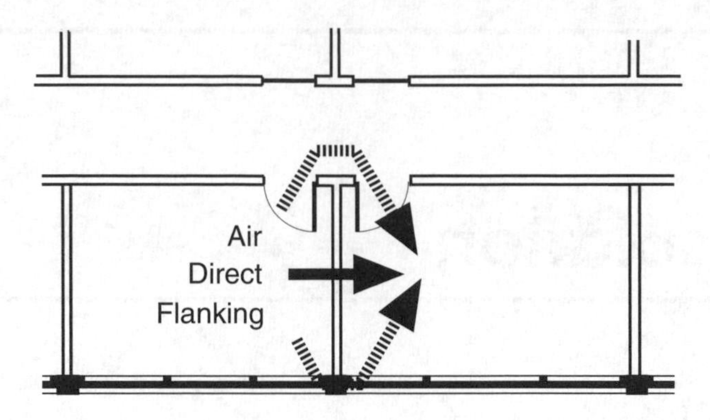

Figure 10.1 Sound can move from one room to the next by three types of room-to-room paths: (1) direct, (2) air, and (3) flanking transmission.

In practice, it is the sound level difference between the adjacent rooms that is of interest. The sound level difference is influenced not only by the sound transmission properties of the acoustically separating wall, or *direct transmission*, but also by adjacent walls and air ducts. This added component is called *flanking transmission* and is discussed later.

10.3 SOUND TRANSMISSION LOSS

Denote the sound power incident on the wall in the sending room, W_{inc}, and the sound power radiated into the receiving room, W_{trans}.

The *transmission coefficient* τ is defined as:

$$\tau = \frac{W_{trans}}{W_{inc}} \tag{10.1}$$

Transmission loss, TL, also called the sound reduction index, R, is given in dB and defined as:

$$R = -10\log(\tau) \tag{10.2}$$

The transmission loss is angle dependent. In the case of an inside wall, the sound field is likely to be diffuse, so there will be an average of the wall's angle-dependent transmission loss, which will be the *effective* diffuse field transmission loss generally referred to.

The difference between the mean sound pressure level due to an unspecified sound source in the sending room L_{pS} and the mean sound pressure level in the receiving room L_{pR} is called the *noise reduction* (NR). The NR will be determined by a few factors. If the sound fields in the rooms are diffuse, the NR will depend on the area of the partition separating the rooms S_p, its transmission loss R_d, and the ability of the receiving room to absorb sound, that is, its sound-absorption area A_R. The sound absorption in the sending room will affect the sound level generated by the source in that room.

The sound transmission through a partition between rooms will lead to an energy exchange between the rooms and the partition, which make up three subsystems. In the frequency range above the Schroeder frequency, the sound fields are diffuse, and we can then use *statistical energy analysis* to study the acoustical interaction between the rooms. If the partition is assumed to be nonresonant (as when the wall behaves as a mass) one can simplify the situation to only involve the resonant energy in the two rooms, as shown in Figure 10.2. Using Equations 4.28, 4.33, and 10.1, the power balance equations can then be written:

$$W_+ + \tau\frac{P_R c S_p}{4} = \frac{P_S c A_S}{4} + \tau\frac{P_S c S_p}{4}$$
$$\tau\frac{P_S c S_p}{4} = \frac{P_R c A_R}{4} + \tau\frac{P_R c S_p}{4} \tag{10.3}$$

If the partition is resonant (as it is above the critical frequency), a third system needs to be included between the two rooms in the figure. If one assumes that the transmission loss of the wall is larger than 10 dB, it is not necessary to take the sound being returned through the partition to the sending room from the receiving room into account. This will be the general case in most buildings in the temperate zones. The power through the partition from room S is simply absorbed by the walls of room R. We then obtain the following equation for the power balance between the reverberant sound fields of the sending and receiving rooms:

$$\tau\frac{P_S c S_p}{4} = \frac{P_R c A_R}{4} \tag{10.4}$$

This gives the ratio between room averaged squared sound pressures as:

$$\frac{<\tilde{p}_S^2>}{<\tilde{p}_R^2>} = \tau\frac{A_R}{S_p} \tag{10.5}$$

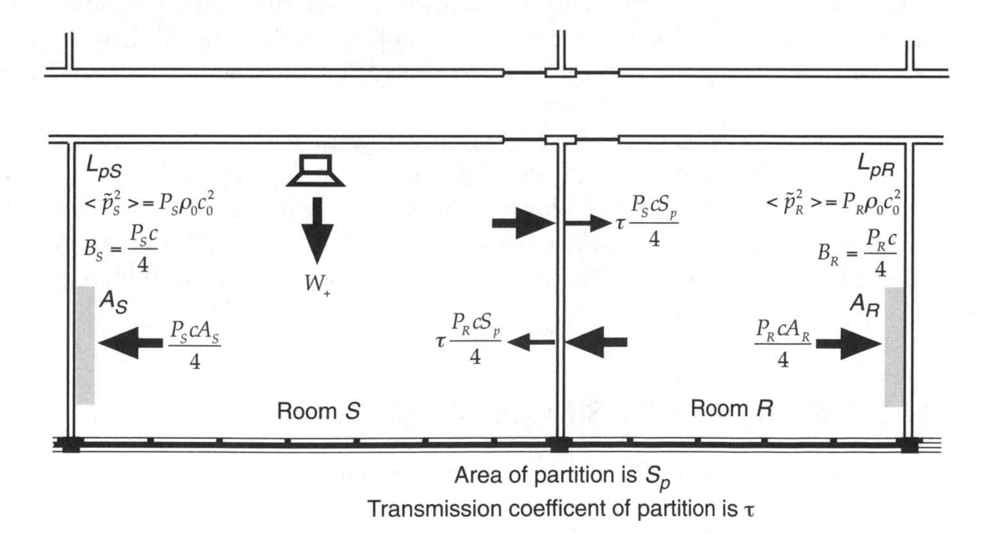

Figure 10.2 The arrows show the power flows from loudspeaker to sound absorbers in the sending and receiving rooms.

Typically, we are interested in the noise reduction (NR) between rooms, that is, the difference in sound pressure level that is obtained for a certain partition and receiving room. Notice that NR is not the same as R; the sound level difference is affected by the *room correction*. Our hearing ultimately senses NR, not the transmission loss. Taking the logarithm of Equation 10.5, we obtain the much-used Equation 10.6:

$$NR = L_{pS} - L_{pR} = R + 10\log\left(\frac{A_R}{S_p}\right) \tag{10.6}$$

Equation 10.6 is generally used in the experimental determination of transmission loss as well as in determination of the necessary sound insulation in buildings.

10.4 SOUND REDUCTION INDEX

Usually the transmission loss is averaged over octave or third octave bands, from at least the 125 Hz octave band (possibly also down to the 50 Hz third octave band) up to the 4 kHz octave band. The reason for not measuring at lower frequencies is the lack of measurement accuracy at low frequencies, where the room sound fields have little diffusitivity because of the low number of modes. However, after specifying the measurement conditions, there is nothing that prevents one from measuring NR—realizing that the data cannot be used with Equation 10.6. Lightweight constructions, such as double-panel drywalls and lightweight concrete block walls, may have poor transmission loss, particularly at low- and mid-frequencies.

It is common to use a sound reduction index R_w to grade the sound transmission loss properties of walls. The calculation of R_w is briefly discussed in the caption to Figure 10.3. The definition of the calculation of the sound reduction index is given in ISO standards. The ISO has published many standards relating to sound insulation and its measurement. The reader is referred to these standards as appropriate. The definition is usually modified nationally, so one should check the national standards applicable. The *Sound Transmission Class* (STC) rating used in the United States to characterize sound attenuation by building elements is calculated in a similar way as R_w.

It is common to differ between results from laboratory measurements R_w and those from measurements in the field R'_w (in the United States, Field STC (FSTC) is used instead). The results from field measurements may often be several decibels lower than those obtained from the laboratory measurements of the same wall type. Sound transmission loss and sound reduction index data may be found at the various websites of building materials manufacturers and national and private building research and testing organizations and institutes.

10.5 TRANSMISSION LOSS OF SINGLE-PANEL PARTITIONS

A single-panel (single-leaf) partition is an impervious, homogeneous building part made of one plane panel (sheet) of material. It is the mass per unit area, the bending stiffness, and the loss factor that determines the transmission loss and its frequency characteristic.

A finite-sized single panel partition must be supported in some way at its edges. The type of support depends on the building construction. For a lightweight partition, the transmission loss behavior varies at different frequencies. At low frequencies, it will behave as a spring (the *stiffness region*) and then show

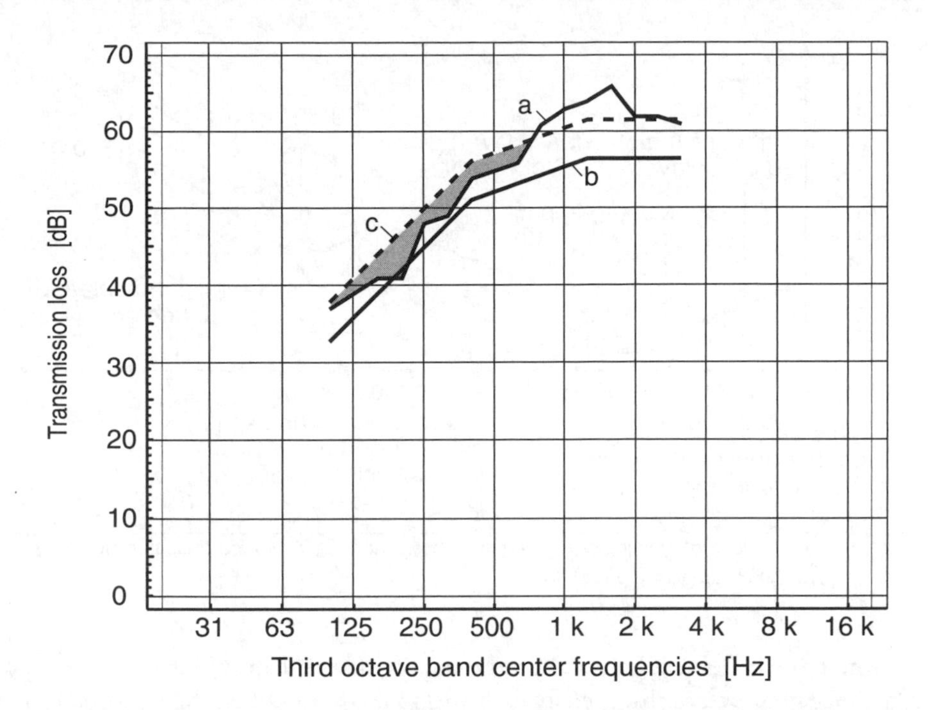

Figure 10.3 Example of calculation of the sound reduction index, R_w: (a) measured sound insulation in third octave bands, (b) reference curve, and (c) reference curve shifted to allow sum of differences above curve a but below the reference curve to be ≤ 32 dB (at the third octave band frequencies from 125 Hz to 3.2 kHz). The R_w value is then determined as 52 + the number of decibels that curve *b* has been shifted upward to fulfill the criterion. (Downward shift subtracts decibels.)

resonant behavior at the low resonance frequencies (the *resonance region*). It will then behave as a mass at higher frequencies (the *mass law region*) because of damping at the edges and in the material. Finally, at extremely high frequencies (the *coincidence region*) and above, there is another type of resonance phenomenon, already discussed in Chapter 9, where the partition may become approximately sound transparent. The transmission loss curve has the basic shape shown in Figure 10.4.

In practice, all of these characteristics are seen only in building acoustics range measurement data for simple, single-pane windows. For most other building partitions, the resonance region is below 100 Hz; it is then that transmission loss in the mass law, the coincidence, and the frequency regions above the critical frequency are usually measured. Single-panel partitions are usually made of poured concrete, concrete building blocks, or bricks. Such heavy, stiff walls have their resonance and critical frequency regions below 100 Hz that results in good sound insulation properties also at low frequencies.

The transmission loss depends on the diffusitivity of the incident sound field. Equation 10.6 was derived under the assumption that the room sound fields are diffuse. The sound field due to sound incident on the facade of a building will not be diffuse—in that case, a different approach is needed.

In Chapter 7, we studied the influence of a limp panel on the sound absorption of an ideal sound-absorbing material (for perpendicular sound incidence). We can study transmission loss of a panel in a similar way for perpendicular sound incidence. The sound power absorbed will be sound power

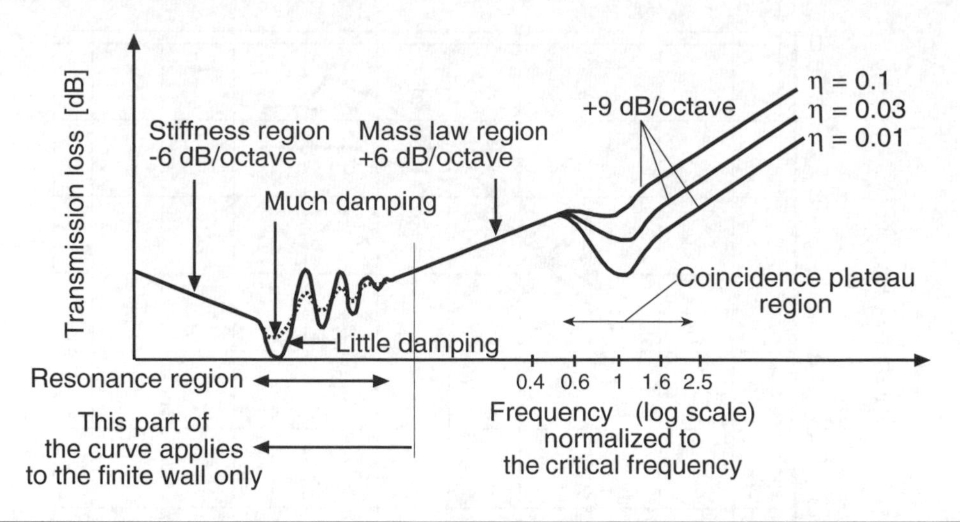

Figure 10.4 The transmission loss of the single-leaf partition shown normalized to the critical frequency for some values of the loss factor η of the material of the partition.

transmitted through the panel. Equation 7.9 gives the ratio of sound absorbed to the sound power incident, that is, α. Since no power is absorbed by the panel in the present case, the absorbed sound is equal to the sound power transmitted to the sound power incident. The transmission coefficient τ_0 will be:

$$\tau_0 = \frac{1}{1+\left(\frac{\omega m''}{2Z_0}\right)^2} \tag{10.7}$$

where m'' is the mass per unit area. From Equation 10.7, we find that the transmission loss for a plane wave, incident perpendicularly on a limp, single panel partition in the mass law region, is approximately given by:

$$R_0 \approx 20\log(m''f)+42 \tag{10.8}$$

In the common case of a diffusely incident sound field, the transmission loss is slightly reduced:

$$R_d \approx R_0 - 7 \tag{10.9}$$

Lightweight, single-panel partitions are usually found only in window constructions and some indoor and old outdoor wall types. The transmission loss in the stiffness region will not be discussed here, but is important in the case of loudspeaker boxes and other types of enclosures.

The critical frequency of a panel can be calculated using Equation 8.20. Table 8.1 lists the critical frequencies f_c of some typical building materials.

The transmission loss will drop considerably in the coincidence region, particularly noted if measured with good frequency resolution. The coincidence phenomenon occurs first for obliquely incident waves that propagate parallel to the partition. As the frequency of sound increases, the coincidence phenomenon will occur at smaller angles of incidence. Around each angle of incidence, there will, however, be only a few modes with similar incidence angles. Since the energy in the sound field is divided between the modes, this means that only a part of the sound field energy is available for transmission at each frequency. This prevents the transmission loss from becoming zero. It is customary to approximate the resulting *coincidence plateau* by a constant transmission loss in the frequency range between 0.5 to 2 times the critical frequency. For the coincidence plateau region, the transmission loss will be approximately:

$$R_{d,f_c} \approx R_d + 10\log(\eta) + 8 \tag{10.10}$$

Over the coincidence plateau frequency region, the transmission loss can be approximated by the expression:

$$R_{d,f>f_c} \approx R_0 + 10\log\left(\frac{f}{f_c} - 1\right) + 10\log(\eta) - 2 \tag{10.11}$$

In summary, one can note that the transmission loss theoretically increases at a rate of 6 dB per octave up to the coincidence plateau and after the end of the plateau by 9 dB per octave. Note that both in the plateau region and above, the transmission loss depends on the loss factor of the panel; this is a result of the resonant behavior of the coincidence phenomenon.

Transmission loss data for some typical single-panel partitions are shown in Figure 10.5. We notice that high transmission loss is provided by massive, heavy materials such as concrete. Concrete and brick have better sound isolation properties than steel, glass, plywood, or gypsum panels. Porous sound-absorbing materials (such as expanded polystyrene and glass wool) have extremely poor sound

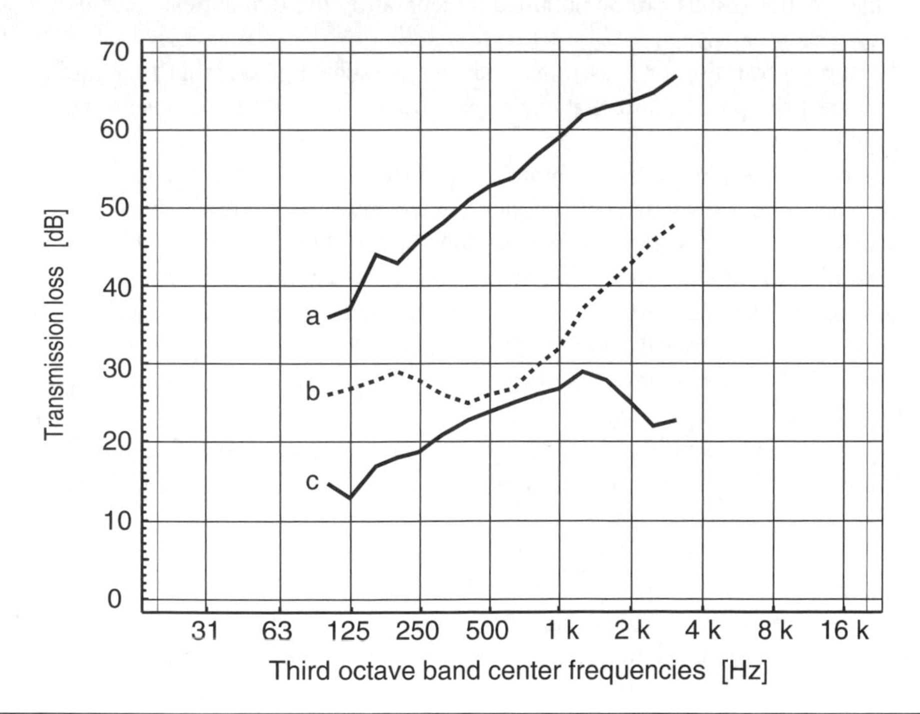

Figure 10.5 Examples of some measured transmission loss curves for single-panel walls: (a) 160 mm concrete, $f_c \approx 100$ Hz, (b) 70 mm lightweight porous concrete, $f_c \approx 400$ Hz, and (c) 13 mm drywall panel, $f_c \approx 2.0$ kHz.

isolation because of the open pores (glass wool) and the low mass of the porous sound-absorbing panel (both). Sound-isolating materials must not transmit air (must be impervious) and must either be rigid or have high mass per unit area. It is important to note that heat-insulating materials are generally poor sound isolators. One should not confuse the NR provided by acoustic absorbers within a room with NR between rooms.

10.6 TRANSMISSION LOSS OF MULTIPLE-PANEL PARTITIONS

The unsatisfactory transmission loss of single-panel partitions can often be increased considerably by the use of multiple panels. If the total panel mass/unit area is small, the transmission loss will still be poor, particularly at low frequencies.

The simplest form of multiple partition uses two panels with a mechanically separating airspace. In practice, it is necessary to use metal or wooden studs to provide a mechanical frame to which the panels can be attached. Attachment is usually done using nails or screws. Because of the mechanical attachment, a practical multiple panel partition has a lower transmission loss than theoretically predicted. Examples of such double-panel partitions are dual- or multiple-pane windows and drywall partitions using gypsum, plywood, or chipboard. Sometimes, when high transmission loss is required, double-panel partitions using concrete or bricks are necessary to ensure sufficient noise rejection.

Studs must be well separated to prevent fastenings from causing too high a stiffness. Steel studs will have lower stiffness than wooden studs that allow the individual panels to behave more according to the mass law. Even better results can be obtained by separating the two panels by using staggered studs (separate studs for each panel).

For the discussion to follow, it is assumed that the mass law applies, that there are no mechanical connections between the panels, and that the critical frequencies of the panels are much higher than any frequency of interest.

The masses of the two panels, in combination with the compliance of the air trapped between the panels, will form a resonant system. The compliance effectively mechanically decouples the panels. The resonance is called the *double-wall fundamental resonance*, and the frequency at which it occurs is the fundamental resonance frequency. Note that this resonance is of a different form than that of the single-panel discussed earlier. (The coincidence resonance phenomenon will be discussed in more detail later, but the higher order resonances will not be discussed here.)

The theoretical transmission loss, as a function of frequency, for an infinitely large double-panel partition with only airspace separating the panels is shown in Figure 10.6. The transmission loss at the resonance frequency depends on the damping of the system. If the panels of the partition are free, the fundamental resonance frequency f_0 of the double-wall for perpendicularly incident waves, is:

$$f_0 = \frac{c_0}{2\pi}\sqrt{\frac{\rho_0}{d}\left(\frac{1}{m_1''}+\frac{1}{m_2''}\right)} \tag{10.12}$$

where m_1'' and m_2'' are the respective masses per unit area of the panels, and d is the depth of the airspace. The fundamental resonance frequency of the double-panel partition is determined by the masses and the air compliance, as if the system consists of two masses joined by a spring. Figure 10.7 shows a double-panel gypsum wall under construction.

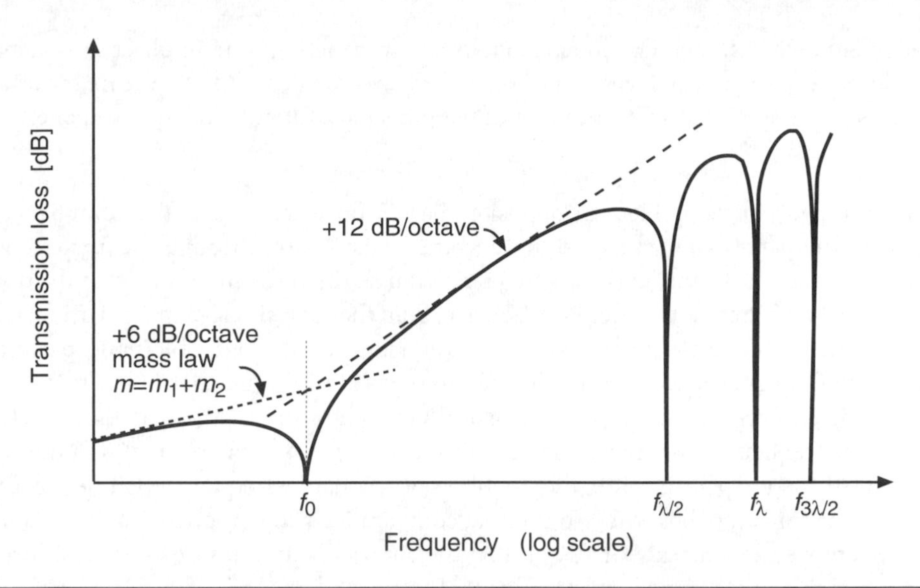

Figure 10.6 The transmission loss, as a function of frequency for perpendicular sound incidence, for an infinitely large double-panel partition (solid line) with only airspace separating the panels. The transmission loss curve for a single-panel partition (dotted line) having the same total mass as the double wall is shown for comparison. Dashed line is the asymptotic line for a double wall not having standing waves in the airspace. The effect of such standing waves is shown at the frequencies where the airspace distance between the two panels is a multiple of one-half wavelength. Curves shown for $f_0 << f_c$.

Figure 10.7 A double-panel drywall under construction showing the back of one panel and the steel studs, waiting for glass wool filling and the second panel to be installed.

At frequencies below the fundamental resonance, both panels vibrate in phase as one joined mass, and the double-panel partition behaves as a single-panel partition according to the mass law but with an equivalent mass per unit area equal to the sum of the mass per unit area for the two panels:

$$m'' = m_1'' + m_2'' \tag{10.13}$$

At resonance, the two panels oscillate in opposite phase. Above resonance, the compliance of the air trapped between the panels will act as a shunt and the transmission through the airspace will diminish as frequency increases. In this mass law frequency range, the transmission loss of the double-panel partition increases at a much faster rate, 12 dB/octave, than that of a single-panel partition, 6 dB/octave.

Sometimes one finds that the measured transmission loss curve of a double-panel partition exhibits sharp dips at high frequencies, which are unrelated to the critical frequencies of the panels. These dips are due to standing waves in the airspace acoustically coupling the two panels tightly. The standing waves occur at the frequencies where the airspace distance between the two panels is a multiple of one-half wavelength. If the airspace is wide, the double-wall resonance frequency will be correspondingly low, but at the same time, the dips will also occur at comparatively low frequencies. The resonances and dips are easily removed by partially or fully filling the airspace with a porous sound-absorptive material, such as a mineral or glass wool isolation blanket. Added advantages of such a porous blanket will be better damping of the fundamental resonance, a drop in the fundamental double-panel resonance frequency, and increased sound transmission loss in this frequency range, as shown in Figure 10.8.

Air space resonances may also occur parallel to the panels and will reduce the transmission loss. Such resonances can be damped by using sound absorption at the edges of the airspace for example in double glazing.

The influence of the coincidence phenomenon will be noted as a leveling of the transmission loss curve. For most multiple-panel partitions, however, this range is at such high frequencies as not to be a problem. If the coincidence effect is a problem, it can often be circumvented by substituting several thinner panels for a thick panel. This technique is often used in drywall construction with good results as shown by the *b* and *d* curves in Figure 10.8. Another application of this technique can be found in the design of high performance sound-insulating windows.

Special types of multiple-panel partitions are the *suspended ceiling* and the *resilient skin*. In both of these cases, the aim is to improve the sound insulation of a heavy and stiff partition, such as a concrete floor slab or wall, or a brick wall. These partitions have low critical frequencies, but, in spite of their mass, the transmission loss is sometimes still not as high as desired. In this case, the addition of an extra layer, effectively building a double-panel partition, is useful to improve the sound insulation, even if the extra layer has low mass. Usually the extra layer takes the form of a panel of gypsum or similar material having a high critical frequency.

In the case of the suspended ceiling, the airspace is usually large, a depth of some 200 mm is typical. In either case, partially filling the airspace by a mineral or glass wool blanket further increases the transmission loss, often by approximately 10 to 20 dB.

The resilient skin is typically an addition of a panel having a high critical frequency on steel or wood studs attached to an existing construction. Fire regulation often prevents the use of wood studs. The air space is usually quite thin (only 2 to 5 cm) and filled with a glass wool blanket. Figure 10.9 shows an example of the effect of such a radiation reducing resilient skin layer to a brick wall.

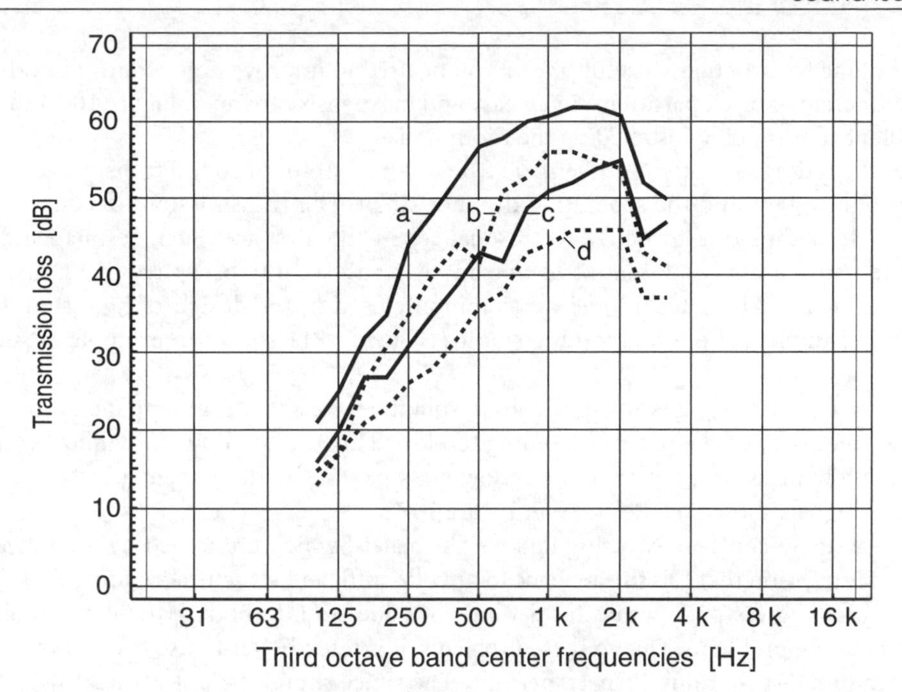

Figure 10.8 Results of some laboratory measurements of drywalls: (a) dual layer double-panel drywall with 45 mm steel studs at 0.6 m distance with 30 mm mineral wool blanket, surface mass approximately 50 kg/m², (b) single-layer, double-panel drywall with 70 mm steel studs at 0.6 m distance with 30 mm mineral wool blanket, surface mass approximately 50 kg/m², (c) dual-layer, double-panel drywall with 45 mm steel studs at 0.6 m distance, no mineral wool blanket, and (d) single-layer, double-panel drywall with 70 mm steel studs at 0.6 m distance, no mineral wool blanket.

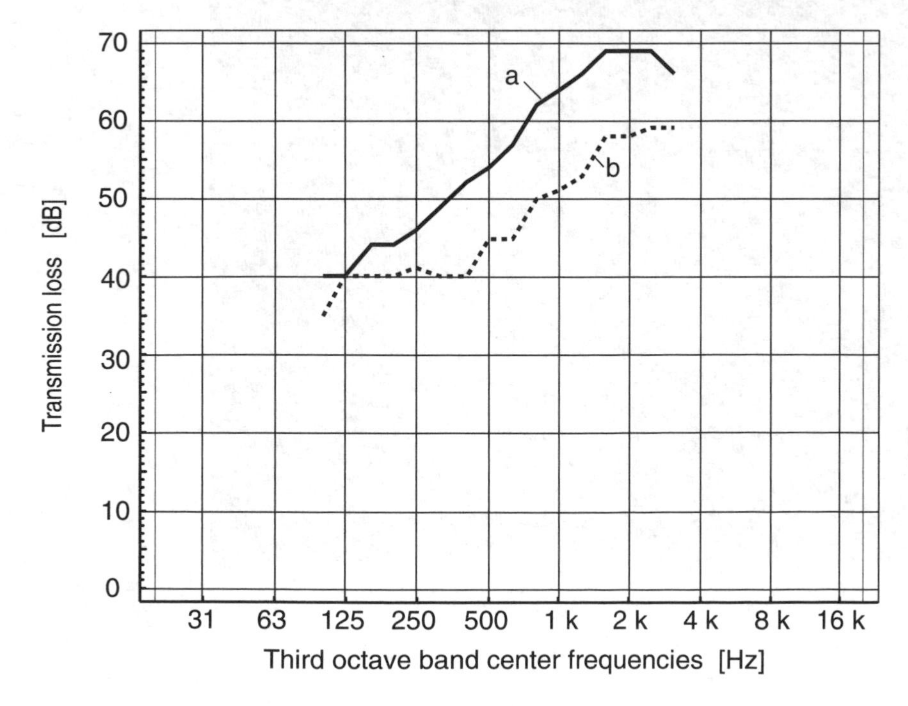

Figure 10.9 Example of the transmission loss of a 200 mm thick brick wall, with (a) and without (b) a resilient skin. The resilient skin was a 13 mm thick gypsum panel on 50 wood mm studs. The air space held a 25 mm thick mineral wool blanket.

The technique is particularly useful since it can be used to improve noise control in existing buildings having concrete or brick partitions, in an easy and inexpensive manner. Figure 10.10 shows a room where a resilient skin is being installed on the room walls.

Doors and windows are often problematic from the viewpoint of sound transmission. The panels or panes are lightweight and the separating distance small. The transmission loss curves of double-glazing often show dips due to half-wave resonances in the airspace. Such resonance dips can be avoided if the panes are angled relative to one another and sound-absorptive materials used at the edges of the air space. These techniques are commonly used in the design of high sound insulation windows—for example, in sound recording studios. Figure 10.11 shows an example of such a studio window.

As mentioned previously, it is advantageous to split the panels into thinner panels if there is danger of the critical frequency of the panels becoming too low. This is often done in window constructions. Because of the low mass per unit area of window glass panes, the fundamental resonance frequency often becomes high as shown by the curve in Figure 10.12.

Doors are usually constructed using lightweight panels, which are joined by a lightweight paper honeycomb construction that fills the airspace to provide sufficient structural rigidity. This type of door construction usually gives poor sound insulation. For good soundproofing, the doors must be heavy, resulting in handling problems. Due to these problems, it is often better to use two doors with an intermediary sluice or short corridor, if space permits. The sluice should be at least 0.2 to 0.5 m long and have sound-absorptive walls and ceiling.

Figure 10.10 A resilient layer being added to a room to improve the sound isolation (photo by Alex Hill, Soundservice, UK).

Figure 10.11 Example of a window between a recording studio and a control room that uses offset (non-parallel) glass panes and sound-absorptive material at the interior air space edges. (Photo by Mendel Kleiner.)

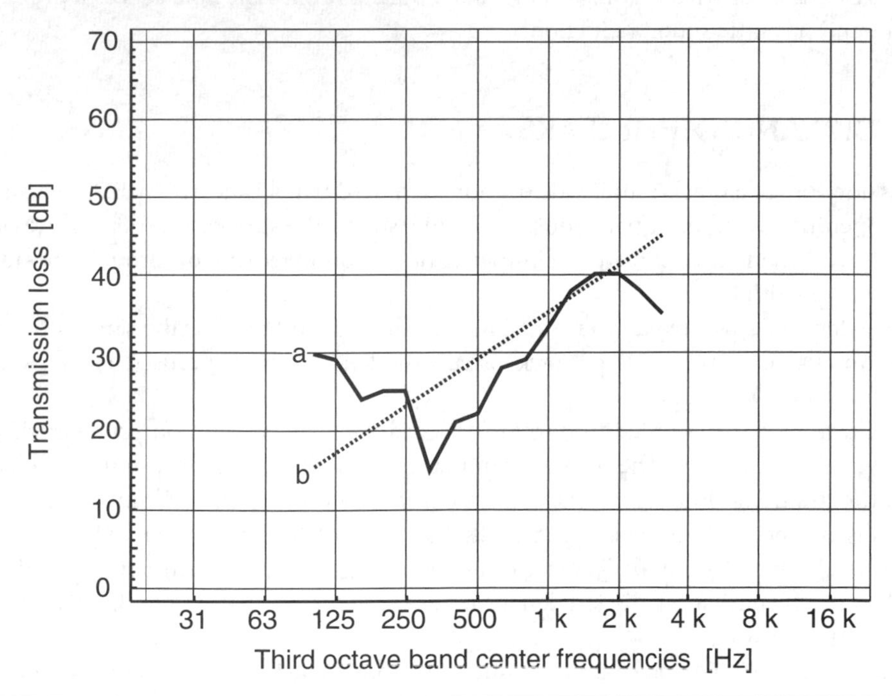

Figure 10.12 Example of the transmission loss of double glazing: (a) glass thickness 3 mm, air gap 10 mm and (b) shows the theoretical mass law curve for single 6 mm glazing having the same total mass.

Another advantage of the latter construction is that most doors and windows have leaks around their perimeters—even good seals will age—and by using a sluice, the impact of poor sound transmission due to leaks is minimized.

Folding doors and walls are particularly difficult to seal. For this reason, two folding units with a rather wide airspace in between are typically used.

10.7 COMPOSITE TRANSMISSION LOSS

Real partitions usually consist of several different parts—wall sections, doors, windows, and cracks or other openings. If one knows the transmission loss of the various parts, the *composite transmission loss* can be calculated using the expression:

$$R_{\text{comp}} = 10\log\left(\frac{\sum_i S_i}{\sum_i S_i\tau_i}\right) = 10\log\left(\frac{\sum_i S_i}{\sum_i S_i 10^{-R_i/10}}\right) \tag{10.14}$$

where S_i are the respective areas of the subelements and R_i are the respective transmission loss values in the frequency band.

Of course, close to a partition subelement having low sound transmission loss, the sound through that particular subelement will dominate, and the local NR will correspondingly be very low. Close to a keyhole, it is easier to hear what is going on in a neighboring room. The hole acts as a probe tube to the sending room and has little sound reduction.

10.8 CRACKS AND OTHER LEAKS

The transmission loss of a crack 1 mm wide or more in a hard panel is nearly 0 dB. One usually considers the crack opening as a separate partition part. By measuring the surface area of a fissure or crack, and using Equation 10.14, one can estimate the influence of the sound leaking in through the small opening on the composite partition.

Long holes tend to have smaller transmission loss than round holes for the same total surface area. Figure 10.13 shows an example of a large crack visually hidden by trim that caused a serious transmission loss problem between two offices.

Sound leakage by fissures around poorly installed drywalls is a common problem since it is easy to visually cover cracks by trim. The trim seldom makes an airtight seal, thus, sound will leak through, even though the fissure no longer attracts visual attention. The slit often used under doors for ventilation purposes is another example of an opening that seriously deteriorates the maximum sound insulation that can be obtained by a door. There are special sound-absorptive ventilation openings available that should be used if ventilation is desired in a similar way. One must also note that fissures are usually at the corners between walls.

As will be discussed in Chapter 14, the position of a sound source determines the coupling of sound from the source to the sound field in the room, since the radiation impedance changes between locations in rooms. A sound source that is located in the corner between three walls usually radiates sound

Figure 10.13 A major leak, between a lightweight concrete wall and the floor, that was hidden by wood trim. (Photo by Mendel Kleiner.)

better than one that is located at the corner between two walls, which in turn radiates more than one in the middle of a wall.

10.9 FLANKING TRANSMISSION

In most buildings, the vibration caused by sound incident on the walls, ceilings, and floors in a room is transmitted through a number of paths to other rooms as shown by Figure 10.14. The resulting transmission loss is determined by areas and acousto-mechanical properties of the various surfaces, and of course the mechanical properties of the joints. The transmission of sound by airborne or structure-borne paths, other than the direct path through the joining wall or floor, is called *flanking transmission*.

The importance of limiting the flanking transmission increases as the direct path transmission loss increases. The presence of flanking transmission is one important reason for the transmission loss of partitions in actual buildings being lower than that measured in the laboratory, where flanking transmission is carefully avoided.

The apparent transmission loss R_{field} may be calculated from:

$$R_{\text{field}} = L_{pS} - L_{pR} - 10\log\left(\frac{A_R}{S_p}\right) \tag{10.15}$$

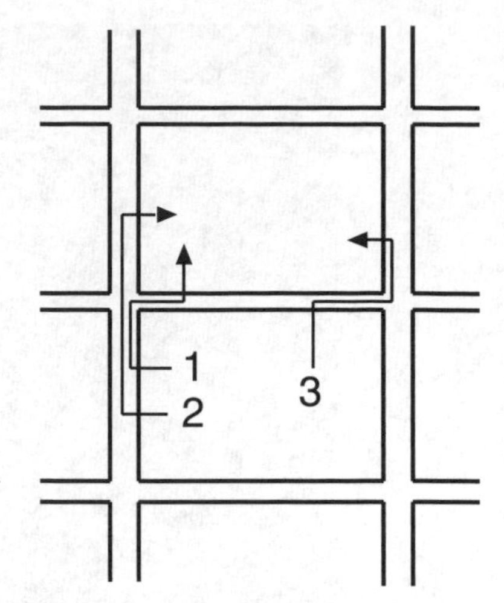

Figure 10.14 Paths 1–3 are the most important flanking sound transmission paths, thus, with four walls there will typically be at least 12 flanking sound transmission paths between two adjacent rooms in a building.

where L_{pS} is the sound level in the diffuse field in the sound source room, L_{pR} is the sound level in the diffuse field in the receiving room, A_R is the sound-absorption area in the receiving room, and S_p is the joint partition area between the two rooms, provided that it is larger than 10 m². If the partition area is smaller than 10 m², or if there is no joint partition area, this area is set to 10 m². Note that one refers to the same partition area irrespective of the path of the sound transport.

If the sound transmission between two rooms is over many paths, and one knows the apparent partial transmission loss for each path, one can calculate the composite transmission loss $R_{\text{field,comp}}$ using:

$$R_{\text{field,comp}} = -10\log\left(\sum_i 10^{-R_{\text{field,i}}/10}\right) \tag{10.16}$$

Since the same transmission area is referred to, this expression is just a simplification of Equation 10.4.

A special form of flanking transmission is the transmission over airborne paths, such as various leaks and crevices, as well as by way of air ducts for ventilation and heating. Occasionally, sound may be transmitted by the water pipes of heating systems.

Flanking transmission is often a problem in buildings using moveable lightweight walls when there are floating floors and suspended ceilings as indicated in Figure 10.15. Because of the mounting method, the structural vibrations in the floating floor will be subject to little *corner* transmission loss.

Flanking sound transmission through the airspace above the suspended ceiling is often considered by adding sound absorption to the suspended ceiling. The best way to limit the transmission of airborne sound is by extending the partition to include the volume over the suspended ceiling. However, even in this case, there may be flanking transmission due to sound pickup by the ventilation duct openings. Once the sound is inside the duct, there will be little attenuation of the sound unless silencers (in the canals or at the vents) are used to prevent transmission through the duct system. Even ventilation ducts without openings into a room can cause flanking transmission since the sound waves will transmit through the lightweight duct walls.

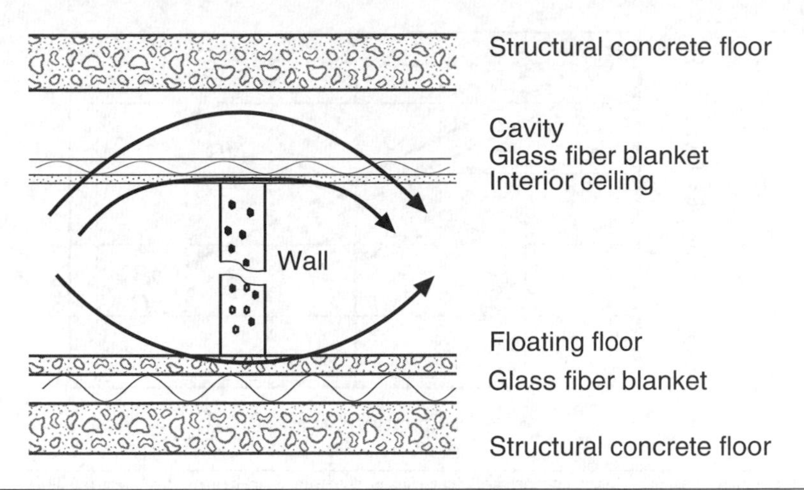

Figure 10.15 Flanking transmission paths for a floating floor and a resiliently suspended ceiling.

A particular problem of this type of flanking transmission is that the sound leaks reduce the privacy. The sound pressure level will be much higher close to a vent or a fissure than in the diffuse field of the room, making it easy to eavesdrop on conversations.

10.10 BALANCED SPECTRUM DESIGN

Any room will have background noise contributions from many sources—HVAC, traffic, neighboring rooms, and facilities. The *balanced spectrum design* approach calls for the contribution of the noise transmitted through a certain partition to contribute negligibly to the level in the receiving room. For continuous, random process noise, such as aerodynamic noise, this is a useful design approach. Figure 10.16 shows the design curves used in the discussion in the following paragraph.

Once a decision has been made on the allowable background noise levels in the receiving room—for example, using the octave band levels specified in Chapter 6—the design can start taking the properties of the sending room noise into account. In the figure, the sound level in the sending room is L_{pS}, and the sound level of the background noise in the receiving room is L_{pB}. The spectral shape of the masking noise will depend on its source; traffic noise (also as transmitted through windows), and ventilation noise (from, for example, air conditioning) have different spectral shapes.

Both noise reduction and transmission loss values typically grow with increasing frequency. The sound level of the background masking noise L_{pB} typically decreases with increasing frequency. However, many noise sources—such as radios, televisions, and musical instruments—have spectra that are not flat. They also have easily recognizable sound features making them irritating to listen to, even when they barely *breakthrough*. For such noise sources, it is advantageous that the transmitted sound level in the reverberant field is at least 10 dB below the masking noise, so as not to be audible.

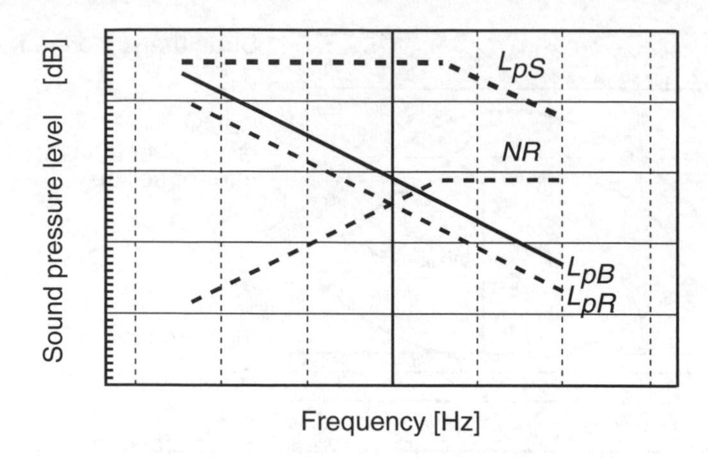

Figure 10.16 Spectrally optimized design approach to privacy by balanced spectrum design. The sound level of the transmitted noise should be below that of the masking noise at all octave band frequencies.

The design curve for the sound level due to the noise through the partition will then be for a sound level curve 10 dB lower, according to L_{pR} in the figure. The difference between curves L_{pS} and L_{pR} will then be the noise reduction target shown by the NR curve.

10.11 INSULATION AGAINST IMPACT SOUND

Often sound-radiating vibration is caused by forces, on the partitions of a building, other than sound. Forces on floors or ceilings are due to footsteps, rolling wheels, machinery, and such. However, heavy-duty, wall-mounted electrical relays, such as those used in elevator systems, and phenomena such as water hammer, may also cause impacts that are disturbing to dwellers.

A floor is usually constructed using several different layers in addition to the layer carrying the structural load. Typical extra layers are floating floors and carpets on the upper side, and sound-absorbing or isolating materials, either ceiling-mounted or resiliently suspended from the ceiling, on the lower side. All of these layers influence both the transmission loss and the sound radiation to the rooms, not only above and below the floor, but also to other rooms by the principle of flanking transmission.

The impact sound transmission is estimated by measuring the *impact sound pressure level* created in an adjoining lower room when using a tapping machine on the floor, as specified by the ISO 140-6 standard. The standardized tapping machine uses five weights, which consecutively drop on the floor 10 times per second, creating vibration in the floor and, of course, considerable sound levels in the upper room. The vibrations on the upper side of the floor are transmitted to the lower side of the floor and radiate sound into the lower, receiving room.

The *normalized impact sound pressure level* is adjusted for the sound absorption of the receiving room and is measured in the third octave frequency bands from 100 Hz to 3.15 kHz since the noise created by the tapping machine varies with frequency depending on the floor and its covering. A high impact sound pressure level indicates an acoustically poor floor construction.

The impact sound pressure level and spectrum created this way may not be relevant to the impact sound mechanisms in some cases, so care should be used when considering sufficient impact sound insulation. This particularly applies to the rating of the impact sound properties of lightweight floors, since the tapping machine uses masses that are much smaller than those of human feet and legs. The method also does not take into account the nonlinearities at the point-of-contact of the impact, which may be due to the elastic properties of, for example, the treatment layer.

Similar to the use of the weighted sound reduction quantity for airborne sound transmission, one uses a single-number quantity to characterize the impact sound transmission properties of floors and additional floor and ceiling layers. The calculation of these quantities, the *weighted normalized impact sound pressure level* $L_{n,w}$ [dB], and the *weighted impact sound improvement index* ΔL_w [dB] are described in the ISO 717–2 standard.

It is important to remember that a high value of $L_{n,w}$ corresponds to a poor construction. Values for $L_{n,w}$ and ΔL_w of typical floor constructions may be found at the various websites of materials manufacturers, research institutions, and testing organizations.

Most floor constructions, whether lightweight or heavyweight, have poor impact noise properties unless acoustically treated in some way. Concrete floor slabs have nearly a frequency-independent normalized impact sound pressure spectra, while lightweight timber and other floor constructions are plagued by high-normalized impact sound pressure at low frequencies (thumping noise).

Attempting to reduce the mass of floors by using concrete floors with cavities or by using precast φ-shaped or similar concrete elements, invariably leads to higher $L_{n,w}$ values than those of massive concrete slabs. Figure 10.17 shows data for a concrete slab and two types of surface treatments.

The low frequency dominant sound spectra obtained for impacts on lightweight floors are difficult to modify without adding more mass to the floor. Figure 10.18 shows an example of the measured normalized impact sound spectra of treated and untreated wood floors. One notes that lightweight wood floors tend to have severe low frequency noise problems unless treated with heavy ballast.

The poor impact sound properties of most floors makes it necessary to use various treatments, such as added layers on top or underneath of the floor, as mentioned previously. In the case of a concrete floor, common treatments consist of adding a resilient layer, such as thermoplastic tiles or carpets or other types of impact-reducing soft layers. More advanced treatments consist of cork or glass wool immediately on top of the concrete slab, covered by a sub-floor made of plywood, parquet, or wood strips. Such a construction is often called a *floating floor*. Since a floating floor is a double-panel partition, it exhibits a fundamental resonance, just as a double wall does. The resonance frequency is typically determined by the surface mass of the sub-floor and the compliance of the resilient layer or trapped air. Good floating floor installations require high-grade workmanship similar to that required in the case of avoiding flanking transmission.

Figure 10.19 shows an example of a multilayer floor characterized by significant high transmission and impact loss. Mechanical short-circuits, vibration transmission paths, as generated by water pipes, for example, must be eliminated for the floating floor to be effective in reducing the impact noise.

The fundamental resonance frequency of a floating floor can be estimated by Equation 10.17. When the frequency is well above the fundamental resonance frequency, the normalized impact sound pressure level improvement is approximately impact sound pressure level:

$$\Delta L = 40\log\left(\frac{f}{f_0}\right) \qquad f_0 \ll f \ll f_c \tag{10.17}$$

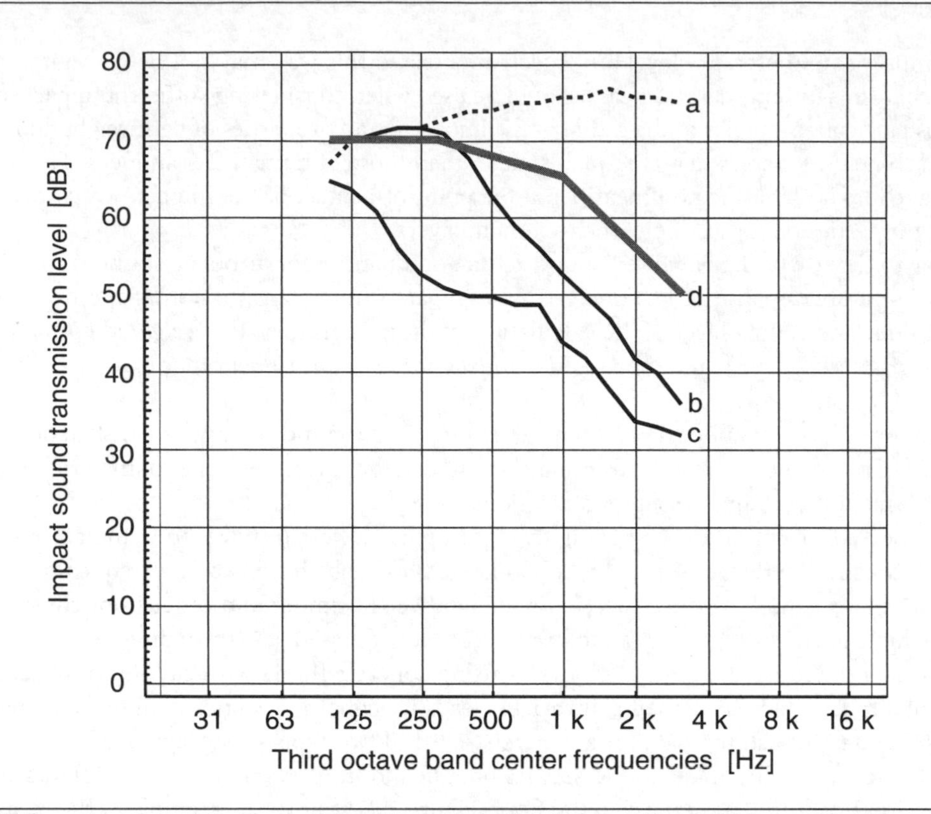

Figure 10.17 Examples of measured impact sound transmission levels for a 140 mm thick concrete slab and the influence of adding a 35 mm floating floor layer of cement on top of an elastic layer: (a) level for bare 140 mm thick concrete slab, (b) level using a 10 mm thick hard plastic elastic layer, (c) level using a 10 mm mineral wool elastic layer, and (d) reference curve used in calculating the weighted normalized impact sound pressure level. The reference curve is shifted in steps of 1 dB to allow sum of differences below measured curve (for example *a*) but above the reference curve to be ≤ 32 dB (at the third octave band frequencies from 125 Hz to 3.2 kHz). The $L_{n,w}$ value is then determined as 68 plus the number of decibels the curve has been shifted upward. (Downward shift subtracts decibels.)

above the *double-wall resonance* frequency given by Equation 10.18:

$$f_0 = \frac{\Lambda c_0}{2\pi}\sqrt{\frac{\rho_0}{m''d}} \tag{10.18}$$

Here d is the thickness of the mineral wool layer and Λ a constant smaller than unity, which is due to the nonadiabatic propagation conditions for the sound waves inside the layer.

It is difficult to obtain sufficiently good impact sound insulation in houses using light floor constructions without adding mass, such as sand, stone, or some other type of ballast. Another common impact sound insulation problem is resonances in the stair landings. These must be isolated from the rest of the building using elastic supports.

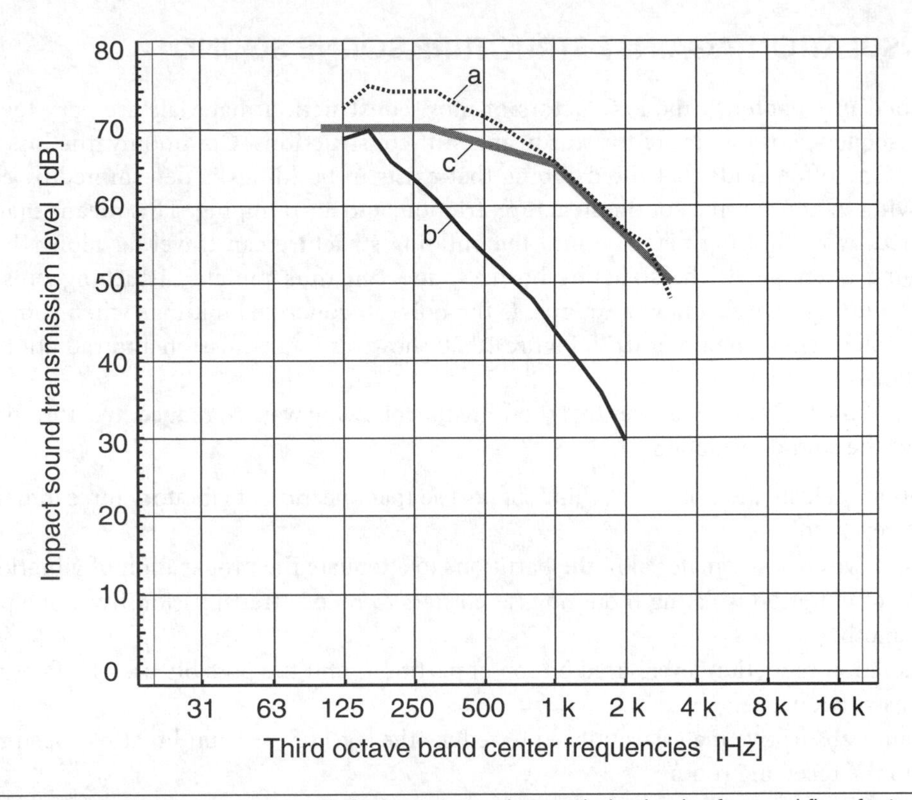

Figure 10.18 Examples of measurement results for impact sound transmission levels of a wood floor for two different treatments: (a) level without treatment, (b) level with treatment, using a particle board floating floor on sand, on top of a mineral wool blanket, and (c) reference curve used in calculating the weighted normalized impact sound pressure level.

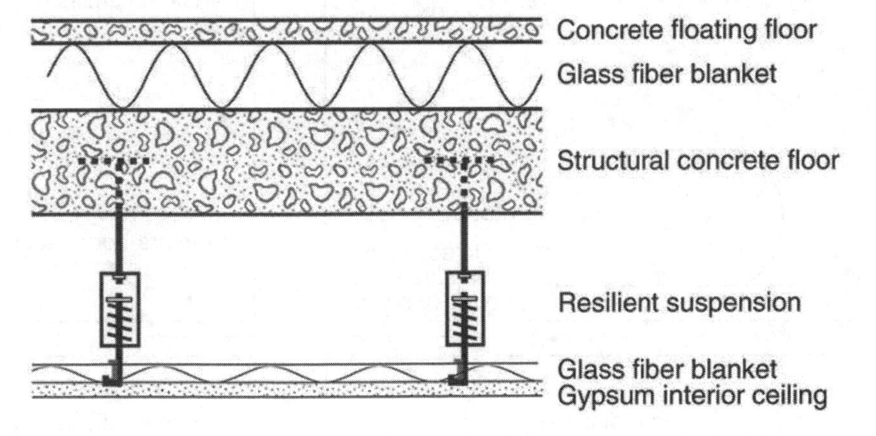

Figure 10.19 Example of a multilayer floor characterized by high transmission loss.

10.12 INSULATION AGAINST STRUCTURE-BORNE SOUND

As mentioned in Chapter 8, the loss factors of most construction materials are very low. This is a natural consequence of the desire for sturdy and stiff constructions. Commonly, the loss factors are below 10^{-2}. One often finds that the damping that exists in buildings is determined by edge losses, energy moving to another part of the structure, friction, and air pumping. This means that the vibrations injected by a vibrating machine into the building structure can travel far along the structure without being attenuated. The structure-borne sound can thus radiate disturbing noise in many places in a building. A well-known example is the noise in concrete buildings when someone is trying to drill holes using a hammer drill. Figure 10.20 shows the transmission and radiation paths of a typical installation.

Looking at the figure, one realizes that there are the following ways to reduce structure-borne sound and improve the sound insulation:

- Vibration insulation of the machine support so that injection of vibratory force into the floor is prevented
- Use of viscoelastic material in the partitions to attenuate the propagation of vibrations
- Use of vibration isolating mountings at corners to reduce transmission from one partition to another
- Radiation reduction layers used on top of partitions and the possible use of a floating floor construction
- Sound-absorptive materials utilized to reduce the level of the sound that has been radiated into the receiving room

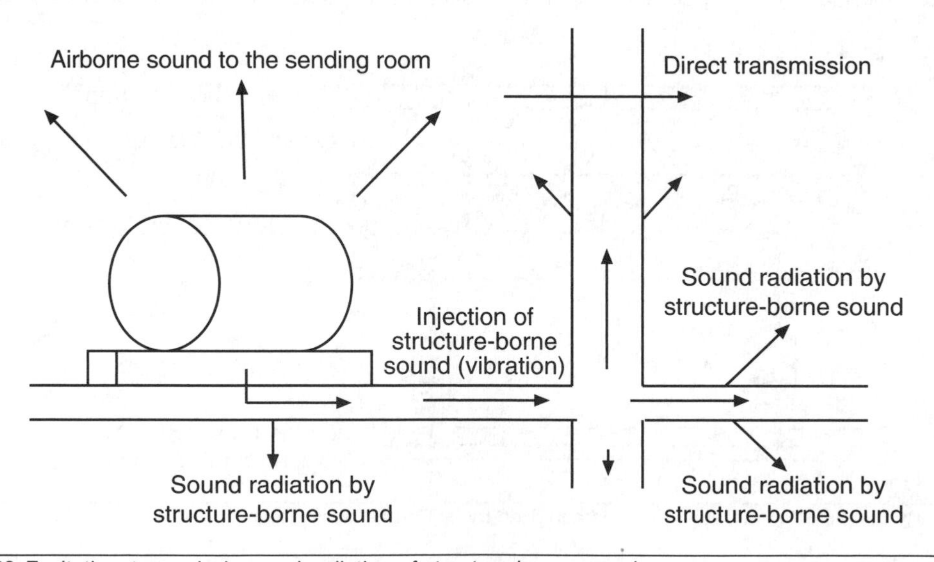

Figure 10.20 Excitation, transmission, and radiation of structure-borne sound.

Usually the sound is radiated by a bending wave field in one or more partitions making up the room. The bending waves have low mechanical impedance and are consequently easy to generate. The prerequisites for using viscoelastic layers to attenuate the propagation of structure-borne sound were discussed in Chapter 8.

It is necessary to differ between resonant and nonresonant waves in the reverberant vibration field in a panel or slab. The nonresonant (direct) wave is primarily attenuated due to geometrical spreading of energy. The vibration level will, in the two-dimensional case, diminish with distance r according to the distance law as $10\log(1/r)$—that is, at a lower rate than in a three-dimensional case. As shown in Chapter 8, the attenuation due to internal losses is approximately -13η dB/λ_B, which should be added to the geometrical attenuation. One notes that the loss factor η must be very large to have any appreciable influence on the propagation of the direct wave.

The resonant bending wave field is dependent on the loss factor. Analogous to the case of airborne sound, one finds that the level of the resonant vibration field is proportional to $10 \log(1/\eta)$ and to the total area of the panel. An increase in loss factor η from 10^{-2} to 10^{-1} will therefore result in much lower resonant bending wave vibration levels over the entire panel, except in the direct field of the vibration source.

In most buildings, it is quite impractical to use single-sided application of viscoelastic materials. Constrained viscoelastic layers are generally more practical than single-sided layers in buildings (see Chapter 8). They can be applied in both concrete and in lightweight fiberboard and gypsum panel constructions. It is often not necessary to cover the entire panel area with the viscoelastic compound or glue.

The best approach is to use the damping layer, or other damping measure, close to the place where the vibrations are induced in the construction. Not only does this result in lower cost, it also reduces the wave conversion at edges and the generation of wave types that are harder to attenuate than bending waves.

In some cases, it is possible to use suspended ceilings or floating floors mounted on special compliant isolators. This is equivalent to the use of a double wall or a resilient skin.

10.13 PROBLEMS

10.1 A wall has an area $S = 15$ m^2, a thickness $h = 0.16$ m, and is made of concrete. There is a door having an area $S_d = 2.2$ m^2 in the wall. The transmission losses of the concrete wall and of the door are 53 dB and 40 dB respectively at 500 Hz.

Task: Calculate the resulting transmission loss of the wall and door combined.

10.2 A television set is radiating noise into a room in the frequency range 125 Hz to 4 kHz. Assume the sound pressure level to be $L_p = 80$ dB in all the octave bands in this range on page 254. The wall to an adjoining bedroom has the transmission loss characteristic shown in the graph. The dimensions of the rectangular bedroom are $3.5 \times 3.5 \times 2.6$ m^3; its reverberation time is $T = 0.5$ s in this frequency range.

Task: Calculate the sound level in dBA in the bedroom due to the television set if there are no other noise sources. Only consider direct transmission.

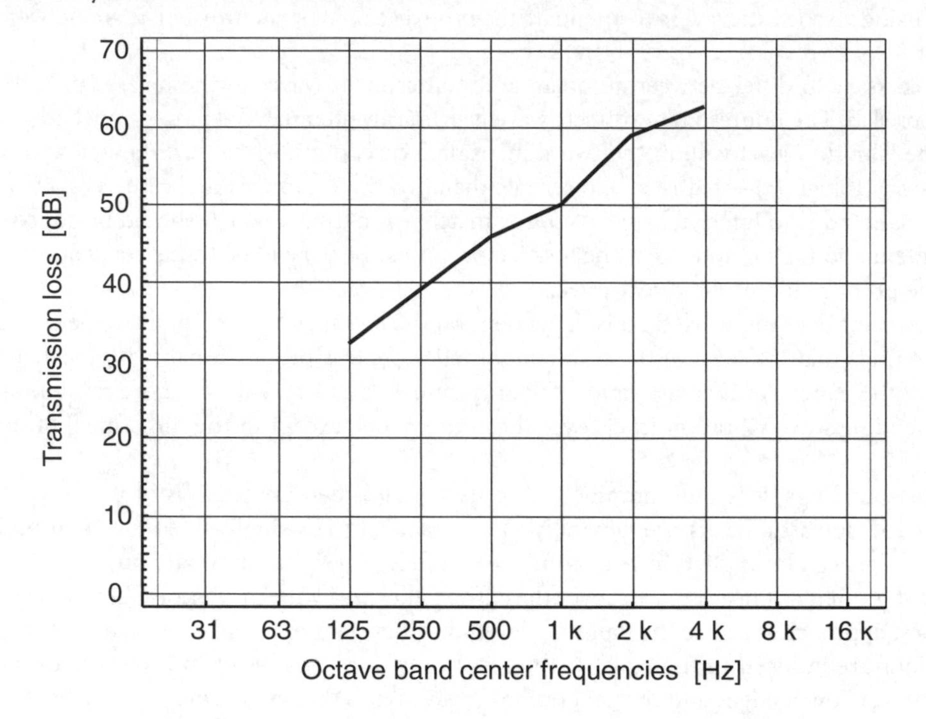

10.3 An apartment consists of two adjoining rooms. The larger room has a volume of 75 m³ and a reverberation time of 0.5 s. The smaller room has a volume of 24 m³ and a reverberation time of 0.5 s. One person, *A*, wants to be undisturbed by the noise from a television set that the other members in the family insist on watching. Assume all persons to be in the reverberant field.

Task: Which solution is best from a noise viewpoint?

a. To put the TV in the smaller room and *A* in the larger room.
b. To put the TV in the larger room and *A* in the smaller room.

Assume that:

1. The TV set's sound power output is the same in both cases 1 and 2 (same volume setting).
2. The TV viewers use the same sound pressure level irrespectively of where they are sitting.

10.4 The transmission loss of a particular drywall is given by the curve in the graph on the next page.
The wall is a double-panel construction using 9 mm and 13 mm gypsum panels. The panels are separated by 75 mm wood studs at 0.6 m intervals. The air gap is filled with glass wool. The panels have the following data: $\rho_{gypsum} = 840$ kg/m³, $K_{c,gypsum} = 32–40$ m/s.

Tasks:

a. Calculate the fundamental resonance frequency and the critical frequency of the construction and explain the appearance of the measured curve
b. Suggest suitable changes to the wall construction to improve its transmission loss.

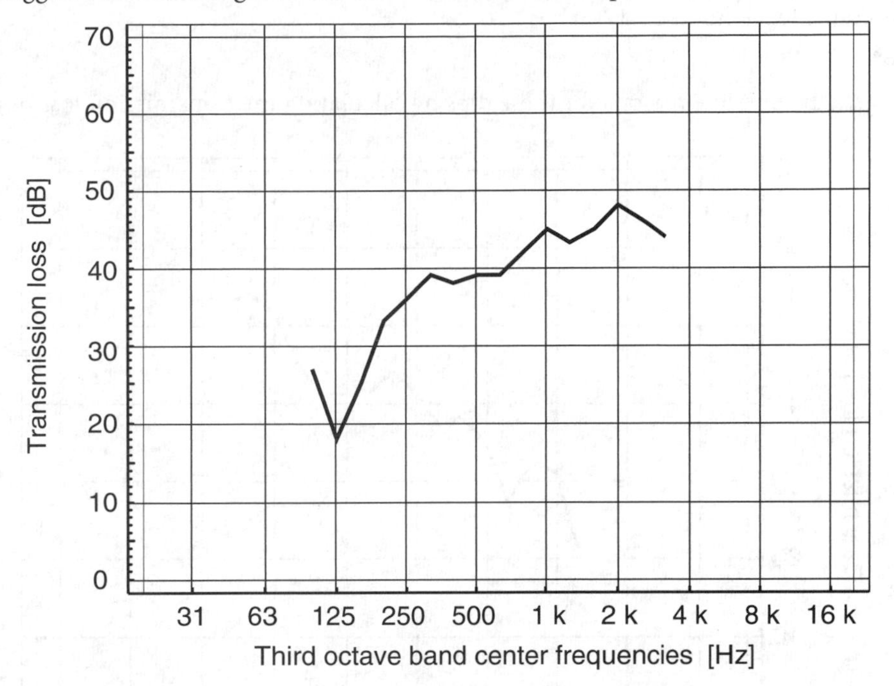

10.5 Installing office equipment often results in the need for cooling. A high volume copier is to be installed in a room (Room 1) neighboring an adjoining room (Room 2). Rooms 1 and 2 have the following data: The reverberation times are 1.5 s and 0.5 s respectively and the volumes are 100 m³ and 40 m³ respectively. The area of the wall between the rooms is 10 m².

The sound power level of the cooling equipment installed in room 1 as well as the transmission loss of the wall, are listed in the table.

Octave band	125	250	500	1 k	2 k	4 k	[Hz]
L_W	74	76	84	86	89	80	[dB] re $1 \cdot 10^{-12}$ W
R'	18	35	40	45	50	40	[dB]

Task: The maximum allowable sound level in the adjoining room is 45 dBA. Calculate the sound level in Room 2 and decide whether this requirement has been fulfilled.

10.6 It is often difficult to avoid resonances that reduce the transmission loss when designing double pane window constructions. The transmission loss curve in the graph on page 256 shows several transmission loss dips at 100, 200, and 400 Hz.

The window has:

Glass pane thicknesses: 10 mm and 4 mm
Glass pane surface dimensions: 0.83 × 1.80 m²
Glass density: 2,500 kg/m³
Glass coincidence number: 12 m/s
Air gap: 56 mm

Task: Investigate the possible reasons for these dips by calculating the transmission loss of the window.

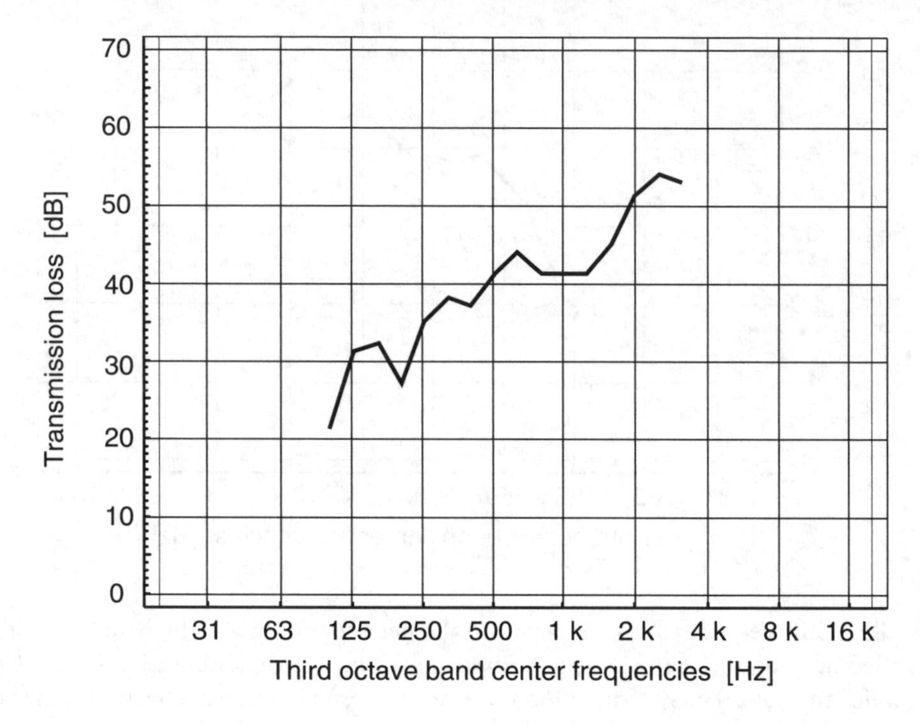

Vibration Isolation 11

11.1 INTRODUCTION

Vibration isolation is the name given to a number of techniques for reducing the transmission of vibration between generating and receiving mechanical structures. In audio engineering, we are interested primarily in two tasks—isolation of force and velocity:

1. *Force isolation* means that we try to reduce the force by which a device, such as a machine or loudspeaker, is acting on its support structure.
2. *Velocity isolation* means that we try to prevent a machine from shaking, in spite of its being mounted on a vibrating base. An example of velocity isolation is mounting hard disk memory or other electronics parts on a vibrating structure, such as a car or floor.

The devices used for achieving isolation are called *vibration isolators*. Vibration isolators need to have high mobility to be effective. Typically, isolators are soft springs of steel or brass, glass fiber pads, or air-filled cushions (for example, air cavities or sheets of glass wool). A simple vibration isolation system used in a phonograph turntable is shown in Figure 11.1.

The metal cross subchassis that will be holding the platter and tone arm is a steel structure suspended by three steel springs and some rubber bushings. The three spring systems are placed on the foundation (for example, a table). The aim is to prevent floor or wall vibration due to footsteps, as well as vibration induced by loudspeaker sound, from being injected into the steel structure.

There are also *alternate ways* of reducing vibrations, such as the *addition of mass* or *stiffness*, as well as the use of *tuned mechanical resonators*, also known as *dynamic absorbers*. These will not be dealt with here.

11.2 CLASSICAL VIBRATION ISOLATION THEORY

Vibration isolation is usually achieved by *creating a mismatch* between the mechanical impedances at the connections of the two structures. The impedances can be those of both translational and rotational motion. The generating and receiving structures generally have a *high mechanical impedance* (low mechanical mobility).

Figure 11.1 A subchassis suspended on steel springs can frequently be used to achieve good vibration isolation for a phonograph at medium and high frequencies.

The impedances of the two structures are often of the same magnitude, which may lead to a strong exchange of vibratory energy between the structures. The exchange can be reduced by inserting an impedance mismatch that will lead to vibration reduction.

For reasons of static stability, most equipment is fastened to its base at several points at some distance apart. There will then be an interaction between the vibratory motion at the various points. In *classical vibration isolation theory*, the influence of such interaction is neglected. Classical theory assumes that there is a discrete impedance, for example, due to a mass attached to a foundation by one or more discrete springs and dampers under the following conditions:

- The motion of the machine is one-dimensional, translational, and perpendicular to the base.
- The foundation has infinitely high mechanical impedance.
- Frequencies are low enough for the source of vibration to be considered a discrete element.
- Frequencies are low enough for the spring elements to be considered as discrete massless springs and for mass elements to be considered inelastic.

These conditions are often approximately fulfilled at low frequencies, and when dealing with small vibratory sources. The theory can then be used to estimate the fundamental resonance frequency of the single degree of freedom vibration isolation system.

It is quite feasible to extend this vibration isolation theory to handle all six degrees of motional freedom (three translational and three rotational directions of motion). This text will, however, only give a short introduction to the topic primarily as it can be applied to audio equipment and associated electronics and will be limited to a single degree of freedom systems.

The theories and methods described here further assume the systems have linear characteristics—that is, only relatively small vibrations are studied. It is common for vibration isolation systems to have *nonlinear properties*. An extreme form of nonlinearity is that used to limit vibration displacement. Many isolation devices, such as rubber pads, have nonlinear characteristics also in their working range. There are also *active vibration isolators* that use electrically variable components to control vibration. These can be designed to interfere with the vibration or to simply change the dynamic properties of the devices in the system, such as effective mass or damping.

11.3 IMPEDANCE AND MOBILITY

For medium and high frequencies, the mechanical systems (machine, isolator, and base) are best analyzed using the *mobility theory*. The *mechanical mobility* $\underline{Y}_M$ is defined as the inverse of the mechanical impedance $\underline{Z}_M$, which in turn was defined in Chapter 2. The theory assumes that the mobilities can be regarded as complex. One can then use the methods of *electromechanical analogies* and electrical circuit theory to find engineering solutions to complicated mechanical systems.

At low frequencies, many devices such as hi-fi/stereo equipment can be regarded as discrete *masses*. This means that the internal mobility can be approximated by $\underline{Y}_M = 1/j\omega M$ where M is the mass of the unit. However, even at a few hundred Hz, internal resonances will lead to the internal mobility having approximate frequency independent characteristics at the equipment supports. These characteristics are similar to those of the foundations.

Springs are the most common isolator component. The mobility of an ideal spring depends on the spring constant k_I that is given by the ratio of applied force $\underline{F}$ to net length change $\underline{x}$ is $k_I = \underline{F}/\underline{x}$. Using the j$\omega$-method, we find that the spring's mobility is $\underline{Y}_M = j\omega/k_I$, since velocity u is related to displacement as $\underline{u} = j\omega\underline{x}$. Any actual spring will be characterized by internal losses. Steel springs have small internal losses whereas rubber, cork, and glass wool springs are characterized by greater losses. Because of the small losses, steel springs have internal resonances that, unless damped, will limit usefulness. At high frequencies, there may even be wave motion in the spring. Figure 8.3 illustrates how such wave motion can be used for special purposes. In many practical cases, it will be necessary to increase losses by adding extra damping to systems using steel springs (for example, the system shown in Figure 11.1). Some spring isolators are made of rubber, which is used in a shear mode, to combine softness and losses.

Components mainly characterized by internal losses are called *dampers*. Viscous dampers, such as air dampers, are common, as are oil-based dampers, but dry friction dampers made of felt are also used. The mobility of dampers can be written as $\underline{Y}_I = r_I$ where r_I is the mechanical admittance of the damper.

In using the classical vibration isolation theory, we often assume that the foundation supporting the system to be vibration isolated has zero mobility. Any real support will have some local compliance at the support points and at medium and high frequencies, the internal mobility will be real. At frequencies, well above the bending wave resonances of a limited size sheet having bending stiffness, the point mobility of the sheet will approach that of an infinite sheet—that is, $\underline{Y}_F = 1/(8\sqrt{(m''B')})$ as discussed in Chapter 8. Equipment to be vibration isolated should be placed over major floor beams or on a concrete slab or solid ground. Lightweight wood floors are problematic from the viewpoint of vibration isolation. In some cases, extra mass, such as sand or stone, can be added to such floors to reduce their mobility. Sometimes stone or cement blocks can be used successfully on the furniture or floor, holding the equipment to be isolated, to reduce the apparent foundation mobility.

11.4 SOME METRICS FOR VIBRATION ISOLATION

Since there are different aims for vibration isolation, several different metrics have been developed to estimate the efficiency of the techniques used. Here we will only study metrics used in applications where linear translational movement is involved.

Two sets of indices will be used, one to denote the structures and one to denote the case. Index *0* is used to denote the free velocity or unblocked force of the source and index *2* is used to indicate the receiving structure. Index u is used to denote the case of no vibration isolation (stiff mounting), and index i is the case of a vibration isolation technique or device being used.

One common vibration isolation metric is the ratio of transmitted force to the force acting on the vibrating device to be isolated, called the *transmissibility* T which is defined as:

$$T = \left|\frac{\underline{u}_2}{\underline{u}_0}\right| = \left|\frac{\underline{F}_2}{\underline{F}_0}\right| \tag{11.1}$$

Another metric is *efficiency* E, defined as:

$$E = \left|\frac{\underline{u}_{2u}}{\underline{u}_{2i}}\right| = \left|\frac{\underline{F}_{2u}}{\underline{F}_{2i}}\right| \tag{11.2}$$

An important metric for the performance of the vibration isolation is *insertion loss* in dB, ΔL, or IL, defined as:

$$\Delta L = 20\log(E) \tag{11.3}$$

Typically, we are more interested in the insertion loss than in the transmissibility since the insertion loss number gives us practical useful information about the success of our vibration isolation efforts. Both efficiency and insertion loss are related to the vibration in the practical case, taking into account the vibratory load on the structure.

11.5 LINEAR SINGLE-DEGREE-OF-FREEDOM SYSTEMS

The Simple Mass-spring System

Figure 11.2 shows a very simple system that idealizes a situation often found in practice. A loudspeaker is mounted in a car door. The loudspeaker consists of a diaphragm and voice coil combination, magnet, and basket (see Chapter 14). In practice, we can consider the diaphragm and voice coil combination as having little mass compared to the basket and magnet; it can be neglected in the further analysis here.

Let us study a device that is to be isolated. It is considered to be a rigid body (mass) M that is acted upon by a force. The vibrating device is connected to the foundation (the floor) by an isolator. Figure 11.3 shows an example of a commonly used practical steel spring isolator having a spring constant k_i. (Note the rubber pad, which increases insertion loss at high frequencies.) The isolator may contain compliance and damping but is assumed to be without mass. The foundation is also assumed to be rigid and to have infinite mass. The system is assumed only to vibrate in one direction. Such a system is called a linear *single-degree-of-freedom system* (often written SDOF system).

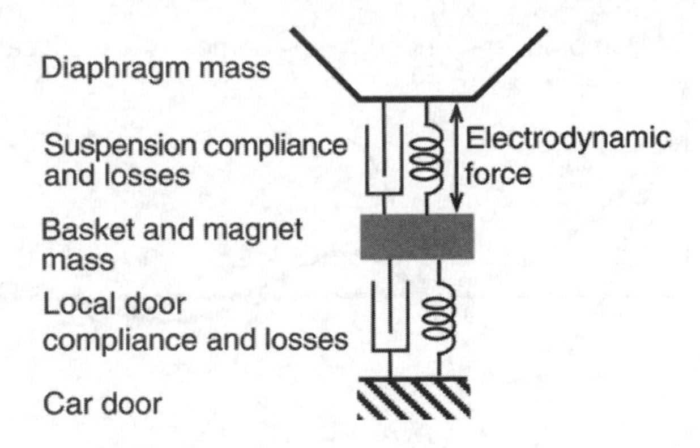

Figure 11.2 An electrodynamic loudspeaker mounted in a car door can be thought of as an example of a simple one-dimensional vibratory system. The diaphragm mass is small compared to the basket mass. The lower three components in the drawing can be considered parts of a single-degree-of-freedom system.

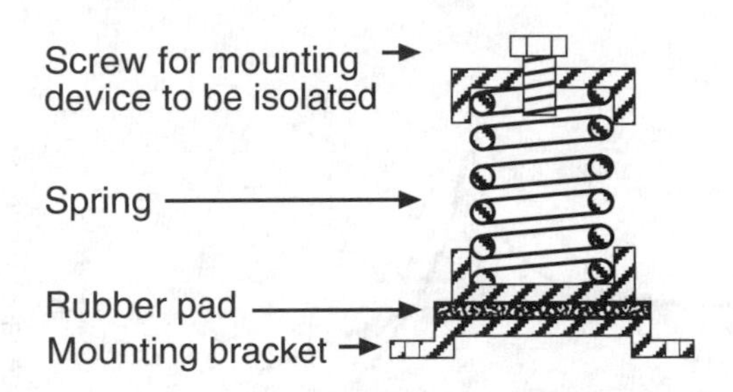

Figure 11.3 A metal-spring machinery isolator combining a steel spring with a rubber pad.

A transient force (or a momentary system displacement) will result in a damped oscillation by the mass as shown in Figure 11.4.

The natural frequency of the resonance for the undamped system will be:

$$f_0 = \frac{1}{2\pi}\sqrt{\frac{k_i}{M}} \tag{11.4}$$

The transmissibility of a simple SDOF system is shown graphically in Figure 11.5—for some different values of viscous damping. (Other types of damping, such as friction, will give different results.) At critical damping, there is no oscillation as the mechanical system loses energy, as shown in Figure 11.4. Note, however, that the transmissibility curve, shown in Figure 11.5, still has some overshoot at the resonant frequency.

The curves in Figure 11.5 are misleading, however. Their simplicity has been achieved by unrealistic simplification. The vibrating device cannot usually be assumed to be a rigid body and thus has internal resonances. The spring has distributed mass, the damper has mass, and the receiving foundation is not solid, but has many resonances.

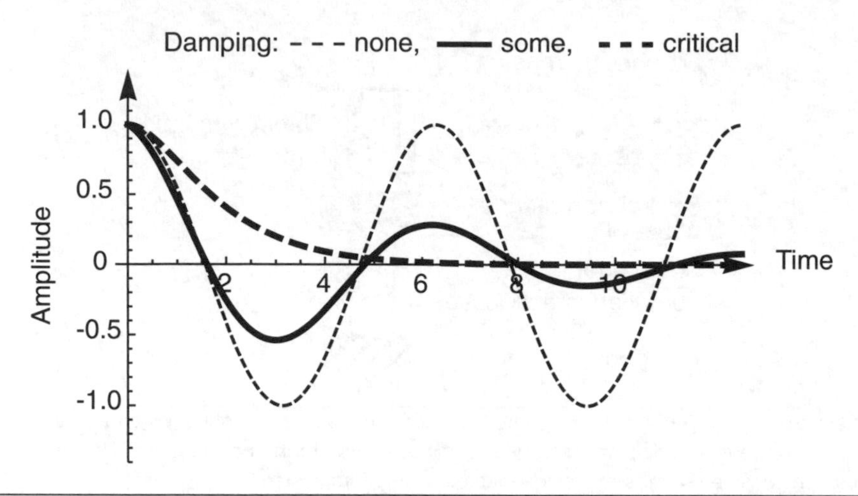

Figure 11.4 Natural vibration of a single-degree-of-freedom system deflected and released without velocity for different cases of damping.

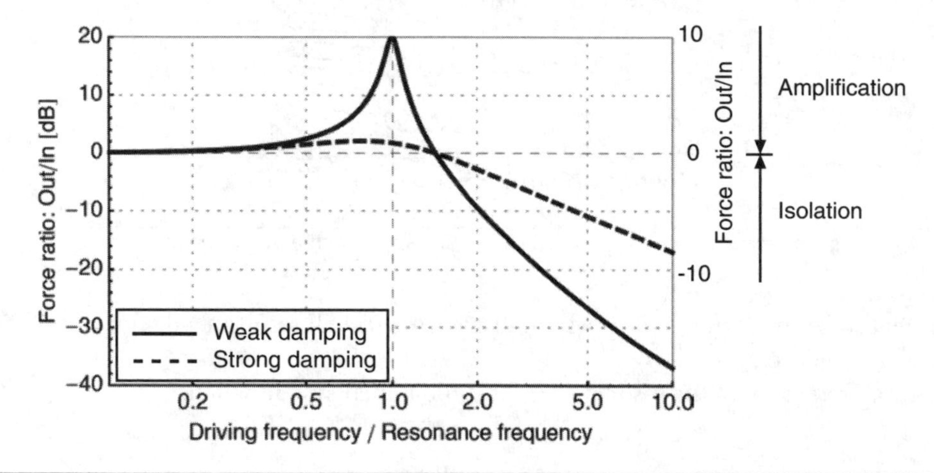

Figure 11.5 Example of the theoretical force transmissibility for a viscous damped single-degree-of-freedom system for two cases of damping.

The attraction of the theory, in spite of these serious problems, lies in that it is intuitive and gives reasonable results for many systems under a certain combination of conditions: a lightweight vibrating body, a relatively high mass foundation (compared to the mass of the vibrating body), and low frequencies, close to and below, the resonance frequency.

Using a more advanced theory, which allows representation of the more exact mechanical properties, we can estimate the response of the system at higher frequencies and for more realistic situations. It also allows us to understand why our vibration isolation efforts seldom lead to more than typically 15 to 20 dB of insertion loss. The more advanced approach uses the theory of electromechanical analogies and can be expanded to multiple-degree-of-freedom systems.

11.6 VIBRATION ISOLATION THEORY USING ELECTROMECHANICAL ANALOGIES

Mechanical and Acoustical Analogies

Electromechanical analogy uses the similarity between the differential equations of electrical circuits and mechanical systems to enable the use of electrical symbols for mechanical components. Applying this approach, one can use the techniques of electrical circuit theory to study such things as transfer functions, impedances, and use circuit theorems, etc., to solve for the desired behavior of the mechanical system. One can also extend the method to include acoustical components. There are two types of commonly used analogies—impedance type and mobility type.

Acoustical analogies are usually of the impedance type and *mechanical analogies* of the mobility type. Techniques exist for the conversion between the two types and for interconnecting electrical, mechanical, and acoustical circuits. Reference 11.6 gives a summary of the principles of electromechanical-acoustical circuit analogies. Table 11.1 shows how the two types of analogies model mechanical systems.

Preventing Force from Acting on a Foundation

We will briefly return to the simple mechanical linear single-degree-of-freedom system discussed previously, but now use the analogy approach.

Start by studying a vibratory system without any vibration isolation, as shown in Figure 11.6. The rigid, vibrating body is represented by a mass M, initially mounted directly on a foundation having high internal impedance $\underline{Z}_f$. We also assume that the mechanical impedance of the mass, over the entire frequency range modeled, is much smaller than that of the foundation, that is, $\omega M \ll |\underline{Z}_f|$. This can also be expressed by saying that the mechanical mobility of the rigid body is much higher than that of the foundation. There is an internal force $\underline{F}_0$ acting on the mass.

Table 11.1 Substitution of mechanical variables by electrical variables for impedance and mobility systems

Impedance type analogy quantity	Mechanical system quantity	Mobility type analogy quantity
Electrical voltage	Mechanical force	Electrical current
Electrical current	Mechanical velocity	Electrical voltage

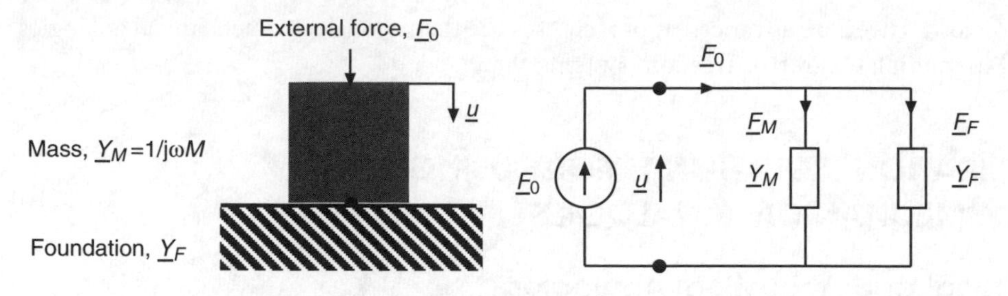

Figure 11.6 A mechanical system consisting of a mass mounted rigidly on a foundation. On the right, the system's electrical circuit analogy using the mobility approach.

Since the mass is mounted directly on the foundation, one realizes that both vibrate with the same velocity. According to Newton's first law, much more force is needed to move the foundation than the mass, since $\omega M \ll |\underline{Z}_f|$. As the mobility of the rigid body is much higher than that of the foundation, the foundation will pick up nearly all of the force internally applied to the rigid body—that is, $\underline{F}_{2u} \approx \underline{F}_{0.}$

We now mount an isolator, in this case just an idealized spring and damper combination, between the rigid body and the foundation as shown in Figure 11.7. One can then draw an electromechanical mobility type analogy as shown.

Now assume that we do not know what the impedances or mobilities are of the rigid body, the isolator, or the foundation. We can still use the same symbolic approach previously used and calculate the resulting force acting on the foundation since the *force currents* will split according to Kirchoff's first law:

$$\underline{F}_{2i} = \frac{\underline{Y}_M}{\underline{Y}_M + \underline{Y}_F + \underline{Y}_I}\underline{F}_0 \tag{11.5}$$

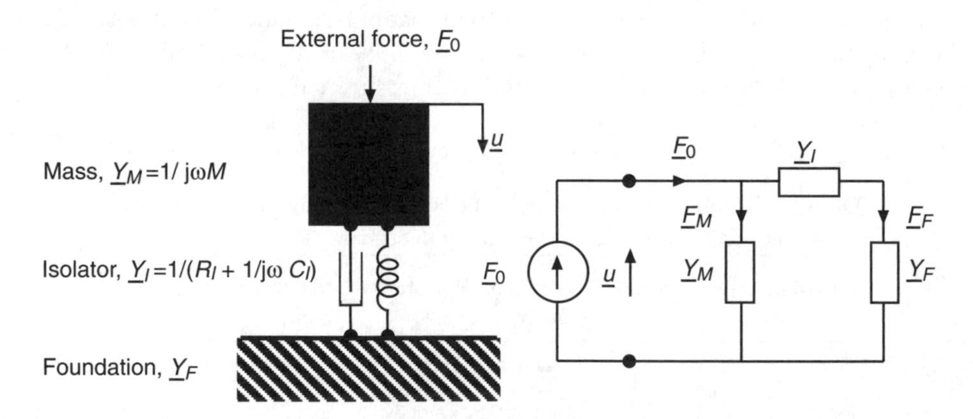

Figure 11.7 A mechanical system with a vibration isolation device inserted between source and foundation. On the right, the system's electrical circuit analogy using the mobility approach. The resulting vibration velocity of the mass is $\underline{u}$.

Using the electrical circuit diagram, we can now calculate the insertion loss as:

$$\Delta L = 20\log\left(\left|\frac{\underline{F}_{2u}}{\underline{F}_{2i}}\right|\right) = 20\log\left(\left|\frac{\underline{Y}_M + \underline{Y}_F + \underline{Y}_I}{\underline{Y}_M + \underline{Y}_F}\right|\right) \tag{11.6}$$

The isolator impedance is $\underline{Z}_I$ and its mobility $\underline{Y}_I$. The mechanical compliance of the spring is $C_I = 1/k_I$, where k_I is the stiffness of the spring:

$$\underline{Y}_I = \frac{1}{\underline{Z}_I} = \frac{1}{R_I + \dfrac{1}{j\omega C_I}} \tag{11.7}$$

One realizes that if the mobility of the isolator is high in comparison to the mobilities of the vibrating body and the foundation, the insertion loss will be:

$$\Delta L \approx 20\log\left(\left|\frac{\underline{Y}_I}{\underline{Y}_M + \underline{Y}_F}\right|\right) \tag{11.8}$$

We also immediately notice from Figure 11.7 that the prerequisite for obtaining a substantial insertion loss is that the mobility of the isolator is high compared to the mobility of the foundation. The resulting force (mobility type analogy current) will then be low.

We further note that another prerequisite for high insertion loss is that the mobility of the foundation be low compared to the mobility of the vibrating body. This is often difficult to achieve. In many real systems, the mobilities of the vibrating body and its foundation are of the same order of magnitude.

Now let us assume that the foundation is rigid (that is, $\underline{Y}_F = 0$), which means that $\underline{Z}_F$ is infinitely high. The equation for the insertion loss 11.6 then becomes:

$$\Delta L = 20\log\left(\left|\frac{\underline{Y}_M + \underline{Y}_I}{\underline{Y}_M}\right|\right) \tag{11.9}$$

We realize that, even in this case, it is necessary for the mobility of the vibrating body to be much lower than that of the isolator to achieve high insertion loss by applying vibration isolation.

The system is at resonance when the imaginary part of the mobility of the vibrating body is the same as that of the isolator but having the opposite sign—that is, if $\mathrm{im}[\underline{Y}_M] = -\mathrm{im}[\underline{Y}_I]$. At resonance, the system can become *matched*, and the insertion loss can, in fact, become negative.

If the real parts of the component's mobilities are low, the system will be poorly damped, resulting in an increased response close to the resonance frequency as indicated in Figure 11.3. High damping results in a poorer insertion loss at frequencies well above the resonance frequency than low damping does.

For a system with a force spectrum containing frequencies both around and over the resonance frequency, the choice of damping will depend on the force spectrum's shape and on other limits, such as the maximum allowable displacement of the vibrating body. For situations involving a force that is swept slowly in frequency across the region of resonance, the choice of damping will be particularly important.

Preventing Body Vibration Due to Vibration of the Foundation

Many practical cases of vibration isolation involve preventing a body mounted on the foundation from vibrating. For instance keeping a turntable for playing vinyl records, a computer hard disk, or a CD/DVD player that is positioned on furniture from picking up floor vibrations due to steps or vibration due to loudspeaker excitation. This is sometimes called velocity isolation in contrast to the force isolation discussed previously.

We now regard the vibration velocity or displacement as given, and we are primarily interested in the resulting velocity or displacement of the attached body. As previously mentioned, we draw the mechanical mobility analogy as shown in Figure 11.8.

We see that the velocity insertion loss achieved by using the isolator will be:

$$\Delta L = 20\log\left(\left|\frac{\underline{u}_{2u}}{\underline{u}_{2i}}\right|\right) = 20\log\left(\left|\frac{\underline{Y}_M + \underline{Y}_I}{\underline{Y}_M}\right|\right) \tag{11.10}$$

We see that in this case the expression for the insertion loss will be the same as that obtained for force isolation with a rigid foundation.

11.7 REAL SYSTEMS

Using measured mobilities for the body, isolator, and foundation, we are in a better position to estimate the practically obtainable insertion loss that can be obtained by vibration isolation of a particular system. In reality, all three components are likely to be multiresonant systems with complex and very frequency-dependent mobilities.

The systems studied until now have been very idealized, also, because they have involved linear one-dimensional motion only. In a real system, there will be vibration, both linear and rotational, in each of the three directions as well as coupling between the attachment points. This makes most vibration isolation tasks difficult if high insertion loss is desired. Some possibilities for wave motion of a phonograph subchassis and the foundation on which it is placed are shown in Figure 11.9. It is important to not overreact to the high loss data shown in Figure 11.3, which applies to the idealized SDOF system case only.

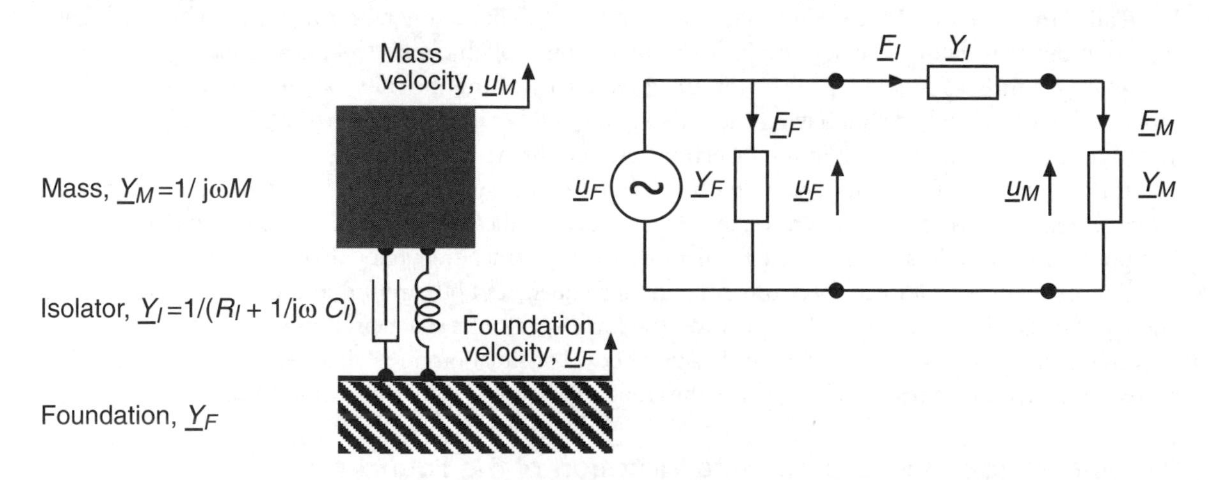

Figure 11.8 *Velocity* isolation and the electrical circuit analogy of the system using the mobility approach.

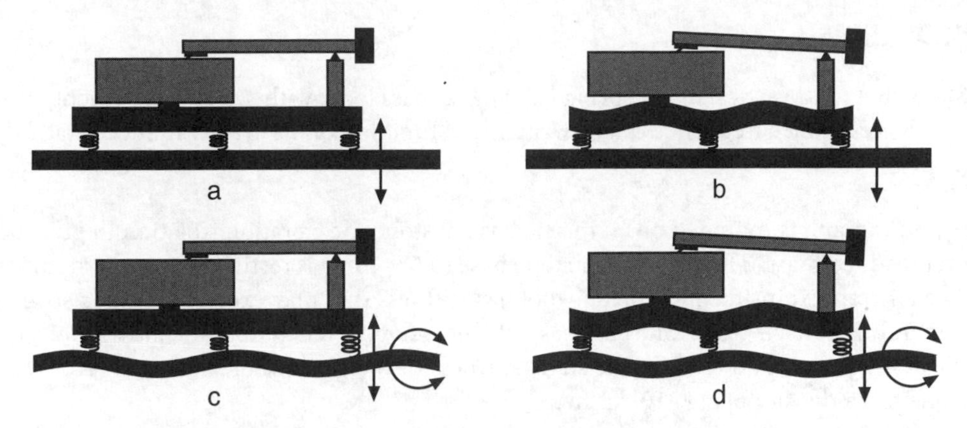

Figure 11.9 Some possibilities for vibration in a foundation and in an attached phonograph subchassis: (a) solid foundation and solid subchassis, very low frequencies, unidirectional translational movement in both, (b) solid foundation but bending wave motion in subchassis, medium frequencies, (c) bending wave motion in foundation but solid subchassis, medium frequencies, and (d) bending wave motion in both foundation and subchassis, high frequencies.

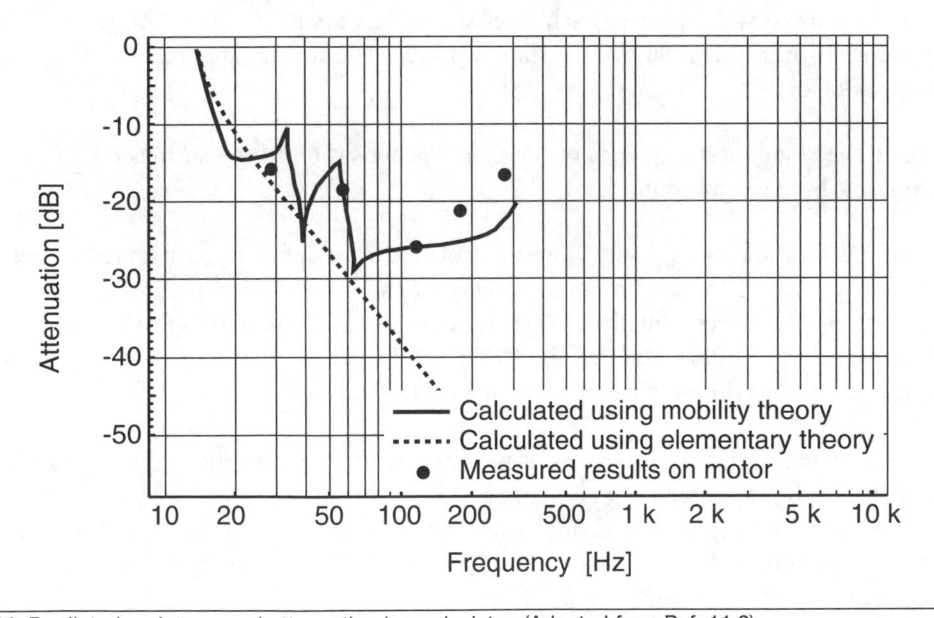

Figure 11.10 Predicted and measured attenuation by an isolator. (Adapted from Ref. 11.2)

At frequencies where the isolator is large compared to the wavelength in the isolator, there will be a possibility for half-wave transmission. This will reduce the insertion loss of the isolator.

Figure 11.10 shows predicted and measured attenuation of a SDOF system using both the simple theory initially described and mobility theory and data. The measurement data shows that the mobility approach gives a much better estimate of the attenuation (see reference 11.2).

11.8 PROBLEMS

11.1 Show that for a vertical linear spring carrying a mass load, with system movement purely vertical, the resonance frequency can be written as a function of the gravity induced static deflection d_{stat}.

11.2 A phonograph is mounted on an inertia base of stone for vibration isolation to give the phonograph added mass. The rectangular stone base (0.5×1 m^2) is resting on a sheet of mineral wool. The air trapped inside the mineral wool has stiffness that may exceed the static stiffness of the mineral wool itself. The stone base has a thickness of 2.5 cm as does the glass wool sheet before the stone base is laid on top. The static stiffness of the glass wool sheet is $1 \cdot 10^4$ N/m and the stone has a density of $2.5 \cdot 10^3$ kg/m^3.

Task: Calculate the resonance frequency for the stone base and the glass wool mass-spring SDOF system, assuming the foundation has low internal mobility.

11.3 The phonograph of Problem 2 is now mounted on a concrete floated inertia base that is supported by three steel springs. Each steel spring has a static stiffness of $12 \cdot 10^3$ N/m and a dynamic stiffness that is 90% of the static stiffness. The springs are placed so that they have the same static load.

Task: Calculate the resonance frequency of the mass-spring system if the springs compress by 10 mm when the concrete base is applied on them.

11.4 It is important that any velocity vibration isolation is designed to have its resonance frequency well below that of the maximum sensitivity of the supported structure—for example, a CD player. The CD-player contains a servo system to control the motion of the lens focusing the laser used to read information off the CD. The CD rotates at most at 10,000 rpm. Estimate the highest allowable vibration isolation resonance frequency.

11.5 In some vibration isolation systems, it may not be possible to control the *Q*-value of the resonant system because damping also reduces the insertion loss above the resonance frequency. The force transmission at resonance can then be reduced by the use of a parasitic resonant system, a *dynamic absorber*. Discuss the dynamic similarities between this and the ported box loudspeaker system shown in Figure 14.16.

Microphones 12

12.1 INTRODUCTION

The ideal microphone is an *electroacoustic transducer* that can convert the waveform of sound pressure or sound pressure gradient to an electrical signal having the identical waveform.

Because of physical and engineering limitations, such conversion is not possible. A practical microphone will feature nonlinearities and self-noise. Because any practical microphone is also a mechanical device, microphones will be sensitive to vibration, which will give electrical output as well. Many microphones also contain electronic signal amplification and other signal enhancement circuits. These also contribute to the nonlinearities and the background noise of the microphone. Digital microphones use analog microphone capsules but incorporate a built-in analog-to-digital converter (ADC) so that the output signal is digitized and coded before being transmitted from the microphone. Such converters can, of course, add all the errors and noise types present in any system (see Chapter 16).

12.2 DYNAMIC RANGE, FREQUENCY RESPONSE, NOISE, AND DISTORTION

Any physical device has a limited *dynamic range*. A microphone's dynamic range is the ratio between the strongest and the weakest sounds that the microphone can faithfully reproduce. This ratio is usually expressed as a level difference in units of dB. It follows that the figure for the dynamic range is somewhat arbitrary; it depends on the background noise and maximum nonlinearities considered acceptable.

All microphone systems are characterized by *distortion*. We differ between two types of distortion, linear distortion and nonlinear distortion. *Linear distortion* is any distortion of the waveform that does not create new frequency components. The frequency-dependent amplification of an amplifier and the frequency-dependent attenuation by an acoustic or electric filter are examples of linear distortion. We usually expect a microphone to have *linear frequency response*, which is the absence of linear distortion in the audio frequency range—between 20 Hz and 20 kHz—often called a *flat* response. Depending on the nature of the acoustic signal and the engineering principle of the microphone, it may at times be necessary to add compensating linear distortion, for example, in the microphone amplifier or mixing console. In addition, the microphone response will be directional, influenced by diffraction around the microphone diaphragm and housing, acoustic resonances in the air behind grilles and such. The irregularities may not be much of a problem in dedicated applications, such as telephone systems, where the frequency response of the microphone can and has to be optimized for criteria such as speech intelligibility or signal-to-noise (S/N) ratio.

Nonlinear distortion is distortion that creates new frequency components being added to the signal. Examples of such distortion are the distortion due to limited movement of the microphone or loudspeaker diaphragm suspension, the distortion due to electric and or magnetic field unevenness in transducers, and the acoustic distortion due to the nonlinearities of air at high sound levels. The distortion that is created by nonlinearities creates harmonics to the waveform—overtones to the fundamental tone. This is called *harmonic distortion*. In addition, the nonlinearities cause *intermodulation distortion*, which is characterized by added tones that have frequencies that are related to the sum and difference frequencies of signal components.

Because of nonlinear distortion, the linear distortion may also change character between high and low sound levels. The frequency response will be different depending on signal level. The system may become saturated.

Most microphones are dependent on the use of preamplifiers. The noise from the preamplifier will sometimes dominate over the microphone's own electrical noise. It is advantageous to express the noisiness of a microphone and preamplifier system by an equivalent room noise level, for example, as an equivalent sound level in dBA. Typically, condenser microphones will have background noise levels in the range of 15–35 dBA equivalent room noise. It is important to note that room noise and microphone noise generally have different character so even at the same dBA level number they will feature very different noise quality. The typical noise spectra for room noise and the electrical noise from microphones are shown in Figure 12.1.

Microphone *self* noise can be due to many effects. One inevitable but negligible noise source is the molecular motion of the air in which the diaphragm exists. This noise is called the microphone *thermal agitation* noise (represented by the real part of the radiation impedance seen by the microphone diagram). A second noise source is the *thermal noise due to the resistance* of the electric circuits. A third noise source

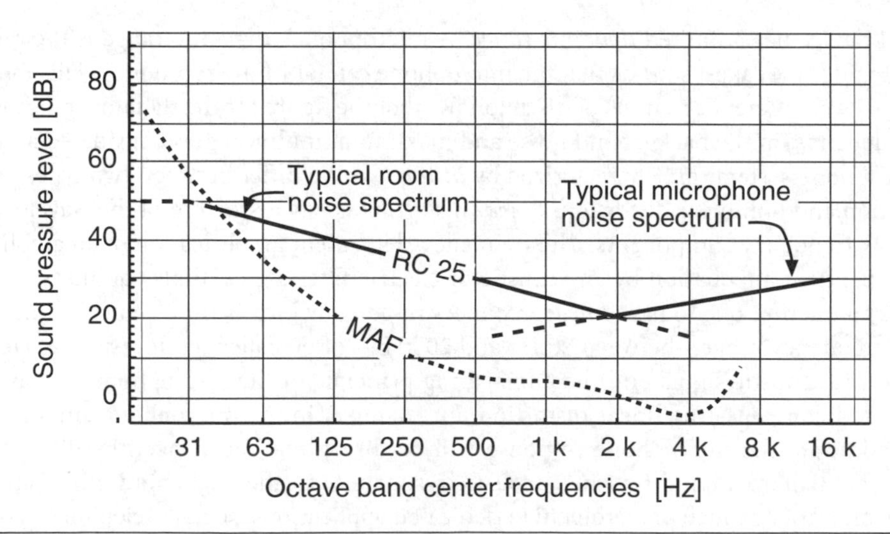

Figure 12.1 Typical spectra for room noise and microphone noise.

is the noise and *hum* induced in the electric circuits of the microphones due to *stray electromagnetic fields* in the microphone environment. A fourth noise source is the noise due to power supplies necessary for microphone electronic circuit operation. Digital microphones will also exhibit *quantization* noise characteristic of analog-to-digital (A/D) conversion systems. Other sources of noise exist as well. Microphone noise may also be due to *mechanical movement of the cables* and connectors by which the microphones are transmitting their signals.

12.3 SENSITIVITY

The *sensitivity* of a microphone is an important property in the choice of a microphone, along with the electric output impedance, and in some cases the electric power supply principle. The sensitivity of microphones usually relates to the output voltage to a certain sound pressure, often 1 Pa. The electric voltage obtained for that sound pressure is then compared to some reference voltage typical of the electric system to which the microphone is connected. Typically, the voltage at the output of an unamplified microphone is about 5–50 millivolts at a sound pressure of 1 Pa.

Many passive, self-powered microphones (microphones not using internal electric amplification) have fairly low *electric output impedance*, typically in the range of 10–200 ohm. Active microphones usually have very low output impedance, except in the case of electret microphone capsules that may need further impedance reducing electric circuits.

Active microphones, such as condenser and digital microphones, need electrical power to operate. The power is usually supplied through the audio output cables using a form of multiplexing. Since the power signal is DC, or very high frequency, it can coexist with the audio signal on the cable. The supply of electric power can be *balanced* or *unbalanced*. The most common type of balanced power supply, *phantom power*, uses an electrically shielded cable with two conductors to feed the microphone. In this way, the electric signal and its noise can be cancelled both at the microphone and at the microphone amplifier input stage. This can be done using electric *balancing transformers* or special signal balancing circuits.

Radio microphones transmit their output signal by radio to a radio receiver connected to the mixing console. Such microphones are commonly used in theatre and musical applications and can be conveniently hidden in the performers' hair or clothing.

Digital microphones often use Bluetooth transmission to connect to a computer or mobile phone. Such microphones still need an A/D conversion process to operate. Studio quality digital microphones will use PCM-based conversion and lossless transmission (see Chapter 16).

12.4 ELECTROACOUSTICAL CONVERSION PRINCIPLES

To obtain an electric signal, an electric, magnetic, or capacitive circuit must be disturbed by a mechanic or acoustic action, for example. Any microphone will also by its presence disturb the sound field in some way.

The sound pressure or the sound pressure gradient act on the microphone diaphragm(s) that is set in motion by the field. As the diaphragm moves in the magnetic or electrostatic field, the mechanical and electrical systems try to block the movement. It is the electrical system that reacts to the diaphragm movement by generating the output signal.

Many microphones work on a principle involving magnetism and inductors (coils) or electrets and capacitors. The reason is that it is practical to obtain the electric signal directly without the need for external power as for resistive microphones.

Because of flexibility, economy, and other advantages, many microphones are designed so that the transducer is housed by itself in a *microphone capsule* that is then immediately coupled, both mechanically and electrically, to a *microphone preamplifier*. The term microphone is then used to describe the entire assembly.

Most microphone capsules are designed to directly give an electric output voltage proportional to the sound. Alternate technologies are available; however, microphones may use a high frequency voltage whose frequency or amplitude is changed by the sound. In such cases, the microphone will need some form of demodulating circuit to yield a signal suitable for further audio signal amplification.

Recently microphones based on the cooling of a heated thin wire have become available. Another type of microphone recently introduced is based on micromachined sensors built directly on silicon blocks. Such microphones can be made very small, but practical microphones will need to be made up of many individual microphone elements to achieve a reasonable S/N ratio.

Microphones are often categorized by their engineering design principle:

- Electrodynamic
- Resistive
- Electromagnetic
- Condenser (electrostatic, capacitive)
- Electret
- Micromachined capacitive
- Piezoelectric

A summary of some advantages and disadvantages of some common microphone types is shown in Table 12.1.

Table 12.1 Some important properties of common audio microphone systems

Type	Advantages	Disadvantages
Condenser (external bias)	Stability High sensitivity Wide frequency range Low vibration sensitivity	Requires external electronics and bias voltage supply Relatively mechanically sensitive Some types sensitive to humidity
Condenser (electret)	High sensitivity Very low cost Small size Small vibration sensitivity	Limited temperature range Relatively high noise level Relatively high nonlinear distortion
Piezoelectric	Self-powered Simple electronics	Sensitive to vibration Limited frequency range
Electrodynamic	Self-powered No electronics needed	Sensitive to dynamic magnetic fields Large physical size Limited frequency range Sensitive to vibration
Variable resistance	High sensitivity Low cost	Limited frequency response High noise level

12.5 SOUND FIELD SENSING PRINCIPLES

A microphone which outputs an electric signal having a waveform proportional to the sound pressure is called a *pressure (sensitive) microphone*. Analogously, microphones where the waveform of the electric signal is proportional to the acoustic pressure gradient or the particle velocity are called *pressure gradient microphones* and *velocity microphones* respectively. The latter two terms are used somewhat interchangeably.

It is also common that microphones are classified by their directional sensitivity; an ideal pressure microphone has an omnidirectional sensitivity pattern and is called an *omnidirectional microphone* or sometimes just *omni* for short. An ideal pressure gradient microphone will have a bidirectional sensitivity pattern and is called a *figure-of-eight type microphone*, or sometimes just *gradient*. The *unidirectional microphone* can be engineered by using a combination of the principles of gradient and pressure microphones.

12.6 DIRECTIVITY

The *directivity* of a microphone describes the microphone's sensitivity to sounds from various directions. The importance and need for various directivity properties was discussed already in Chapter 3 where various alternatives of microphone directivities were considered for stereo recording and reproduction. Since all microphones have a finite size, all microphones will become directional at sufficiently high frequencies.

Directivity can be obtained in two different ways. One type of microphone attains its directivity by long or wide sound pickup areas. Examples of such microphones are array microphones and microphones that use parabolic reflectors. An array microphone uses sound pickup at multiple points; there are passive and active array microphones. Both of these types rely on the size and geometry of the pickup area to give the desired directivity.

A second way of obtaining directivity is using multipole techniques. The pressure gradient microphone has a sharp null in its directivity characteristic, the same as that of a dipole. Quadrupole and higher multipole patterns can also be realized, both passively and actively. The advantage of multipole techniques is that the microphone can be small. The disadvantage is low sensitivity and associated high background noise.

The directivity of a microphone is usually described by a *directivity function*. The directivity function $F(\theta,\varphi)$ is the relationship between the microphone output signal at angles (θ, φ) (see Figure 12.2) compared to its output in the *reference direction*, usually $\theta = 0$ at which $|F(\theta, \varphi)| = 1$. The directivity function is usually frequency-dependent and its magnitude ≤ 1.

Directivity plots are used to graphically display the directional properties of the microphone, for example, in the case of measurement of the directivity characteristics of real microphones. The directivity plots are usually based on $20\log|F(\theta, \varphi)|$.

The directivity function is usually defined with relation to some *reference direction* of the microphone. Usually the reference direction is chosen to be the normal to the diaphragm surface. Since most microphone diaphragms are circular, the center of the coordinate system is chosen as the center of the diaphragm. The directivity function and the directivity plot are usually defined or measured for plane

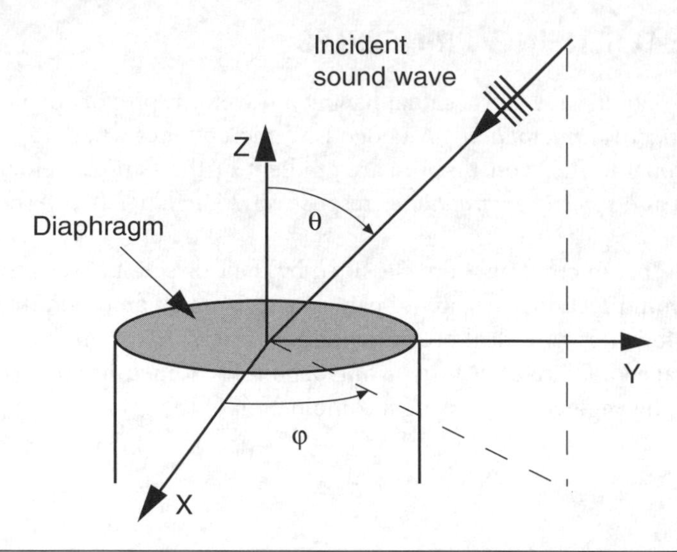

Figure 12.2 Definition of the coordinate system for the discussion of microphone directivity characteristics. The microphone diaphragm is in the plane $z = 0$. The z-axis is in the center of the microphone diaphragm.

incident waves. The direction of incidence is, in principle, specified by two angles, θ and φ. However, since so many microphones essentially have cylindrical symmetry, using θ only is usually sufficient for definition of the directivity function $F(\theta)$.

Since an *omnidirectional microphone* will have equal sensitivity to sound from any direction, its directivity function will be $F(\theta) = 1$.

A *bidirectional microphone* is characterized by a directivity function:

$$F(\theta) = \cos^n(\theta) \tag{12.1}$$

The number n specifies the *gradient order*. A first order gradient microphone has $n = 1$ and is sometimes said to have a figure-of-eight directivity since graphically the directivity pattern is a simple cosine.

There are several *unidirectional* microphone directivities. The most common unidirectional microphone is the *cardioid microphone*, which usually has the directivity function:

$$F(\theta) = \frac{1 + \cos(\theta)}{2} \tag{12.2}$$

A similar type is the *hypercardioid* directivity function:

$$F(\theta) = \frac{1 + 3\cos(\theta)}{4} \tag{12.3}$$

By adding the outputs from coincident microphones (microphones that are essentially at the same point in space and that have different gradient orders and angles), one can achieve desired directivities. Some common audio microphone directivities are shown in Table 12.2.

One sometimes needs sharper directivity patterns than those that can be obtained by using gradient or multipole microphone techniques. This can be done using various forms of array microphones as mentioned previously.

Table 12.2 Some important audio microphone directivity characteristics

Characteristic	Omnidirectional	Cardioid	Hyper-cardioid	Super-cardioid	Second order cardioid	Bidirectional	Second order bidirectional
Polar response pattern							
$F(\theta)$	1	$(1 + \cos(\theta))/2$	$(1 + 3\cos(\theta))/4$	$0.37 + 0.63\cos(\theta)$	$(\cos(\theta) + \cos^2(\theta))/2$	$\cos(\theta)$	$\cos^2(\theta)$
−3 dB pickup arc	360°	131°	105°	115°	77°	90°	60°
Random energy efficiency [dB]	0	−4.77	−6.02	−5.70	−8.70	−4.77	−6.99
Distance factor	1	1.73	2	1.93	2.74	1.73	2.24

One usually thinks of microphones as small devices; a typical electret microphone has a cylindrical shape with length and diameter typically about 6 mm. In spite of this, over frequencies of a few kHz, most microphones will exhibit directivity properties that are a function of their finite, although small, size.

The *directivity factor D* is a power-related metric that describes the ratio between the microphone power output for sound incident in its most sensitive direction and the mean power output over all angles. The directivity factor is defined as:

$$D = \frac{4\pi}{\int_0^{2\pi}\int_0^{\pi} |F(\theta,\varphi)|^2 \sin(\theta)\, d\theta\, d\varphi} \tag{12.4}$$

A common metric of the microphone's directivity is the directivity index (DI), that is expressed in units of dB and is defined as:

$$DI = 10\log(D) \tag{12.5}$$

The DI tells us how much more dB sensitive the microphone will be in its nominal 0°-direction than an omnidirectional microphone, for example, when recording sound in a reverberant and noisy room.

A more practical metric is that of the *random energy efficiency*, which gives the relative output of the microphone in a diffuse sound field (for example, random incidence noise or reverberation) relative to an omnidirectional microphone having the same sensitivity in its nominal 0°-direction. The reverberation energy efficiency is by definition equal to the negative number of the DI.

The *distance factor* shown in Table 12.2 shows the increase in reverberation radius that is the result of the suppression of random incidence reverberation. Figure 12.3 shows theoretical and measured data for the frequency response of a cylindrical microphone to an incident plane wave characterized by flat frequency response. We see that the output voltage level varies by more than 10 dB in the high frequency region because of the diffraction effects due to the finite microphone size. The microphone's directivity will become very directional at these frequencies as well.

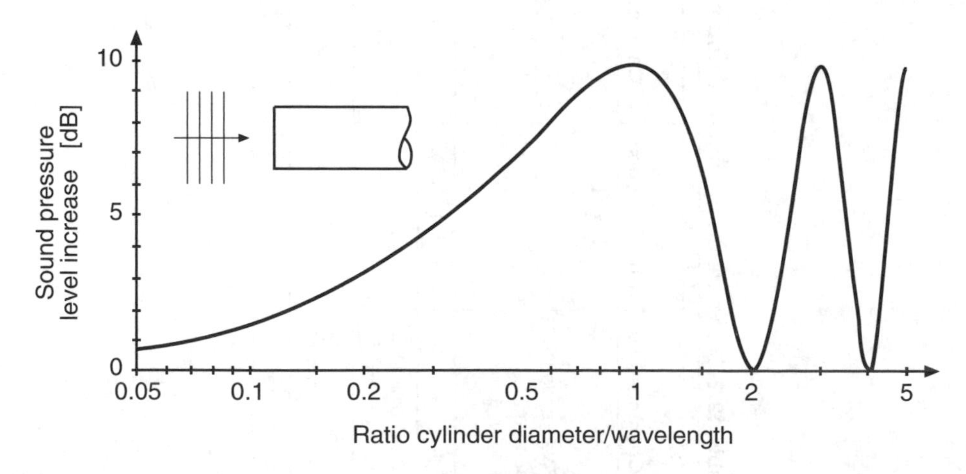

Figure 12.3 Influence of the finite dimensions of a cylindrical microphone. Theoretical and measured values for the sound pressure increase at the diaphragm end of the cylinder relative to the free field sound pressure.

12.7 DIAPHRAGM MECHANICAL PROPERTIES

The diaphragm and its suspension form the basic dynamic system of the microphone. The equation of motion for the mechanical system can approximately be written as follows:

$$\underline{F}_A = M_M \underline{a}_D + R_M \underline{u}_D + k_M \underline{\xi}_D = \left(j\omega M_M + R_M + \frac{1}{C_M} \right) \underline{u}_D \tag{12.6}$$

where

$\underline{F}_A$ = force
M_M = diaphragm mass
R_M = mechanical resistance
$C_M = 1/k_M$ diaphragm suspension compliance that is the inverse of the spring constant k_M
$\underline{a}_D$, $\underline{u}_D$, $\underline{\xi}_D$ = diaphragm acceleration, velocity, displacement.

The equation of motion shows that there will be a resonance at a frequency f_0.

$$f_0 = \frac{1}{2\pi} \sqrt{\frac{1}{M_M C_M}} \tag{12.7}$$

The microphone diaphragm's resonance frequency can be designed to be in the desired part of the audio spectrum. The choice of frequency will depend on the properties of the electromechanical transduction mechanism chosen for the microphone as well as on the choice of pressure or pressure-gradient acoustic sensitivity desired. Pressure sensitive microphones based on electrodynamic transduction will need to have their vibrating system controlled mainly by mechanical resistance and have the resonance frequency in the center of the audio frequency range—*resistance control.* Those pressure sensitive microphones that are based on capacitive sensing of diaphragm motion need to have their vibrating system controlled by the system's mechanical compliance—*compliance control.* Pressure gradient sensitive microphones need to have their vibrating system controlled by the system's mass—*mass control.* Figure 12.4 shows the basic frequency response of three alternatives available.

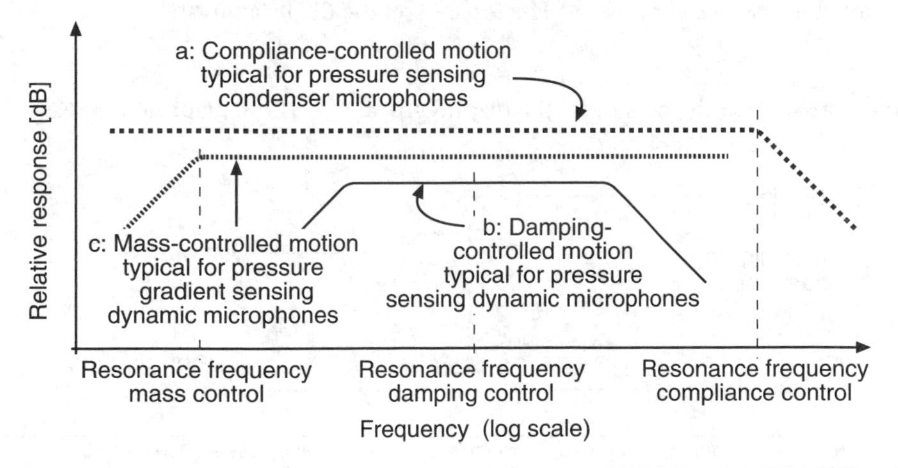

Figure 12.4 Relative frequency response for microphones using various diaphragm motion control: (a) dashed line pressure sensitive condenser microphone (compliance-controlled), (b) solid line pressure sensitive electrodynamic microphone (resistance-controlled), and (c) dotted line pressure gradient sensitive electrodynamic microphone (mass-controlled). Curves displaced to improve legibility.

By designing the mechanical system in such a way that its resonance frequency is much higher than the highest frequency of interest f_{max}, it is possible to give the microphone capsule a flat frequency response up to frequencies slightly below f_{max}. This requires the mechanical system to have its mechanical impedance $\underline{Z}_M$ dominated by the compliance term. We say that the system is *compliance-controlled.*

A *mass-controlled* diaphragm will have its resonance frequency below the audio frequency range. Typically, this design is only used with pressure gradient sensing microphones.

12.8 RESISTANCE MICROPHONES

The classical variable *resistance carbon microphone* was long the dominant microphone in telephone systems. The function of a carbon microphone is based on dynamic compacting of electrically conductive granules between two electrodes in a cavity, as shown in Figure 12.5. Since the granules are only loosely packed, variations in the degree of compaction will cause a variable resistance and thus a varying DC current in a simple electric, battery-operated circuit.

Like all resistive transducers, the resistive microphone needs an external power supply to operate that contributes toward its high sensitivity. Of course, the compaction of the granules is a nonlinear process, which means that the nonlinear distortion of the carbon microphone is high, so the carbon microphone was only used for speech purposes.

The electrical output of the microphone $\underline{e}_{out}$ is proportional to the compacting of the carbon granules and consequently to diaphragm displacement, that is:

$$\underline{e}_{out} = K\,\underline{\xi}_D \tag{12.8}$$

where K is a constant determined by the microphone design, and $\underline{\xi}_D$ is the displacement of the diaphragm from the rest. In this discussion the stiffness of the air in the cavity is neglected compared to that of the carbon granules.

This means that a pressure sensitive resistance type microphone must have its diaphragm displacement proportional to the sound pressure. The force $\underline{F}_A$ on the diaphragm is:

$$\underline{F}_A = S_D \underline{p}_D \tag{12.9}$$

where $\underline{p}_D$ is net pressure on the outside of the diaphragm, and S_D is the diaphragm area.

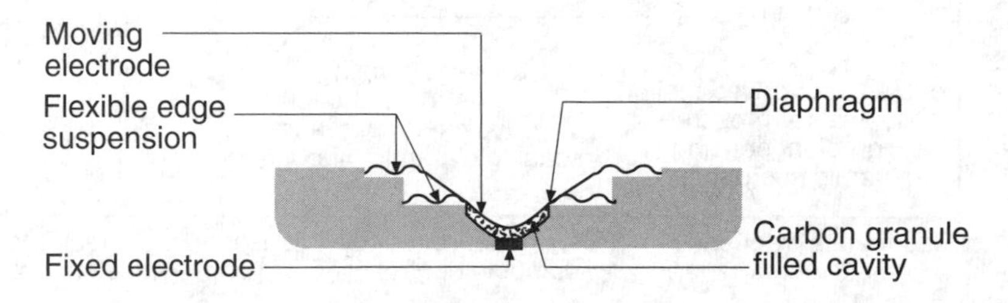

Figure 12.5 A microphone operating, using a sound modulated, carbon granule resistance cartridge.

The pressure sensitive, variable resistance microphone must use a compliance-controlled diaphragm unless frequency response equalization takes place in an associated filter or amplifier. The output voltage is then proportional to sound pressure since:

$$\underline{e}_{\text{out}} \propto \underline{\xi}_D \propto \frac{\underline{u}_D}{j\omega} \propto \underline{p}_A \quad (12.10)$$

Modern variations on the variable resistance microphone exist as micro-machined piezoresistive microphones. Since these are usually built on a monocrystalline substrate, the amplifier filter may be easily integrated into the function. Such microphones are now often found in communications equipment such as mobile phones.

12.9 PIEZOELECTRIC MICROPHONES

Piezoelectric microphones (also sometimes referred to as crystal or ceramic microphones) are based on the force of the diaphragm acting on a piezoelectric element. A piezoelectric element will become electrically charged if acted upon by an external force. Examples of materials that exhibit piezoelectric properties are crystals, such as quartz and Rochelle salt, as well as ceramic materials, such as lead-titanium-zirconate and polymers like polyvinyldifluoride (PVDF). The design of a typical piezoelectric *bender* microphone is shown in Figure 12.6.

The electric charge buildup causes a voltage over the piezoelectric element electrodes. The piezoelectric element's output voltage is given by:

$$\underline{e}_{\text{out}} = K\,\underline{\xi}_D \quad (12.11)$$

where $\underline{e}_{\text{out}}$ is the output voltage without electrical load, $\underline{\xi}_D$ the deformation, and K a constant depending on the piezoelectric material and the engineering design.

The piezoelectric element is an electric charge generator, and its electrical characteristic is typically that of a fairly low capacitance, typically a few nanofarads. Unless used with a microphone preamplifier, usually an *impedance converter*, to lower its output impedance, the microphone output impedance is very high, making its output voltage sensitive to cable capacitance and leakage resistance.

Since the electric output is proportional to displacement, the mechanical system of the piezoelectric microphone must be designed to the same principles as the resistive microphone. The microphone will typically have a flat frequency response, limited at the high frequency end by the mechanical resonance of the mechanical system, and at the low frequency end by the design of the electric circuit. Within this

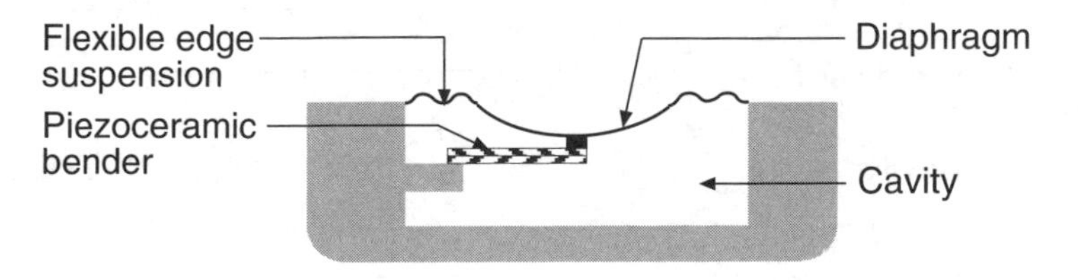

Figure 12.6 A piezoelectric microphone capsule operating, using a ceramic bimorph element.

frequency range, its frequency response can be made very flat, making the microphone suitable also for acoustic measurements.

12.10 CAPACITIVE MICROPHONES

Capacitive microphones are also known as electrostatic microphones or, more often, *condenser microphones.* The primary advantage of condenser microphones over other microphone types is their low linear and nonlinear distortion. Condenser microphones, because of their very low output voltage and their high capacitive internal electric impedance, always need a microphone preamplifier, typically an impedance converter. The internal capacitance is typically in the range of 5–50 pF.

Condenser microphones are available as two types, with or without electrically prepolarized diaphragms. The mechanical design of both types follows the same rule. The advantage of the prepolarized type is primarily a simpler power supply. The design principle of the condenser microphone is shown in Figure 12.7. The microphone capsule, the transducer, is a simple plate capacitor in which the distance between the plates varies due to sound pressure. The capacitance C_E between two parallel, equally large, metal plates placed close to one other, is:

$$C_E = \frac{\varepsilon\varepsilon_0 S}{d} \tag{12.12}$$

where ε, ε_0 is dielectricity constants, S is plate area, and d is distance between the opposing plates.

If the distance between the plates varies, the capacitance will vary as well. This capacitance variation can be used in several ways to produce an electric output signal proportional to the sound pressure.

The traditional, passive condenser microphone uses a flexible, electrically conductive diaphragm and a *back electrode* to form the two electrodes. In measurement and studio microphones, the capacitor is usually housed in a microphone capsule. Passive condenser microphones have to be supplied with a *polarization (bias) voltage* to generate a static charge between the plates.

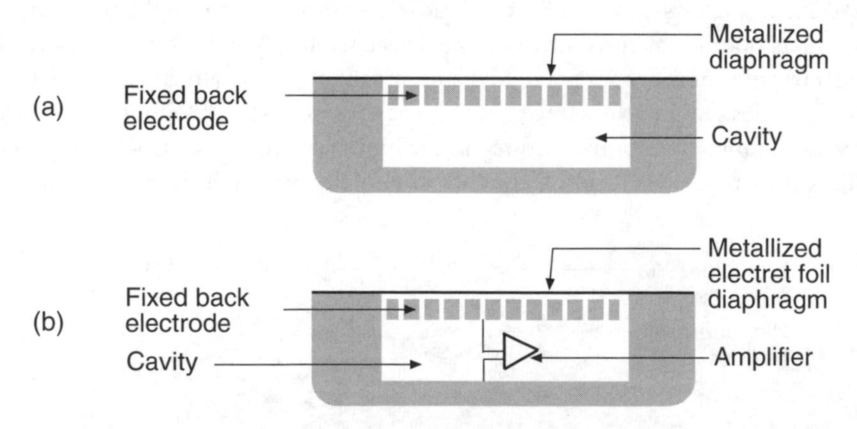

Figure 12.7 Two varieties of condenser microphone capsules: (a) traditional high voltage charged diaphragm, (b) pre-charged electret foil diaphragm.

The bias voltage is supplied to the condenser microphone capsule by way of a large electric resistance, typically 200 MOhms or higher. If the so-called time constant of the microphone, the product of the resistance and the capacitance of the microphone capsule, is large enough, the charge Q_E over the capacitor will be approximately constant over more than one cycle of sound pressure variation. Usually the time constant is chosen to be larger than 1 s. Fast variations in capacitance, such as those due to audio sound pressure variations, will, because of the constant electric capacitor charge, lead to voltage variations that follow the sound pressure variations.

For a charge Q_E, the voltage U_E over a capacitor is related to the capacitance C_E as:

$$U_E = \frac{Q_E}{C_E} \tag{12.13}$$

This means that the audio voltage $\underline{e}_{\text{out}}$ over the constant charge capacitor will be related to the diaphragm displacement $\underline{\xi}_D$ as:

$$\underline{e}_{\text{out}} = K\ \underline{\xi}_D \tag{12.14}$$

where K is a constant that depends on the microphone design and polarization voltage. Since the audio voltage is proportional to the diaphragm displacement, this microphone is also used in its compliance-controlled mode.

The low capsule capacitance and the low audio output voltage make the microphone sensitive to electrical interference. In practice, the microphone capsule has to be close to the preamplifier.

Because of the low capsule capacitance, the microphone circuit is sensitive both to *leakage resistance* and to *stray capacitance*. Leakage resistance is often due to the finite parallel resistance that appears between the electrode connectors due to dirt and fungus growth in a humid atmosphere. Even small leakage currents will usually lead to noise at the microphone output.

The stray capacitance is the capacitance that appears between the electrical conductors on the signal's path to the preamplifier. The preamplifier will also contribute parallel resistance and capacitance. Stray capacitance will lead to reduced microphone sensitivity.

In addition, the leakage of sound to the back side of the diaphragm will result in a further drop of response at low frequencies. Since it is usually desired for the microphone to have flat frequency response, the mechanical circuit of the microphone capsule has to be compliance controlled, that is, have its resonance frequency much over the highest audio frequency of interest.

Most condenser microphones manufactured today are *electret microphones*. The electret microphone uses an electrostatically charged, prepolarized device, an *electret*, usually a plastic film, to permanently charge the microphone capsule so that no bias voltage is necessary. PVDF is a plastic that can be electrostatically charged so that it can be used in microphones. The film can be used both as a diaphragm, after metallization, and as a layer on the back electrode.

By using the precharged electret film, there is no need for a bias voltage to be applied to the microphone capsule to provide the charge necessary to produce an audio output voltage. Together with a preamplifier using a field effect transistor, the electret microphone can be made very small. It can be used directly where probe microphones would otherwise be necessary or where it is important not to disturb the sound field. It also has the advantage that it can be manufactured at low cost.

Another way of using the capacitance variations produced by the sound pressure on the diaphragm is to employ the capsule capacitance as an element in an electric resonance circuit. The shift in resonance frequency can be detected in many ways. An advantage of this method is that there is no noise due to leakage currents.

The diaphragm is usually made of thin nickel or plastic film. Plastic films have to be made electrically conductive, typically by application of a thin gold or aluminum layer on the film by sputtering or electrochemical deposition.

The use of plastic films for condenser microphone capsules is now widely accepted even for high quality measurement microphones for laboratory use. Although plastic films do not have as good a resistance to high temperature and humidity as nickel film, accelerated tests have shown the sensitivity of, for example, electret films to be virtually unchanged over many years. One advantage of nickel films is that the diaphragm can be tensioned much more tautly, which results in a higher resonance frequency that gives a wider frequency range response. Of course, the sensitivity of the microphone capsule to sound suffers from this and the S/N ratio of the microphone will be lower. One important advantage of the low mass diaphragm of any capacitive microphone is that the microphone has little sensitivity to vibration.

The condenser microphone can easily be made so that the surface area of the diaphragm covers the entire end of a cylinder. This design results in the microphone having little influence on the sound field, and having comparatively low directivity. As shown in Figure 12.2, there is still considerable pressure buildup in front of the microphone capsule in such designs. The pressure buildup can be compensated for by strong damping of the microphone resonance. This damping is equivalent to a frequency response filtering by −6 dB over some limiting frequency. The microphone will then have fairly flat frequency response for a plane wave incident on the axis of symmetry of the microphone. Such a microphone will not, however, have flat frequency response in a diffuse sound field.

Micro-electric machined systems (MEMS) microphones form a recent addition to the available range of microphones. Capacitive MEMS microphones may be made by etching a grill, with a cavity below it, on a monocrystalline silicon substrate (that may already carry a CMOS electronics region). The grill is then covered by a polymer coating. The cavity will need to carry a gas—for example, air—at the same pressure as the surrounding air. This is accomplished by venting the cavity through a (meandering) channel etched into the substrate. There also are MEMS microphones based on piezoelectric and piezoresistive technologies. Yet another MEMS microphone technology uses a flexible diaphragm that works as the moving gate of the field effect transistor.

In practice, any microphone capsule needs a protective grille. This grille can be designed to minimize but not remove the frequency response difference between frontal incident sound and diffusely incident sound. Most laboratory measurement condenser microphones are designed as such pressure-sensitive omnidirectional microphones.

The bidirectional response can be obtained by using the diaphragm together with an acoustically transparent back electrode and no rear cavity.

The principles used in the design of capacitive microphones also make it relatively easy to design microphones having switchable omni, bidirectional, and unidirectional directivity as shown in Figure 12.8.

The unidirectional response is achieved by giving the back electrode a suitable acoustic resistance and making the rear side of the back electrode open to the sound. The bidirectional characteristic is achieved with making the capsule symmetric by having two diaphragms and combining the output voltages of both.

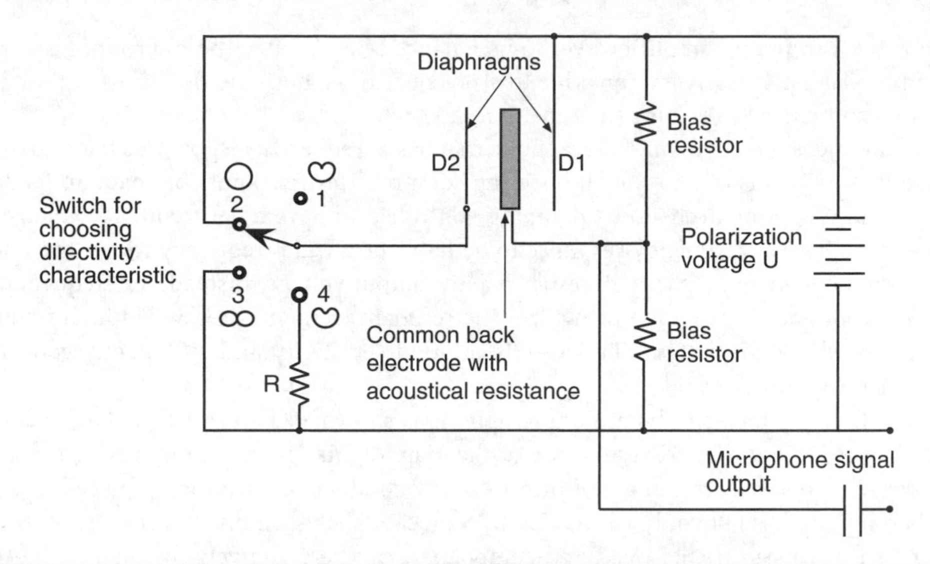

Figure 12.8 Electrical circuit for a capacitive microphone with variable directivity. Switch positions: 1) cardioid, 2) omnidirectional, 3) bidirectional, and 4) hypercardioid directivity. Note that the back electrode has to have a special flow resistance for the system to operate as intended. Microphone electrical connection is between back electrode and ground.

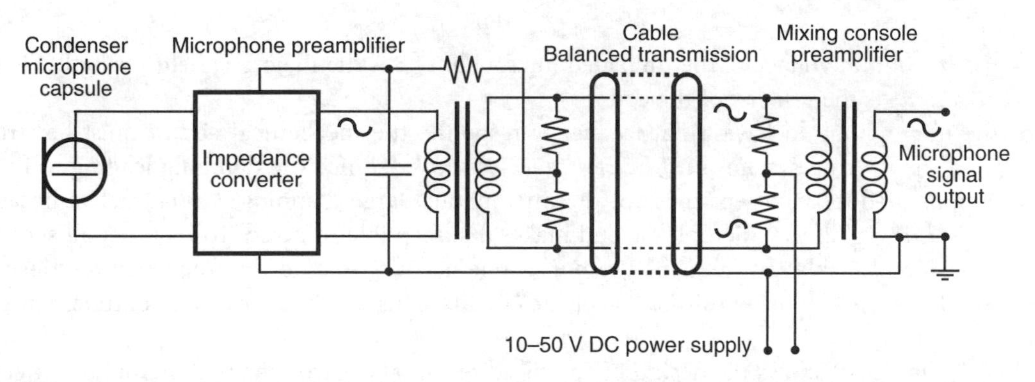

Figure 12.9 The phantom feed principle often used for capacitive microphones.

Professional condenser microphones are often powered using the phantom feed principle. This means that the audio and the supply current are multiplexed on the same wires, as shown in Figure 12.9.

12.11 ELECTRODYNAMIC MICROPHONES

The *dynamic microphone* works on the electrodynamic principle—a voltage will be induced in an electrical conductor moving in a permanent magnetic field. Since the magnetic field can store much energy, the microphone type can be made fairly sensitive. Because of its design, it is typically self-powered and does not require an impedance converter. The low impedance makes it possible to typically connect it to the preamplifier over long, balanced lines. Since the microphone works by the electrodynamic principle, it is very sensitive to stray electromagnetic fields (hum), although a compensation coil can be used to reduce the sensitivity to such fields.

It is difficult to design a dynamic microphone that has a frequency response as flat as that of a condenser microphone. This is due to the engineering compromises that must be made in its design. An electrodynamic microphone needs to be damping-controlled to have a flat frequency voltage response to sound pressure. For the frequency response to be flat over a wide frequency range, the microphone will then need to be damped, which will result in a low output voltage. Instead, the frequency response of dynamic microphones is extended using acoustic resonant elements at low and high frequencies in front of and behind the diaphragm. This results in good sensitivity and sufficiently good frequency response for many applications.

A section of a simple electrodynamic microphone is shown in Figure 12.10. The incident sound pressure works primarily on the external side of the diaphragm. The diaphragm is firmly attached to a paper or plastic cylinder carrying a coil of electrically conductive wire that is placed in a magnetic field. The diaphragm is usually only suspended by a mechanically compliant structure at its edge. The resonance, which is caused by the mass and compliance, can be effectively damped both acoustically and mechanically.

The movement of the diaphragm and voice coil can be described by the same equation as previously, Equation 12.7. The voltage generated by the moving coil is:

$$\underline{e}_{\text{out}} = Bl\underline{u}_D \tag{12.15}$$

where $\underline{e}_{\text{out}}$ is the output voltage of the unloaded microphone, B is the magnetic field strength, l is voice coil length, and $\underline{u}_D$ is the velocity of the coil.

For the microphone to have a flat frequency response, the mechanical circuit must be strongly damped, and the moving system must be *resistance-controlled.* Since the damping leading to flat frequency response reduces the sensitivity of the microphone, large damping is unattractive because it reduces the sensitivity of the microphone and makes it susceptible to noise. To retain good sensitivity and still have a reasonably flat frequency response, one needs to extend the frequency response both upward and downward by using resonant acoustic circuits. This results in less than optimal reproduction of transient sounds.

While Figure 12.10 shows the design of an omnidirectional electrodynamic microphone, dynamic microphones can also be designed for a bidirectional or unidirectional directivity characteristic.

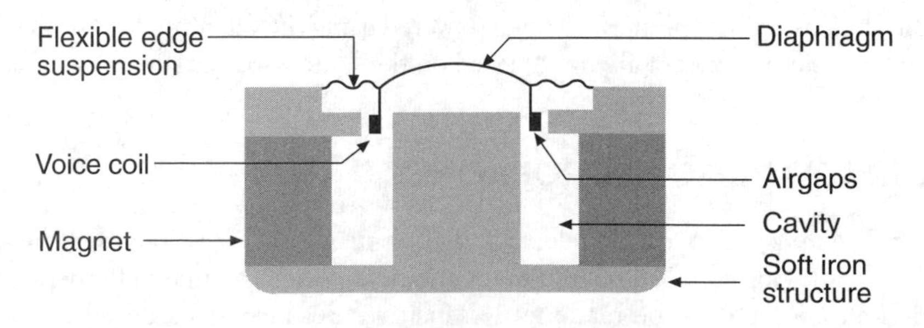

Figure 12.10 The electrodynamic omnidirectional microphone.

Such an alternate electrodynamic microphone is the *ribbon microphone* shown in Figure 12.11. The ribbon microphone is best designed as a pressure gradient microphone and usually has a bidirectional response because of this front/back symmetry. On occasion, the bidirectional response is desired for the balance between direct sound and reverberation, as well as noise reduction.

The ribbon microphone uses a lightweight electrically conductive ribbon in a magnetic field, as shown in the figure. The pressure difference between the two sides of the ribbon causes the ribbon to move. Because of the magnetic field in which the ribbon moves, a voltage will be generated across the ends of the ribbon. The voltage is low but is transformed to conventional microphone output voltages by a transformer.

The pressure sensitive ribbon microphone senses the sound field in the following way. Assume that the microphone has its ribbon in the $z = 0$ plane. Further assume the sound wave to be incident at an angle relative to the $z = 0$ plane, as shown in Figure 12.1. The resulting net force on the diaphragm $\underline{F}_A$ will be:

$$\underline{F}_A = S_D\left[\left(\underline{p}(z) - \frac{dp}{dz}\Delta l\cos(\theta)\right) - \underline{p}(z)\right] = S_D\frac{dp}{dz}\Delta l\cos(\theta) \tag{12.16}$$

where $\underline{p}(z)$ is sound pressure at a point z, $\cos(\theta)$ is angle of incidence relative to the z-axis, Δl is effective path length between the microphone's front and back sides, and S_D is the area of one side of the diaphragm.

The diaphragm is assumed to be very thin. Since the pressure gradient is proportional to the particle velocity in a plane wave, the net force can also be written as:

$$\underline{F}_A \propto S_D\,\omega\,\underline{u}_D\,\Delta l\cos(\theta) \tag{12.17}$$

Because of this property, the pressure gradient sensing microphone type is also sometimes called a *velocity microphone.*

The mechanical action of ribbon can be considered a mass-spring system that is damped by the radiation impedance and the viscosity of the air in the air gaps between the edges of the ribbon and the sides of the magnet's pole pieces. The ribbon behaves somewhat like a string and is characterized by many resonances; the higher order resonances are very damped, however.

We can use the same equation of motion as for the previously studied microphones because of the simple mass-spring analogy. Since the force is proportional to the pressure gradient $\partial\underline{p}/\partial z$, which

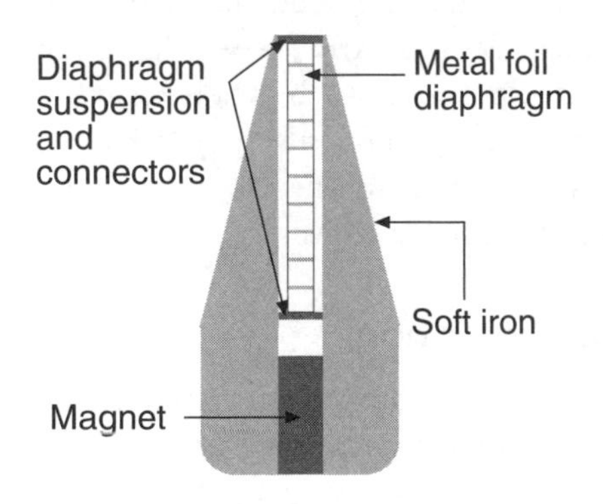

Figure 12.11 A pressure gradient ribbon microphone using a limp metal foil diaphragm. The soft iron magnet pole pieces are shaped so that the magnetic field in the air gap will be uniform.

becomes stronger as the frequency increases, the microphone must be made so that the mechanical impedance of the mass compensates for this frequency increase.

The microphone diaphragm must be *mass-controlled*—the M_M term must dominate in the equation of motion, Equation 12.7, to obtain a flat frequency response. This means that the fundamental resonance frequency of the mechanical system must be very low, well below the lowest audio frequency of interest.

The voltage response of the ribbon microphone to an incident plane wave can then be written as:

$$\underline{e}_{\text{out}} = K\, \underline{p}_A \cos(\theta) \tag{12.18}$$

where K is a constant that depends on the microphone design.

Note that the frequency response of the microphone is essentially flat over the resonance frequency, but that the microphone has a cosine-shaped directivity. At high frequencies, the microphone will have an even sharper bidirectional response than that of a cosine. This is due to the microphone diaphragm becoming large compared to the wavelength. The magnet's poles' pieces will also cause slight acoustic resonances leading to increased directivity and amplification.

Since the microphone is mass-controlled, it will also act as a vibration sensor; the ribbon microphone must be softly suspended to reduce the vibration sensitivity. This makes the microphone impractical to use.

Since the magnet of a ribbon microphone is usually made having large magnetic flux, a ribbon microphone can have about the same sensitivity as a conventional dynamic microphone. The need for a large magnet, a transformer, and high mechanical precision makes the ribbon microphone expensive to manufacture.

12.12 SUPER-DIRECTIONAL MICROPHONES

As mentioned previously, directional microphones can be designed according to two principles, multipole (gradient) designs and array designs.

Gradient designs are usually relatively small whereas *array microphones* are large since their function requires the microphone baseline to be large compared to the wavelength of the sound. Gradient designs can have directivities that are fairly frequency-independent over at least two decades, whereas array type microphones have directivities that are frequency-dependent. Most directional microphones used for studio recording are gradient designs. Array type microphones are sometimes used for measurement purposes and for specialized purposes, such as voice and birdsong pickup at large distances.

Since the directivity of an array microphone is also dependent on the microphone capsules used, one can use gradient microphones as elements in the array to achieve some low frequency directivity along with the directivity offered by the array action.

Noise Canceling Gradient Microphones

Gradient microphones base their function on the pressure gradient of the incident wave. In some cases, one can use this property to make microphones that have an improved S/N ratio.

The sound field close to a small sound source is different from that of a plane wave as studied in Chapter 1. The output signal of a velocity microphone placed close to a monopole sound source—that is, in a sound field that has spherical symmetry—will increase toward low frequencies in a sound field that has a flat frequency sound pressure response. If the gradient sensing microphone is designed so that it will compensate for this frequency increase, the microphone will correspondingly reject incoming plane waves (besides the rejection due to directivity). Many types of noise—for example, in office environments—are typically characterized by random incidence because of reverberation, and this noise canceling property will then function very well.

This type of microphone is common in various types of helmets and headsets for close pickup of sound; sometimes the microphone type is called *close-talking microphones*.

Array Microphones

Examples of popular array microphones are the parabolic reflector microphone, the passive or active line microphone, and the slit microphone. The directivity of these microphone types is based on the constructive and destructive interference between added sound components from various directions because of phase differences. More advanced arrays may use inhibition digital signal processing circuits, such as those used in cocktail-party processors, to reduce side lobes in the directivity pattern.

The directivity of an array microphone can be predicted from simple addition of its location dependent output signals. By using signal processing one can turn the sensitivity of the array microphone to the desired direction by manipulation of the phase (delay) and amplitude of the microphone signals, as indicated in Figure 12.12.

Figure 12.13 shows the design of a simple passive line array microphone. It functions by the passive addition of the sound at various tube openings. The sound is transmitted from the opening to the sound collecting cavity in front of the microphone diaphragm by tubes that act as *delay lines*. Active line and surface arrays give much better results since the addition can be optimized to give approximately frequency independent *main lobe* directivity or *side lobe* free directivity. The direction of the main lobe can also be easily adjusted.

Array microphones that have linear extension can only achieve toroidal or shotgun type directional patterns. To obtain receiving patterns that are better controlled, one can use two-dimensional arrays or

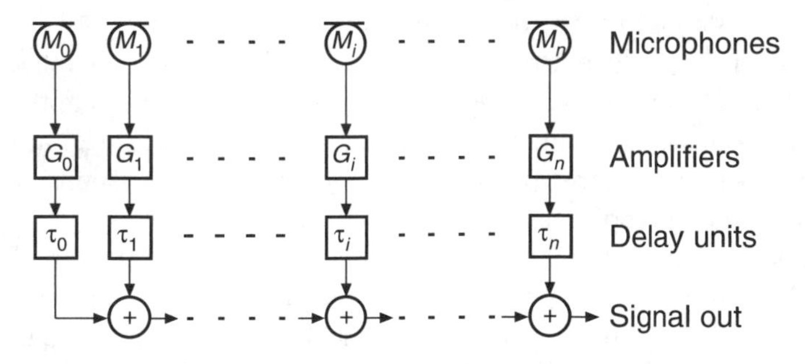

Figure 12.12 The basic operating principle of array microphones using beam-forming electronic control of pickup directivity pattern.

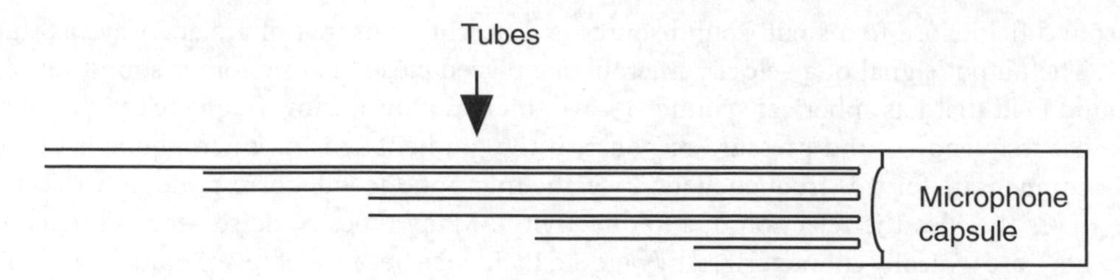

Figure 12.13 An old-style passive array microphone design using acoustic delay lines to feed the signals from the pickup points to the microphone.

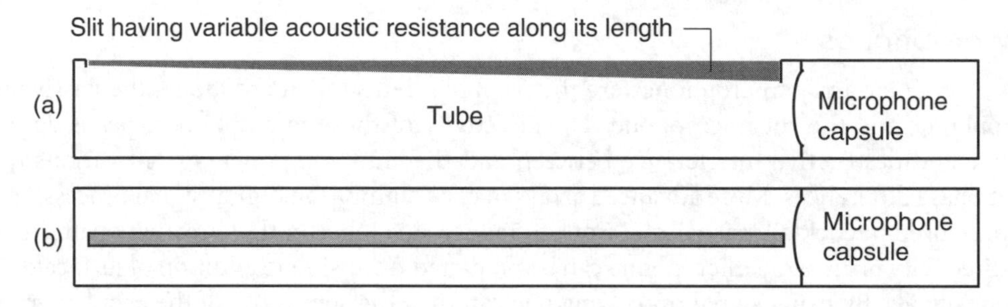

Figure 12.14 A slit microphone made using a cylindrical tube: (a) section through the microphone tube, and (b) top view of the microphone tube.

a parabolic reflector microphone, which essentially behaves as a passive two-dimensional continuous array.

The two most common types of line arrays are the *broadside array* and the *endfire array*. If a number of microphones are positioned along a line, and their output voltages added, the resulting directivity pattern will take the form of a toroid with the line at its symmetry center; this is the broadside pattern. If instead all microphone signals are added up by means of appropriate delay lines so that all signals sum in phase for a sound wave in the direction of the line, the pattern becomes directed toward one end of the line; this is the endfire pattern. The design shown in Figure 12.13 is a passive array using an endfire design. The acoustic signals add up at the microphone capsule.

The slit microphone is an endfire array, where all the pickup points have been replaced by a slit (Figure 12.14). The tube in front of the microphone can be thought of as a continuous distribution of pickup points. The slit is covered or contains an acoustic resistance that varies along the length of the slit. The sound that travels longest in the tube is most attenuated so the resistance to the incoming wave has to be highest close to the microphone.

The parabolic reflector microphone uses the circular pickup area of its parabolic reflector as a *continuous surface array*. The sound is picked up by a pressure or uni-directional microphone at the focal point of the parabola. The directivity can be changed by modifying the parabola, offsetting the microphone from the focus, and by choice of microphone directivity. For the parabolic reflector to work efficiently, the reflector diameter must be very large compared to the wavelength of sound, typically at least three times the wavelength. The *amplification* of the parabolic reflector is frequency-dependent

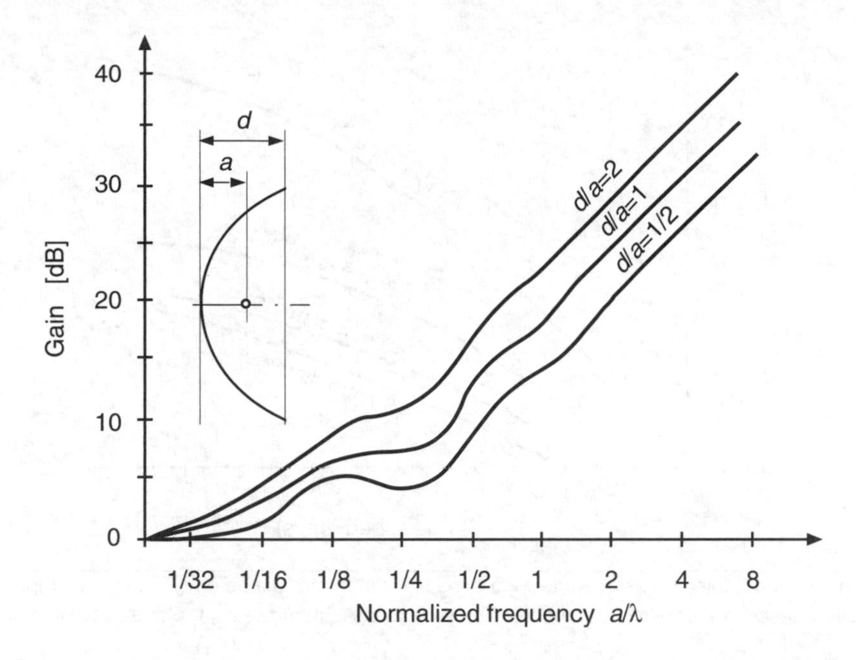

Figure 12.15 The pressure amplification characteristics of a parabolic reflector for some different choices of design variables (adapted from Reference 12.3).

so parabolic reflector microphones should be provided with a frequency response correcting filter to have a flat frequency response. The frequency response is irregular in the frequency region where the reflector diameter is of about the same size as the wavelength. Figure 12.15 shows the amplification of sound pressure at the focus of a theoretical parabolic mirror as function of frequency for varying reflector diameters.

12.13 WIND NOISE

The turbulence around microphones in wind causes local pressure fluctuations that are picked up by the microphone as noise. Such *wind noise* is a common problem when measuring and recording sound outdoors. The turbulence occurs close to the microphone, which disturbs the flow of air.

Wind noise can also occur when recording voice, due to breath. When voice is a problem, a *pop screen* is used to reduce the pops, which typically occur when plosives such as *p* and *b* are spoken or sung. The microphone may also be overloaded (buffeted) electrically and/or acoustically by the low-frequency wind gusts.

By moving the region of turbulence away from the microphone, one can reduce the noise picked up because of the strong decrease of pressure with distance. Typically, wind shields take the form of an acoustically transparent layer of cloth or plastic foam on a wire sphere surrounding the microphone at some distance. Even though the area of turbulence increases in this way, the increased distance between the microphone and the zone of turbulence more than compensates for the increased volume of turbulence.

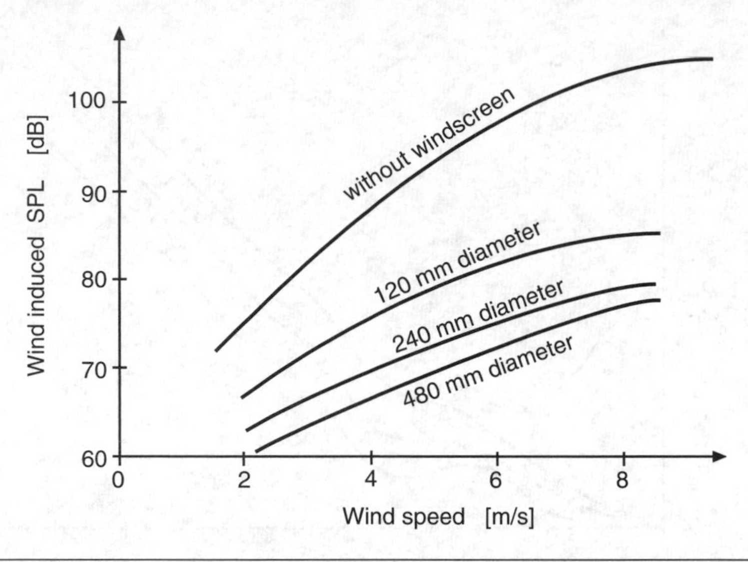

Figure 12.16 Wind induced noise levels at some wind speeds for a 25 mm diameter cylindrical microphone at the center of spherical windscreens of different diameter. The wind is incident in the $z = 0$ plane according to Figure 12.1. (Adapted from Ref. 12.1)

Figure 12.16 shows some results for the level of noise picked up by a microphone in wind, with and without windscreens of different diameters.

12.14 PROBLEMS

12.1 For a microphone, small compared to the wavelength, one may assume that the same force is acting over all of the microphone diaphragm's surface area; that is, the force is the product of sound pressure and diaphragm surface area.

Tasks:

a. Draw a simple sketch showing which parts of the microphone diaphragm correspond to mass, spring, and damper in a simple mechanical model. Draw this mechanical model and derive the equation of motion for the membrane, the mechanical impedance of the system, and its resonance frequency.
b. Show how one can obtain a constant ratio between rms sound pressure and rms output voltage above, below, or at the resonance frequency of the moving system of a microphone in which the output voltage is proportional to the diaphragm's.

   i. Excursion (which type of microphone?)
   ii. Velocity (which type of microphone?)
   iii. Acceleration

12.2 A dynamic microphone having the following characteristics is placed in a sound field:

Bl-product $5 \cdot 10^{-2}$ N/A
Diaphragm mass $1 \cdot 10^{-3}$ kg
Diaphragm resonance frequency $3 \cdot 10^{2}$ Hz
Diaphragm surface area $2 \cdot 10^{-4}$ m²

Tasks:

a. Calculate the microphone's output voltage as a function of sound pressure level and the *Q*-value of the mechanical system. Assume the simplest possible mechanical model for the microphone. Hint: Q_M is defined as $Q_M = \omega_0 M_M / R_M$.
b. Draw the frequency response curve for the following values of Q_M: 1, 0.1 and 0.01. Plot the output voltage level as a function of frequency using a logarithmic frequency scale. Which value will the output voltage have at 1 kHz for a sound pressure level of 94 dB?

12.3 A dual-diaphragm condenser microphone is assumed to have the following directivity function:

$$F(\theta) = \frac{Q_1}{2}(1+\cos(\theta)) + \frac{Q_1}{2}(1+\cos(\theta+\pi))$$

where Q_1 and Q_2 are constants proportional to the charge on the two diaphragms respectively.

Task: Calculate, using this expression, how the polarization voltage should be chosen to obtain the following directivity characteristics:

a. Omnidirectional, $F = C$
b. Cardioid, $F = C_1 + C_2 \cos(\theta)$
c. Bi-directional, $F = C \cos(\theta)$

12.4 A bi-directional microphone has the directivity function $F(\theta) = \cos(\theta)$.

Task: Calculate the microphone's DI.

12.5 A dynamic, pressure sensing microphone is positioned close to the mouth of a speaker. The microphone has the following characteristics at 1 kHz:

Bl-product 1 N/A
Diaphragm mass $1 \cdot 10^{-2}$ kg
Diaphragm suspension mechanical resistance $6.8 \cdot 10^{3}$ Ns/m
Diaphragm surface area $3 \cdot 10^{-4}$ m²

Task: Calculate the output voltage of the microphone for sound having sound pressure levels shown in the table.

Frequency	500	1k	2k	[Hz]
SPL	85	88	80	[dB] re 20 μPa

12.6 An intercom uses an electrodynamic loudspeaker unit both as loudspeaker and microphone. The microphone/loudspeaker has the following characteristics:

Bl-product 1 N/A
Effective loudspeaker unit diameter $3 \cdot 10^{-2}$ m
Diaphragm mass $1 \cdot 10^{-3}$ kg
Voice coil resistance 5 Ω

Task: Calculate the unit's S/N ratio when used as a microphone above its resonance frequency in the following case. The sound pressure level L_p at the membrane is 60 dB and the frequency of the sound is 1 kHz. The rms noise voltage in a resistor is given by $e = (4\ RkT\Delta f)^{1/2}$, where R is the resistance [Ω], k is Boltzmann's constant, T is the absolute temperature [K], and Δf is the noise bandwidth [Hz]. Assume the system bandwidth to be 4 kHz.

Phonograph Systems 13

13.1 INTRODUCTION

A phonograph system uses electromechanical means to engrave and subsequently track a groove containing signal information onto a disc.

The principles of recording and playback systems are shown in Figure 13.1. A *cutting lathe* is used to hold the original *lacquer covered aluminum disc*. The audio signal is amplified and fed to an electrodynamic *cutting head*, with a *stylus* moving in proportion to the applied current. The heated stylus is used to cut the groove in the thin lacquer layer on the surface of the rotating aluminum disc. The signal to the cutting head will modulate the *stylus tip* position both *laterally* and *vertically*, creating a *modulated groove*. The groove modulation is tracked by the pickup cartridge stylus on playback and converted to an electric signal. The cutting head travels on a radially positioned bar so that the cut groove is tangential to the cutting head at the point of contact. Using an electroplating process, the engraved disc is used in a process for making a number of stampers used for the pressing of vinyl disc copies. These copies are then played back using a *turntable*.

In addition to the turntable, the playback system has two main components:

- A tone arm that holds the pickup cartridge, making it possible to track the groove but that does not move at audio frequencies.
- A pickup cartridge with a stylus arm to track the audio frequency signal. The cartridge also has an electromechanical system to convert the stylus arm movement to an electrical signal.

It should be emphasized that while this section only discusses cutting heads and cartridges, the resulting sound quality is determined to a large extent also by the other components in the chain, such as amplifiers, *tone arm*, phonograph motor, *turntable*, and bearings. Noise from turntable bearing, motor vibration, dirt, and dust particles in the groove will result in annoying noise. Feedback due to vibrations induced in the system supporting the playback turntable may also be annoying.

13.2 DISC CUTTING

Since the groove position contains the recorded information, it is necessary to also discuss the groove-cutting process. The disc cutter is based on the use of a turntable rotating at constant rotational speed, typically 45 or 33.33 rpm, although discs made using *half-speed cutting* is also commercially available.

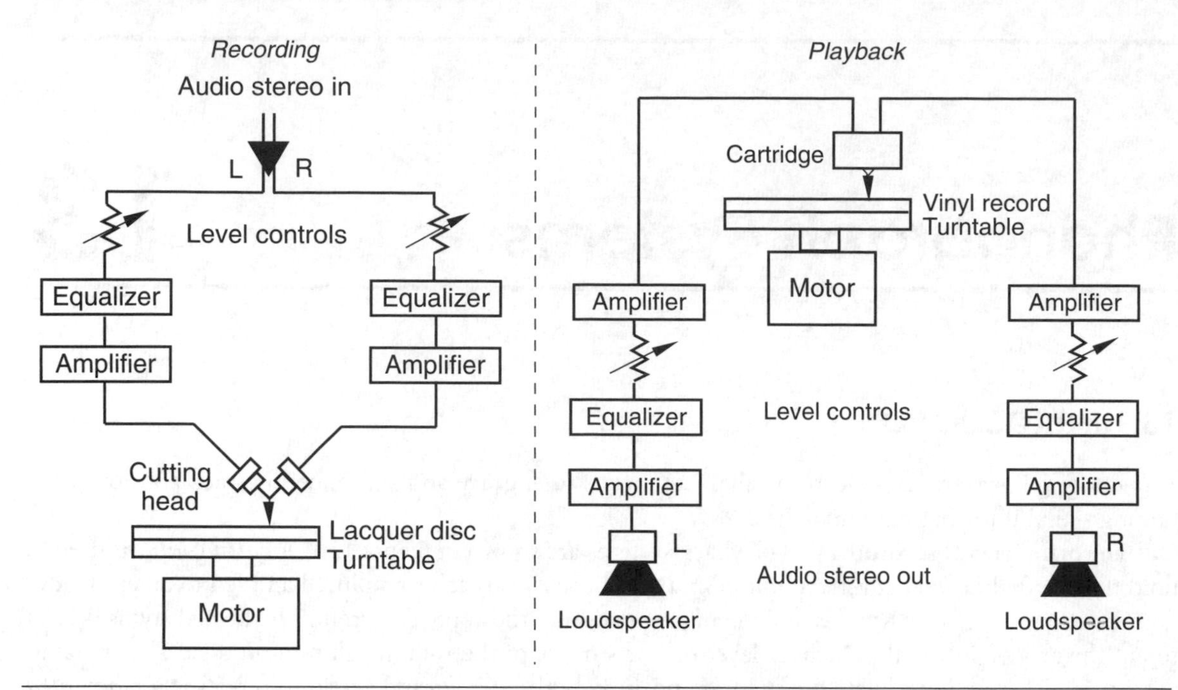

Figure 13.1 The phonograph sound recording and playback systems (after Ref. 13.1).

Guides and a screw carry the cutting head radially across the disc while the disc is spinning, cutting a spiral track in the lacquer on the original disc. The radial speed of the cutting head is dynamically changed according to the need for high excursions by the cut groove. This is done by sensing the signal to be recorded a short instance in advance to know the need for cutting head movement. A conventional stereophonic disc is cut using both lateral and vertical cutter stylus tip movement.

Cutting Heads

The basic design of a stereo disc cutting head is shown in Figure 13.2. The electromechanical system typically consists of two electrodynamic drivers that move the stylus. The drivers are mounted at right angles to one another and generate stylus tip movement in a plane almost perpendicular to the groove. Using the so-called 45/45 system, grooves can be cut in combinations of both lateral and vertical movements. The left and right channel information is carried by the groove walls at ± 45° angles to the disc surface.

The cutting head is usually tilted so that the movement plane of the cutting stylus is at an angle of 15° relative to the vertical axis. In this way, the cartridge will be able to track the groove with little nonlinear distortion.

The cutting head's motional system, in fact, consists of two first-order resonant systems, one for each transducer channel shown in Figure 13.2. These have their resonances in the middle of the audio frequency range.

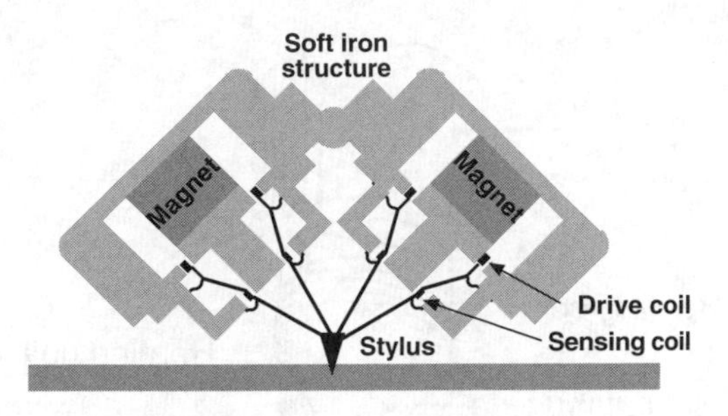

Figure 13.2 An example of the construction of a stereo cutter head.

Each electrodynamic transducer carries two coils. One coil is the main cutting coil and the second coil is the sensing coil. The output of the sensing coil is a voltage proportional to the tip movement. This voltage is used as feedback to adjust the driving current to the main coil so that the cutting head stylus moves according to the input voltage. The feedback system makes the frequency response correct and reduces the nonlinear distortion.

Examples of typical frequency response curves, with and without feedback applied, are shown in Figure 13.3. Special recording pre-emphasis filters are used that, along with filters in the playback preamplifier, achieve the goals of low noise, low distortion, and flat frequency response in the overall system. The standardized RIAA playback response curve is shown in Figure 13.5.

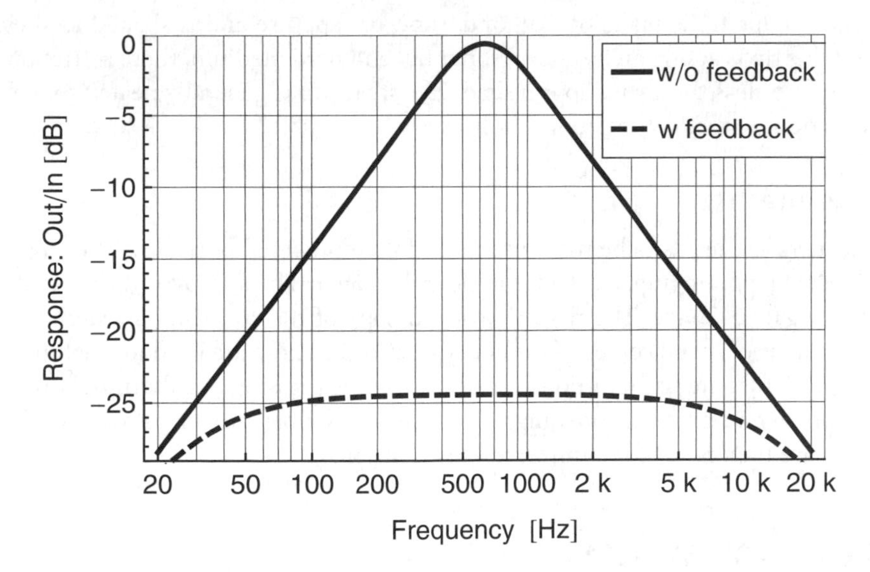

Figure 13.3 An example of the typical frequency response characteristics of a cutter head without A and with B feedback.

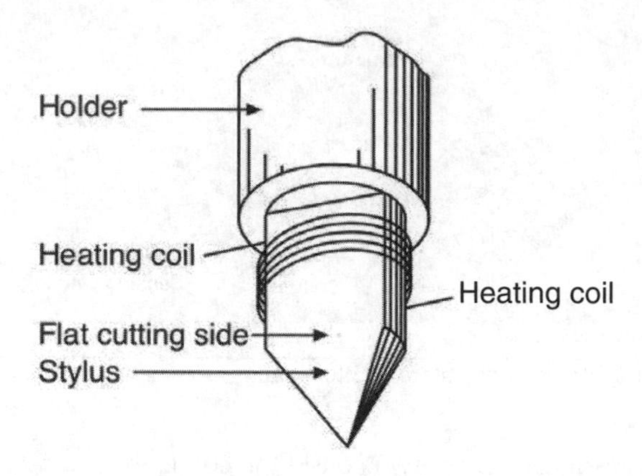

Figure 13.4 An example of a cutter head stylus equipped with a heating coil.

The amplifier driving the cutting head must typically be able to deliver at least 100 W of electrical power to each cutting head coil. The main cutting coil and the cutting head both become very hot in the cutting process. Cutting heads often use cooling systems to remove the heat using fluidic systems.

Cutting Styli

The cutting head's stylus tip is made of diamond, ruby, or sapphire and is shaped as shown in Figure 13.4. Since the stylus tip is acting on lacquer, a softer but still hard medium, there is friction, resulting in noise and distortion unless the stylus tip is heated. The stylus tip is generally heated to melt the lacquer to reduce this friction. Direct cutting to metal disks is also possible.

Disk Manufacture

The original lacquer disk needs to be mass-produced for commercial purposes. The engraved disk can be played back using a phonograph cartridge as described later. The approved disk is electroplated and a metal backed "negative" made. The negative, ridged copy of the cut disk ("master") is then used to manufacture a number of metal copies of the cut lacquer disk. These are used for making metal backed negatives, "stampers", that are used for double-sided hydraulic pressing of a thermoplastic material. The stampers wear out after a number of pressings. The resulting double-sided plastic ("vinyl") copies of the two lacquer disks are then ready for commercial distribution.

13.3 THE PLAYBACK SYSTEM

The playback turntable must be suspended by quiet bearings and have high mass so that any speed irregularities induced by the playback system and the turntable motor drive system (direct, pulley, or belt drive) are eliminated. The vinyl record should rest snugly on the turntable top mat so that any resonances in the plastic disc are damped by the turntable and its mat.

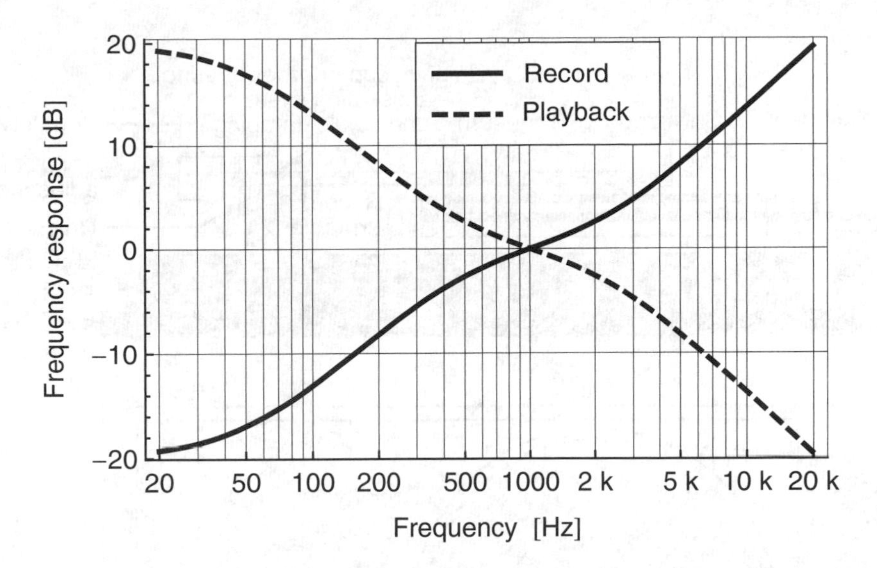

Figure 13.5 The curves shows the RIAA velocity frequency response recording characteristics used in the production of commercial vinyl recordings.

Since the playback equipment is generally placed in the same room as the loudspeakers, the sound pressure levels at the record playback equipment may become very large. The sound will induce vibrations both in the playback equipment and furniture carrying the equipment, as well as in floors and walls. These vibrations will arrive somewhat later than the direct sound from the cartridge and effectively add coloration to the sound from the record. It is important that the tone arm and turntable are constructed in such a way as to reduce such vibrations to a minimum. A rigid stand for the turntable system is the best.

Tone Arms

There are essentially two types of tone arms, those that have the cartridge to travel a curved path across the record, and those that allow the cartridge to travel a radial path, mimicking the way that the cutting head moves across the original cut. Most tone arms are the first type, as shown in Figure 13.6. Such pivoted tone arms must be shaped so that the *angular error* between the groove cut by the recording equipment and the cartridge needle system is minimized. The angular error will depend on tone arm position, length, and shape. It will also depend on cartridge mounting and the geometry of the stylus arm and tip. The linearly tracking tone arm is a modified pivoted arm that rests on a moveable base that can move along the side of the turntable so that the cartridge will move radially acoss the record. Such arms need a servo system to control the position of the moveable base so that the tracking error is kept within some small range.

Any tone arm must have bearings, allowing the arm free movement in two planes without skew or friction. Some tone arms do not use conventional bearings but rest on a *needle tip*, so called unipivot arms. The friction in the tone arm suspension and the compliance in the signal cables between the pick up and the tone arm base, which usually runs through the tone arm, introduces further forces on the cartridge's mechanical tracking system. Since the tone arm cannot be made ideally stiff, its mechanical resonances must be well-damped and preferably above the audio frequency range.

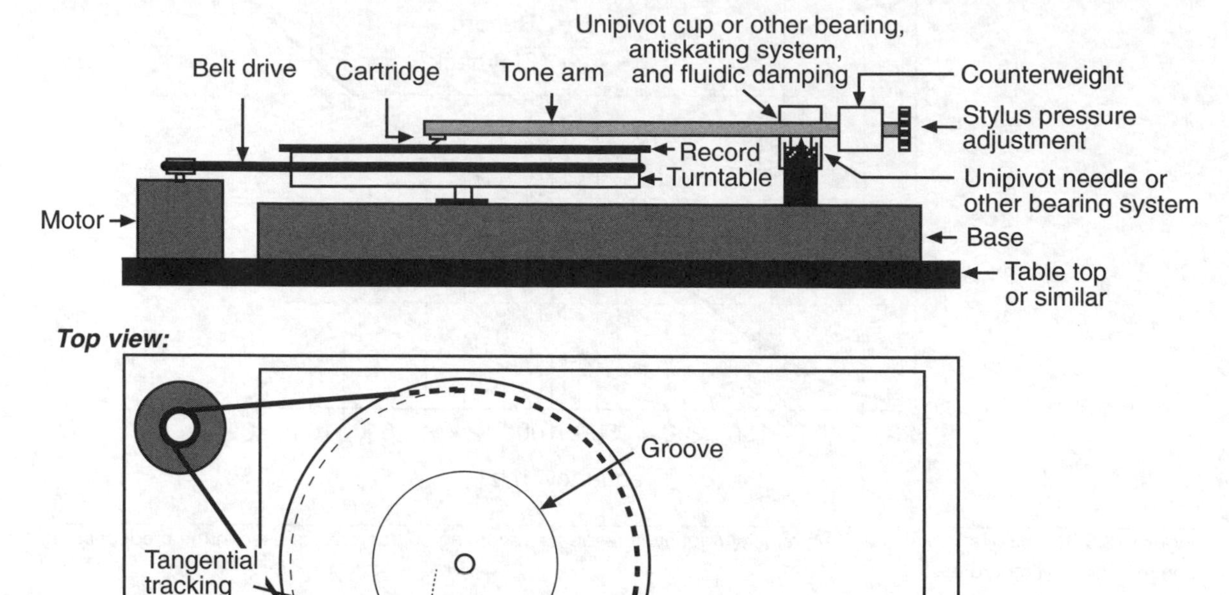

Figure 13.6 Outline of a modern belt-driven turntable with external motor. Special tone arm geometry is needed to reduce the tracking angle error for a pivoted arm since the recording is done with a tangentially mounted cutter head. Since the cartridge is offset by an angle to reduce the tracking angle error, there is an additional *bias* force acting to draw the cartridge toward the turntable center as the disk is rotating. Most arms will have an arrangement for compensation of this bias force. In the case shown, the counterforce is assumed to be achieved magnetically.

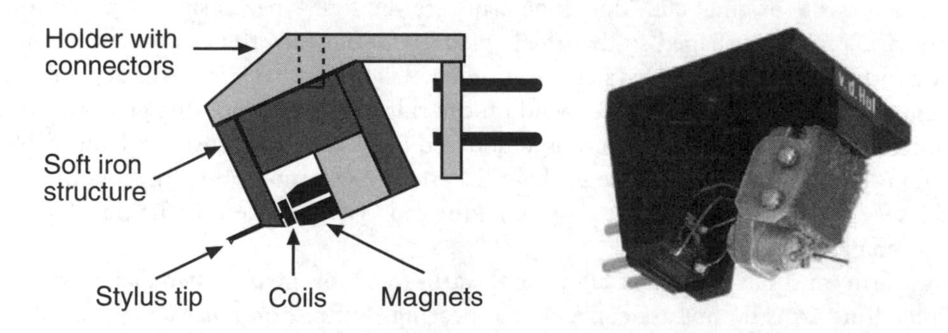

Figure 13.7 Basic components of a moving-coil cartridge (left) and a practical implementation by Van der Hul, Netherlands (right) (photo courtesy of A.J. van den Hul b.v., The Netherlands).

13.4 CARTRIDGES

The playback quality depends on the mechanical properties of the record groove, the stylus tip and arm, the transduction mechanism, the cartridge body, and the tone arm, along with the factors mentioned previously. A high-quality cartridge is shown in Figure 13.7.

At low frequencies, the response of the cartridge mounted in the tone arm will depend on the compliance of the stylus arm suspension and the inertia of the tone arm forming a second order high pass filter. The resonance due to these two components generally has a frequency in the range of 5–20 Hz. It is advantageous to have the resonance at approximately 15 Hz since this will reduce the stylus arm movement due to the vinyl record not being ideally flat and centered. Typically, the disc rotates at a speed of 33 rpm, which means that the lowest frequency induced by record unevennesses will be about 0.5 Hz. The tone arm/stylus resonance may be damped by possible viscous tone arm damping but more often, the resonance is damped because of the mechanical damping in the stylus arm suspension. It is advantageous to keep the Q-value of this resonance low.

The high frequency limit is given by the filter characteristics of the resonant system, consisting of the stylus tip mass and the compliance of the vinyl groove walls, forming a mechanical low-pass filter.

Stylus Tips

To reduce friction and associated disk wear, the *stylus tip* is highly polished. It also needs to have low mass so that the groove wall will not be deformed by repeated playback. The walls of the vinyl groove are compliant but not ideally elastic. There is a pressure limit above which the stylus groove wall will permanently yield and not recover its former shape.

The technical requirements regarding the stylus tip are conflicting. On one hand, the tip needs to have a small radius of curvature to follow the groove well at high frequencies. The cutting stylus is diamond-shaped, so the cartridge stylus needs to have a shape closely approximating the cutting stylus. On the other hand, the cartridge stylus tip must not apply so much pressure on the groove wall while accelerating that the wall yields. The wall will only endure a limited number of playback cycles until it deforms unacceptably. The stylus tip can have a conical, elliptical or other cross-sections. A low pressure requires a spherical tip that has a large radius of curvature. Most current stylus tips have an elliptical lateral cross-section to better track the groove at high frequencies.

The compliance of the groove walls together with the mass of the stylus tip form a resonant system, having a low-pass filter characteristic that removes high frequencies. Typically, this resonance occurs at 20–25 kHz.

Stylus Arms

The *stylus arm* (cantilever) is an elastic beam, typically hinged in a rubber collar. It has the stylus tip at one end and a magnet or coils at the opposite end. The cantilever will have bending wave resonances at various frequencies, preferably only at frequencies above the audio range. Hollow tubes made of synthetic semiprecious materials, such as ruby, make good stylus arms.

The movement of the beam at each resonance frequency will be characterized by a mode shape. The mode shape and mechanical properties of the stylus arm, along with the stylus tip mass, will determine the *effective mechanical impedance* at the stylus tip. The modal movement of the stylus arm may result in an uneven frequency response. The mode shapes will depend on the shape of the stylus arm and the way it is suspended, the mechanical impedances at the stylus tip (including the groove wall impedances), and the magnets or coils attached to the arm.

The effective stylus mass (given by the effective mechanical impedance) can be regarded as a primary quality metric for cartridges since this mass will be directly proportional to the forces that act on the groove walls. These forces determine the abrasion and other groove deformation processes that occur on playback of the vinyl record. The higher the effective stylus mass, the more difficult it will be for the stylus tip to track the groove. In the extreme case, the tip will jump out of the groove. This can be circumvented by higher static pressure, again resulting in more groove wear.

Electroacoustic Systems

The main problem in designing the electroacoustic system is retaining linearity over a wide range of static stylus tip positions. Different users will use the cartridge at different stylus pressures, which means that the static position of the stylus arm will vary quite a lot. Another problem is obtaining sufficient output voltage so that the preamplifier noise will be negligible.

Some Common Cartridge Types

Most cartridges are *moving-coil* and *moving-magnet* designs, although other designs such as piezoelectric mechanisms were very common some decades ago.

Moving-coil cartridges work according to the electrodynamic principle, having electrically conductive coils moving in a static magnetic field. An outline drawing of a moving-coil cartridge is shown in Figure 13.7. The two coils are attached orthogonally to the rear end of the stylus arm. This results in comparatively low output voltage since there can only be a few coil windings. The mass of moving-coil cartridges is generally higher than that of moving-magnet cartridges, resulting in a lower tone arm resonance frequency, making the system more sensitive to record unevenness.

In a moving-magnet cartridge design, a magnet, positioned at the rear end of the stylus arm, creates time-varying magnetic fields as the arm moves (see Figure 13.8). The magnetic field induces voltages in

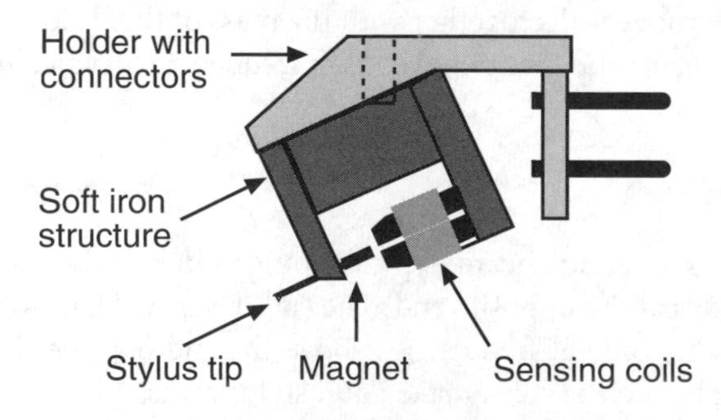

Figure 13.8 The basic design of a moving magnet cartridge.

two coils of electrically conductive wire. The design is quite rugged and has several advantages. Since the static coils can have many turns, the sensitivity can be high, and the output impedance relatively high. The design allows the user of the cartridge to replace the stylus, which is practical since styli need to be replaced after some thousand hours of use.

13.5 AMPLIFICATION

Since the output power of the cartridge is very low, the playback system needs a preamplifier to make the signal level high enough for the audio power amplifier that drives the loudspeakers. The preamplifier/equalizer must have very low noise level. The preamplifier will contain the frequency response correction network discussed previously.

Typically, a preamplifier for moving coil type cartridges needs to be preceded by a step-up transformer that may have a 1:40 winding ratio to increase the output voltage of the cartridge by a factor of 40. This results in an increase in signal-to-noise ratio by about 30 dB. Because of the transformer, the amplifier will also see the cartridge as having a different impedance. Most moving coil cartridges have an internal impedance that is primarily resistive and in the range from 2 Ohms to 50 Ohms. Since the impedance transformation ratio is the square of the voltage ratio, the resulting impedances will be in the range of about 3 kOhm to 80 kOhm. Most audio preamplifiers have an input impedance of 100 kOhm or more in parallel with some small capacitance. With the resulting high impedances, the frequency response of the cartridge may be affected by the filter action of the resistance and capacitance combination, particularly since the transformer will also feature some inductance.

For magnetic cartridges, the noise problem is smaller. On the other hand, the high inductance of the cartridge in combination with the cable capacitance also leads to low pass filter behavior, the damping of which is primarily determined by the preamplifier input resistance. It is always advisable to have short cables between the cartridge and the preamplifier for minimum capacitance and also some possibility for adjusting the load impedance. Since electromagnetically-induced noise *hum* is always present, it is advantageous to use a preamplifier having balanced signal input so that the hum is cancelled.

Loudspeakers 14

14.1 INTRODUCTION

A *loudspeaker* is an electroacoustic device for converting an electric signal into an acoustic pressure signal retaining the waveform. The loudspeaker's diaphragm moves the surrounding air driven by a force from an electromechanical transducer.

A loudspeaker generally consists of a *loudspeaker driver* (sometimes also known as *motor*) and a *loudspeaker box*. The term loudspeaker is popularly used to describe both the loudspeaker driver and the complete loudspeaker, including the loudspeaker box. The term driver is sometimes used to specifically describe the type of loudspeaker motor used for medium and upper frequency range speakers, known as *horn loudspeakers*.

It is convenient to group loudspeakers according to how the (front) side of the loudspeaker diaphragm radiates sound into the surrounding environment:

- *Direct radiator loudspeakers* are loudspeakers in which at least one side of the diaphragm of the loudspeaker radiates directly (sometimes called direct firing).
- *Horn loudspeakers* are loudspeakers in which the diaphragm radiates through an acoustic filter in the form of an *acoustic transformer*, a *horn*.
- *Nondirect radiating loudspeakers* are loudspeakers in which the diaphragm radiates by way of an *acoustic filter*. These are sometimes called *bandpass* box designs.

The radiation from the second side (rear side in direct radiating loudspeakers) is usually, but not always, prevented by a *box*, *baffle*, or similar construction to prevent an aerodynamic short-circuit of the volume velocity generated by the diaphragm.

The shape and acoustic design of the loudspeaker box contribute to how the loudspeaker driver radiates sound. In addition, there is usually radiation of sound by the vibrations of the box walls. These can be excited both by the loudspeaker driver and the sound field inside the box.

There are many types of drive principles that can be used to generate the force needed to move the diaphragm. The most common designs use electrodynamic, piezoelectric, or electrostatic transduction principles. Loudspeaker drivers have been constructed using modulated ionization of air as well as modulated release of pressurized air. The latter are used for various industrial purposes.

In-depth discussions of the many issues involved in loudspeaker design and use may be found in References 14.1 and 14.2, and in the general references for this chapter.

14.2 RADIATION AND DIRECTIVITY

A simple and instructive way to analyze the radiation by a loudspeaker diaphragm is to use an approximation, where the diaphragm is assumed to be a *vibrating plane piston*. This approximation is relevant for *electrodynamic loudspeakers* in the frequency region above the *fundamental resonance frequency*, due to the *diaphragm mass* and *suspension compliance* of the loudspeaker. For *electrostatic loudspeakers*, it is valid well above the fundamental resonance frequency of the membrane. Most electrodynamic loudspeakers use dome- or cone-shaped diaphragms. The following analysis is reasonably valid also for such diaphragm shapes, provided the dome or cone is reasonably shallow (cone depth much smaller than λ).

In Chapter 9, we studied the sound radiation for the case of a monopole in free space and on a hard surface. In Chapter 1, we found that the sound pressure at a distance r from a monopole having a volume velocity $\underline{U}_D$ is:

$$\underline{p}(r) = j\omega \underline{U}_D \rho_0 \frac{e^{-jkr}}{4\pi r} \tag{14.1}$$

The sound at some distance will be the linear sum of the sound from the monopole and its mirror image, and will be twice that of one free monopole.

Now assume that the loudspeaker diaphragm is a circular piston, which, besides being stiff, is mounted, airtight, in an infinite, hard, immovable plane baffle at $z = 0$, as shown in Figure 14.1. The piston is assumed circular, in the plane $z = 0$, with its center at the origin.

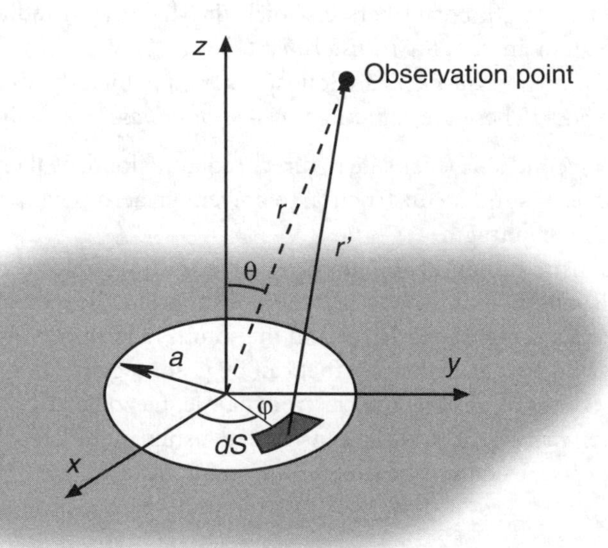

Figure 14.1 The coordinate system for calculation of the sound radiation by a circular piston in an infinite baffle, both in the $z = 0$ plane.

The volume velocity is generated by the piston vibrating in the z-direction. When the piston is vibrating, each element on the piston surface area (on both the + and $-z$ sides of the piston) can be thought of as an individual, in-phase monopole, generating volume velocity.

Because of linearity, we can again use the principle of superposition of pressure to determine the total sound pressure at the observation point. Since each monopole is on a hard surface, the principle of doubling of pressure due to mirror images also applies.

Each monopole will contribute to the total sound pressure as if it has a part of the total volume velocity $d\underline{U}_D = \underline{u}_D dS$, due to the vibrational velocity of the diaphragm $\underline{u}_D$ and the size of the surface element of each surface element dS. The total pressure can then be obtained by summing the sound pressure contributions $d\underline{p}$ from each surface element dS of the entire piston surface:

$$d\underline{p}(r') = j\omega(\underline{u}_D dS)\rho_0 \frac{e^{-jkr'}}{2\pi r'} \tag{14.2}$$

where

$$r' = \sqrt{r^2 + \sigma^2 - 2r\sigma\sin(\varphi)\cos(\theta)} \tag{14.3}$$

For points far away from the piston, the integration results in:

$$\underline{p}(r,\theta) = j\omega(\underline{u}_D \pi a^2)\rho_0 \left[\frac{2J_1(ka\sin(\theta))}{ka\sin(\theta)}\right]\frac{e^{-jkr}}{2\pi r} \tag{14.4}$$

where J_1 is a Bessel function (due to the cylindrical symmetry). Replacing $\underline{u}_D$ in this expression with $\underline{U}_D/\pi a^2$, one can see that the expression for the total sound pressure at r is the same as that for a monopole on a hard surface, having the volume velocity $\underline{U}_D$, except for a directivity term:

$$\frac{2J_1(x)}{x} \approx 1 - \frac{x^2}{8} + \tag{14.5}$$

One finds:

$$\underline{p}(r,\theta) = j\omega\underline{U}_D\rho_0 \left[\frac{2J_1(ka\sin(\theta))}{ka\sin(\theta)}\right]\frac{e^{-jkr}}{2\pi r} \tag{14.6}$$

As mentioned in Chapter 9, the radiated acoustic power can be estimated by integration of the sound intensity for the far-field of the piston, where particle velocity and sound pressure are in phase. One incident for far-field conditions is that the sum pressure diminishes with distance as $1/r$, which is the case at long distance.

The function $F(\theta, \varphi)$:

$$F(\theta,\varphi) = \frac{2J_1(ka\sin(\theta))}{ka\sin(\theta)} \tag{14.7}$$

is a *directivity function* describing the directional sound radiation in the far-field of the loudspeaker. Because of the symmetry, there is no variation due to the angle φ, for a point source $F(\theta, \varphi) = 1$. (Note that the coordinate system in Figure 14.1 corresponds to that of the microphone studied in Chapter 12.)

The sound pressure close to the surface of the piston has a complicated structure for frequencies over those in which the piston has a size comparable to that of the wavelength. Figure 14.2 shows isobars near a piston for a case where $ka = 10$; that is, the piston diameter is about 3 λ.

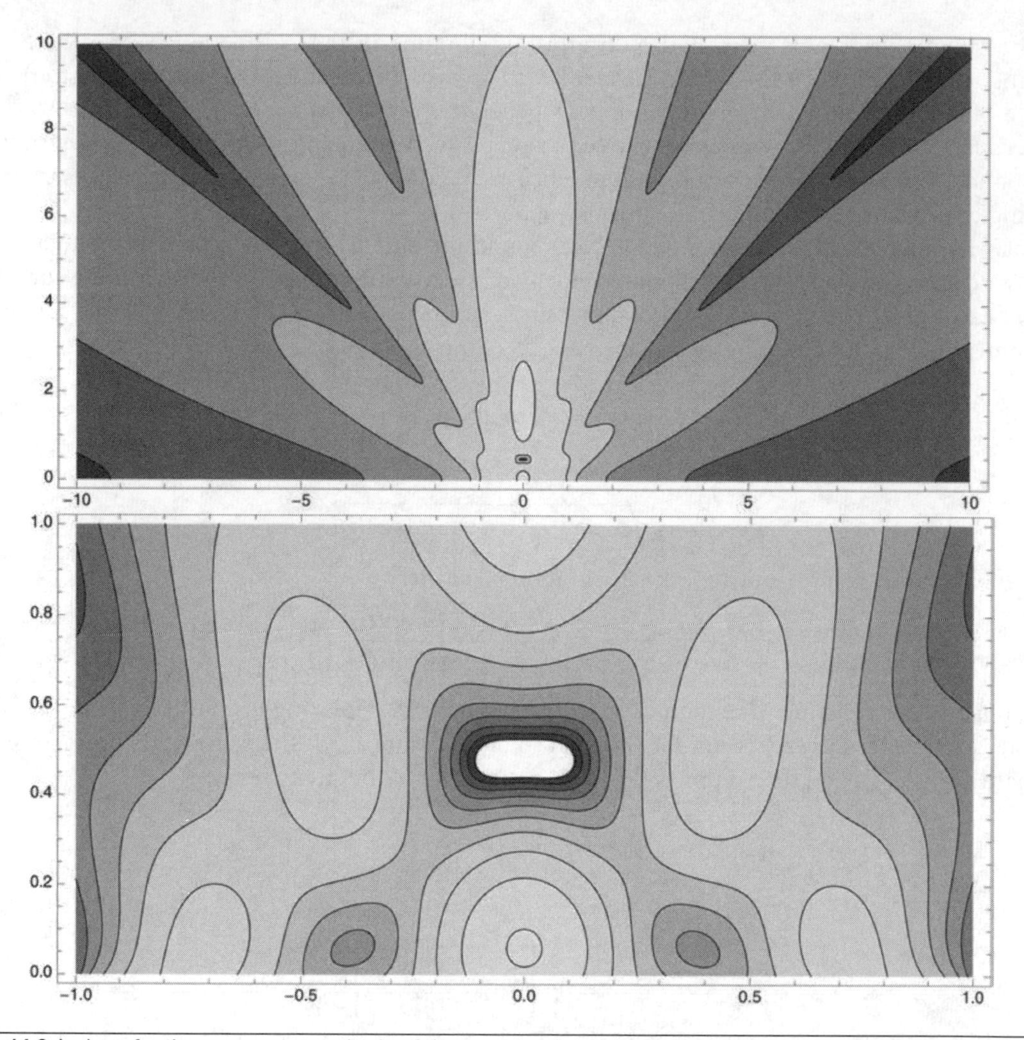

Figure 14.2 Isobars for the pressure magnitude of the sound radiated by a circular piston in a hard baffle for *ka* = 10. The wave number is *k*, and *a* is the piston radius. Near- and far-field regions for *z* from 0 to 10 and *a* from −1 to 1 (top) and near-field region for *z* from 0 to 1 and *a* from −1 to 1 (lower).

The sound pressure on the normal of the circular piston is obtained as:

$$\underline{p}(r,0) = -Z_0\underline{u}_D\left(e^{-jk\sqrt{r^2+a^2}} - e^{-jkr}\right) \tag{14.8}$$

A commonly used criterion to estimate the shortest far-field distance $r(\lambda)$ is:

$$r(\lambda) > 4\frac{a^2}{\lambda} \tag{14.9}$$

where a is the piston radius and λ the wavelength of sound.

The *directivity D* describes the intensity of the sound in the direction of maximum radiation relative to the intensity at the same distance from a monopole having the same radiated sound power. This is directly analogous to the case of the microphone. The directivity is given by the ratio:

$$D = \frac{4\pi}{\int_0^{2\pi}\int_0^{\pi} |F(\theta,\varphi)|^2 \sin(\theta)\, d\theta\, d\varphi} \tag{14.10}$$

Also note that it is more common and practical to use the *directivity index* (DI) expressed in dB units as a metric for the directionality of the loudspeaker's radiation rather than the directivity. The directivity index is defined as:

$$DI = 10\log(D) \tag{14.11}$$

The radiation pattern from a loudspeaker having a shallow circular cone, mounted in the large flat surface of a loudspeaker box, is approximately described by the directivity function for a circular flat piston in an infinite, hard baffle given in Equation 14.7.

Some directivity patterns for a vibrating plane piston in a hard baffle, given by Equation 14.7, are shown in Figure 14.3 for some values of *ka*. The patterns correspond well to those of practical conical loudspeaker diaphragms having the same effective diameter. It is important to note that different cone profiles have different directivity patterns. The directivity patterns of nonplanar diaphragms are best obtained numerically, for example, by the boundary element method, or more approximately, by numerical integration as outlined initially in this section.

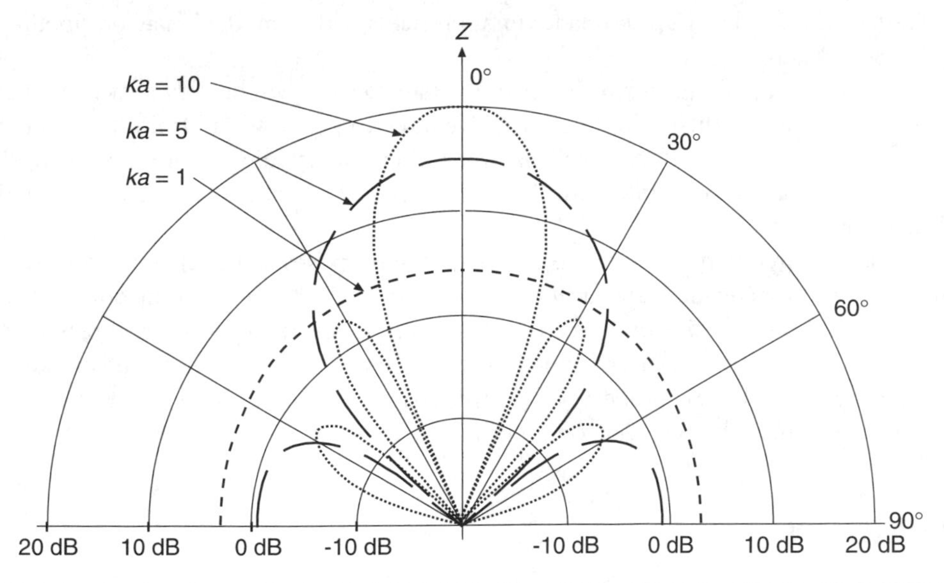

Figure 14.3 The sound radiation by a vibrating circular plane piston is highly directional when the piston diameter becomes large compared to wavelength. This graph shows the far-field directivity patterns for a piston having radius *a* in a baffle for three values of *ka* as a function of the angle θ relative to the *z*-axis in Figure 14.1. The curves show gain in dB over that of a free monopole having the same radiated power as the piston.

The directional behavior of the loudspeaker determines the effective area of good listening, the *sweet spot* in stereo and surround sound reproduction. The directivity of the loudspeaker is in practice affected by how the loudspeaker is mounted, how many loudspeaker drivers are being used, how the signal is distributed between the loudspeaker drivers, and additional factors. The directivity of a loudspeaker is also influenced by the diffraction of sound by the edges of the loudspeaker box.

The directivity is particularly difficult to control in the case of loudspeakers where the entire frequency range is handled by one driver only. Control of directivity by using several loudspeaker drivers is also problematic. For domestic loudspeakers, one usually wants to have a gradual change of directivity over the frequency range, from omnidirectional to more directive. The more reverberant the space, the more important the control of loudspeaker directivity becomes. In extremely reverberant spaces, such as churches and sports arenas, the directivity of the loudspeaker must be very high in the speech frequency range. It can then be necessary to eliminate frequencies where the loudspeaker has insufficient directivity by using suitable filters.

Horn loudspeakers are particularly advantageous when there is a need for high directivity. Horn loudspeakers and horn drivers use comparatively small diaphragms with more piston-like behavior than conventional direct radiating electrodynamic loudspeakers. To have the same radiating area and directivity as a horn loudspeaker, one must use many direct radiating loudspeakers. It is, however, difficult to obtain drivers that are uniform enough to obtain the desired array properties.

Most electrodynamic loudspeakers have very erratic phase and amplitude performance at medium and high frequencies. This is due to the diaphragm no longer vibrating with the same amplitude and phase over its area. At medium and high frequencies, the wave propagation in loudspeaker diaphragms will be as bending and torsional waves, the diaphragm will no longer vibrate as a rigid piston. The bending and torsional waves will set up resonant vibration fields with a modal behavior, popularly known as *breakup*.

In case of breakup, the frequency response of the loudspeaker will be very peaky, characterized by peaks and sharp dips due to the positive and negative interference of sound from the different parts of the diaphragm. The frequency response will also vary when the listening position is changed since the directivity pattern will be poorly controlled. The directivity pattern will often be characterized by many lobes and sharp nulls.

One can use the directivity of loudspeakers for special purposes. All electrodynamic, piezoelectric, and electrostatic electroacoustic transducers are reciprocal—they can function both as transmitters and receivers. Also, since the directivity is a result of interference, the transmit and receive directional responses are reciprocal. The directivity function will be the same, whether a loudspeaker is used to transmit or pick up sound. This property of electrodynamic loudspeakers is often used in intercommunication devices, where the transducer has both roles.

14.3 EFFICIENCY

The sound power required from loudspeakers varies considerably. Typically, the maximum acoustic power available from conventional electrodynamic loudspeakers is about 1 W. Higher radiated powers can be obtained by using a horn or other transformer-coupler, typically 10 W. Loudspeaker efficiency—acoustics power for a certain electrical input power—may be an important parameter in some applications.

Loudspeaker sensitivity is usually given as sound pressure at some distance from the loudspeaker for a certain input voltage, usually as sound pressure level on-axis at 1 m distance for an input voltage of 2.83 V in an anechoic environment.

There are several ways to express the sensitivity and efficiency of a loudspeaker system. One common way of writing the electroacoustic efficiency η is to use the ratio between acoustic output power and electric input power as:

$$\eta = \frac{W_{\text{acoustic}}}{W_{\text{electric}}} \tag{14.12}$$

The sensitivity can be expressed by the particular sound pressure level that can be obtained in a certain direction for a given input voltage or power.

Most direct radiating electrodynamic loudspeakers have efficiencies in the range of 0.1 to 5%. Using horns or other impedance transformers, one can increase the efficiency substantially. Horn loudspeakers often have efficiencies in the range of 5 to 15% in commercial designs.

Electrostatic loudspeakers have nominal efficiencies close to 100%, but the losses in transformers and amplifiers are substantial because of the reactive capacitive electrical load of the loudspeaker.

14.4 FREQUENCY RESPONSE

The acoustic load on a small sound source by sound field impedance, the *radiation load*, is primarily reactive and has a mass-type character for distances from the origin where $kr \ll 1$, as shown by Equation 2.27. For distances such that $kr \gg 1$, the sound field impedance is primarily resistive. Since the radiation load by the air influences the mechanical system (it adds to the mass and resistance in the equation of motion), and since the radiated power depends on both the vibrational velocity of the loudspeaker diaphragm and the real part of the acoustic load on the diaphragm, the loudspeaker driver must be designed with this in mind.

Most high-quality electrodynamic drivers are designed to operate mainly in the region where the diaphragm radius, a, is much smaller than the wavelength, so that $ka \ll 1$—that is, the radiation load is primarily reactive and its resistive component increases proportionally to frequency squared. As long as the diaphragm is small compared to wavelength, the mass load due to radiation will be constant.

The mass component of the radiation load may be as large as the mass of the loudspeaker diaphragm itself. This puts a limit to how thin and weak the diaphragm can be. If the diaphragm is weak, it may buckle under the radiation load. This, of course, leads to considerable nonlinear distortion.

In the $ka \ll 1$ frequency region, the sound intensity, as a function of frequency, in the radial direction in the far-field will be:

$$I = \frac{p^2}{Z_0} = \frac{\rho_0 a^4}{8c_0}\left|F(\theta,\varphi)\right|^2 \frac{\omega^2}{r^2} u_D^2 \tag{14.13}$$

As long as the magnitude of the directivity function $F(\theta, \varphi)$ is close to unity, the vibrational velocity must diminish proportionally to the frequency for the sound intensity at the listening point to be frequency-independent.

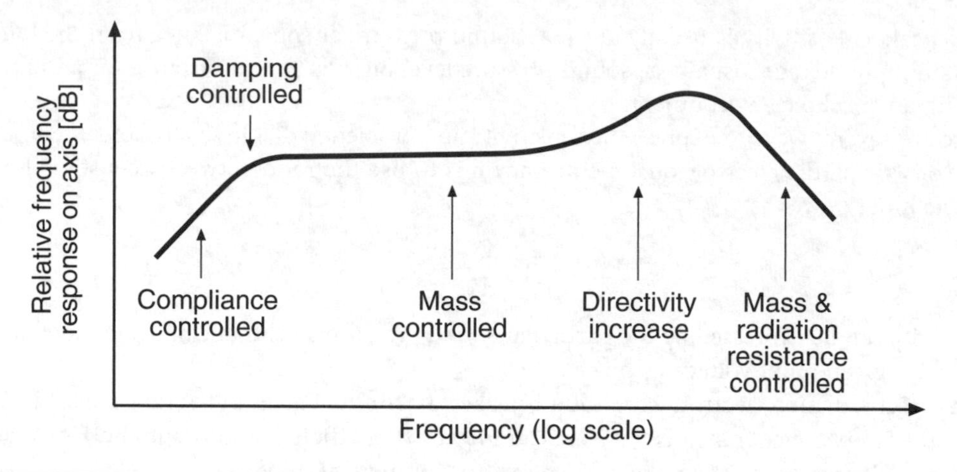

Figure 14.4 Idealized frequency response curve for an electrodynamic loudspeaker.

This, in turn, means that the vibrational velocity must be determined by the vibrating mass, which is the sum of the diaphragm mass and the mass caused by the radiation load. *Mass-controlled* operating conditions can only be achieved if the resonance frequency of the mechanical system is well below the lowest frequency at which the loudspeaker is designed to operate.

At high frequencies, the intensity will diminish because the radiation load no longer compensates for the decrease in vibration velocity with increasing frequency.

The directivity, however, will start increasing already for frequencies at which the diameter of the diaphragm is smaller than the wavelength, which will result in an increase of sound intensity on the symmetry axis of the loudspeaker.

The different factors will influence the frequency response of the loudspeaker driver so that its typical frequency response will be that shown in Figure 14.4 (see also Problem 14.4). Various ways of adjusting the frequency response to be more linear will be discussed later in this chapter.

14.5 ELECTRODYNAMIC DRIVERS

The electrodynamic loudspeaker driver can be given a robust, simple, nondangerous, and inexpensive design. These advantageous properties, over other designs, such as piezoelectric and electrostatic designs, in combination with sufficient acoustic power output, reasonably good frequency response, and good directivity, have contributed to the almost total market dominance by electrodynamic drivers.

An *electrodynamic driver* uses the force generated by an electric current in a wire placed in a magnetic field. The design is usually circularly symmetric, as shown in Figure 14.5, using a coil of metal wire, the *voice coil*, in the magnetic field provided by a *permanent magnet*.

Besides the coil, the driver also consists of a *diaphragm* firmly attached to the coil, both mounted in a stiff *basket* (usually made of stamped sheet metal), which also holds the magnet. The diaphragm is attached to the basket using two concentric compliance elements, the *suspension*.

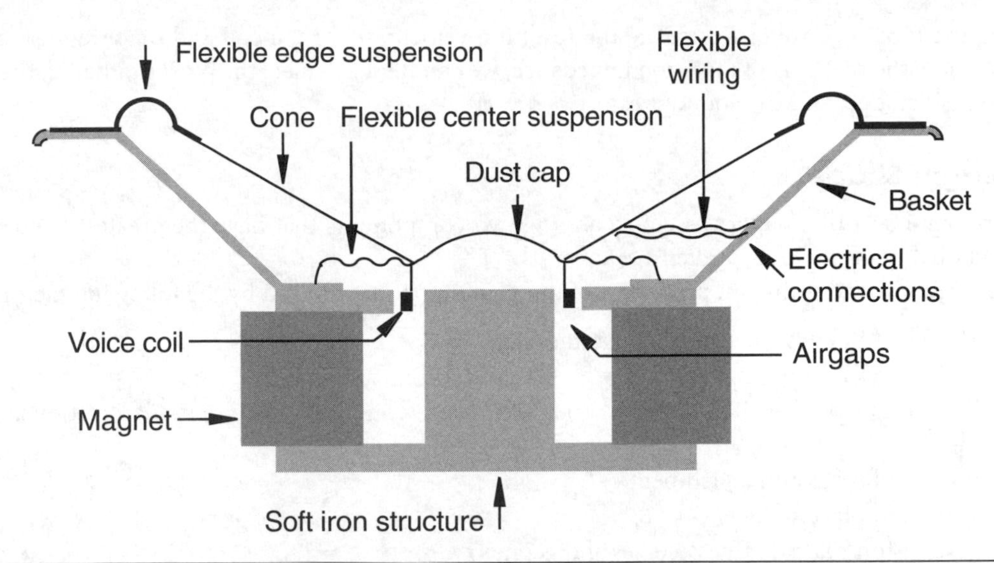

Figure 14.5 A cross-section through a typical electrodynamic driver having a concave conical diaphragm.

The force $\underline{F}_{vc}$, due to the interaction between the dynamic magnetic field of the current $\underline{i}$ through the voice coil wire and the permanent magnetic field is:

$$\underline{F}_{vc} = Bl\underline{i} \tag{14.14}$$

where B is the flux density of the magnetic field and l the length of the voice coil wire. The Bl-product is an important property.

The electric current through the coil is determined by the voltage applied to the coil, the electric resistances, inductances, and capacitances in the wiring and filters connected to the loudspeaker terminals and by the voltage induced in the voice coil due to its motion in the magnetic field. The field tries to keep the coil still.

This force is used to set the mechanical and acoustical system in motion. If the acoustic system is included in the mechanical system, the force equation is:

$$\underline{F}_{vc} = M_M\underline{a}_D + R_M\underline{u}_D + \frac{\underline{\xi}_D}{C_M} = \left(j\omega M_M + R_M + \frac{1}{j\omega C_M} \right)\underline{u}_D \tag{14.15}$$

where $\underline{a}_D$, $\underline{u}_D$, $\underline{\xi}_D$ are the acceleration, velocity, and displacement of the voice coil. Further, M_M is the total mass of the voice coil, the diaphragm, and the air load (the mass part making up the reactive part of the radiation impedance). The losses due to radiation are included in the mechanical resistance R_M along with the viscous losses in the air gaps and losses at the loudspeaker suspension. The compliance C_M includes compliance of the suspension and air inside the loudspeaker box. The compliance C_M is the inverse of the system's spring stiffness k_M:

$$C_M = \frac{1}{k_M} \tag{14.16}$$

Knowing the force $\underline{F}_{vc}$, we can calculate the resulting velocity $\underline{u}_D$ of the coil and diaphragm assembly. Since we know the real part of radiation impedance, we can then calculate the power radiated, the intensity at some distance, and the sound pressure $\underline{p}(r, \theta, \varphi)$.

Diaphragm Shape

The diaphragm and the magnet assembly are the two components that have the greatest influence on the sound quality of the loudspeaker.

The vibration and radiation properties of the diaphragm are affected by the following factors:

- Material thickness, density, Young's modulus
- Geometry and inhomogeneities
- Diaphragm profile (flat, straight cone, hyperbolic cone, corrugated circular sections, and such)
- Voice coil mass and attachment
- Voice coil placement
- Suspension mass, stiffness, losses and geometry
- Air load on both sides of the diaphragm

The shape of the diaphragm, its geometry, density, and elasticity influence the sound reproduction properties so much that many loudspeakers are similar except regarding the diaphragm. It is also important to note that inner and outer suspension rings of conventional electrodynamic loudspeakers also radiate sound.

Some high quality loudspeakers have diaphragms that are ideally stiff over most of the desired frequency range, so the diaphragm will move in phase over its entire surface. These loudspeakers tend to become very directional as the wavelength of sound is comparable to, or larger than, the diaphragm radius. The diaphragms of such loudspeakers are usually shaped as deep cones. Because of the depth, there will be considerable interference between the radiation from the inner, center, and the outer parts of the cone, resulting in frequency response dips already at quite low frequencies.

There are also high-quality loudspeakers where the diaphragm is designed to vibrate as a membrane with well-damped bending wave motion (see Figure 8.14). By controlling the diaphragm modes and their damping, one can achieve a fairly uniform frequency response and directivity, particularly if one is only concerned with the response in the reverberant sound field of a room. Loudspeaker manufacturers using this technique usually design diaphragms that are shallow cones or have flat circular or rectangular shapes.

Flexible cones have long been popular in the construction of high quality loudspeakers. Using flexible diaphragm materials (featuring suitably damped bending wave motion), one can achieve nonresonant behavior, and the effective cone area will shrink as the frequency is increased. This simplifies the use of the loudspeaker in multi-loudspeaker systems, where different loudspeakers are used for different frequency ranges. To control the directivity properly, one wants the loudspeakers to have about the same effective area in the crossover frequency region.

Some common cone shapes and profiles are shown in Figure 14.6. The use of elliptical cones makes it possible to increase the cone area in some applications where a circular cone would not use the available area optimally. The elliptical shape can also be used to give the loudspeaker different directivity

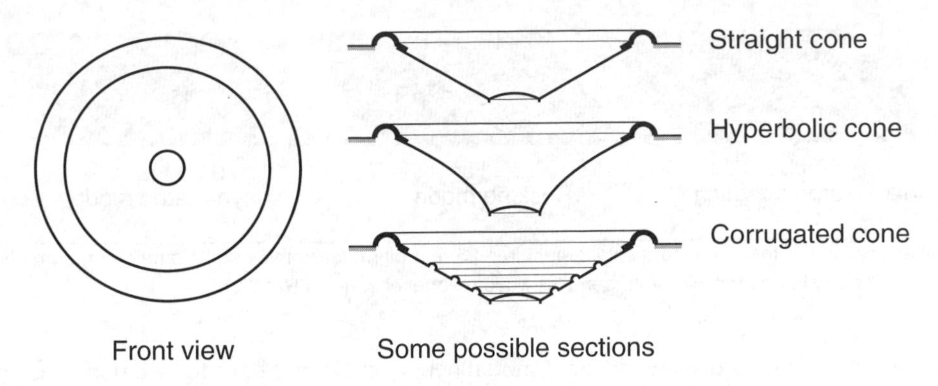

Figure 14.6 Some examples of typical diaphragm profiles.

characteristics in the horizontal and vertical planes. If the loudspeaker diaphragm moves as one unit, its directivity characteristics will be similar to those of the plane piston discussed in Section 14.1. A conical diaphragm will show similar but smaller directivity than the piston. The theory in Section 14.3 can be used to calculate the directivity of cones and other diaphragm shapes that are not too deep. For deep cone shapes and when the exact motion and directivity pattern is desired, one can use finite element or boundary element modeling to design the diaphragm and the associated directivity pattern. Finite element modeling can be used also to model possible buckling and other nonlinearities.

The choice between convex or concave diaphragms is determined to some part by the maximum allowable thickness of the loudspeaker. Concave diaphragms give less directional radiation than do convex ones, but convex diaphragms can sometimes be used to house and hide the magnet in shallow loudspeakers.

Many midrange and high frequency loudspeakers use dome shaped diaphragms. Looking at Figure 14.7, one can note that the directivity is much less dependent on dome depth or height for concave domes than for convex domes (that bulge outwards).

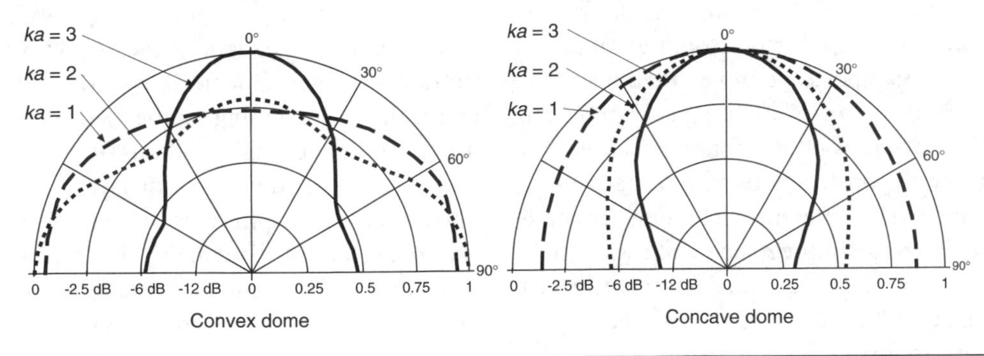

Figure 14.7 Directivity patterns, $F(\theta)$, for the sound radiation by vibrating convex and concave dome diaphragms for various values of ka. The wave number is k, and a is the dome radius. For the case shown, the dome radius and the dome height/depth ratios were constant. The dome vibration was in the z-direction shown in Figure 14.1. Note the linear scale. (Adapted from Ref. 14.4)

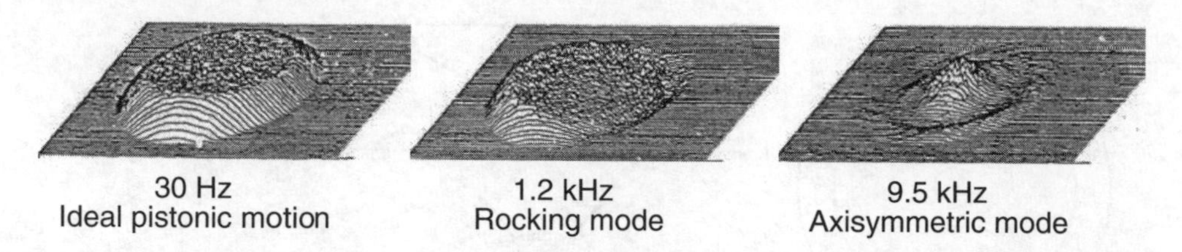

Figure 14.8 Laser scans of the vibration displacement in a 25 mm diameter convex soft-dome diaphragm showing pistonic motion and mode shapes for two of its bending wave resonances. (from Ref 8.2)

Domes, as well as cones, display resonant modal behavior. Domes have to be damped appropriately, the same way as conical diaphragms. Examples of mode shapes at some resonances of a loudspeaker dome diaphragm are shown in the laser vibrometer scans in Figure 14.8.

The mechanical compensation of improper loudspeaker diaphragm behavior is the economically most attractive and can take many forms, such as:

- Progressive vibration isolation of the cone
- Use of small, concentric extra cones in combination with vibration isolation of the main cone
- Use of soft, flexible cones with high damping

All of these actions have the goal of reducing the vibrating and radiating area at medium and high frequencies.

Diaphragm Materials

Most simple loudspeakers use conical diaphragms made of thermoformed plastic out of sheet stock or paper pulp that has been dried to conical shape. Such loudspeaker cones have a poor rigidity and yield poor frequency response.

Lightweight, thin cones cannot be stiff unless they are shaped as deep cones. The original cone shape was due to the cones being made from rolled paper. Today, paper cones are usually made by shaping and drying paper pulp with additives, such as animal hair, glue, and softeners. This technology makes it possible to design cones in various geometries with curved or flexible sections. The main advantage of using cones made of paper and felt mixtures is that the damping can be made much higher than in homogenous plastic cones. Because of the reasonably high Young's modulus of paper, one can make sandwich constructions of two sheets of paper with a constrained viscoelastic layer.

A somewhat different technique is applied when making diaphragms of woven materials. Traditionally, resin-impregnated cloth was used to cover the center of the cone so that dirt would not enter the voice coil gap. Dome diaphragms have long been made of impregnated cloth since it was possible to make soft domes in that way. Coarsely woven Kevlar cloth, impregnated using soft plastics, has been used successfully in diaphragm design. The coarse cloth can be made to use the viscous damping properties of the plastic in a good way.

An advantage of plastics is that the material can be formed by injection molding or thermoforming processes, which are much faster and simpler than the paper pulp process. Since plastics are more homogeneous between batches than paper, one can have more similar properties and higher production yield. Typical plastics are polystyrene and polypropylene. The disadvantage of these plastics is their low internal damping that cannot be increased much by added external damping compounds. A better plastic for loudspeaker diaphragms is talc or mica-filled polypropylene. The addition of solid compounds helps in keeping the damping high and constant over a wide temperature range. Conventional plastic-based viscoelastic materials only function over a narrow temperature range and are not suitable for loudspeakers for automotive use where the temperature range is typically −40 to 80°C.

Over some frequency, the poorly damped bending wave motion of the cone will lead to resonant behavior due to the reflection of bending waves from the edges where the diaphragm meets the suspension. This leads to uncontrolled resonances and modal behavior—*breakup*. Many diaphragm constructions are also characterized by noncircular modes excited by asymmetrically mounted wire leads, and such (see Figure 8.1).

The resonant behavior can be minimized by using various damping compounds on the diaphragm or by using diaphragms made of materials with high inner damping.

Diaphragms made of sandwich-type materials can also be used for loudspeakers. A typical sandwich panel is made of two outer layers of stiff material, such as plastic, with an intermediary honeycomb structure. The honeycomb structure has low bending stiffness. Such materials have long been used in ship and aircraft design to build stiff, lightweight walls and structures.

If the diaphragm material is correctly designed, the wave speed in sandwich materials can be tailored to follow a different rule than for solid plates. This makes it possible to design sandwich panels having an increased modal density compared to plates having bending waves.

Diaphragm Supports and Surrounds

The diaphragms of most electrodynamic low- and midfrequency range drivers need to make large displacements to radiate sufficient low-frequency energy. Maximum diaphragm displacements of large long-throw driver designs are approximately ± 1 cm. For closed loudspeaker boxes, the maximum diaphragm displacement can be calculated from knowledge of the diaphragm area and the desired sound level at the lowest frequencies the loudspeaker is supposed to reproduce. Various types of tricks can be used to reduce the need for long diaphragm travel, such as the use of horn loading and acoustically resonant boxes (for example, bass-reflex loudspeaker boxes).

High-frequency driver diaphragms move much less than those of low- and mid-frequency range drivers. This makes two different types of suspension designs necessary. In long throw designs, the diaphragms are usually quite heavy; diaphragm mass of about 10 to 100 grams is common.

It must be possible to ship and handle commercial loudspeakers. For conventional drivers using conical diaphragms, there will be an outer suspension and an inner suspension (see Figure 14.5). Unless the diaphragm is supported by two concentric suspensions, it will not be possible to design the loudspeaker driver in such a way as to avoid the vibrating voice coil being displaced off-center in the magnetic field. In the worst case, the voice coil will be touching the walls of the magnet in the air gap. If the voice coil touches the gap walls, there will be nonlinear friction, since the voice coil walls are usually quite roughly machined. The effects of such nonlinearity are heard as *buzz*, not as regular harmonic distortion.

Typically, short-throw designs are found in treble drivers, mobile phone loudspeakers, and in earphone drivers. Such drivers are characterized by low-mass, dome-type diaphragms and are designed to have resonances in the medium to high-frequency range. This makes it possible to use single-ring suspensions that are stiff enough to keep the voice coil from being catastrophically statically displaced in the magnetic field.

The compliance characteristics of the suspension must be both linear and nonlinear. In the linear operating range of the suspension, it must have the correct compliance to give the desired resonance frequency and not show any hysteresis. The suspensions must also prevent the voice coil from leaving the magnet air gap; this is usually the task of the inner suspension. It is difficult to design loudspeaker suspensions with just the right amount of nonlinearity so that the voice coil movement will be linear in the main operating range.

The inner suspension is usually made of impregnated cloth so that the air trapped between the inner suspension and the magnet assembly is free to escape. The inner suspension must have high radial static stiffness so that the voice coil is centered in the air gap. In some experimental designs, the inner suspension has been made of metal rods, which makes it possible to achieve higher linearity and less hysteresis. Some disadvantages of this type of approach are that dirt can enter the air gap and that the rods need to be provided with viscoelastic damping to remove bending wave resonances in the rods themselves.

The outer suspension, the *surround*, has two tasks. It should keep the cone in place and introduce sufficient damping to the cone rim so that resonances due to radial and circular bending wave propagation in the cone are prevented.

In guitar loudspeaker drivers, the outer rim is usually made as corrugated continuation of the cone. This type of surround has comparatively high stiffness and low losses, which causes severe bending wave reflections at the rim. It is also a common suspension design in single suspension earphone and treble drivers.

High-quality, low-frequency and midrange drivers using cone diaphragms usually have the type of outer suspension shown in Figure 14.9, often called *roll surround*. Roll surround allows much larger and more linear movement than the corrugated surround. The roll surround is usually made of nitrile rubber, impregnated cloth, foam plastic, or foam rubber. The choice of material is critical for the durability of the loudspeaker. Surround *rot* is a common cause of loudspeaker failure and can be caused by ozone and exposure to sunlight.

An advantage of the roll surround is that it can be combined with vibration damping at the outer rim of the cone. A suitable flat section of the surround is allowed to cover the outer end of the cone. This section, if designed properly, can be made to act as an anechoic termination for the radial bending waves

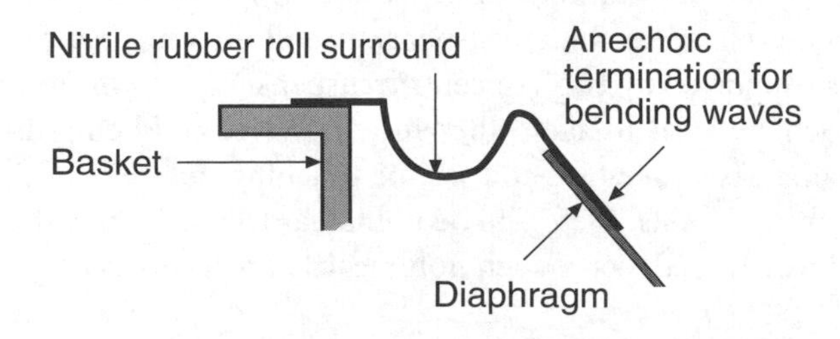

Figure 14.9 An example of a *roll surround*-type suspension designed to act as a vibration absorber.

in the cone (see Figures 8.12 and 8.14). This will prevent the formation of resonances due to reflection of bending waves at the rim.

Voice Coil and Magnet Air Gap

The magnetic system in an electrodynamic loudspeaker consists of the dynamic magnetic field induced by the voice coil and the static magnetic field in the magnet's air gap, which interact to give the force that moves the voice coil. Some modern loudspeaker drivers use multiple voice coils and air gaps to improve the force linearity.

The *voice coil* is the center of the loudspeaker driver. A stiff *voice coil former* is attached to the diaphragm and carries the voice coil wire windings. It is important that the voice coil former and leads are attached in a particular way to the diaphragm so that they do not excite unwanted bending waves in the diaphragm. Specially formed copper, silver, or aluminum wire is usually used for the voice coil. Typically, the wire cross-section is hexagonal or rectangular so that the electrical resistance of the loudspeaker can be reduced. The target is to have a voice coil with low mass and low electrical resistance that uses little volume.

The former must contribute to the cooling of the windings, thus, metal voice coil formers are preferred for high-power applications. In high-power, low-frequency drivers, the voice coil is often made of aluminum or copper and is cooled by incorporating special perforations in the voice coil former to increase turbulence and air movement. A common problem in loudspeakers is that the voice coil overheats, which makes the voice coil lacquer soften and lose its grip on the voice coil wire. This results in rub and buzz and ultimately makes the loudspeaker fail.

The voice coil winding must have a height that is suited to the design of the magnetic field. In principle, it is possible to compensate for magnetic field and diaphragm suspension nonlinearities by special voice coil designs, as shown in Figure 14.10. Typically, the voice coil has a height that is slightly less than that of the air gap. Low-frequency drivers sometimes have long voice coils, as shown in Figure 14.10c, since these drivers are less sensitive to high voice coil mass and need to be able to generate greater volume velocity.

The voice coil resistance is determined by the coil winding. An electrical resistance in the range of 3 to 7 ohm is typical. The resistance must be chosen so that the power losses in the wires leading to the loudspeaker are small compared to the losses in the voice coil winding, which encourages the use of high resistance circuits. At the same time, a high voice coil resistance is a disadvantage from the viewpoint of amplifier design, particularly in automotive applications.

The voice coil electrical impedance is a result of both the voice coil resistance and the inductance of the coil and the magnetic circuit, and typically behaves as shown in Figure 14.11. The inductance will

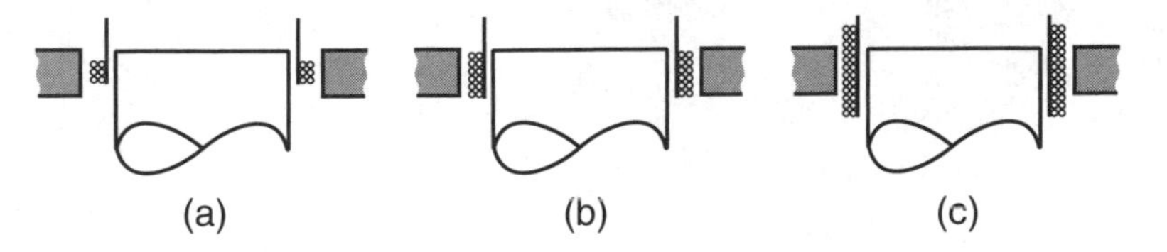

Figure 14.10 Some examples of short, medium, and long voice coil designs.

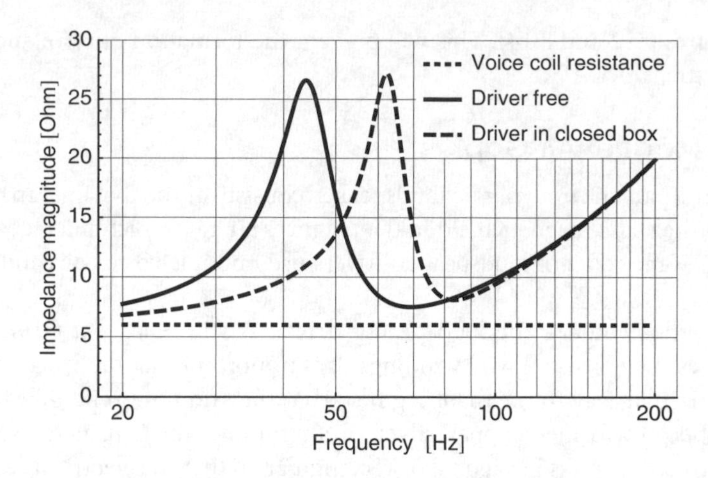

Figure 14.11 Typical behavior of the modulus of the electric input impedance Z_{el} of an electrodynamic loudspeaker.

contribute to a rise of the impedance at high frequencies. At low frequencies, the back electromotoric force (emf) of the voice coil movement will result in one or several impedance peaks. The frequencies and levels of these peaks can be used to trim the acoustic design of the loudspeaker box.

The static magnetic field in the air gap of the electrodynamic loudspeaker driver is provided by a magnet and a soft iron magnetic circuit. Figure 4.12 shows some magnet circuit designs. The desired magnetic field is determined by the mechanical system, the electrical system, and the loudspeaker box. All of these factors influence the loudspeaker sensitivity, the resonance frequency, the damping at resonance, and the suitability of the loudspeaker for a particular loudspeaker box construction. The magnetic circuit of high quality drivers must be designed so that the field in the air gap can be kept within small tolerances.

To have sufficient magnetic flux density in the air gap, the gap has to be narrow. Typically, the voice coil is cooled by the magnet structure, and a narrow air gap is necessary to be able to cool the voice coil properly. If the air gap is too narrow, the unavoidable sideways movement of the voice coil will lead to contact between the air gap walls and the voice coil resulting in rub and buzz sounds. This is a common problem in loudspeakers for mobile phones. A wide air gap will result in a weak magnetic field and low loudspeaker sensitivity. By adding a magnetic fluid consisting of a suspension of ferrite particles in a viscous *ferrofluid*, one can have a somewhat wider air gap and still reasonable flux density. This is a com-

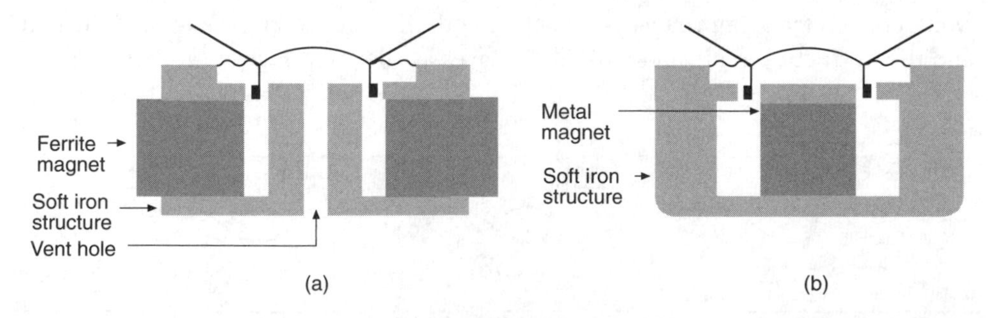

Figure 14.12 Two different magnet systems for electrodynamic loudspeakers: (a) ferrite magnet typical for low-frequency drivers and (b) metal magnet (for example, Neodymium) typical for high-frequency drivers.

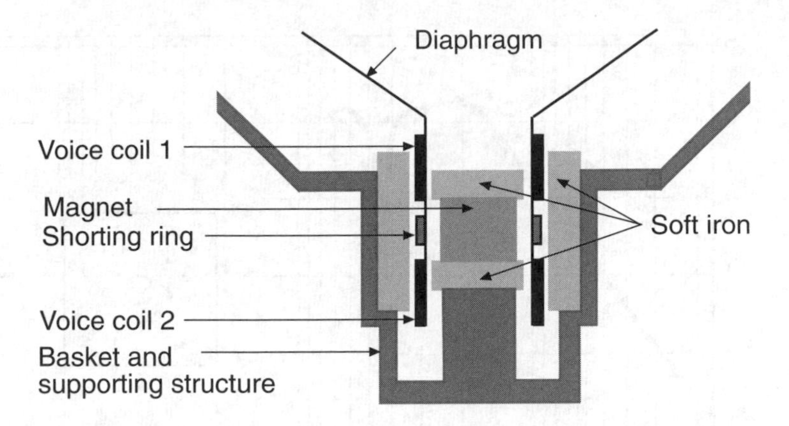

Figure 14.13 The basic design of a *dual drive* driver combining the action of two voice coils to reduce nonlinear distortion due to magnet field asymmetry. (After Ref. 14.7)

mon approach used in midrange and treble drivers. The method is unsuitable for low-frequency drivers because of the viscosity of the fluid. The ferrofluid also contributes to better cooling of driver voice coils. A disadvantage of this approach is that the oil emulsion-based magnetic fluid has a tendency to dry up.

An important addition to the magnet system is the shorting ring that acts as a secondary transformer winding, and reduces the magnetic field modulation in the magnet system induced by the voice coil's magnetic field. The position of this shorting ring is critical for reduction of nonlinear distortion in the loudspeaker. The shorting ring also reduces the inductance of the voice coil.

Another approach to reducing the nonlinear distortion is to use dual magnetic drives. The aim in using dual drives is to provide drivers with less mass, reduced stray magnetic fields, and that are more linear—that is, have less nonlinear distortion. Figure 14.13 shows the basic construction of such a driver. Note the shorting ring shown in the figure.

14.6 LOUDSPEAKER ENCLOSURES

Any finite vibrating plate will have a frequency below which the diffraction of sound from the rear side of the panel will interfere to reduce the sound pressure at a distance. This phenomenon is known as *aerodynamic short circuit*. Since it is impractical to build electrodynamic loudspeaker drivers with diameters larger than 30 cm, the low-frequency limit would become unacceptably high. This makes some form of acoustic isolation between the front and back side of the diaphragm necessary.

The isolation can be obtained in several ways, depending on design compromises. Three basic approaches exist: the use of *baffles*, *boxes*, and *horns*.

Baffles

The purpose of a *baffle* is simply to prolong the path that the rear radiated wave has to travel. For a baffle to be effective, it must be about as large as the wavelength at the lowest frequency one wishes to reproduce. The frequency response of a loudspeaker mounted in two different positions in a square baffle is shown in Figure 14.14.

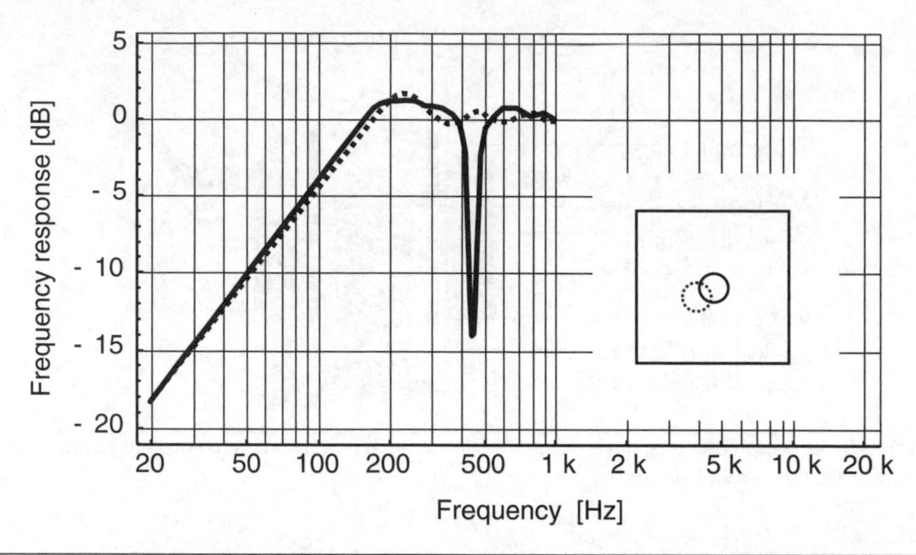

Figure 14.14 Two examples of frequency response curves for a loudspeaker mounted on center (solid line) and asymmetrically (dashed line) in a square baffle having 1.2 m long sides. (After Ref. 14.3)

Since one is usually listening at a distance of a few meters in homes, a large baffle will, in addition, contribute toward making the rear wave more attenuated than the front wave due to the longer travel distance.

The diffracted wave can be attenuated to some extent by providing a sound-absorptive layer on the rear side of the baffle. A sound-absorptive layer on the front side of the baffle would somewhat reduce the sound pressure at distance because of less pressure doubling due to the acoustically softer surface. Sound-absorptive treatments of this type are common to reduce the front side diffraction on loudspeaker boxes.

The baffle must have sufficient mass and rigidity so that it is not excited by the loudspeaker. This requires quite heavy baffles. Baffle resonance will lead to reduced sound insulation between the front and rear side of the baffle and to baffle sound radiation that might interfere with the direct sound from the loudspeaker.

An advantage of the baffle approach is the absence of acoustical resonance in the air. This is a particularly useful advantage when using loudspeakers that have light membrane-type diaphragms and thus lack sound insulation. Electrostatic loudspeaker panels, which use low mass, flexible membranes, are often mounted in baffles for this reason.

Loudspeaker baffles require particular attention when positioned in rooms due to sound cancellation by the rear radiation. In contrast to conventional loudspeaker boxes, which usually obtain the best low-frequency coupling to the sound field in the room when mounted in corners, baffle loudspeakers are best mounted with the transducer close to a side wall.

Loudspeaker Boxes

Loudspeaker boxes are designed to contain or control the rear radiation of the loudspeaker, so that it does not interfere, but couples optimally, with the front side radiation of the loudspeaker driver. Some common box designs are shown in Figure 14.15.

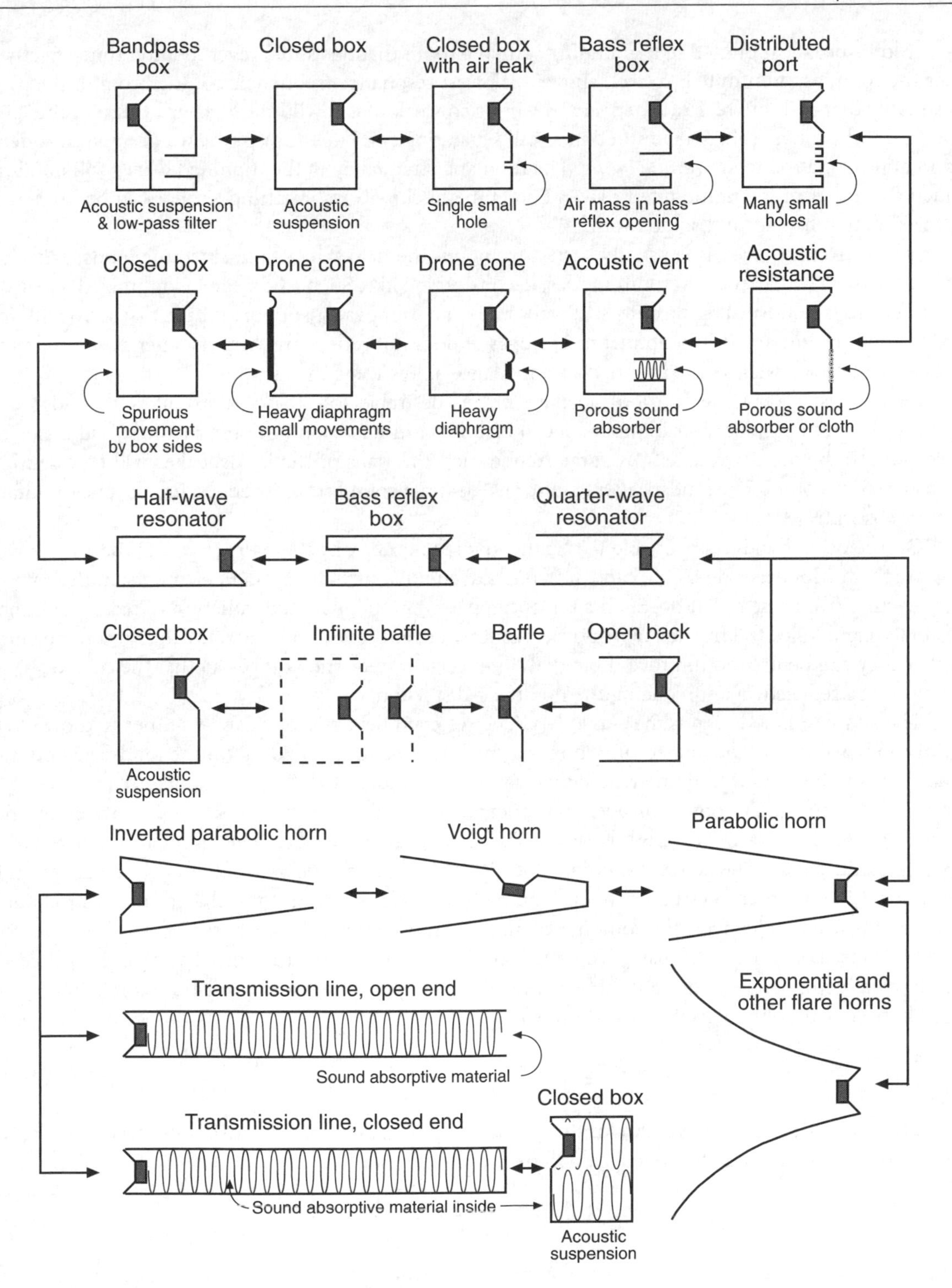

Figure 14.15 Some common loudspeaker box designs and their interrelation. (After Ref. 14.5).

A loudspeaker mounted in a bounding wall of a room that has short reverberation time effectively sees the room as an infinite box, well above the lowest resonance frequencies of the room. If the room is suitably damped, the rear radiated sound will be so weak that it will not be heard through the loudspeaker. If the loudspeaker is mounted in a wall separating two large rooms, the wall can be considered as an approximation to an *infinite baffle*. The resonance frequency of the mounted driver will not differ much from that of the unmounted driver. Sometimes a closet or something similar can be used as an approximation to a second, large room.

In practice, however, it is usually easiest to mount the driver in a box, since the loudspeaker can then be moved with respect to room modes, decor, and the like. Such a box can be optimized in various ways. The ideal loudspeaker box has stiff walls and is an immovable support for the driver. In addition, it does not give off any diffraction from its edges, it does not reflect any high-frequency sound, and it provides a known, well-defined acoustical impedance at low frequencies.

Real loudspeaker boxes can only behave in this desirable way to some extent. The loudspeaker box also has limited mechanical impedance; it can be set in vibration not only by the sound inside the box, but also by the driver itself. At some frequencies, the walls of the loudspeaker will be essentially transparent to sound. The loudspeaker exterior will diffract sound from its edges, unless it has a suitable sound-absorptive front.

On the other hand, it is possible to use the loudspeaker box in a constructive way, particularly for low-frequency loudspeakers. One can use acoustical circuits, such as resonant elements, in the paths of the rear and front side radiation. Such circuits can passively attenuate the radiation to reduce nonlinear distortion and tailor the frequency response. Another advantage of such passive acoustical impedances is that they can be used to improve the radiation in certain frequency ranges so that the cone displacement can be reduced, again reducing the nonlinear distortion.

The simplest box design is the *closed box*. The resonant behavior of closed volumes is known to us from room acoustics. The loudspeaker box is a small room and, depending on the loudspeaker design tradeoffs, the interior volume may be either damped or undamped. The resonance frequencies of an undamped rectangular room can be easily calculated. At very low frequencies, well under the first resonance frequency, any closed loudspeaker box will appear as an acoustical compliance element, irrespective of its shape. The acoustical compliance is determined by the volume of enclosed air given by Equation 1.77. In order to estimate the influence of the enclosed air on the behavior of the mechanical system, it is useful to convert the compliance into an equivalent spring stiffness.

We must convert the acoustical compliance C_{Abox} into mechanical compliance C_{Mbox}. Since the mechanical compliance is the ratio between displacement and force and acoustical compliance is the ratio between air volume displaced and pressure, we find:

$$C_{Mbox} = \frac{C_{Abox}}{S_D^2} \tag{14.17}$$

Since the total stiffness k_M is the sum of the stiffness of the diaphragm suspension k_S and the stiffness of the volume V of air inside the loudspeaker box k_{Abox}, we find:

$$k_M = k_{Mbox} + k_S \rightarrow \frac{1}{C_M} = \frac{1}{C_{Mbox}} + \frac{1}{C_S} \tag{14.18}$$

where k_S and C_S are the stiffness and the compliance of the diaphragm suspension respectively.

Thus, the mechanical compliance C_M of the loudspeaker and closed box is:

$$C_M = \frac{1}{\dfrac{1}{C_S} + \dfrac{\rho_0 c_0^2 S_D^2}{V}} \tag{14.19}$$

Sometimes manufacturers publish the driver compliance in the form of its equivalent acoustical volume. This is useful information since, if the driver is used in a closed box having that volume, the resulting driver resonance frequency will be about 1.4 times that of the unmounted driver. This volume is typically the minimum recommended for a closed box loudspeaker design.

Filling the interior of the loudspeaker box with sound-absorptive material will increase the compliance. This is due to the nonadiabatic behavior of sound as it moves through a medium that has heat exchange with a secondary structure—in this case, the fibers of the sound-absorptive material. The heat exchange can lower the speed of sound to about 80% of its value in free air. A consequence of the reduced propagation speed is an apparent increase of box volume. If the volume is entirely filled by the material, the increase in apparent volume can be up to about 50%.

There are several approaches to closed box design. The choice of approach depends on the type of loudspeaker used and the desired behavior at the loudspeaker's resonance frequency. Since the mass moves against the compliance of both the air trapped in the box and the compliance of the loudspeaker suspension, the *mounted resonance* frequency is going to be higher than the *free resonance frequency* of the loudspeaker.

The traditional approach was to let the mounted resonance frequency only be minimally higher than the free resonance frequency. This is a reasonable choice for a loudspeaker that has a stiff suspension. The typical applications are loudspeakers mounted in quasi-baffle-type enclosures, that is, large enclosures. This approach requires loudspeaker drivers having strong magnets or mechanical resistance high enough to give reasonable damping. There is otherwise a risk that the voice coil may travel so far at low-frequency transients that it moves out of the air gap, which is likely to damage the voice coil.

A modern approach is to make the box so small that its compliance is much smaller than that of the loudspeaker. This *acoustic suspension* design results in a marked increase of the mounted resonance frequency, compared to the free resonance frequency. The addition of the compliance of the trapped air results in better distortion characteristics for loudspeaker drivers that have poor suspension linearity. In such a case, the air may be more linear than the mechanical diaphragm suspension. Acoustic suspension designs need loudspeakers having rigid cones and large diameter voice coils.

The closed box design always results in an increased resonance frequency, unless a suitable sound-absorptive material is used in the box as discussed previously. This, in turn, leads to an increase in the Q-value of the mechanical system leading to a more resonant character of the high pass properties of the transfer function of the loudspeaker system.

To have optimal properties, it is necessary to adjust the properties of the box to those of the loudspeaker driver. At high frequencies, where the inner dimensions of the box start to become close in size to one-half wavelength of the sound, the box will become resonant. This means that it will be necessary to apply sound-absorptive materials to the inner box surfaces. A small air leak is necessary to compensate for changing atmospheric pressure.

The *bass-reflex loudspeaker* uses an extra resonant circuit to decrease the lower operating frequency of the loudspeaker. Below this cutoff frequency, the loudspeaker will have a faster drop in frequency

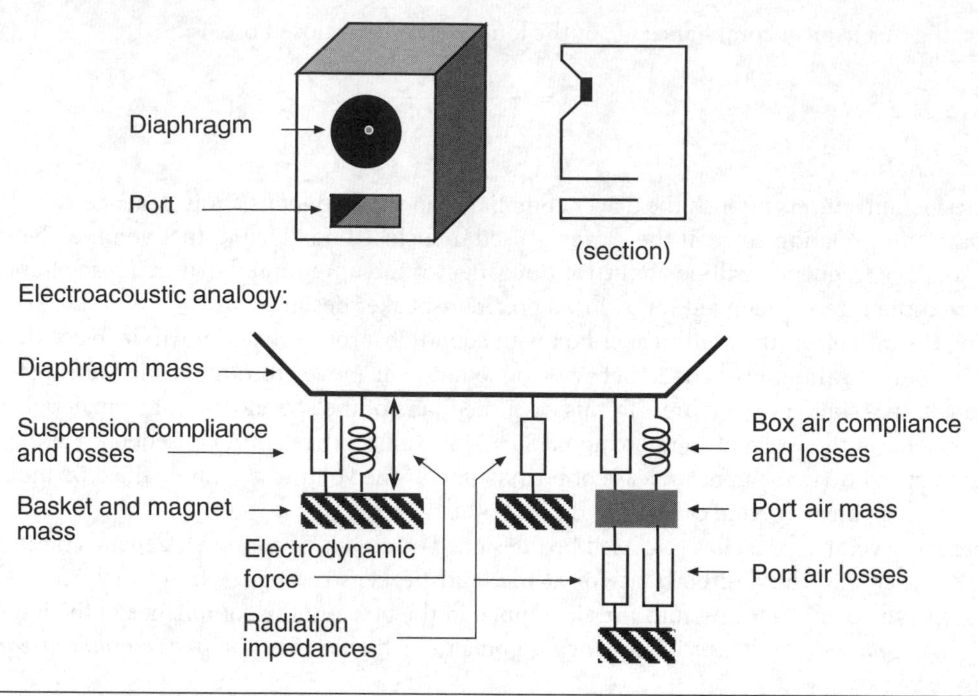

Figure 14.16 A ported box—bass-reflex—loudspeaker and the mechanical analogy of its acoustical system.

response than will the closed box loudspeaker. The mechanical system that is analogous to that of the acoustical system in the bass-reflex loudspeaker is shown in Figure 14.16. It is easy to design the system for various types of response if one knows all the necessary electrical, acoustical, and mechanical properties of the loudspeaker and the bass-reflex or closed box in question. The most common approach is to give the system a *Butterworth-type response*, in analogy to that of electrical filters. Such a response is sometimes also called a *maximally flat* response.

14.7 HORN LOUDSPEAKERS

By using a horn in front of the loudspeaker, it is possible to dramatically increase the efficiency and the sound power output of the loudspeaker. This is not due to confinement of the radiation but because the horn acts as an acoustic impedance transformer. A practical finite horn, however, introduces resonances and dispersion. Only an infinitely long horn can be resonance free. Bends and constrictions in the horn also cause resonances.

The horn is characterized by an expansion constant known as the *flare rate*. The flare rate determines the lowest frequency at which it is possible for sound waves to propagate through the horn. The propagation is often assumed to be as a plane wave, but this is a theoretical simplification. The waves inside the horn are seldom plane.

The resonant characteristics of the horn are due to sound being reflected at the mouth of the horn. Reasonably reflection-free conditions are obtained when the diameter of the mouth of the horn is larger than approximately 0.3 times the wavelength at the cutoff frequency.

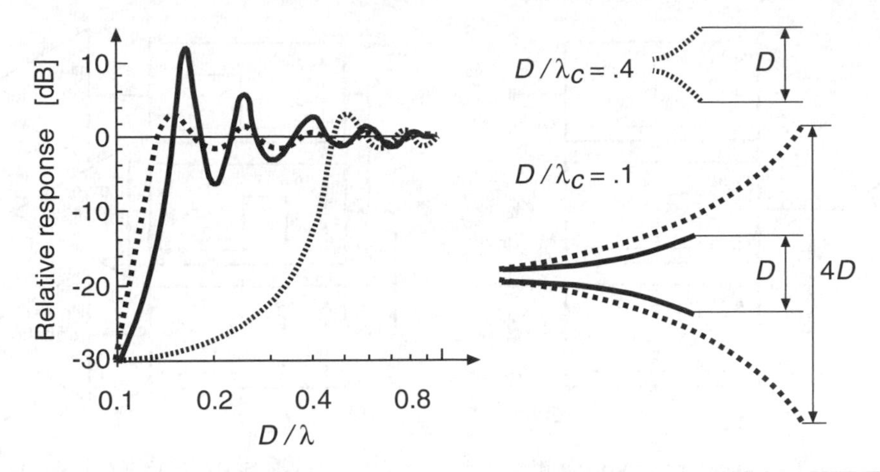

Figure 14.17 Typical frequency response of some exponential horn loudspeakers with different flare (expansion) rates, mouth diameters (*D*), and cutoff frequencies ($\lambda_c = c_0/f_{cutoff}$). (After Ref. 14.3)

Clearly, this limits the use of large exponential horns to places such as outdoor arenas, theaters, and cinemas, although designs for domestic use exist as well—for example, those shown on the cover of this book and in Figure 9.3. Short horns are common for midrange and treble loudspeakers since for these frequency ranges the horn can be made large compared to the wavelength in the working range. Figure 14.17 shows the response characteristics for some exponential horns having different flare rates and lengths.

14.8 MULTIPLE DIRECT RADIATOR LOUDSPEAKER SYSTEMS

The frequency response problems of electrodynamic loudspeakers have been outlined in Section 14.3. These problems result in a *high fidelity* loudspeaker driver often being usable only over a frequency range of 3 to 4 octaves. To circumvent some of these problems, one tends to subdivide the frequency range so that different loudspeaker drivers are responsible for different frequency ranges.

These designs are called two-way and three-way designs (or more if appropriate), according to how many loudspeaker drivers that are used in a single loudspeaker box to cover the desired frequency range of the loudspeaker. These designs require different types of electrical low-pass and high-pass filters, usually called *crossover filters*. The filters may be passive, using only resistors, inductors, and capacitors to divide the frequencies of signal from the amplifier to reach the appropriate loudspeaker driver. Figure 14.18 shows how *crossover networks* can operate on both low and high level signals. Crossover networks (and most other things concerning loudspeakers) are discussed in detail in Reference 14.7.

It is difficult to design loudspeakers using multiple drivers. Some of the difficulties concern the frequency and phase responses, the power handling, the directivity, and the balance between the frequency response in the main lobe direction and that in other directions.

There are now many types of computer software available designed to model the response of multiple driver direct radiator loudspeakers. Without such software, the development time for new loudspeaker designs would be much longer and slower. An example of the frequency response characteristics of a

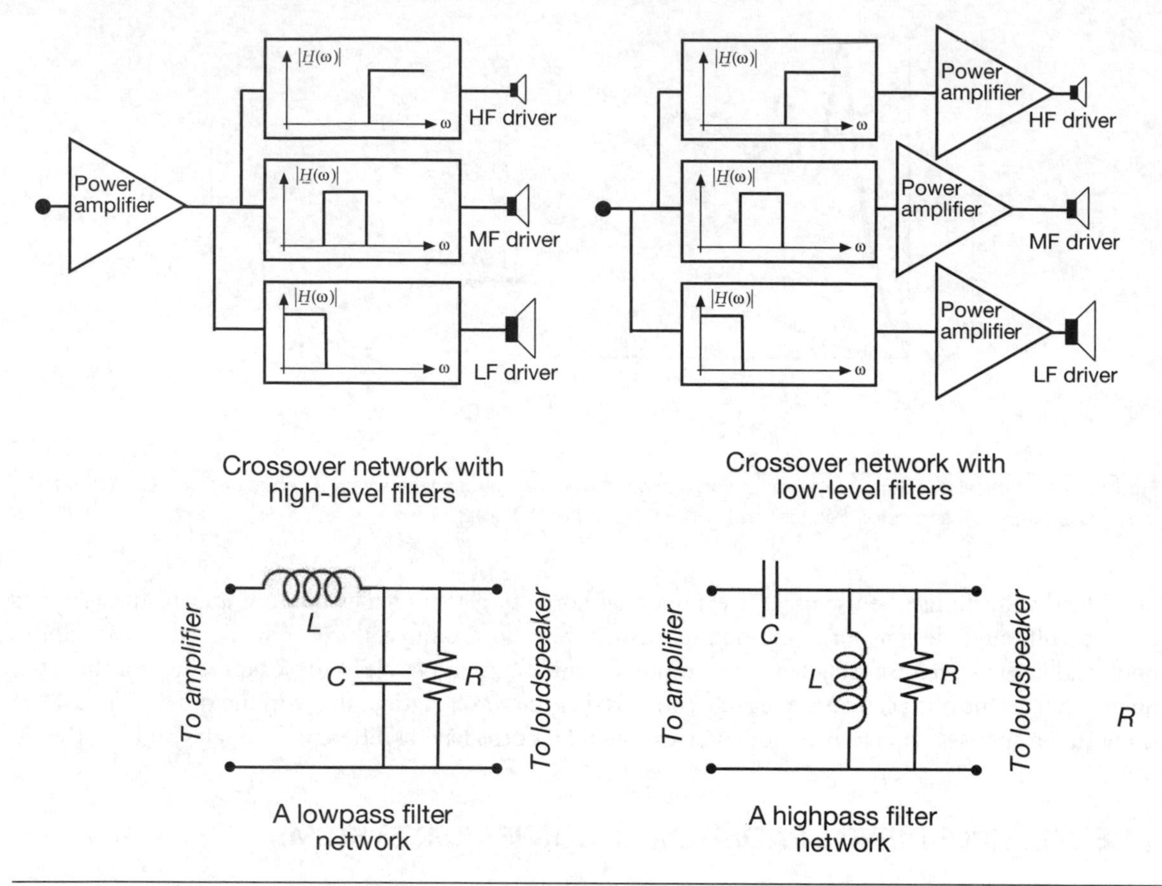

Figure 14.18 Two types of crossover networks. The lower part of the figure shows two simple electrical circuits similar to those used in practical crossover networks. For example, a bandpass network can be accomplished by connecting a lowpass and highpass filter in series.

commercial two-way system is shown in Figure 14.19. Obtaining flat frequency response in the crossover region may be difficult since the sound pressure and the sound power characteristics are different. The direct sound from the loudspeaker will usually not have a flat frequency response off the nominal axis of the loudspeaker system.

14.9 ARRAY LOUDSPEAKERS

Many sound reinforcement applications—such as churches, theaters, cinemas, and sports arenas—require loudspeaker systems that have a high- and frequency-independent directivity.

A high directivity requires that the sound radiator have large dimensions compared to wavelength. As previously noted, it is impractical to design loudspeaker diaphragms having diameters larger than wavelength. If high directivity is needed, it is necessary to replace the large diaphragm loudspeaker with an array of smaller diaphragms or a horn, as mentioned earlier.

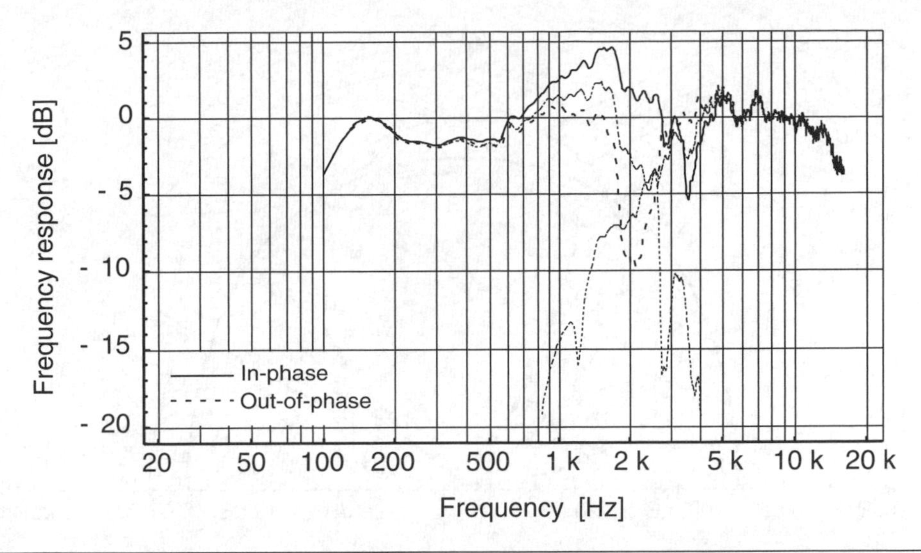

Figure 14.19 An example of the on-axis frequency responses of the individual drivers of a two-way Yamaha NS10 loudspeaker system (thin dotted lines) and the resulting responses for correct and incorrect (out-of-phase) interconnection (anechoic measurements).

Arrays of loudspeakers are often used in place of single large loudspeakers. A single one-dimensional *array loudspeaker* is usually called a *column loudspeaker*. Its directivity will depend on the directivity of the individual drivers, the distance between the drivers, the length of the array, and the way that the signals are distributed between the loudspeaker drivers. Since there are always unavoidable production tolerance differences between the drivers, the results are seldom as good as predicted by simple theory. Horn loudspeakers often have better directivity performance than arrays.

The basic directivity pattern of array loudspeakers can be designed approximately by adding the contributions of the individual driver elements at the receiving points. As an example, consider a linear array of four (small) drivers that has the following directivity function:

$$F(\theta) = \frac{\sin\left(2kb\sin(\theta)\right)}{\sin\left(\frac{kb}{2}\sin(\theta)\right)} \tag{14.20}$$

Some examples of directivity patterns for this array are shown in Figure 14.20. One notes that the directivity function is frequency-dependent. This problem can be solved for large arrays by *tapering* the response of the individual drivers.

It is also important to note that one-dimensional arrays only give high directivity in one plane. Sometimes two-dimensional arrays are necessary to control the directivity. Such two-dimensional arrays are occasionally used to confine the noisy area in discotheques and nightclubs.

Note that the far-field where the level will drop at −6 dB per distance doubling does not start until a considerable distance from a large array. At close distance to the array, the level will vary rapidly with location as discussed for large loudspeaker diaphragms earlier in this chapter.

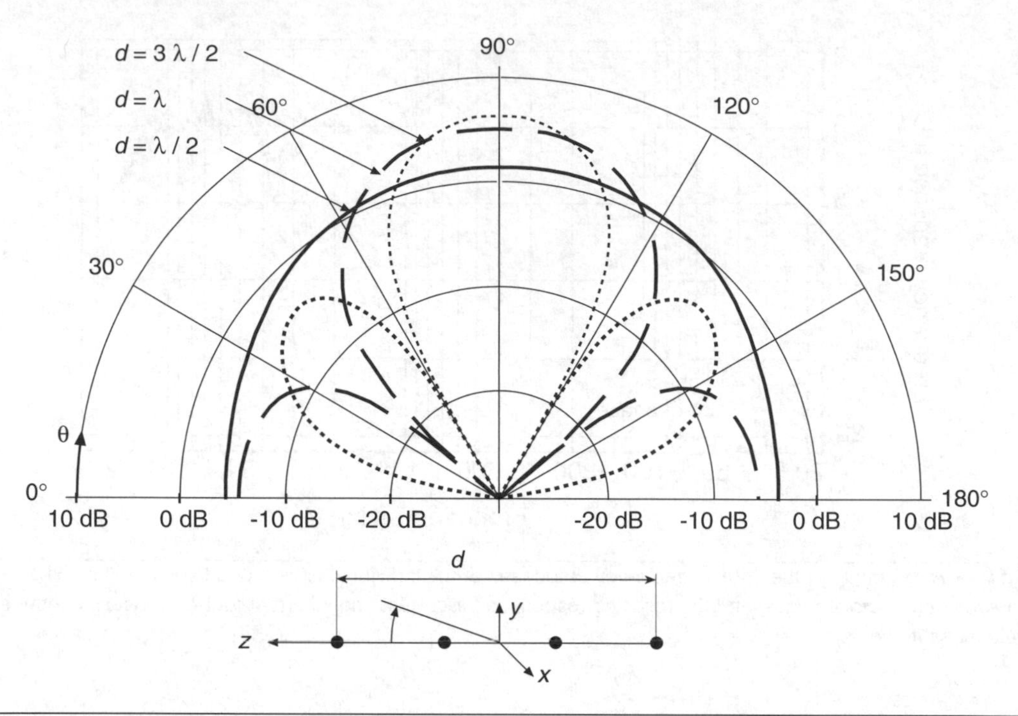

Figure 14.20 Directivity patterns of a simple, linear, four monopole array for three ratios of interelement distance d and wavelength λ. The curves show gain in dB over that of a free monopole having the same radiated power as the piston. Note that the array is along the z-axis shown in Figure 14.1 and that only one half of the toroidal pattern around the z-axis is shown.

Commercial loudspeaker systems for sound reinforcement often make use of several arrays to shape the directivity function over their working frequency range. Ribbon-type dynamic loudspeakers for domestic use have directivity characteristics similar to those of array loudspeakers.

14.10 ROOM EFFECTS

The listening room influences the response of the loudspeaker system in many ways. It is convenient to separate the low- and high-frequency effects. The low-frequency effects typically extend up to around 200 Hz in rooms of up to about 100 m^3 of volume. The high-frequency effects have already been discussed in Chapter 5. The discussion here will be confined to the low-frequency range, under the Schroeder frequency given by Equation 4.67.

The radiated power by a loudspeaker is influenced by the radiation impedance, which changes as the loudspeaker is moved close to sound reflecting surfaces. The effect will vary depending on such things as the properties of the loudspeakers, the position of the drivers, the surface reflection coefficients, and the number of (intersecting) reflecting surfaces.

The simplest case to handle is that of loudspeakers having a high internal impedance compared to the radiation impedance, which is often the case at low frequencies. For a constant volume velocity

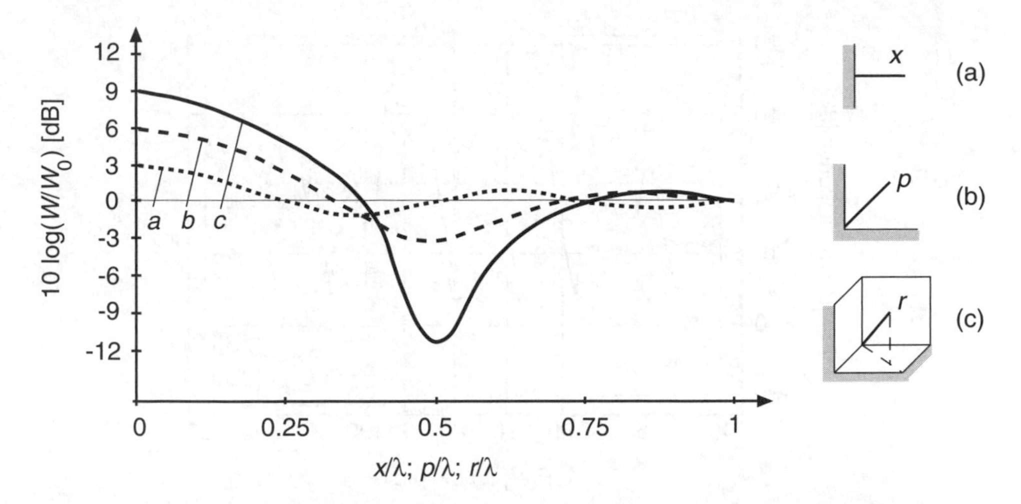

Figure 14.21 The sound power W radiated by a monopole close to rigid reflecting surfaces, relative to free field W_0, as a function of the ratio between distance and wavelength. Three cases shown: (a) plane, (b) on the diagonal between two orthogonal planes, and (c) on the space diagonal between three orthogonal planes. (After Ref. 14.6)

loudspeaker sound output, the power output of the loudspeaker will vary with frequency as shown in Figure 14.21. The sound pressure response will behave similarly.

For the loudspeaker to behave as if under free field conditions, the loudspeaker has to be at least a quarter to half a wavelength away from the sound reflecting surfaces. In practice, this often corresponds to a few meters. Note that this discussion concerns omnidirectional loudspeakers with constant volume velocity. Most loudspeakers are omnidirectional at the frequencies of interest in this discussion.

In a resonant sound field, such as that obtained in a rectangular room (discussed previously in Chapter 4), the position of the loudspeaker will influence the radiation impedance felt by the loudspeaker and, in this way, the radiated sound power. It is therefore important to test various positions for the loudspeaker, to find which position gives the best response in a particular room. This testing can be considerably simplified by using the reciprocity between source and receiver position indicated by Equation 4.38.

The principle of reciprocity can be used by putting the loudspeaker at the desired listening position and then moving an omnidirectional microphone between alternative loudspeaker positions to find the one that has the smoothest response. Figure 14.22 shows the result of such a pair of measurements in a rectangular room having a volume of approximately 25 m^3. The differences between the curves are small compared to the peaks and dips due to the modes of the room.

14.11 TRANSIENT RESPONSE

Human hearing functions as a complex time and frequency analyzer. At high frequencies, the time constant of hearing is only a few milliseconds.

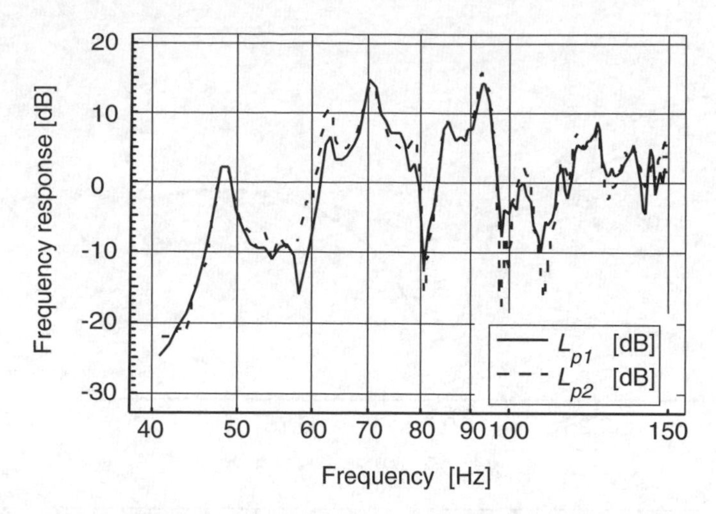

Figure 14.22 Two frequency response curves for a pair of points in a rectangular listening room having a volume of 90 m³ obtained by switching microphone and loudspeaker positions. Both microphone and loudspeaker were small and omnidirectional in the frequency range shown. The effects of reciprocity are clearly visible, up to a frequency of 150 Hz.

The transient response of a loudspeaker is determined by the electrical circuit, the mechanical properties of the diaphragm and suspension, and by the geometrical extension of the loudspeaker. It can be very complex, showing considerable ringing, particularly at the loudspeaker resonance frequencies and in the crossover frequency regions in multiple driver direct radiating loudspeaker systems.

An additional complication is that a seemingly almost flat frequency response can be the result of many closely spaced resonances.

14.12 NONLINEAR DISTORTION

The first noticeable result of the presence of nonlinear distortion in loudspeaker systems is the generation of frequency components in the reproduced sound that were not present in the electrical signal supplied to the loudspeaker driver. The nonlinear distortion can come from several sources.

Tonal distortion components are created by the nonlinearities in the magnetic field and in the diaphragm's mechanical system, particularly the suspension. Tonal distortion is also created by the nonlinearity of the trapped air in the loudspeaker box and when sound propagates long distances at high intensity in horn-type loudspeakers. Distortion, such as turbulence noise, is created when air moves at high speed around the voice coil and in acoustic resonator components such as the port opening in bass-reflex loudspeaker boxes, as discussed in Chapter 9. Another source of nonlinear distortion is the buckling that can occur when driving the loudspeaker diaphragm with high force. An example of finite element modeling of such buckling is shown in Figure 14.23.

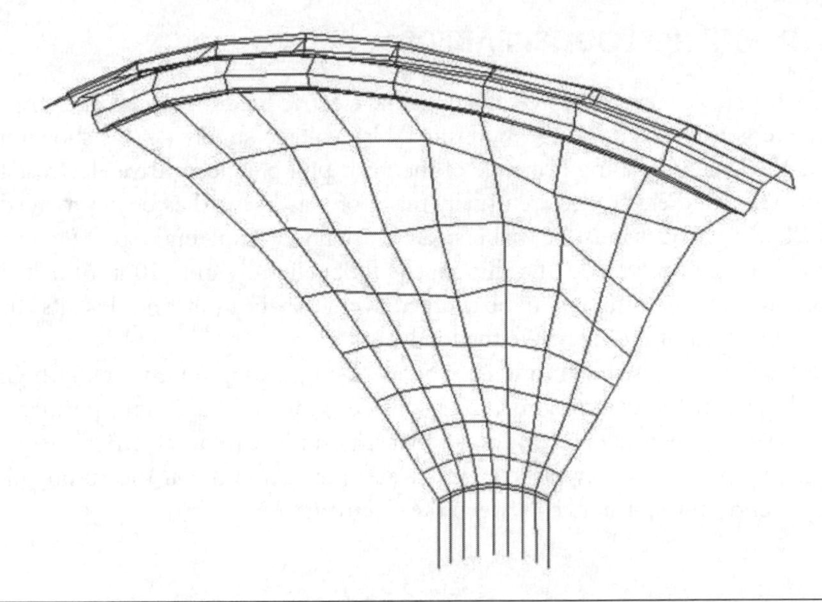

Figure 14.23 Buckling of a loudspeaker diaphragm calculated using finite element modeling.

Some of these types of nonlinear distortion can be minimized by using balanced designs, for example, in the magnetic field and in the diaphragm suspension. Nonlinear distortion can also be reduced by replacing mechanical components by simulation using various types of electronic feedback circuits. Other types of distortions, such as the distortion due to air nonlinearity and turbulence, can only be minimized by reducing sound pressure and particle velocity. A particular type of tonal distortion is the Doppler distortion that is the result of movement by the diaphragm at several frequencies at the same time, particularly high displacement movement at low frequencies.

Reduction of nonlinear distortion can also be done using both analog and digital signal processing if an accurate model of the distortion generation mechanism is available or if the nonlinearity can be sensed in some way.

14.13 ELECTRONIC COMPENSATION OF NONLINEARITIES

Since loudspeaker design is still mostly focused on loudspeakers to be used with amplifiers having linear frequency response, one must correct for the frequency and directivity responses in electrical, mechanical, or acoustical ways. This can only be done in an incomplete way by passive methods, such as those described in this chapter.

To achieve good loudspeaker system properties, one must regard the loudspeaker as a part of a complete sound reproducing system that includes amplification and signal processing of various kinds. This systems approach gives much larger design freedom. The signal processing can include compensation for both linear and nonlinear distortion, voice coil power, and diaphragm displacement overload protection, and, in the case of array loudspeakers, frequency-dependent adjustment of the directivity characteristics.

14.14 ELECTROSTATIC LOUDSPEAKERS

The *electrostatic loudspeaker* uses the force of a dynamic electric field on electrical charge to move a diaphragm. The charge is usually introduced by using a bias voltage supply e_{bias}, as shown in Figure 14.24. The figure shows the basic operating principle of the push-pull balanced drive electrostatic loudspeaker (and headphone). The fixed electrodes are usually made of plastic that has been provided an electrically conductive layer. The electrodes must be well insulated, usually by applying a special paint or insulator.

The diaphragm is typically made of a thin plastic foil, generally only 10 μm thick. The diaphragm should have low conductivity, sufficient to be charged over a day but must not lose its charge in the case of arcing—local ionization of the air inside the loudspeaker.

Modern electrostatic loudspeaker systems operate using a diaphragm having high surface resistivity, so that each part of the diaphragm can be considered to hold a certain permanent charge. The charged foil is placed between the electrodes, so that it is subject to an electrical field $\underline{E}_{audio}$. The field strength between two planes, closely placed, large electrodes depends on the audio voltage applied to the electrodes $\underline{e}_{audio}$ and on the distance between the electrodes h.

$$\underline{E}_{audio} = \frac{\underline{e}_{audio}}{h} \tag{14.21}$$

The electrodes need to be perforated so that sound can escape the loudspeaker. This will cause the field to be somewhat uneven because the foil has some mass and stiffness. The electrostatic force will tend to average over the surface of the foil. The force per unit area $\underline{F}''$ is ideally:

$$\underline{F}'' = q''\underline{E}_{audio} \tag{14.22}$$

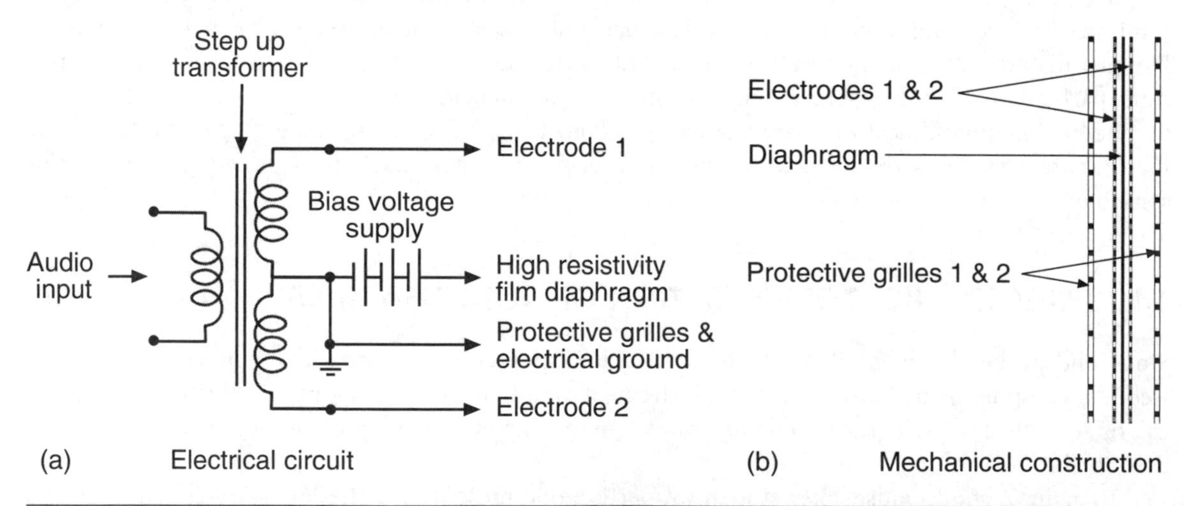

Figure 14.24 The electrical circuit for a *push-pull* balanced drive electrostatic loudspeaker. The electrical input impedance of the electrostatic loudspeaker is primarily capacitive making it a difficult load for power amplifiers (a). The electrostatic loudspeaker uses the force of an electric field on electric charges on a thin polymer film diaphragm to move the diaphragm. The electrodes are fixed and have little acoustic resistance but need to be isolated to withstand the maximum voltages between the electrodes, ground, and film. The film must have high electrical resistivity so that charges remain evenly distributed over the film as it moves and if there is arcing to an electrode (b).

where q'' is the charge per unit area on the foil. The charge is ideally constant as a function of time on each patch of the foil since the foil has such high surface resistivity. The charge is determined by the area of the foil S_D, the voltage to which the foil is charged e_{bias}, and the capacitance C_E between each side of the foil and the corresponding electrode, the same as with the condenser microphone (see Equations 12.11 and 12.12).

$$q''S_D = e_{bias}C_E \tag{14.23}$$

In contrast to magnetic fields, electrostatic fields may ionize the air if the field strength is large. On ionization the air forms a conductive path between electrodes and foil, this is called arcing. Arcing can burn holes in the foil and in the electrode insulation layer causing the loudspeaker to cease operating. Typically air will become ionized if the field strength exceeds 3 to 6 kV/mm, depending on air humidity. The field strength inside the loudspeaker depends both on the bias voltage applied to the foil and on the audio voltage applied to the electrodes. In practice, the applied audio voltage is limited to a few kV. Once the air is ionized, it will burn a hole in the plastic foil and carbonize a part of the foil. This will reduce the acoustic output of the loudspeaker. High resistivity foils are less susceptible to this type of damage than foils that have been metalized and have high conductivity.

Because the time constant of the charge circuit cannot be infinitely large, the charge will start to flow to and from the bias supply for very low frequencies. The position of the foil becomes unstable, and the foil may even be attracted to, and stuck to, an electrode. For this not to happen, the foil needs to be mechanically tensioned so that most of the displacement available can take place without the foil going into one of the two unstable regions close to the electrodes.

The foil tension, together with the mass and Young's modulus of the foil, determine the lowest resonance frequency of the electrostatic loudspeaker. The frequency should be at the low end of the spectrum, since the frequency response of the loudspeaker is essentially frequency-independent above this resonance frequency and up to the Helmholtz resonance frequency determined by the compliance of the air between the foil and the electrode, as well as by the air mass in the holes of the electrodes.

Electrostatic loudspeakers are used as dipoles without boxes but could be mounted in a baffle. The reason is that the loudspeaker diaphragm area needs to be large to generate sufficient volume velocity to radiate much sound power and achieve high sound pressure at the listening position. The diaphragm movement is controlled by the radiation resistance. Because the diaphragm has such low mass per unit area, this leads to most diaphragm vibration modes, except the lowest frequency ones, being extremely well damped.

The frequency response of an experimental full frequency range electrostatic loudspeaker strip under anechoic conditions is shown in Figure 14.25. The electrostatic loudspeaker strip (0.15×1.8 m^2) was mounted in a rigid baffle and the measurement point was at 1.5 m distance from the strip, on-axis, and symmetrically positioned. Low-frequency diaphragm resonances are clearly visible as is the low-frequency sound pressure cancellation by the back side radiation and also the frequency response peak at about 18 kHz of the Helmholtz resonance of the air between the diaphragm and the electrodes and the holes in the electrodes.

Since the sound radiating surface is typically large, even small diaphragm movements lead to satisfactory sound pressure in home settings. The large diaphragm results in high directivity that reduces the influence of the room on the reproduced sound. The small diaphragm movement and the balanced drive push-pull operation results in electrostatic loudspeakers being characterized by low nonlinear distortion.

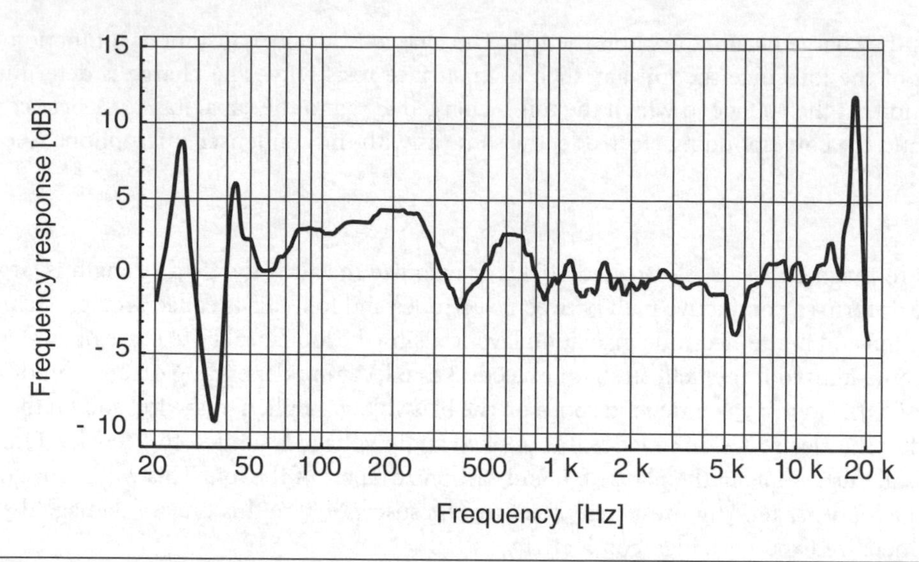

Figure 14.25 Anechoic measurement of the on-axis frequency response at 1.5 m distance of an electrostatic loudspeaker strip (0.15 m by 1.8 m) mounted in a rigid baffle 1.2 m by 2.2 m. Low-frequency membrane resonances are clearly visible, as is the sound pressure cancellation by the back side radiation (at about 300 Hz), and also the frequency response peak of the Helmholtz resonance of the air between the diaphragm and the electrodes and the holes in the electrodes (at about 18 kHz).

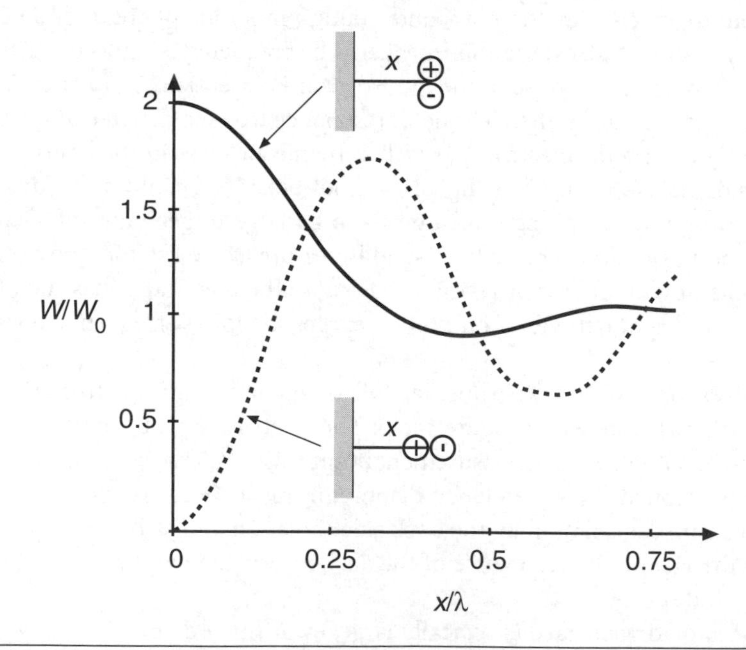

Figure 14.26 The power radiated by a dipole close to a rigid plane depends on its direction, relative to the plane. This graph shows the power W radiated by a dipole for two different directional alignments of the dipole, relative to the free field power radiated W_0. The center of the dipole is at a distance x from the reflecting plane. (After Ref. 14.6)

An interesting psychoacoustic effect in using full frequency range electrostatic vertical strip loudspeakers is that the sound in listening appears to come from the part of the loudspeaker at the height of one's head.

In contrast to loudspeaker box designs, electrostatic and other open baffle mounted loudspeakers are best placed away from the walls of the listening room. Because the dipole is effectively two different sources, the alignment of these sources relative to the reflecting surface is important, as indicated by the power radiation curves for dipoles shown in Figure 14.26. For good low-frequency sound power generation, the dipole operating open baffle loudspeakers should be positioned so that their dipole action is parallel relative to the close reflecting wall.

14.15 PROBLEMS

14.1 An electrodynamic transducer has the following electrical and mechanical data:

Diaphragm mass	$M = 10^{-2}$ kg
Suspension stiffness	$k = 2 \cdot 10^3$ N/m
Suspension loss resistance	$R_M = 2$ kg/s
Air gap flux density	$B = 1$ N/A
Voice coil winding length	$l = 7.5$ m
Voice coil electrical resistance	$R_E = 10\ \Omega$
Diaphragm diameter	$d = 0.1$ m

Tasks:

a. Calculate the *free* resonance frequency f_0.
b. Calculate the rms value of the cone displacement and velocity as a function of frequency.
c. Draw a curve showing the diaphragm displacement and velocity as a function of frequency for a rms drive current of 2A.

Use a logarithmic frequency axis and express diaphragm displacement and velocity as levels, using 1 mm and 1 m/s, respectively, as reference levels.

14.2 Assume that the electrodynamic transducer given in Problem 14.1 is mounted in a *closed box* arrangement. The box volume (after removing the volume occupied by the loudspeaker) is $2 \cdot 10^{-2}$ m^3.

Task: Calculate the resulting resonance frequency f_{mount} for the loudspeaker when mounted in the box.

14.3 The acoustic power radiated by the electrodynamic transducer vibrating with the velocity $\tilde{u}$ and box described in Problems 14.1 and 14.2 can be written as:

$$W \approx sSZ_0\tilde{u}^2$$

Here, S is the radiating area, and s is the radiation ratio.

When the dimensions of the radiating surface are much smaller than the wavelength of sound, the radiation ratio can be approximated as:
Here, a is the loudspeaker diaphragm radius

$$s \approx \left(\frac{\omega a}{c_0}\right)^2$$

Tasks:

a. Calculate the radiated sound power at the following frequencies: $1/4 f_{mount}$, f_{mount}, and $4 f_{mount}$ for a rms electrical drive current of 2A.
b. Calculate the resulting sound power level as a function of frequency using a logarithmic frequency axis.

14.4 An electrodynamic transducer mounted in a *closed box* was measured in free field. The frequency response given in the graph was measured *on axis*.

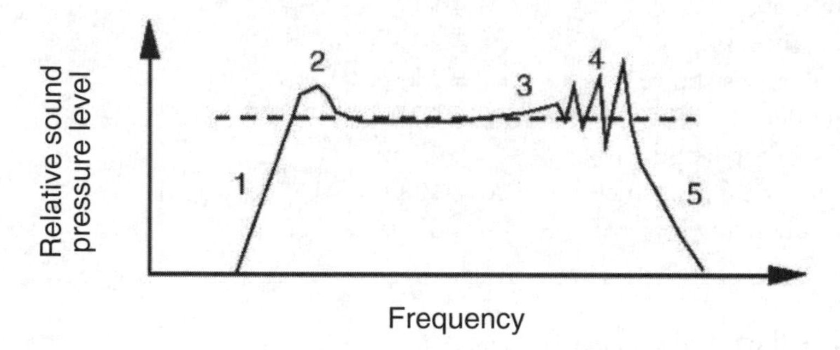

Tasks:

a. Explain the reasons for the deviations 1 to 5 in the curve, relative to *flat* frequency response.
b. Explain how the frequency response problems may be changed/altered/removed by modifying the construction of the loudspeaker drive unit and box.

14.5 An electrodynamic transducer, having a free resonance frequency of 50 Hz, is mounted against the end of a tube closed at the opposite end. The length l of the tube is 2 m, and its cross-sectional area S is $3 \cdot 10^{-2}$ m². The mass of the loudspeaker's diaphragm M_M is $1 \cdot 10^{-2}$ kg.

Task: Determine the resulting resonance frequency using a graphic or numerical solution if the loudspeaker's effective cone area is equal to the cross-sectional area of the tube.

14.6 One can use a loudspeaker having an electrodynamic transducer as a resonance absorber. Determine an expression for the resonance frequency expressed as a function of a loudspeaker driver's diaphragm area, diaphragm mass, suspension stiffness, suspension losses, and effective box volume. Also, derive an expression for the absorption of the loudspeaker.

14.7 An electrodynamic transducer having the following specifications is mounted in a closed box arrangement. The maximum displacement of the diaphragm results in a volume change of $3 \cdot 10^{-4}$ m^3. The sound power output is also limited by the maximum power dissipation allowed in the loudspeaker drive unit's voice coil.

The drive unit's specifications are:

Resonance frequency (fitted)	50 Hz
Effective diaphragm area (circular)	$8 \cdot 10^{-2}$ m^2
Efficiency ($W_{\text{acoustical,out}}/W_{\text{electrical,in}}$)	3%
Maximum voice coil power dissipation	20 W

Assume that the loudspeaker radiates as if the diaphragm were an ideal sphere having a radius equal to one-half of the diaphragm's radius.

Task: Calculate the maximum sound power available from the loudspeaker as a function of frequency for frequencies from 0.5 times the resonance frequency up to the frequency corresponding to ka equal to 0.5 where k is the wave number and a is the radius of the cone.

14.8 We wish to construct an intercom using an electrodynamic transducer as both loudspeaker and microphone (see Problem 12.6). The electrodynamic transducer is mounted in a *closed box*, having an effective volume of $5 \cdot 10^{-4}$ m^3.

The transducer's specifications are:

Resonance frequency (free)	250 Hz
Effective diaphragm diameter	$3 \cdot 10^{-2}$ m
Diaphragm mass	$1 \cdot 10^{-3}$ kg
Bl-product	1 N/A
Voice coil resistance	5 Ω

Task: Calculate the sound pressure level, L_p, at a distance of 1 m from the transducer at a frequency of 1 kHz for a voice coil driving voltage of 1 V.

Hint: Assume that the diameter of the transducer is much smaller than the wavelength of sound and that the fitted transducer radiates as a monopole having a volume velocity of $\underline{U}_{\text{diaphragm}} = \underline{u}_{\text{diaphragm}} S_{\text{diaphragm}}$.

Headphones and Earphones 15

15.1 INTRODUCTION

Headphones are small personal loudspeakers that radiate sound close to the ear canal opening. The distance between the ear canal opening and the headphone driver is about 5 cm or less. *Earphones* are smaller devices designed to be inserted into the ear canal entrance, typically forming a sealed cavity.

Headphones and earphones enjoy wide usage, extending into areas far beyond general high fidelity audio sound reproduction. Many forms of military and civilian telecommunications rely on the use of headphones, for example, telephones, protective gear, or diver and pilot helmets.

Since headphones and earphones are used in mobile applications and are often handled carelessly, they are subject to stringent mechanical requirements. They must also fulfill demanding hygienic requirements, since they are close to the body, become dirty, and are exposed to comparatively high temperatures and humidities.

Many earphones are used in hearing aids. A hearing aid contains the microphone, filters, (digital) signal processor and compressor, amplifier, and headphone in one single small unit, possibly insertable into the ear canal and nearly filling the concha. Sometimes it is advantageous to use bone conduction transducers when the middle ear is damaged or when the ear canal entrance cannot be closed.

The most common headphone and earphone designs for high-quality sound reproduction use are electrodynamic, electromagnetic, and piezoelectric drivers.

15.2 HEADPHONES/EARPHONES VS. LOUDSPEAKERS

The audio requirements for headphones are different from those of loudspeakers. Headphone and earphone transducers need only generate small volume velocities, thus, they may operate in a more linear range and have less nonlinear distortion. Since the listener's room is eliminated from the audio chain, the response characteristics of the room are eliminated.

Do headphones and earphones need *flat* frequency response? When we listen to sound reproduced over loudspeakers in rooms, the aural impression is formed by the total response of the direct sound, early reflected, and reverberant sound. Since most living rooms have short reverberation times, direct and early reflected sound will be dominant. The sound pressures at the listener's ears will depend on such things as the loudspeakers, their direct sound frequency response, the response in other directions, the reflection characteristics of the room surfaces, and scattered sound from objects. But the sound pressure will also depend on the reflection of sound by the listener's body and head.

We have already discussed the concept of the head-related transfer function (HRTF) in Chapter 6. Figure 15.1 shows an example of the frequency response of the HRTF for frontal sound incidence, as well as for diffuse field incidence, for one ear on a manikin head. Clearly, the frequency responses for both cases are far from linear.

Examples of frequency response curves, measured using a simple coupler, for some high-quality headphones are shown in Figure 15.2. One notices that even high-quality headphones have non-flat frequency responses. The response curves for the semi-open supra-aural headphone and the insert earphone show relatively resonance free behavior. The response curve of the circumaural phone shows the influence of the modes inside the cavity formed by the headphone. Note, however, that our hearing is used to these types of resonances, which, in any case, are associated with the external ear.

The peak/rms relationships in sound pressure between headphone, earphone and loudspeaker listening are also going to be different. Significant musical peak levels can have a duration as short as a few milliseconds, particularly with popular amplified instruments, such as electric guitars. Looking at the dashed curve in Figure 3.18, which shows the reported level difference between sine-wave tone bursts and continuous sine waves at 1 kHz, we note that for a 5 ms burst, the perceived level *deficiency* is approximately 20 dB. When listening to headphone sound, which, of course, has not been subject to the room impulse response, the peak/rms ratios are thus likely to be 10 to 20 dB higher than in loudspeaker listening.

Since it is likely that listeners adjust the average reproduced sound level to about the same loudness for both types of listening, it is likely that people will listen to peaks that are relatively much stronger than while listening to loudspeaker reproduced sound. This will increase the risk of hearing damage.

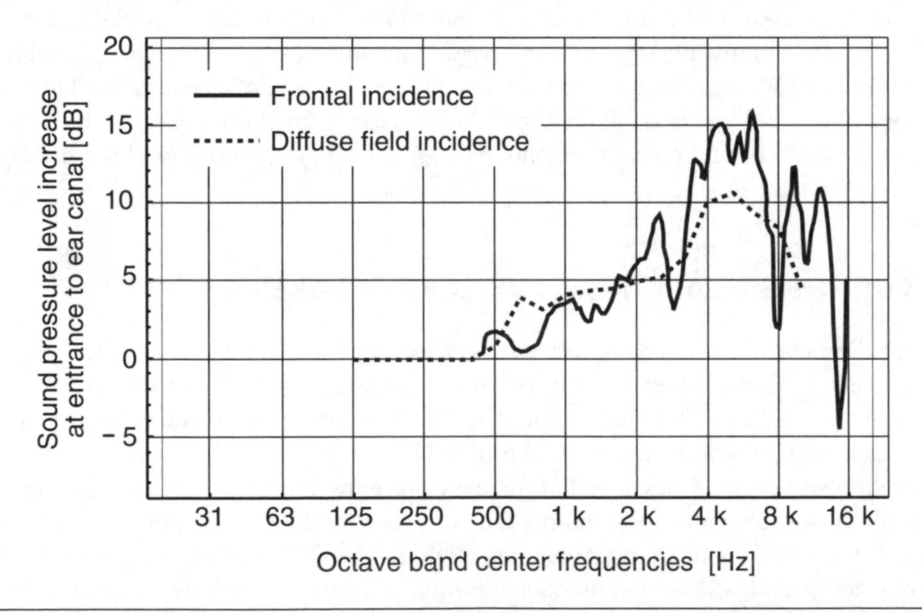

Figure 15.1 The log magnitude (frequency response) of the HRTF for an ear on a manikin head measured in an anechoic chamber for frontal sound field incidence and in a reverberation chamber for diffuse field incidence. No torso was used with the head for this measurement—if used, it would have increased levels in the 250 Hz range.

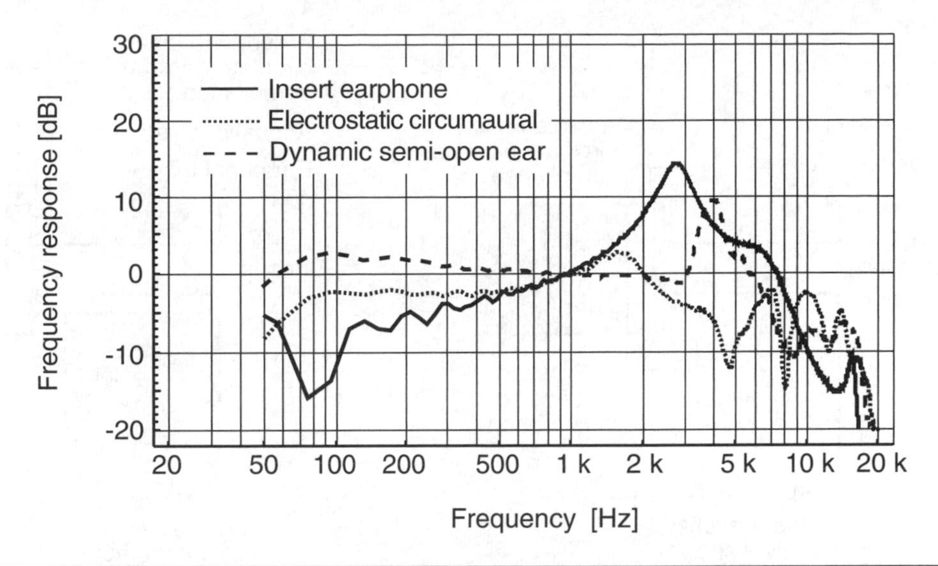

Figure 15.2 The frequency response of some high quality head- and earphones, measured using a simple small volume coupler.

Note that this also is a serious problem in mobile phone use, where the phone often is also used for music reproduction over headphone or earphones. Additionally, people generally adjust monaural sound reproduction to be louder than binaural sound reproduction.

By including a convolver in the amplifier, headphone or earphone electronics, having impulses responses of (or representative of) a real or fictive room, the convolution of the recorded audio and the room impulse responses will eliminate this risk.

15.3 THE ACOUSTIC ENVIRONMENT

The *acoustic environment* in which the headphone operates is very different from that of general loudspeakers. Some are made to operate directly coupled to the ear while others are designed to be used at some small distance from the head so that there is an air path or leak between headphone and head. The use of a correctly designed coupler is essential for the measurement of the response of headphones and earphones. Many couplers, however, are designed for telephony and will not adequately model the acoustic properties of the ear to be useful for full frequency range audio. Such a coupler typically has a volume of about a few cubic centimeters. The anthropometric head of the manikin shown in Figure 17.6a, however, has soft rubber pinnae, as well as a concha, ear canal, and eardrum simulator. Its design is shown to the left in Figure 15.3. On the right is a simple coupler that still allows the concha and ear canal modes to be similar to those in actual head (see Reference 15.4).

The direct coupling headphones can use an ear-surrounding tight cushion (in the case of *circumaural headphones*) or an *ear bud* inserted into the ear canal with little leakage (insert earphones).

Those headphones that are used away from the ear are designed so that they are primarily affected by the acoustic impedance of the opening between the headphone and the head, but not by the ear and the ear canal. These are often called *supra-aural headphones.*

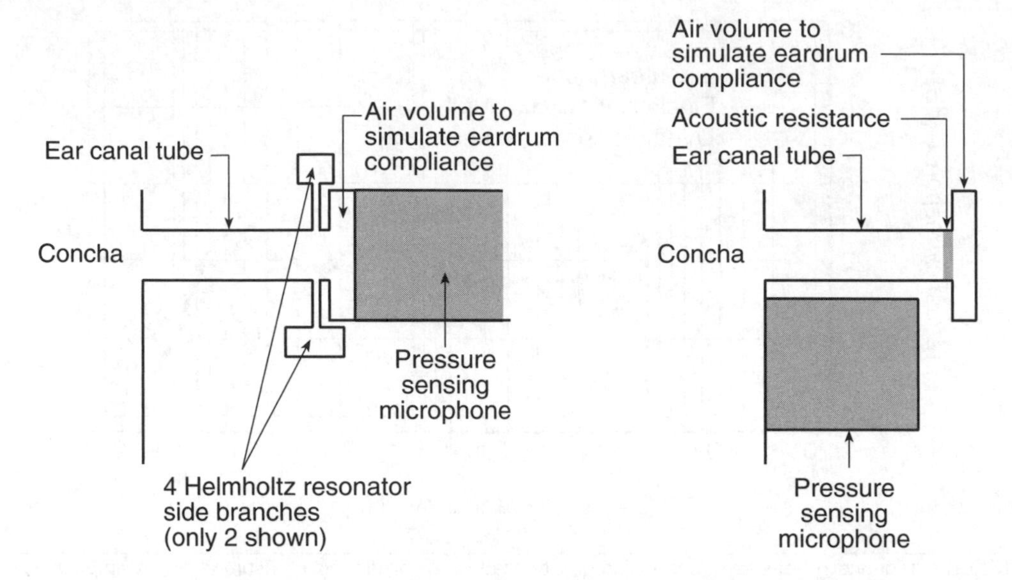

Figure 15.3 Two types of couplers to be used in headphone/earphone measurement and calibration as well as binaural sound recording. To the left is shown the Zwislocki coupler design. On the right is shown a simple coupler. Both allow the concha and ear canal modes to be relatively similar to those in a real head. The latter type is easier to equalize for use in binaural sound recording.

Usually headphones are much smaller than the wavelengths of sound over much of the audio frequency range. This means that the headphone (depending on construction) will act as a nearly ideal monopole or dipole sound source. The radiation impedance at frequencies up to 0.5 kHz can be described by the expressions given in Chapter 1. Since the headphone operates close to the head, the sound reflection by the head will considerably influence the impedance at higher frequencies. Any headphone design will need to be optimized regarding the operating distance between head and headphone. Figure 15.4 shows the large influence of leakage on the resulting frequency response of a headphone.

One must also note that the contact pressure will influence the compression, and, thus, the flow resistance, of any foam plastic pads used to seal the leak between head and headphone. Such variations will also influence the resulting frequency response, as shown by Figure 15.5.

Just as in the design of loudspeaker systems, it is necessary to have the intended use of the transducer clearly focused in its design. Since headphones are generally used to simulate the experience obtained when listening to loudspeakers, it is important to equalize the frequency response of the headphones so that it is similar to that of loudspeakers, at the ear canal entrance. Simulation of the diffuse and direct field loudspeaker frequency responses will be different.

15.4 ELECTROMAGNETIC HEADPHONES

There are several different electroacoustical operating principles used in headphone design. Electromagnetic and piezoelectric headphones are used in telephones and hearing aids. Electrodynamic and electrostatic headphones are used for high-quality audio sound reproduction.

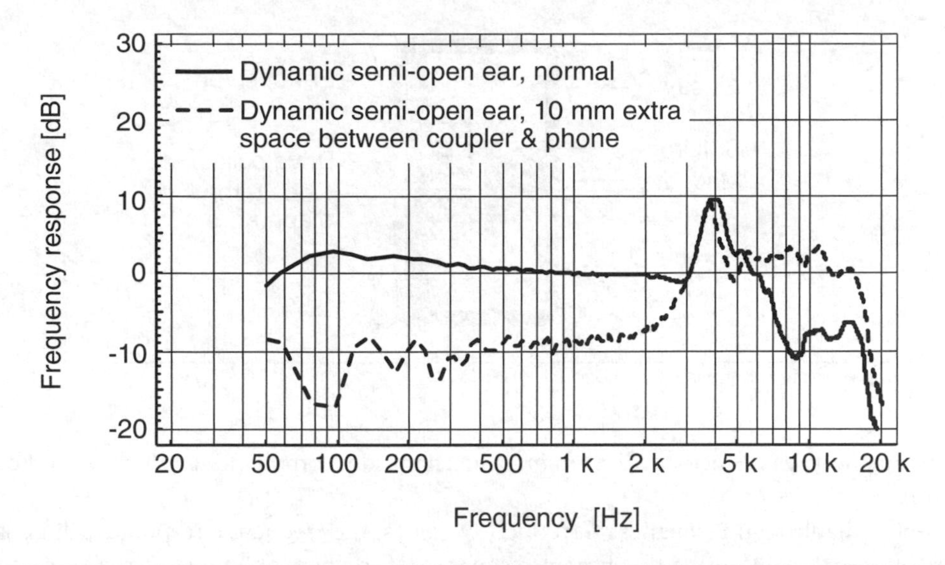

Figure 15.4 The leakage between the headphone and the body/ear influences the frequency response drastically. The solid line shows the frequency response of a particular headphone with little leakage and the dashed line shows the frequency response with large leakage.

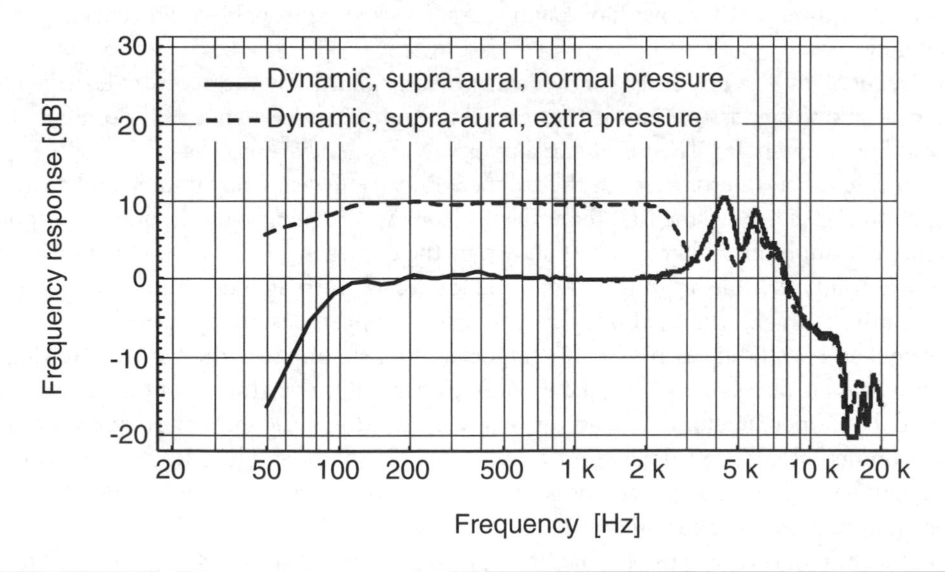

Figure 15.5 An example of the influence of leakage due to different contact pressures on the frequency response of a supra-aural headphone using a foam plastic seal.

The *electromagnetic headphone* (one design shown in Figure 15.6) has high sensitivity, a simple and rugged construction making use of a permanent magnet, a small air gap, a magnetically conductive diaphragm, and a coil of electrically conductive wire to enable modulation of the magnetic field. When audio voltage is applied to the coil, the magnetic field strength will vary, and so will the attraction

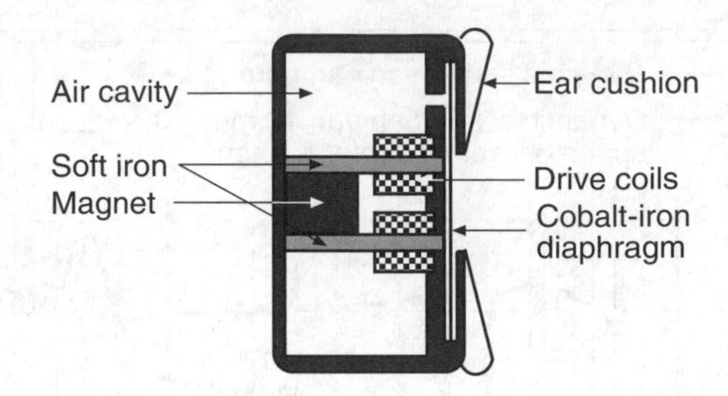

Figure 15.6 A simple electromagnetic headphone.

between the magnet parts. Since the diaphragm is mechanically compliant, it will move and change the air gap width.

Because the diaphragm resonances have high *Q*-values, the frequency response will be jagged for frequencies at and above the first diaphragm resonance. If the headphone has a tight seal against the ear, the acoustic impedance will be primarily capacitive at frequencies up to approximately 2 kHz. This means that the diaphragm stiffness and the stiffness of the trapped air will act as two series-coupled springs. This requires the diaphragm stiffness to be low, to generate sufficient sound levels, which, in turn, leads to a requirement for low diaphragm mass to have a reasonably high first diaphragm resonance frequency.

At low frequencies, the response will be determined primarily by the acoustic leakage around the seal of the headphone against the ear. If the seal is not tight, its impedance will dominate that of the trapped air at low frequencies. This results in a poor, low frequency response.

At higher frequencies, resonances in the trapped air, various acoustic circuits, and such will affect the frequency response. Above the first diaphragm resonance frequency, the frequency response of the electromagnetic headphone will drop off, as shown by the curves in Figure 15.7. Due to the asymmetry of the force-extension relationship of the air gap and the rest of the magnetic circuit, the electromagnetic headphone is inherently nonlinear, although symmetric, balanced designs are possible.

Lightweight and occluding earphones are important for listening to audio while exercising and traveling. The low mass is necessary so that the earphone does not move around in the ear canal entrance, causing it to fall out or create noise. The occlusion is desirable to have good low-frequency response and to keep environmental and transportation noise out so that the best signal-to-noise ratio is obtained. An advantage of having low background noise is, of course, that it removes the necessity for listening at sound levels that may lead to hearing damage.

Different technologies are in use for insert earphones. One technology that lends itself to miniaturization, and that has been in use a long time in hearing aids, is the balanced armature design shown in Figure 15.8.

The frequency response of such an earphone in a small coupler is shown in Figure 15.2. Since an insert earphone only needs to produce sound inside the ear canal, a cavity that is small, minimal volume velocities need to be generated by the diaphragm for high sound pressure at the eardrum. The small volume also results in a high acoustic impedance seen by the diaphragm, so the diaphragm suspension

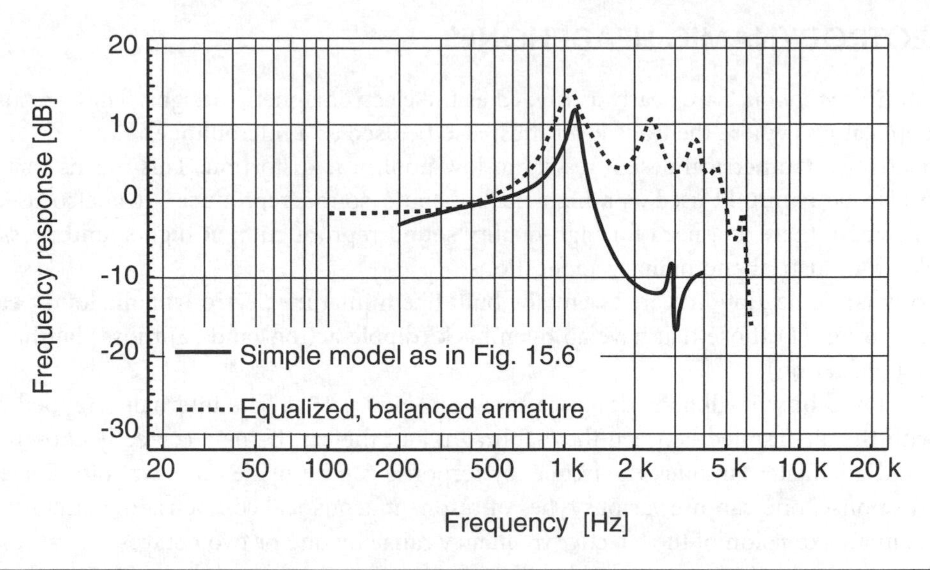

Figure 15.7 Frequency response curves for some electromagnetic headphones. Data from Knowles BK (balanced armature) series. Unknown couplers.

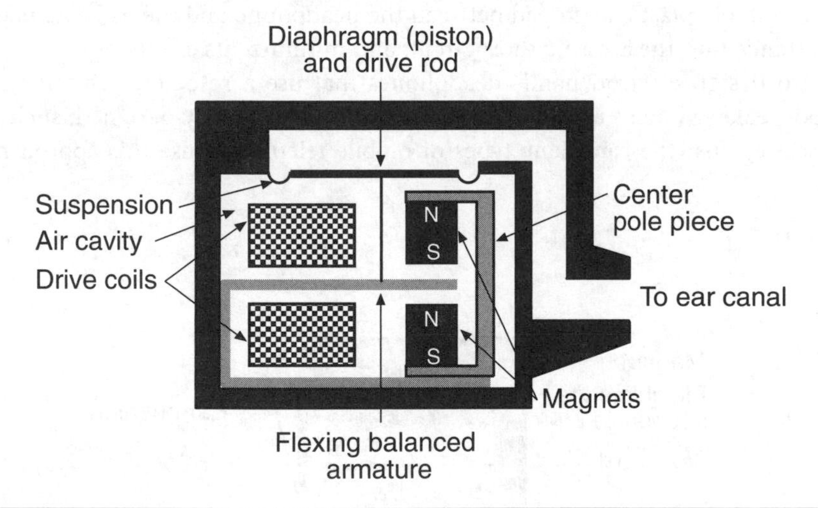

Figure 15.8 A simple electromagnetic bridge-type insert headphone using a flexible balanced armature that drives a small piston. Center pole piece and armature are magnetically conductive.

may be quite stiff. As with other electromagnetic transducers, the response is essentially stiffness controlled, leading to a drop-off of −12 dB/octave for frequencies above the resonance. This, in turn, results in the need for several transducers to cover the entire audio range with high sensitivity. The frequency response curve for the insert earphone shown in Figure 15.7, clearly shows the resonances and the associated high frequency drop.

15.5 ELECTRODYNAMIC HEADPHONES

Electrodynamic headphones are nearly as rugged as the electromagnetic designs. They are also suitable for those applications where the transducer must also be used as a microphone.

The relatively symmetrical design results in low nonlinear distortion. This means that the headphone can also be gainfully used as a high-fidelity audio sound reproducer. It is also used in other applications where there is a need for high-quality sound reproduction at high sound pressure levels, for example, in industrial and military applications.

Electrodynamic headphones are essentially built like miniature electrodynamic loudspeakers. Two types are common: (1) those that have an open back (dipole action) and (2) those that have a closed back (monopole action).

A basic closed box headphone design is shown in Figure 15.9. The internally trapped air will act in series with the air trapped between the diaphragm and the ear. Using a correctly chosen resonance frequency and adequate damping, the frequency response can be made quite smooth. For even better frequency response, one can use various types of resonant acoustical compensating circuits. This typically results in an extension of the effective frequency range by one or two octaves.

Any headphone having an *open back* will suffer from aerodynamic short circuit—that is, dipole cancellation. This leads to poor low frequency response and sensitivity unless the headphone is worn close to the ear. The response can be improved by a careful choice of diaphragm resonance frequency and the use of a suitable plastic foam pad between the headphone and the ear. The pad will introduce an acoustic resistance into the acoustic short circuit and minimize its effects.

One can also design electrodynamic headphones that use a resonance circuit similar to that of ported box loudspeaker systems as shown in Figure 15.10. If correctly designed, such headphones do not require a seal against the ear. Many types of mobile telephones use this approach to headphone design.

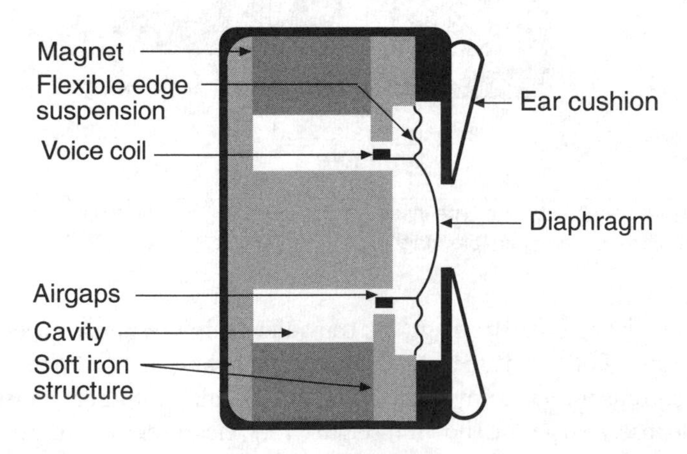

Figure 15.9 A simple electrodynamic headphone

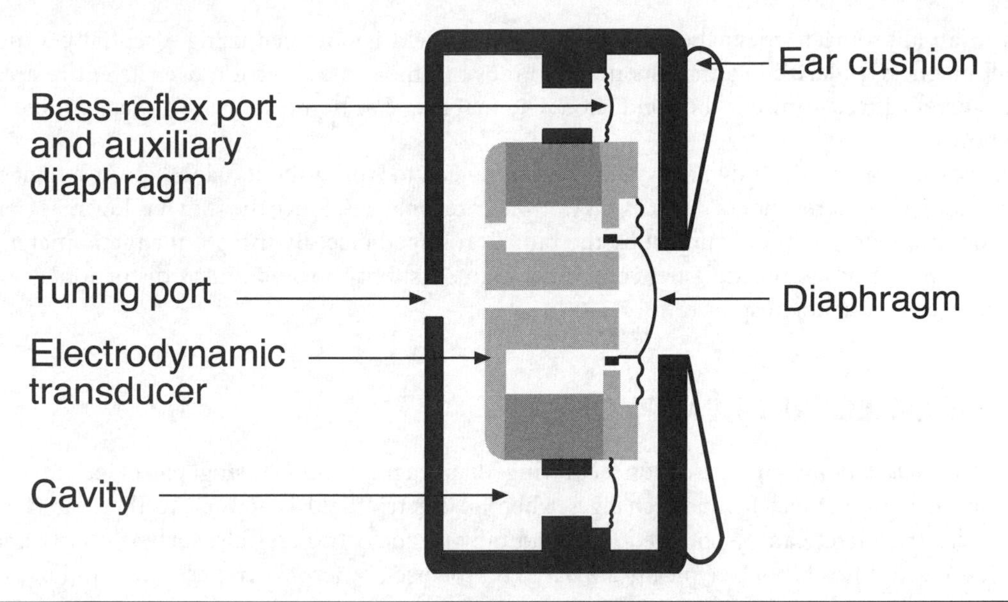

Figure 15.10 An example of a modern electrodynamic headphone using a bass reflex system ("ported box design").

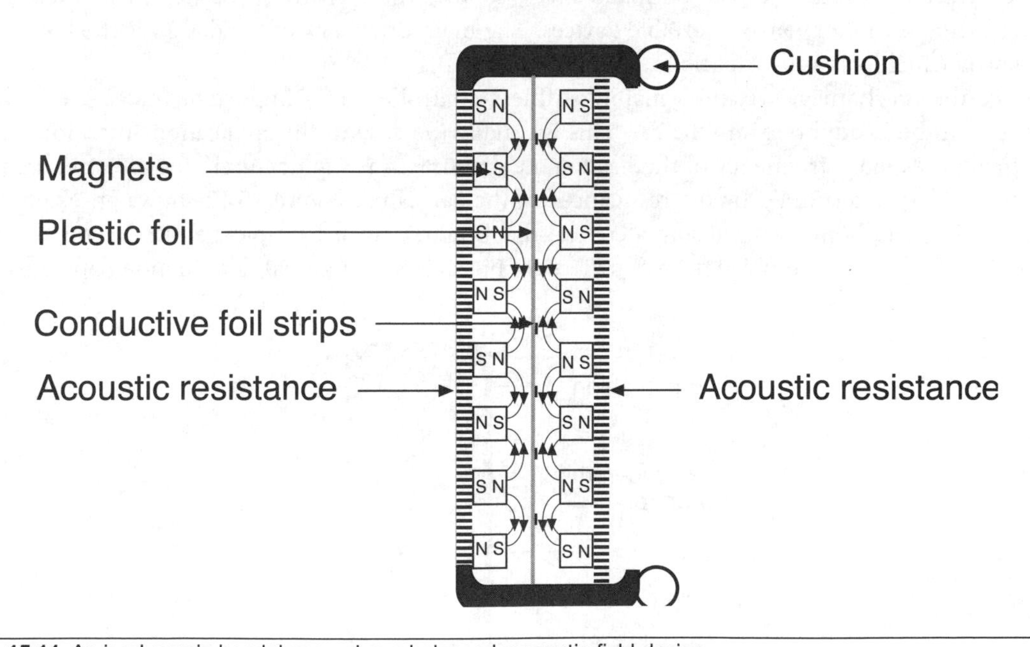

Figure 15.11 An isodynamic headphone using a balanced magnetic field design.

A modern version of the electrodynamic headphone is the *isodynamic* headphone, a dipole design, shown in Figure 15.11. Its design is reminiscent of that of electrostatic headphones, in that the entire surface area of the diaphragm is driven with about the same force per unit area. This is approximately achieved by applying a meander of an electrically conductive foil path onto a thin plastic film that is placed in an appropriate magnetic field. The magnetic field is obtained using essentially a multitude of small magnets, polarized so that the film will move in the same direction over its entire area when electric current is fed through the conductor. The magnets can be made using a plastic mixed with a ferrite powder.

The advantage of this design is that one does not need to worry about diaphragm resonances other than the fundamental resonance, such as *rocking mode* resonances. Since the film has low mass, any high frequency resonances will be damped by the radiation impedance. By using a magnetic material with high flux, one can design headphones combining high sensitivity, low nonlinear distortion, as well as a wide and even frequency response.

15.6 PIEZOELECTRIC HEADPHONES

Piezoelectric headphones are based on achieving diaphragm motion using piezoelectric materials. Such materials are subject to shape changes when an electric field is applied to the material. While the piezoelectric effect can be obtained from naturally piezoelectric crystals, such as quartz and some salts, piezoelectric headphones typically are based on the use of piezoelectric ceramics and plastics (see Chapter 13).

Piezoelectric headphones can be made small, robust, and sensitive. The design is often used in hearing aids and in inexpensive portable devices. The main drawback of the design is the high electric impedance, which is mainly capacitive.

Since the mechanical system is mainly stiffness-controlled, it is important that there be a tight seal between the headphone and the ear. This is usually the case in the applications mentioned. If the fundamental resonant frequency of the headphone diaphragm is high enough, the effective frequency response will be determined by the resonances of the ear canal. Figure 15.12 shows an example of a piezoelectric headphone design using a conical diaphragm driven by a piezoelectric ceramic bender. The ceramic bender is assembled from two pieces of piezoelectric material (length mode operation) that

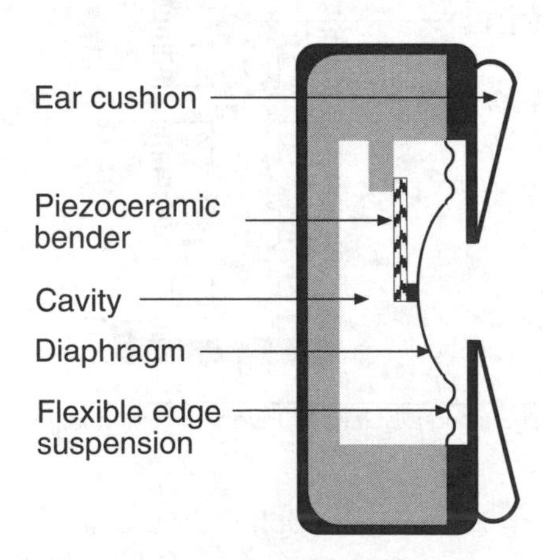

Figure 15.12 A piezoelectric headphone using a ceramic bender.

are combined in a *bimetal* fashion to achieve large excursions. Insert earphones may use a pile of thin piezoelectric disks operating in their thickness mode.

As mentioned, some modern piezoelectric headphones operate using piezoelectric plastic films. Superficially similar to electrostatic headphones, their mode of operation is, however, more similar to that of the bender design described previously.

15.7 ELECTROSTATIC HEADPHONES

Electrostatic headphones are generally used when the highest possible sound quality reproduction is desired. Most modern implementations feature an open-back design, as shown in Figure 15.13. Electrostatic headphones operate on the same principles as the electrostatic loudspeakers discussed in Chapter 14. The electrostatic headphone is, in principle, designed as an electrostatic loudspeaker with a comparatively large diaphragm. The problem of achieving sufficient loudness, often encountered when using electrostatic loudspeakers, is not as serious when using electrostatic headphones, since the latter are placed close to the ear.

There are two types of electrostatic headphones—those that use a bias voltage supply to provide charge for the membrane suspended between the electrodes and those that use *precharged* electret film to provide the charge that the audio voltage field acts on.

The mechanical system of the electrostatic headphone consists only of the film, which is mounted between two electrodes for push-pull operation. The film has low mass and is tuned for a low resonance frequency. The motion of the film is mass-controlled in the case of an open-back headphone. The resonances of the pinna and the ear canal will remain unaltered by the presence of the headphone since the film has such low impedance. If a sealing pad is used between the headphone and the head, the resulting cavity will exhibit acoustic resonances detrimental to the audio quality.

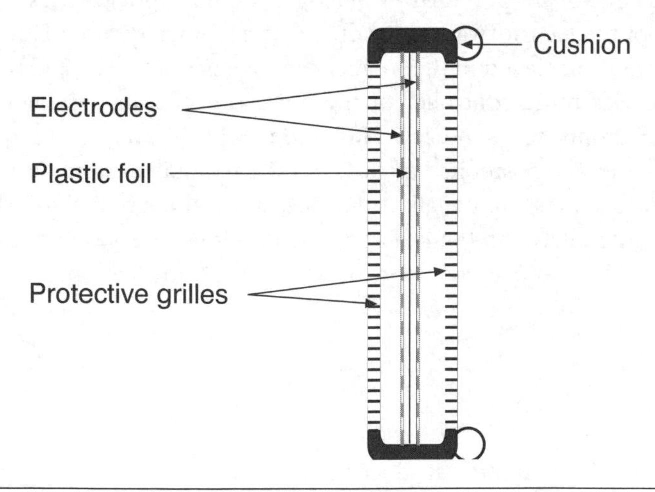

Figure 15.13 A circumaural headphone using a push-pull electrostatic transducer.

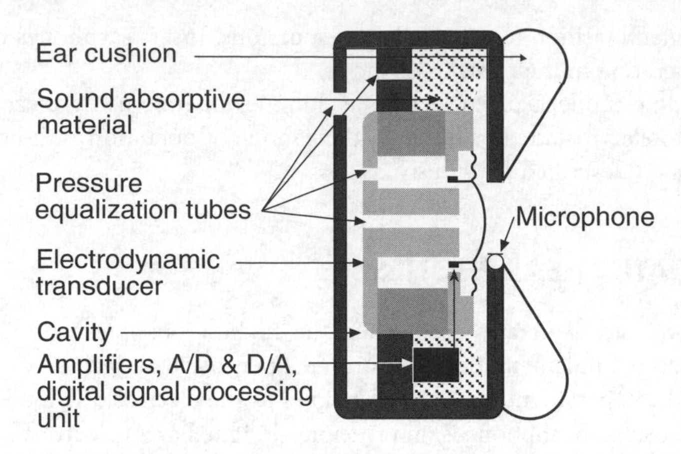

Figure 15.14 A typical headphone design using active control for noise reduction over a limited bandwidth.

As with most open-back headphones, sound insulation is nearly nonexistent; music listening, using these headphones, is best done in quiet surroundings.

15.8 NOISE-CANCELING HEADPHONES

Headphones are also used as noise-cancellation devices. The sound insulation of traditional headphones is mainly due to the mass of the headphone cups, their seal against the head, and the stiffness of the headband. Current signal processing techniques allow the use of active noise cancellation to reduce the environmental noise leaking in through these paths.

Two types of active control are provided by analog or digital signal processing. The noise-cancellation technique can be of the feed-forward or feedback type. Typically, attenuation of 10 to 20 dB over that of passive devices can be achieved in the frequency range of 100 Hz to 2 kHz.

To be effective, the electronic controller of the noise cancellation circuit must use an adaptive approach and an error microphone, as indicated in Figure 15.14.

The microphone is generally positioned as close to the ear canal entrance as possible. In the design shown, it is necessary to use pressure equalization tubes since the ear cushion seal must be effective to block out sound leaking in between the head and the cushion. The active circuit is usually effective in the range below 1 to 2 kHz—above this frequency, the noise cancellation generally relies on passive attenuation.

15.9 PROBLEMS

15.1 An electrodynamic headphone transducer of a hearing protector is coupled to a small sound coupler. Assume that the transducer current is determined by the resistance of its voice coil.

Task: What will the frequency response be of the sound pressure in the coupler (below any acoustic mode)?

15.2 An electrodynamic circumaural headphone that has an open back is to be used in reverse as a microphone when it is not being used for transmission. The *Bl*-product of the transducer is 0.01 N/A. The resonance frequency of the transducer is 2 kHz when it is attached to the head. For an rms input voltage of 1 V at 250 Hz, the headphone generates a sound pressure level of 94 dB inside the air cavity that has a volume of $3 \cdot 10^{-6}$ m³. The voice coil resistance is 1000 Ohms.

Task: Calculate the output voltage of the headphone for a surrounding sound pressure of 94 dB at 250 Hz.

15.3 An electrodynamic supra-aural headphone is designed to give a flat frequency response at the head. The transducer can be regarded as a small dipole system.

Task: What will the approximate frequency response characteristic be of the sound pressure at far distance in an anechoic chamber?

15.4 An insert earphone is designed using a balanced armature stiffness controlled transducer that generates a volume velocity of $2 \cdot 10^{-8}$ m³/s for an input voltage of 1 V at 400 Hz. Assume the mechanical system to have its resonance frequency at 8 kHz.

Task: Calculate the approximate frequency response at the eardrum for frequencies below 2 kHz.

Digital Representation of Sound

16

16.1 INTRODUCTION

Digital representation of sound is commonplace. Digitization of sound, digital signal processing, storing, transmission, and conversion to analog signals are all used both for measurement and for communication purposes.

Most communication uses of digitized sound signals incorporate some type of signal information reduction, usually called *coding*, to reduce the transmission and storage requirements. Coding can be further categorized into *lossless and lossy coding*. Lossy coding may be optimized for the properties of human hearing; such coding is called *perceptually optimized*. Transmission technologies such as the *pulse code modulation* (PCM) used in personal computing, *compact disc* (CD), and *Digital Audio Tape* (DAT) use various forms of lossless coding. Most other modern sound distribution media and formats all use some form of *perceptual coding*.

This chapter only discusses the fundamental principles of some digitization and lossy coding techniques. More in-depth discussion at a fundamental level can be found in References 16.1–16.5, in journals, such as the *Journal of the Audio Engineering Society*, in standards and recommendations issued by organizations, such as ISO, IEC, ITU, EBU, as well as at many websites.

16.2 SAMPLING AND DIGITIZATION

Most digital audio systems will use the processing blocks shown in Figure 16.1. The incoming analog signal is sampled in the time and amplitude domains to convert its history into a *digital format*. Generally, sampling is done at regular intervals in time determined by the *sampling frequency*. The sampling frequency needs to be well above any audio signal frequency. In most basic consumer oriented analog-to-digital conversion systems for audio, the sampling frequency is 44.1 kHz or higher. A number of standard audio sampling frequencies exist—common are 48, 96, and 192 kHz. For speech and low quality music, lower sampling frequencies such as 22.05, 11.025 kHz, or lower, are also used.

To digitize the amplitude of the incoming audio signal, it is first band-limited by low-pass filtering to remove unnecessary, and possibly detrimental, high-frequency sound information as discussed later.

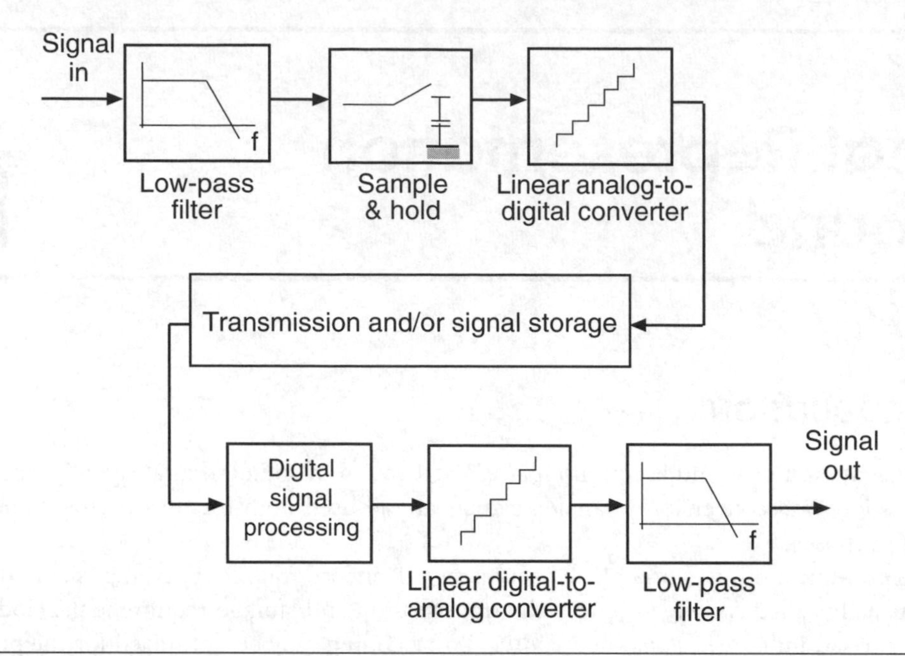

Figure 16.1 Basic digital signal sound reproduction system blocks.

Analog-to-digital Conversion

To provide time for the analog-to-digital converter (ADC) to convert the analog signal's voltage to a number, the voltage at the sampling moment is *remembered* by a sample-and-hold (S/H) network until the next sampling instance. The digitized momentary voltage can then be stored, delayed, coded, transmitted and reassembled into an analog electrical signal by a *digital-to-analog converter* (DAC, D/A) and a low-pass filter. This signal must then possibly be decoded, amplified, and fed to an electroacoustic transducer to become listenable. If there is no lossy coding and all technical components are perfect, the only signal degradation will be the low-pass filtering and the digitization processes at the input and output of the system. Any quantization system will introduce bandwidth-limiting errors, round-off errors, and clipping (or overload) errors.

Let us study an analog signal varying in time as shown in Figure 16.2a. If one samples the signal at equal time intervals, as indicated by the drop lines in Figure 16.2b, one will obtain momentary voltage spikes having the values indicated by the filled dots.

Such signals are difficult to work with; time is needed for the *analog-to-digital converter* to convert the voltage at the sampling instance into an equivalent number, so the voltage is kept constant to give the ADC the time necessary for the conversion as shown in Figure 16.2c. The building blocks of an ADC operation using a common summation and comparison method is shown in Figure 16.4. Note the DAC that is necessary for the operation and is shown in more detail in Figure 16.3.

The time between sample moments t_s gives the *sampling frequency* f_s as $f_s = 1/t_s$. When sampling at this rate, the signal spectrum will be uniquely determined up to the *Nyquist frequency* f_N, which is half the sampling frequency—$f_N = f_s/2$. One can show that if one filters the signal shown in Figure 16.2b by a

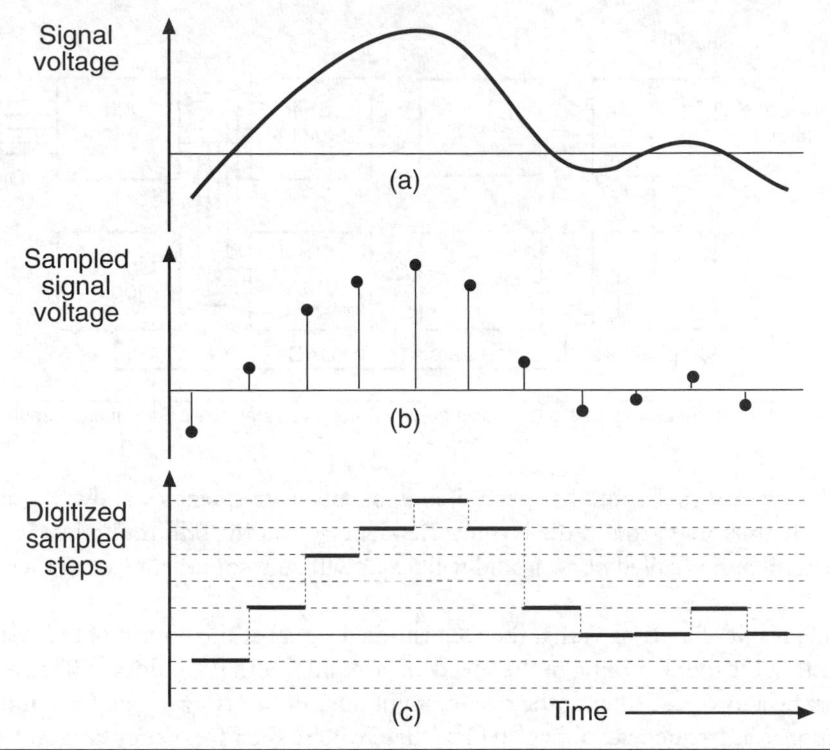

Figure 16.2 Principle of equal time interval sampling: (a) audio signal, (b) instantaneous samples, and (c) step samples.

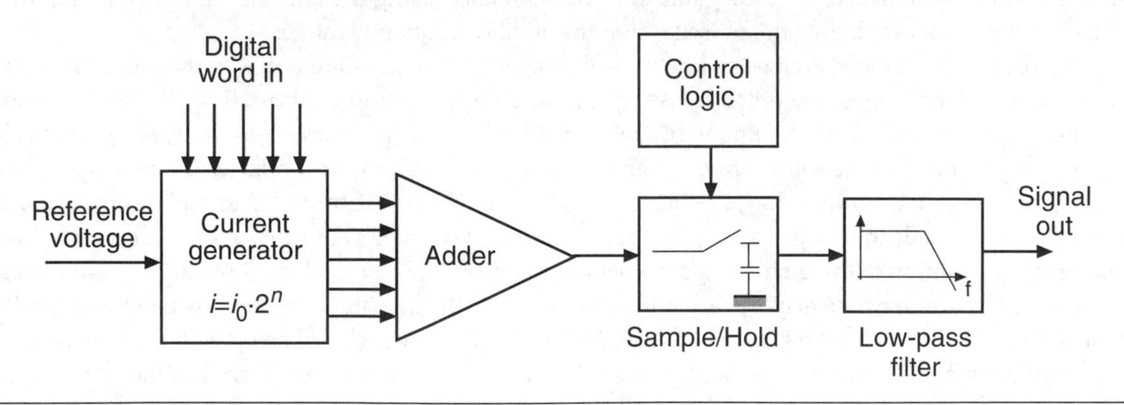

Figure 16.3 The basic building blocks of a digital-to-analog converter.

phase linear low-pass filter (such a filter has the same time delay at all pass band frequencies), one will obtain the signal in Figure 16.2a. Working on the signal shown in Figure 16.2c, there will be a slight drop in high-frequency content in the filter output that can be compensated for by digital or analog signal processing. The recovered signal will not have any added noise, linear or nonlinear distortion.

The sampling has not introduced any errors, provided the time instances of sampling are exact, the filter is linear, and there are no frequency components above f_N in the spectrum of the analog signal.

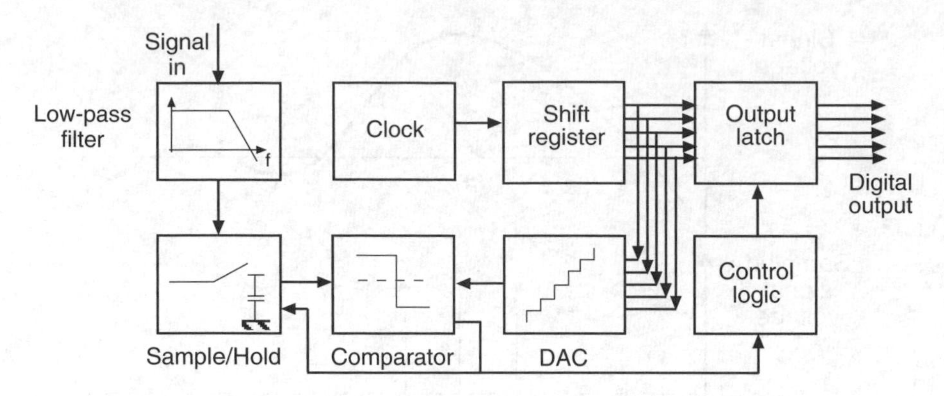

Figure 16.4 The basic building blocks of a successive approximation register analog-to-digital converter.

If, however, the input signal contains spectral components at frequencies above f_N, such components will be repeatedly mirrored around the Nyquist frequency f_N and its odd multiples, as shown in Figure 16.5. This phenomenon is called *aliasing* and will occur with any sound component having a frequency higher than f_N.

The audible result of aliasing is that the reconstituted signal at the output of the system will contain noise and whistles due to the folding of the spectrum around f_N and its multiples. This means in practice that there must be a low-pass filter at the system input (an *anti-aliasing filter*). Even if the microphones have poor response at frequencies above 20 kHz, there will be such frequency components in the signal at the digitizing system's input due to noise, nonlinear distortion, residuals from FM-stereo receivers and other sampling systems, etc. These signals must be attenuated using the anti-aliasing filter so that their aliased components are below the noise level in the audible frequency range.

There are two ways of solving the low-pass filtering requirement. One way is to use analog low-pass filters having steep slopes, the other to use "gentle" analog filters and oversampling. To avoid aliasing the filters need have 40 to 60 dB attenuation per octave above the *low-pass cutoff frequency* f_{LP}, typically 15–20 kHz. (Some filter action is provided already by the microphone and mixer.) A low sampling frequency will require corresponding low-pass filters having low cutoff frequencies and steep slopes. By necessity, there needs to be a guard-band between the low-pass filter's cutoff frequency and the Nyquist frequency. To minimize the guard-band one can use, for example, so-called elliptical filters; these can be designed to also have low ripple in the pass band. Typically, the ripple in the pass band will need to be less than 0.1 dB. Such filters will need to be *high order filters*, which, if analog, will need many reactive components such as capacitors and possibly inductors. Any high order filters will have resonances having high Q-values. Such analog filters will be costly, have poor aging, and poor transient response properties. The low-pass slope of the anti-aliasing filter (the *roll-off rate* in dB/octave) should be chosen according to the sampling frequency and the signal's spectral content.

Because of the mentioned drawbacks, another way of solving the low-pass filtering requirement is modern digital sound systems that use so-called *oversampling*—sampling at a frequency much higher than any anticipated signal frequency. The analog anti-aliasing filter can be of low order and inexpensive, while the *digital filter* used after sampling may use steep slopes to remove all frequency components

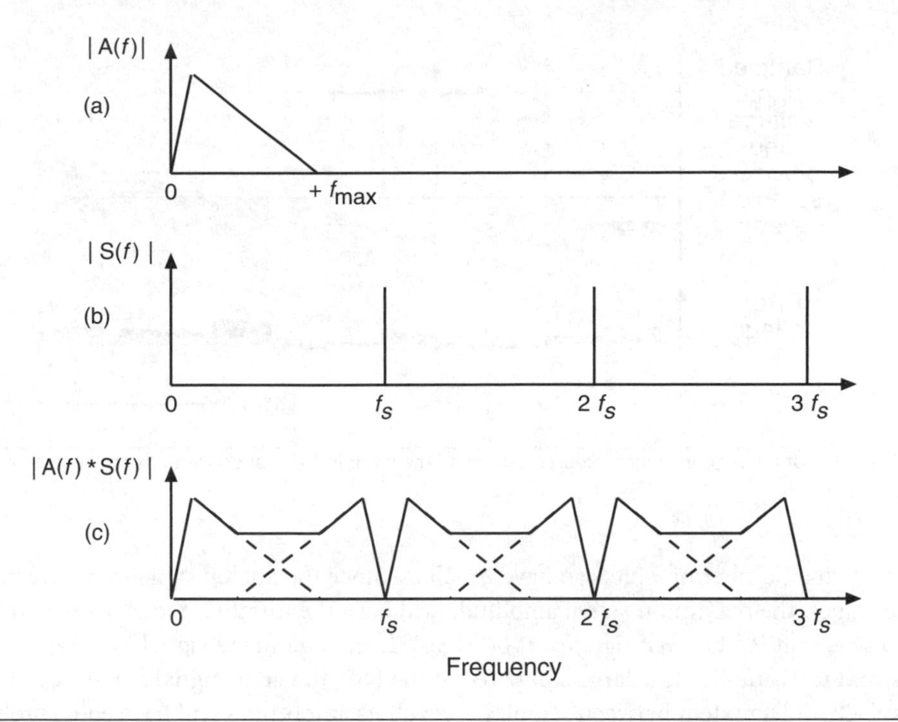

Figure 16.5 Principle of aliasing ($f_{max} > f_s/2$): (a) audio signal spectrum, (b) sampling signal spectrum, and (c) resulting sampled signal spectrum showing aliasing.

above the audio range. The digital filters do not change with age and are of course easily reproducible. The oversampling frequency is then reduced to the desired operating sampling frequency, for example 44.1 kHz, by *downsampling*. The downsampling is done using an interpolation process. The audio quality of downsampling routines can differ.

16.3 QUANTIZATION

Resolution and Quantization Noise

The instantaneous values of the amplitude of the analog signal at the times of sampling are digitized by the ADC in a quantization process. The accuracy of the numbers is determined by the sample and hold circuit, the ADC, and the sampling clock. Because of the limited resolution of any number scheme, there will always be an error, the *quantization error*—the difference between the true value and the number obtained in the quantization process. Figure 16.6 shows an analog signal's instantaneous value and its digitized sampled amplitude values; the gray bars show the difference between these values. One should note that the amplitudes and signs of the errors differ between samples. Figure 16.6 shows how the difference between the analog and digital signal results in quantization errors while the signal is being sampled.

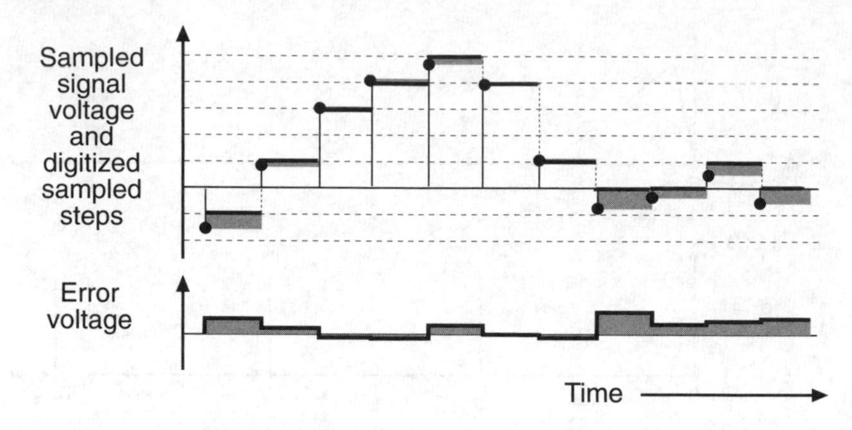

Figure 16.6 Influence of finite quantization step values on a time sampled signal showing the quantization errors at different time instances.

A binary digital number of *n* bits can have 2^n values. Since the analog signal can have both negative and positive values, the maximum signal amplitude will have the number $\delta \times 2^{n-1}$, where δ is the quantization step size, that is, the *least significant bit* (LSB). If the quantized signal has a wide spectrum of frequencies, and if its amplitude is large compared to the LSB, the error signal will have a random character (statistically independent between samples), and all its amplitudes will have equal probability over the interval ± 0.5 LSB. The error signal is sometimes inappropriately called *quantization noise* because of its near random character, in the case of high amplitude, wideband quantized signals. The RMS value of the error signal (quantization noise) will then be (see Reference 16.1):

$$\tilde{e}_{\text{noise}} = \frac{\delta}{2\sqrt{3}} \tag{16.1}$$

If the analog signal is a sinusoid having a peak value of $\delta \times 2^{n-1}$, the error signal will have a RMS value of $\delta \times 2^{n-1.5}$. The signal-to-quantization noise ratio will then have a maximum value of:

$$\left.\tfrac{S}{N}\right|_{dB} = 20\log\left(\tilde{e}_{\text{maximum}}\right) - 20\log\left(\tilde{e}_{\text{noise}}\right) = 20\log\left(\sqrt{\frac{3}{2}}2^n\right) \approx 2 + 6n \quad \text{dB} \tag{16.2}$$

From this expression, one could incorrectly infer that a perfect 16-bit linear quantizing system has a signal-to-noise (S/N) ratio of approximately 98 dB.

In reality, however, the audio signal being digitized usually needs about 20 dB of *headroom*; that is, the difference between the peak and long-term RMS values of the audio signal, corresponding to approximately 3 bits. In addition, the music audio quality becomes poor if less than 10 bits are used. This means that the dynamic range, in a sense, can be said to be only 3 bits—that is, 24 dB. This explains the need to move from 16-bit to 24-bit PCM for high-quality music recording and reproduction. The reasoning is illustrated in Figure 16.7 that shows the relationship between minimum digital resolution and dynamic range for three types of audio.

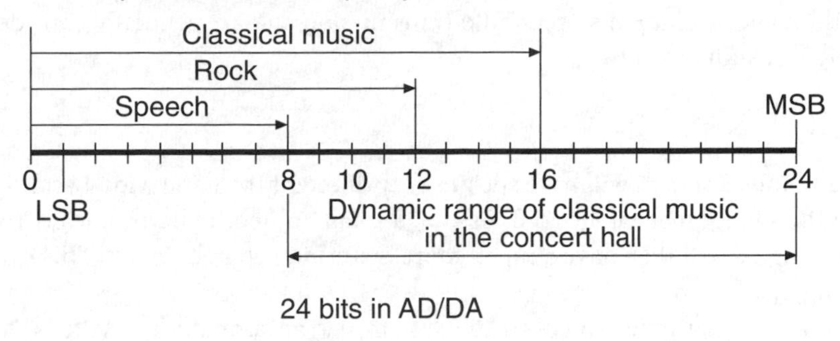

Figure 16.7 The graph shows the minimum number of bits needed for acceptable transmission of various audio materials using linear PCM. Below a resolution of 8 bits there is serious degradation of the signal due to quantization noise and distortion. These do not seriously affect speech intelligibility at resolutions above 4 bits, however.

This explains why the main advantages of the 16-bit CD format over the analog vinyl format for the reproduction of classical music are the absence of pops and clicks in the playback and the ease of copying and handling, rather than audio quality.

Some of the power of the quantization error signal is outside the audio spectrum $f < f_{LP}$. The power spectrum of the quantization error signal, under the assumptions made earlier, can be shown to be (see Reference 16.1):

$$G_{\text{noise}} = \tilde{e}^2_{\text{noise}} t_S \frac{\sin^2(\omega t_S/2)}{(\omega t_S/2)^2} \tag{16.3}$$

The power spectral density can be seen to drop toward higher frequencies; at the Nyquist frequency f_N, it is nearly 4 dB below the low-frequency value. Its behavior is, therefore, nearly the same as for white noise, which has a spectral density independent of frequency.

Since the frequency dependence of the power density spectrum of the sample-and-hold type pulse is the same as that of quantization error signal, frequency equalization of the sampled signal will lead to the quantization error signal also having a frequency-independent power spectrum. The quantization error will then sound like white noise. The level of the quantization noise within the audio band can be reduced by using oversampling. In oversampling, the sampling frequency is increased from the marginal values of 44.1 kHz or 48 kHz for 20 kHz bandwidth audio to, for example, 96 kHz or 192 kHz. The quantization noise will then be spread out over a bandwidth two or four times larger in these later cases. Because of the almost white noise character of the quantization noise, the level of the noise in the audio band will drop by 3 or 6 dB, respectively.

A traditional analog way of achieving perceptual noise reduction in audio systems is to use spectral pre- and de-emphasis. This means using a high-frequency boost at the system input and a high-frequency cut at the system output, usually as simple shelving filters as shown in Figure 2.4. Provided the input signal has a power spectrum that drops toward high frequencies, there will be a *transmission margin*; that is, the high frequencies in the input signal are not likely to force an overload on the input.

This margin can then be reduced by preemphasis of the high frequency part of the spectrum before quantization. Similar methods have been used for many years in gramophone recording, analog tape recording, television audio, and FM stereo radio transmission and are generally considered transparent from the viewpoint of audio quality.

Dither

Most real voice and music signals will have spectra very different from the wide-band signal assumed in this discussion. Often the signal will be composed of a small number of harmonically related tones. The quantization error signal will then have a much more disturbing characteristic. This can be understood from the following discussion.

Assume that the analog input signal to the system has an amplitude smaller than 0.5 LSB. Two extreme cases are shown in Figure 16.8. If the input signal is centered over one quantization step, the output will be a square wave as shown to the left in Figure 16.8. If, on the other hand, the analog signal is offset relative the quantization step, there can be a case of only a static output signal, a steady-state constant voltage, as shown to the right in Figure 16.8.

In either case, one must consider the signal as seriously distorted. A linear ADC and DAC transmission system has its highest S/N ratio when the (wide bandwidth) input signal has maximum rms value. In practice, however, one needs to keep the long term average of the signal at least 2 to 40 dB below its peak value, and the properties of a linear 16-bit PCM system are then similar to those of a high-quality analog system. The main advantage of the digital system is its robustness and the possibility of perceptual coding to reduce its transmission bandwidth and storage requirements.

The system performance can be subjectively improved by adding an extra random noise signal, called a *dither* signal, to the signal being digitized. Figure 16.9 shows the influence of one type of dither on the case shown in Figure 16.8. One notes that in both cases the introduction of dither results in a noisy pulse-width-modulated signal. If this signal is synchronously integrated over several time segments, the result will be that of the sinusoid input signal but with noise added as shown in Figure 16.10 for different numbers of synchronous averaging. Hearing does this type of integration, and dither can then subjectively improve the system performance, making it possible to hear tones having amplitudes

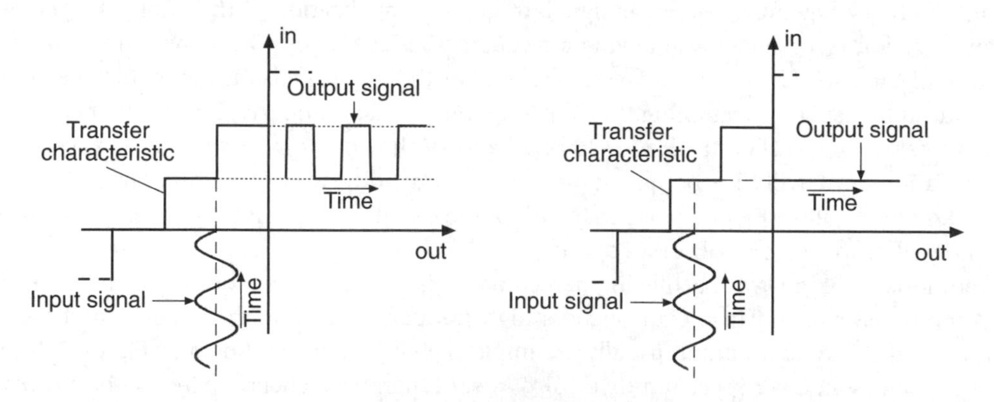

Figure 16.8 Two cases of digitization of an audio signal having amplitude of less than 1 LSB.

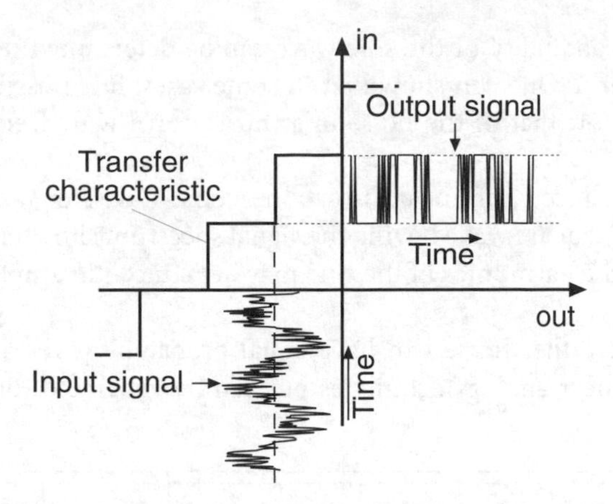

Figure 16.9 Influence of dither on the transfer characteristics shown in Figure 16.8. With dither noise the square wave becomes modulated.

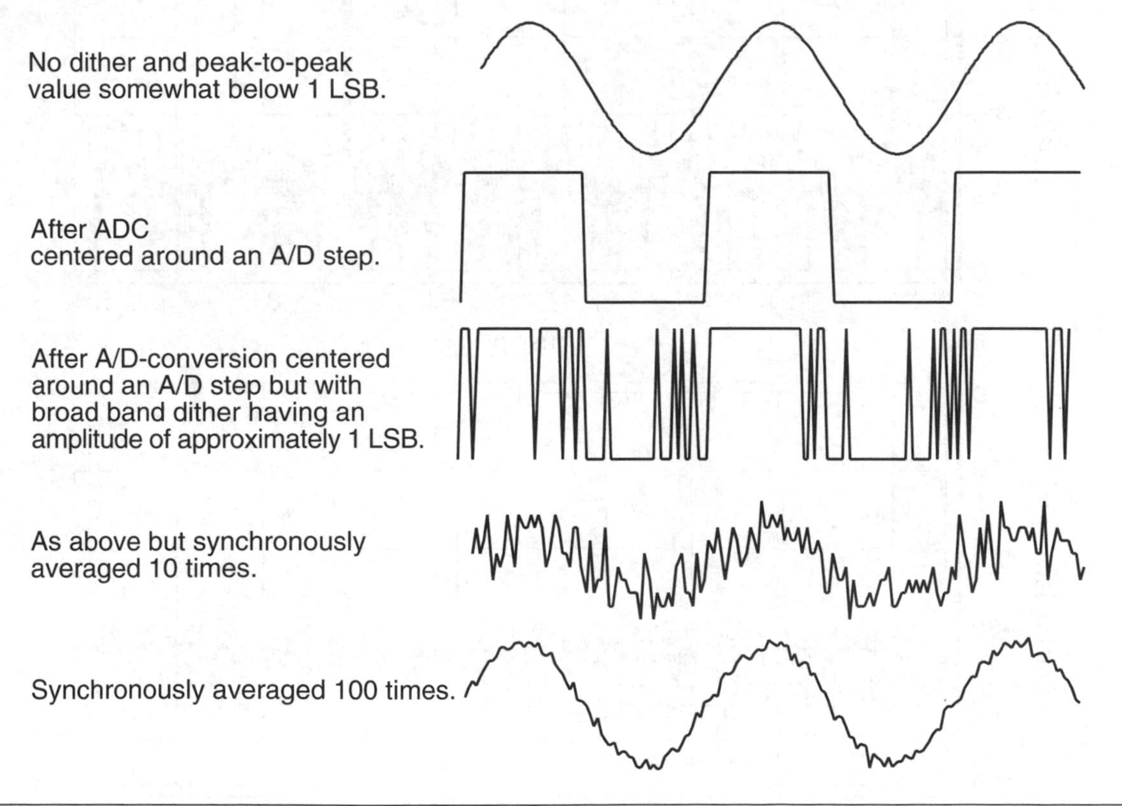

Figure 16.10 Influence of dither on a sine-wave signal when the signal has been D/A-converted and synchronously averaged over a number of periods.

smaller than the LSB. The audibility of the sine wave can be determined using the discussion of critical bands found in Chapter 3. One can show that in some cases, it is possible to perceive sines having a power of 10 to 15 dB below that of the noise in a third octave-wide frequency band (see Reference 16.12).

Dither also results in reduced nonlinear distortion. A sine-wave signal having a peak-to-peak value of 1 LSB will quantize into a square wave having the signal spectrum shown in the upper graph in Figure 16.11. Using dither noise, the harmonics of the sine may be reduced in amplitude as shown in the lower graph in Figure 16.11.

For example, a simple dither noise can have equal probability over a ±0.5 LSB interval. This is difficult to obtain using linear analog techniques but can be generated digitally using a DAC. High-

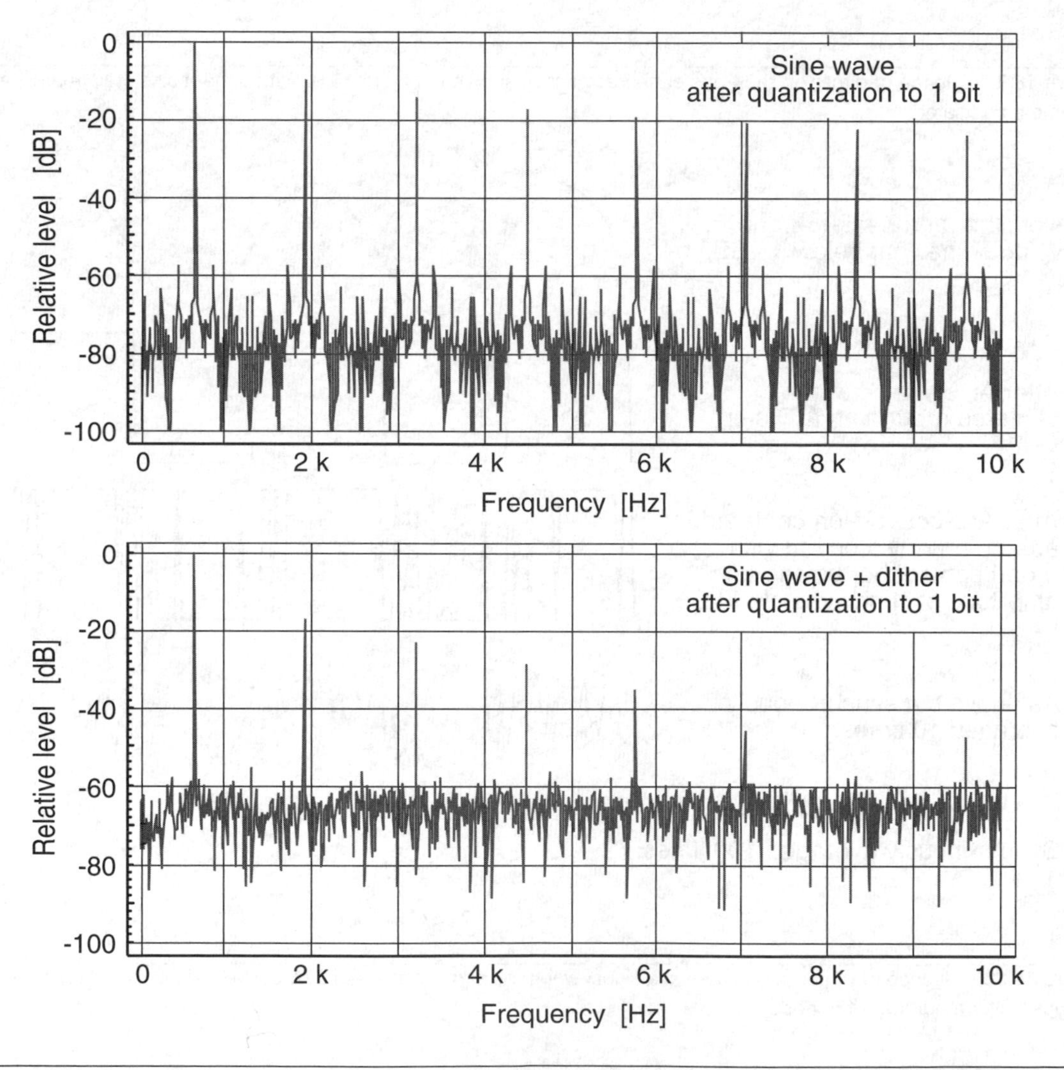

Figure 16.11 The spectrum of a sine-wave after quantization to 1 bit, and of a sine wave + dither signal after the same quantization.

frequency noise present in the input signal is often advantageous since it acts as dither. Different audio signals need different optimum dither noise characteristics. In many cases, however, a dither noise having a triangular probability density over the ±0.5 LSB interval turns out to be a good choice.

Quantization Methods

There are many ways to achieve the task of converting the analog signal value at the sampling instance to a number. It is reasonable to expect a faster conversion to be more coarse than a slower conversion, simply because there is not enough time for the electronic digital circuitry to convert with higher resolution. The power consumption by the ADC is also an important issue.

There are two extreme cases. The first is to convert to the desired resolution within the time window t_S accorded to the *sample-and-hold circuit* (SHC). Usually t_S is set as $1/f_S$; this is the method used in PCM converters using the *successive approximation register* (SAR) approach, as shown previously in Figure 16.4. Such a converter has inside it a precision DAC that outputs the instantaneous value of the conversion process (see Figure 16.3). The digital input to the DAC is numerically ramped in a positive or negative direction until the DAC output is the same as the analog input voltage within the resolution of 1 LSB. At this point, the digital input value is output to the buffer at the ADC output for further processing. Typically, such ADCs output a binary code, 6, 8, 12, 16, or 24-bits wide.

At the time of writing, most personal computers have a built-in sound card having a 16-bit ADC. Such a converter, in the worst case, has to step through all possible 2^{16} values to reach its settling point (where the output of the internal D/A converter is the same as the input voltage). Usually, however, the final value is reached in a series of binary comparison steps.

In conventional audio CDs, the audio data is represented as two channels of 16-bit wide data words at a sampling frequency of 44.1 kHz, whereas the old DAT format uses a 48 kHz sampling frequency. Modern ADCs use 24-bit resolution at a sampling rate of 96 or 192 kHz.

The sampling rates can be converted between one another by interpolation or decimation processes, as required. Such sampling frequency conversions may generate errors that can be detrimental to the subjectively determined sound quality.

A completely different approach is that used in *delta modulation* (DM) and *sigma-delta modulation* (SDM) converters. These operate at very high speed but with low resolution; in the extreme case the resolution is only just one bit. The principles and basic operation of 1 bit DM and SDM converters are shown in Figures 16.12 and 16.13. The output of the one-bit conversion process is integrated, and the integrated value is continuously compared to the analog input value. The bit sign is then changed when the appropriate value has been reached. The number of bits at the required output sampling frequency (for example, 48 kHz) is accumulated into a corresponding 16-bit word. Because the converter must, in the worst case, pass through all 2^{16} values, the 1-bit sampling frequency must be 2^{16} times higher—that is, 3.1 GHz. Such a high sampling frequency is currently impractical, and various ways are used to reduce the sampling frequency.

SDM converters are similar to DM converters but use the integrator block before the quantizer. An obvious advantage of the DM and SDM converters is that only a simple integration circuit is necessary to retrieve the analog signal. This integrator, in a sense, does the same task as the DAC used in SAR converters.

As discussed previously, the quantization depth—that is, the number of bits used, determines the quantization noise. The noise of the DM or SDM converter is much higher than that of the PCM con-

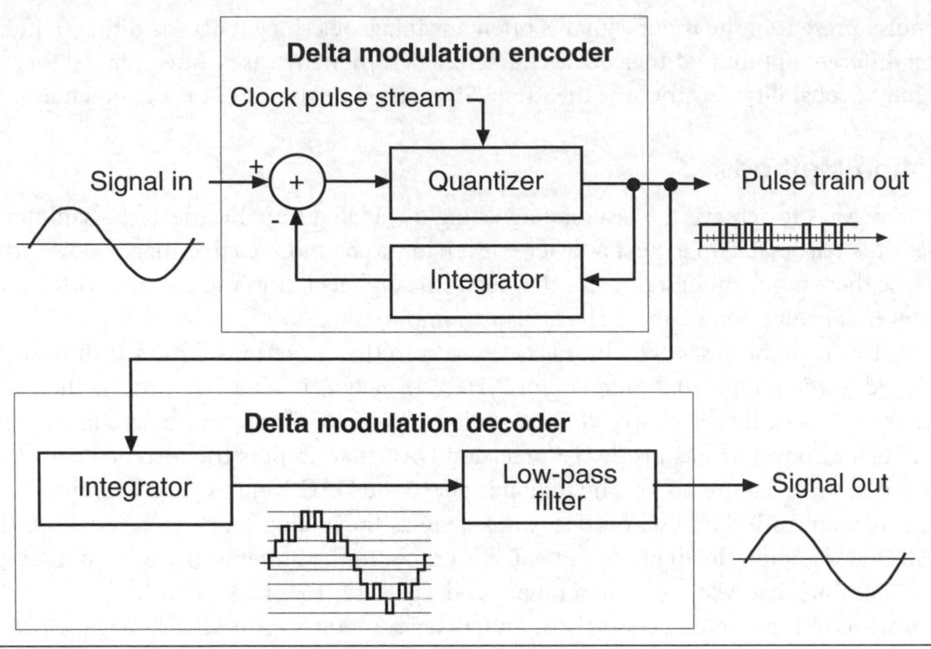

Figure 16.12 Basic operating principles of DM encoders and decoders.

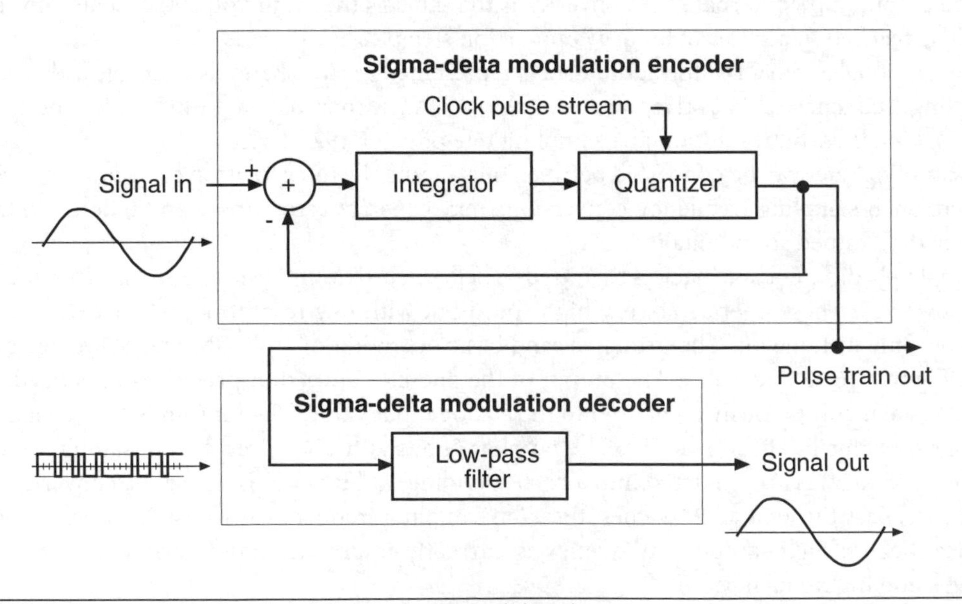

Figure 16.13 Basic operating principles of SDM encoders and decoders.

verter because of the small number of bits. Because of the high sampling frequency, however, the quantization noise power is spread out over a much wider frequency range than is typically the case in PCM converters, so the noise can still be low at audio frequencies. A particular problem with DM and SDM converters is the *idling noise*; that is, the noise when the input signal is close to zero. This idling noise

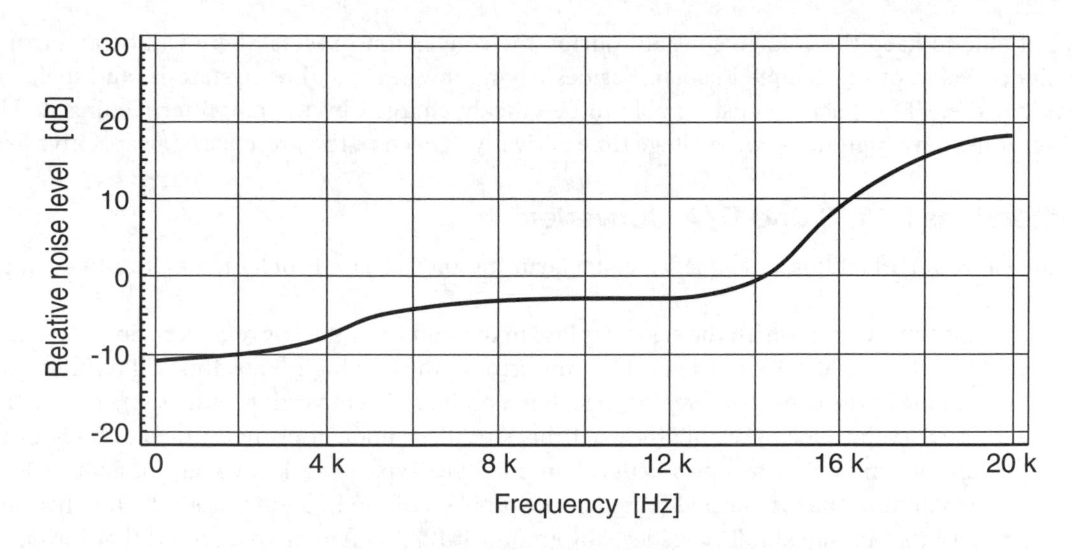

Figure 16.14 An example of a noise-shaping contour that moves noise power from the 1 to 5 kHz frequency region where the ear is sensitive to the higher frequency region of the audio spectrum. (Adapted from Ref. 16.5)

tends to take the form of tones. Consequently, there is a need for dither noise addition in these applications as well. The noise of the SDM is further characterized by high correlation from sample to sample.

The quantization noise can be regarded as an essentially white noise signal that is injected at the quantizer. This noise is affected by the feedback loop of the converter and appears as a high-pass filtered noise having a power spectrum that increases with frequency. The term *noise-shaping* is often used to describe the action of the system. Even more radical reduction of the audio band noise of the SDM converter can be achieved by using so-called higher order noise-shaping. The noise-shaping can take the masking characteristics of hearing into account; this is so called *perceptually optimized noise-shaping*. In this case, one finds that the oversampling should be at least six times the highest audio frequency of interest. It is perceptually advantageous to remove noise power from the 1–5 kHz frequency region and put the noise in the frequency region above 15 kHz where it is less noticeable. An example of such a noise-shaping frequency response is shown in Figure 16.14.

Other types of quantization schemes are *pulse-width modulation* and *pulse-density modulation*. These quantization schemes are often used in the DAC section of audio units. *Flash converters* are characterized by extremely short conversion times but feature comparatively fewer bits, that is, coarser resolution.

16.4 ADDITIONAL PROBLEMS IN A/D AND D/A CONVERSION

Sample-and-hold Circuits

It was mentioned previously that some time is required for the ADC to calculate the value of the waveform at the sampling instant. This time is provided by the S/H capacitor that ideally holds the sampled voltage of one sampling instance constant until the next sampling instant. If the S/H capacitor is not able to keep the voltage constant during the conversion process, there will be an error in the assigned value of the sampled signal. Besides having low leakage, low hysteresis, and dielectric absorption, the S/H capacitor must be able to be quickly charged by the amplifier feeding it. This amplifier must have high *slew-rate* (voltage time derivative), even in the presence of a capacitive load.

Nonlinearities in A/D and D/A Converters

From the viewpoint of audio signal quality, equal quantization step size is of larger interest than amplitude accuracy.

For an 8-bit converter in which the signal is close to the middle of the conversion region, the binary value must switch from 10000000 to 01111111. If the error in the first bit is large, this will result in large errors in the middle of the conversion region, which is a problem in converting audio signals since they do not carry a DC voltage component. To avoid this so-called modulation noise the accuracy of the most significant bits must be 1 to 2 bits better than one LSB. Typically a DC voltage of about 10% of the full scale maximum value is injected to reduce the problem of modulation noise, which only results in a reduction of the AC signal full range level by around 1 dB. A different approach is that zero input voltage is in the middle of a quantization interval so that no input will result in essentially no active bits.

In any digitizing system, it is important that the input signal be suitably amplitude-limited before, within, and after the anti-aliasing filter. Any limiting in or after the filter will result in high-frequency harmonics that will be mirrored around the Nyquist frequency and that will fall into the audio range. It is also important to consider the transient properties of the anti-aliasing filter so that even a square wave input does not cause clipping (and associated harmonics) due to ringing in the filter. This is particularly a problem in active analog filters. Hum and small variations—noise—in the timing of the sampling (so-called *jitter*) may also cause nonlinearities in the sampling of high-frequency audio signals.

To improve DAC properties, a SHC is often used at the output of the converter. The purpose of this SHC is to sample the converter output voltage after it has settled to its proper value. The SHC is followed by a low-pass filter to remove high-frequency components that may interfere with recording equipment, causing aliasing. The *slew-rate* properties of the low-pass filter influences the output as shown in Figure 16.15. A low slew-rate capability will cause distortion in high-amplitude, high-frequency sounds.

16.5 CODECS, UNIFORM, AND NONUNIFORM QUANTIZATION

Most commercial high-quality digital audio recordings are at the time of writing issued on CDs using 16-bit PCM with a sampling frequency of 44.1 kHz. Digital tape recording on DAT uses 48 kHz. If recordings are to be edited and manipulated using digital signal processing, PCM with a resolution of 24 bits and a sampling frequency of at least 96 kHz must be used to avoid audible artifacts once the signals have been downsampled to CD characteristics. A typical audio CD will contain approximately 45 minutes of stereo audio that requires about 650 Mb of data. For transmission by radio, television, or the Internet, the CD data rate is too high and must be reduced. This is done by using various processes called compression.

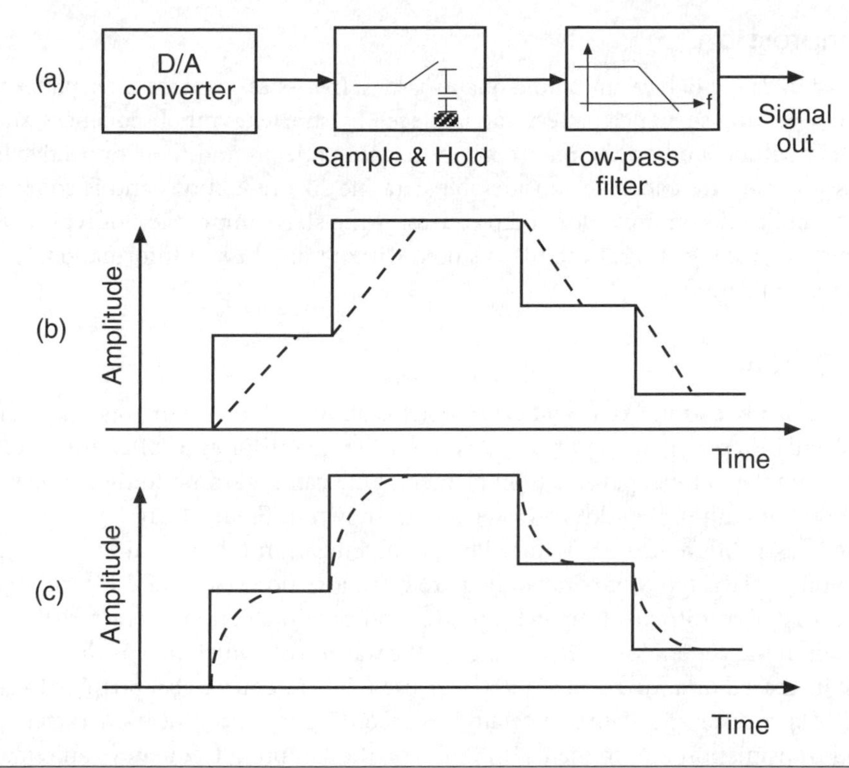

Figure 16.15 (a) Basic building blocks of the output section of a D/A-system, (b) slew-rate limited slope between settling levels, and (c) exponential slope between settling levels.

One usually distinguishes between *lossless* and *lossy compression*. Lossless compression is such compression that results in smaller data storage requirements but that allows exact data recovery. Lossy compression allows much smaller data storage requirements than lossless compression, but data cannot be recovered exactly. This, however, may not be a problem if the receiving system, such as human hearing, cannot detect the loss. This type of compression is called *perceptual coding*.

The file format MPEG II layer III, often just called MP3, which is used for music storage and transmission, is based on the use of lossy compression. The same applies to the PCM technologies used in digitizing voice in telephone applications such as GSM; here the quantization is typically done to 8 bits and at a sampling frequency of 13 kHz (see Reference 16.7). One does not need to be a trained listener to hear the difference between ADC and DAC conversions at 16-bit 48 kHz and 24-bit 96 kHz; it is quite easy to hear the differences at low audio signal levels. The differences are particularly easy to hear on long, reverberant piano tones.

Compression schemes of various types are consequently needed to reduce the amount of space necessary to transmit and archive sound data recorded at high bit rates. Some of the important issues in compression are fidelity to the original, the speed of compression and decompression (data rate and delay), and complexity. Audio data compression is a discipline in its own, and the reader is referred to the references for this chapter as well as to timely internet resources for further information (see References 16.5, 16.6, 16.8).

Lossless Compression

Lossless compression does not have any audio quality issues. In lossless compression, patterns in the data are targeted, and repeating sequence patterns are replaced by shorter symbol sequences, thus achieving a compression that will depend on the recurrence of certain patterns and their respective lengths. This type of compression can often achieve considerable data file size reductions and is comparatively fast, both to compress and to decompress, for audio codecs. A file size compression of typically 50% can be achieved. The main advantage is that the file is stored without any loss of information. This is used to compress software and data files.

Nonuniform Coding

The use of nonuniformly coding ADC and DAC systems allows a larger dynamic range, since the S/N ratio with signal will be less dependent on level. Nonuniform quantization is often used in speech applications since it is simple and inexpensive to implement. One can regard nonuniform quantization as a linear quantization but with pre- and post-distortion, as shown in Figure 16.16.

The relative height differences are reduced before quantization (companding) and increased after the DAC (expanding). Two types of companding are common: power-law and logarithmic companding. In practice, it is often difficult to match the pre- and post-distortion characteristics. In addition, the methods result in the generation of harmonics of the waveform, which means that the sampling frequency must be increased or a further low-pass filter used. It is of course also possible to use a conventional linear PCM quantizer and then just retain bits according to a nonlinear characteristic to reduce data storage and transmission requirements. In this case, the sampling frequency can remain the same.

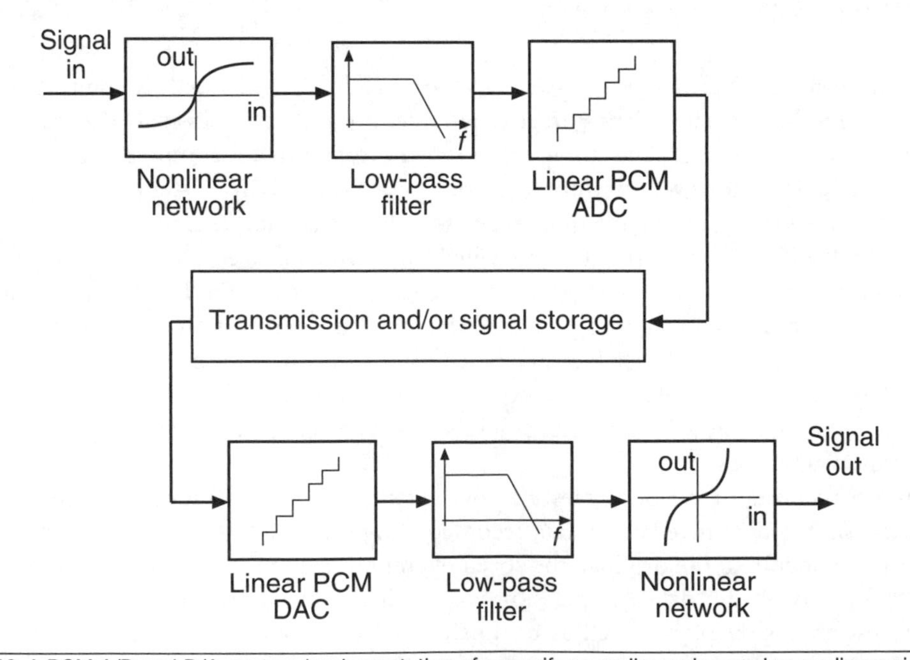

Figure 16.16 A PCM A/D and D/A-system implementation of nonuniform coding using analog nonlinear circuits.

Floating-Point Quantization

Human hearing features a wide dynamic range and *gain-riding*. Unless a signal is extremely quiet, loud, or contrasted, we usually do not consider its loudness. Gain-riding in this context means that hearing continually adjusts the perceived level to be similar; the hearing mechanism effectively reduces the apparent dynamic range. To reproduce sounds as we hear them in classical music, a dynamic range of about 100 dB is needed, for voice about 50 dB (the level difference between whispers and shouts). Over a short time, however, the dynamic range does not need to be as large as 100 dB; it takes some time for hearing to recover its sensitivity after it has been exposed to louder sounds.

In floating-point quantization, one tries to give the same resolution to both weak and strong signals. The aim is to reduce the number of bits used for storing and transmission, compared to linear PCM, but with little subjectively perceived audio quality reduction. This is possible if the bit scheme is chosen appropriately. The basic layout of a floating-point quantizer is shown in Figure 16.17.

One forfeits accuracy of strong signals by giving them the same number of bits as weaker signals and instead describes the *strength* by a scale factor. The sample value is quantized into a word having two parts, the mantissa and exponent. The mantissa and exponent are used to describe the basic quantization and gain adjustment factor respectively. A system using 4 bits for the exponent and 12 bits for the mantissa is denoted E4M12.

When dealing with floating-point quantizers, it is particularly important to separate between dynamic range and S/N ratio.

Block Floating-Point Quantization

In *block floating-point quantization*, one looks at the average strength of the signal over a number of samples making up one block of data. The principles of a block floating-point quantizer are shown in Figure 16.18. The choice of block length is critical and will depend on the audio source.

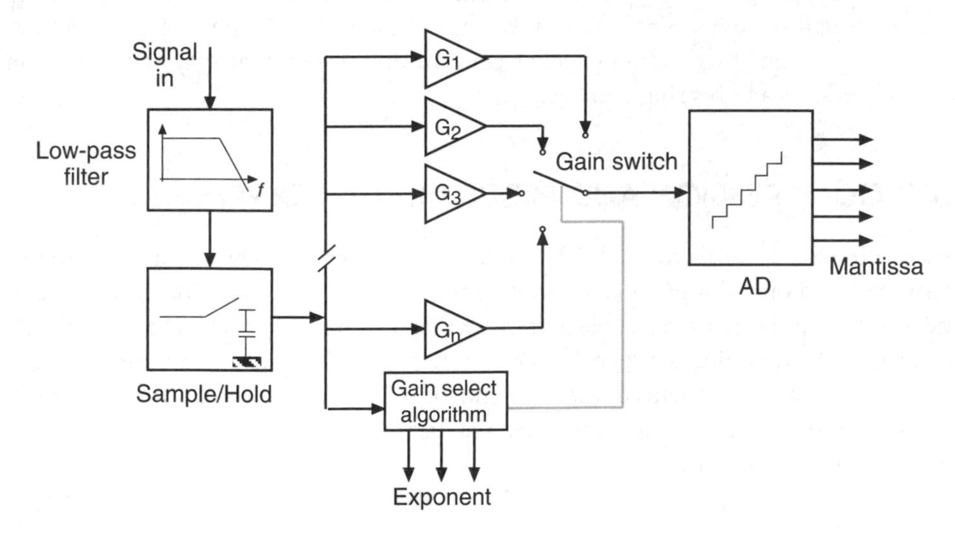

Figure 16.17 Operating principles of a floating-point ADC.

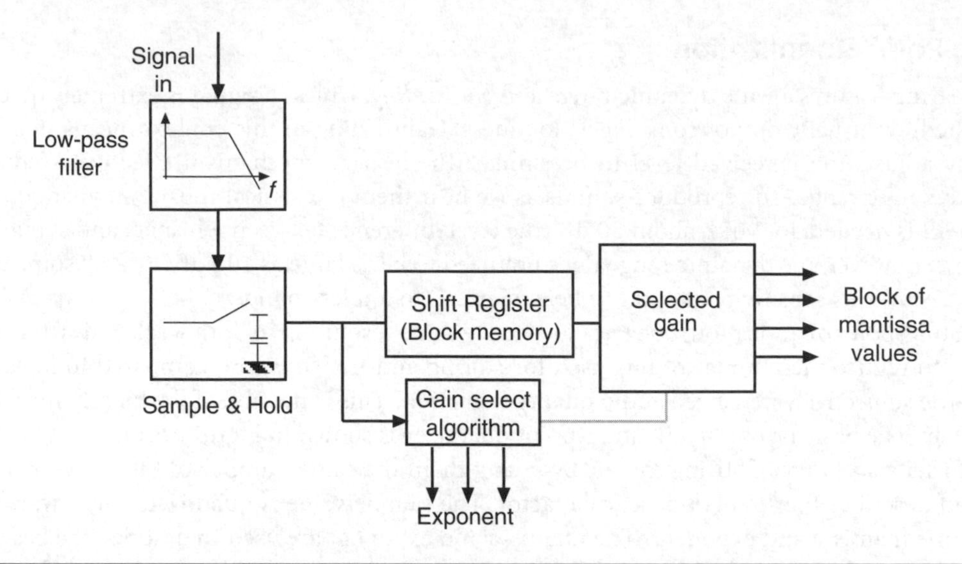

Figure 16.18 Operating principles of a block floating-point A/D converter using a linear PCM A/D converter and subsequent calculation of exponent and mantissa to set block gain.

The masking of pure tones by noise was subjectively determined and results show that a 15-bit mantissa was sufficient (see Reference 16.10). Further experiments using music have shown that in using a block size of 50 ms (approximately 2,200 samples at a sampling frequency of 44 kHz), one could use E3M13 format with exponent gain steps of 6 dB giving a range of 66 dB + 48 dB. The maximum ratio between signal and noise without signal (S/N_{ns}) will be approximately 114 dB if 1-bit dither is used, but the maximum ratio between signal and noise with signal (S/N_{ws}) will be only 66 dB (see Reference 16.9).

If S/N_{ws} is too low, or if the block length is unsuitable to the audio content, one can experience *modulation noise* or gain-riding artifacts. One advantage of the system is that the exponent bits can be neglected if needed, and one then has an effective companding system that may be useful if one wants to avoid hearing tiredness and hearing damage.

16.6 LOSSY COMPRESSION AND PERCEPTUAL CODING

At the time of writing most commercial ADC and DAC systems for high quality sound reproduction use the CD quality standard of 16-bit PCM at a sampling frequency of 44.1 kHz. Sometimes it is necessary to reduce the quality requirements to be able to transfer the information with less bandwidth or memory requirements. Using PCM coding for speech, one can show that 6–8 bits and a sampling frequency of 10–12 kHz is sufficient for good quality speech transmission from the viewpoint of speech intelligibility. Often, however, even this reduction of transmitted data is not sufficient, for example, in multichannel applications and mobile telephony.

Time-to-Frequency Mapping

Because of the perceptual properties of human hearing, it is often advantageous to transfer the signal from being represented in the time domain—the way we normally hear it—to being represented by spectra of suitably long blocks. This allows signal manipulation taking into account frequency masking, for example. It also allows efficient coding techniques, particularly regarding speech. Two alternatives for time-to-frequency mapping are filter bank methods and transform methods.

Filter Bank Methods

In using *filter bank methods*, one applies digital filters to the signal. These subdivide the audio frequency range into a number of frequency sub-bands, as shown in Figure 16.19. The output of each sub-band filter is then resampled at the optimum bit resolution for quantization noise in that band to be inaudible. The available bits can then be distributed to the various sub-bands by, for example, perceptually relevant methods. At playback, the different bands are reconverted from the digital to analog domain and recombined to become a reasonable replica of the original signal. The method is difficult to use as the number of frequency bands is increased. The principle of frequency band subdivision of the audio signal is, in a sense, similar to that used in the Dolby A analog compander system.

Transform Methods

The operating principles of a transform coder and decoder are shown in Figure 16.20. *Transform methods* are similar to the filter bank methods; it is mainly the approach to filtering that is different. The transform methods may use Fourier, wavelet, and other transforms. Fourier transforms are often used since they are easy to make relatively computationally efficient.

The Fourier transform is applied to a short block of audio data with a spectrum that should not change much over the time interval. As discussed in Chapter 3, speech can be considered as noise in

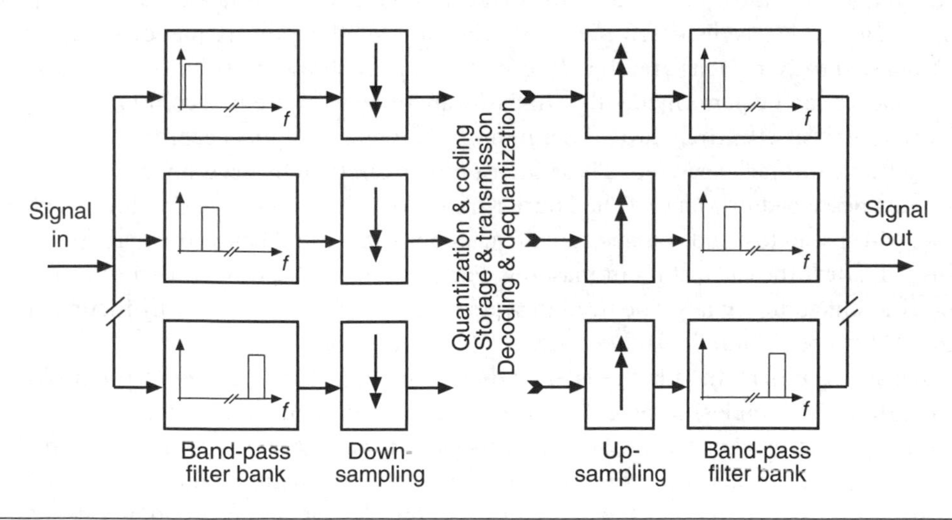

Figure 16.19 Operating principles of a coder & decoder using the filter bank method.

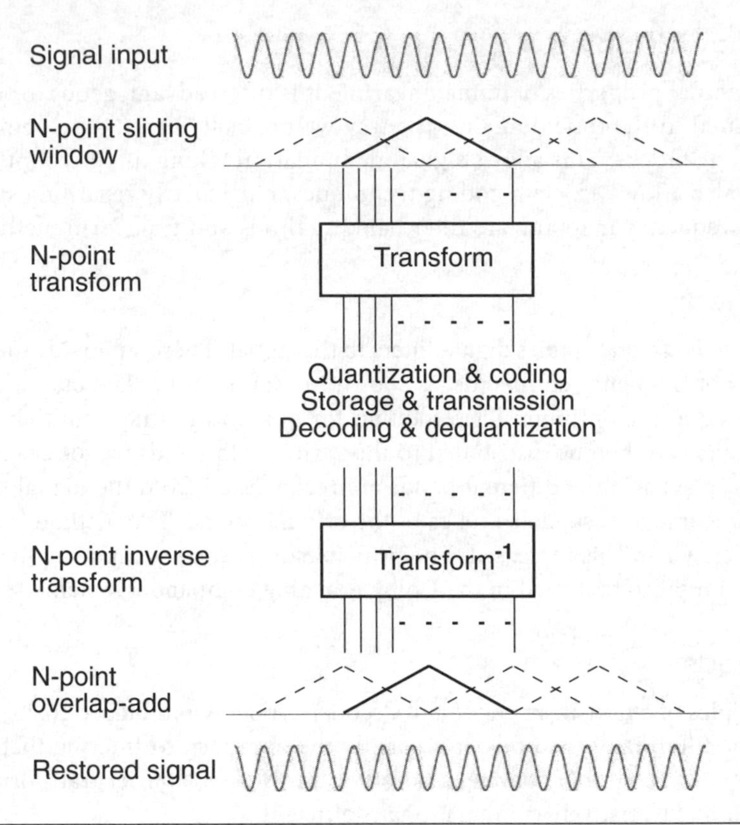

Figure 16.20 Operating principles of a transform coder & decoder using windowing and overlap-add methods.

a number of frequency bands (octave or third octave), modulated at different rates (compare to the speech transmission index as discussed in Ch. 3). The basic modulation frequency is approximately 7 Hz. Audio blocks 10 to 20 ms long are, therefore, comparatively short and suitable for Fourier analysis. The Fourier analysis can be performed either in hardware or by software in a computer.

The Fourier analysis effectively provides a number of filtered outputs according to the number of samples in the frame or block analyzed. These filter outputs can then be used much in the same way as the filter outputs described previously. The Fourier transform filters will have side lobes—energy from the main lobe will spread to a wide range of frequencies both above and below the main (lobe) filter frequency. This will affect the calculation of masking. For optimal results, different window functions are required for different audio signals. The need to switch window functions *on the fly* further complicates the coding and may create audible artifacts between blocks of data.

One drawback of this method is the latency due to the need for an entire N-point block of data (typically at least 1,024 samples) to be collected before the transform can be executed; however, any frequency filter will exhibit signal delay. Some coders use a combination of the transform method with the filter bank method. Because an entire block of data is processed simultaneously, the quantization noise of the process is spread out over the entire block. If the block length is too long, this may result in audible noise, as shown in Figure 16.21.

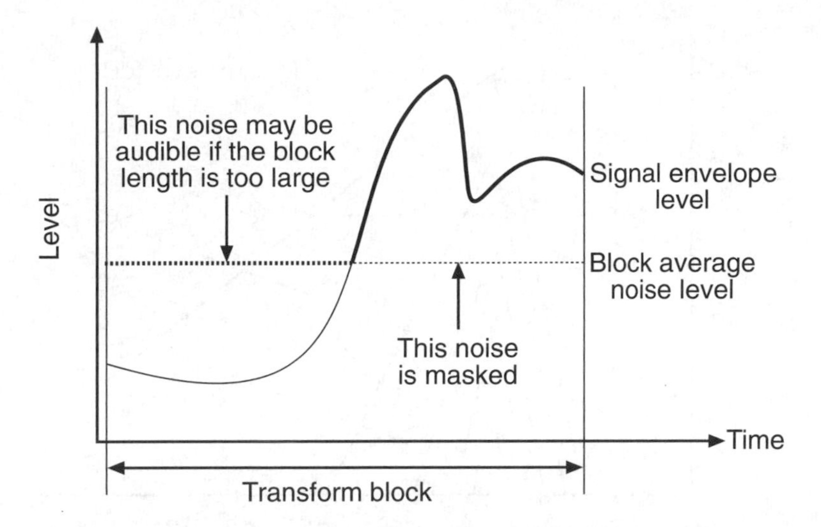

Figure 16.21 Quantization noise can become audible if the block length is excessively large. (After Ref. 16.4)

Perceptual Coding

Perceptual coding uses filtered versions of the input signal, sampled at high resolution and high sampling frequency. The filtered signals are then analyzed from the viewpoint of masking. It is well known that masking takes place both in the time domain and in the frequency domain. However, most perceptual coding schemes use only the principles of masking in the frequency domain, also known as simultaneous masking (see Figure 3.20). The masking will differ between individuals and signal levels. It turns out, however, that effective use of simultaneous masking can be done even with simple approximations to the actual masking properties of human hearing.

It is important to remember that most investigations of masking curve properties have been done with simple maskers and maskees, such as pure tones and narrow-band noise. In perceptual coding, one assumes that these measured characteristics are applicable to general audio signals and that tones and narrow-band noise have the same masking characteristics.

The excitation patterns of narrow-band noise signals, at the same sound level, are fairly similar, irrespective of frequency, if shown on a bark band frequency scale. The patterns will vary with sound level; lower masker levels lead to less prominent masking. Many coding schemes assume this effect to be negligible and use a triangular *spreading function* for calculation of masking effects. An example of spreading functions is shown in Figure 16.22. A further complication is given by the fact that audio signals, typically, will contain many maskers and maskees. This leads to the problem of how to account for the overlap of spreading functions.

If a sound would be inaudible in the recording situation, should it then be coded? The hearing threshold is defined by tones or band-limited noise measured at certain sound pressure levels. Microphone, mixing console, and other noise sources at recording will usually have noise spectra stronger at most frequencies than the hearing threshold of a normally hearing person. The standard assumption for 16-bit PCM is to assume that its maximum digital value corresponds to a sine having a sound pressure level of 96 dB. It is important to remember, though, that one does not know the playback sound pressure level of the decoded signal.

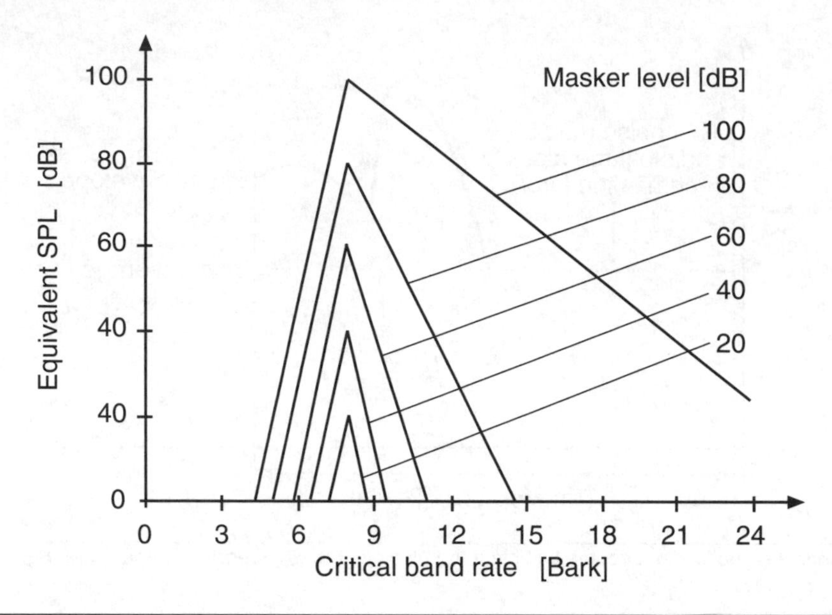

Figure 16.22 Examples of slope-spreading functions approximating the masking curves shown in Figure 3.20.

After sampling the signal, a block of its data will be analyzed from the viewpoint of masking. This can be done in various ways. A straightforward way is to look at the level of each Fourier analysis frequency and calculate its contribution to the total spreading function. This looping method is slow since it requires that all (typically thousands) frequencies be analyzed. Another way is to bundle the contribution from all frequencies into appropriate critical bands. Still another method is to find the *tonality* of a signal and only take into account the dominant tones.

Having decided on the influence of masking, the final step in coding is to decide on how the available bits should be allocated; strong frequency components will be represented by many bits and those that are masked by no bits. The quantization noise can be controlled for each frequency band and shaped so that it will be inaudible. Note, however, that there now is a quantization of the signal in the frequency domain as well.

In summary, one can say that different audio signals will perform differently in various coding/decoding schemes, and further that, even within a standard, different software implementations of the standard can sound differently.

Stereo and Multichannel Coding

The previous discussion of perceptual coding strictly concerned only a single channel of audio data. In practice, most audio is represented as stereo (2 channels) or multichannel (5.1 surround or higher, such as in DVD film and on Super Audio CDs). Holographic techniques such as wave field synthesis (WFS) using hundreds of channels may become common in the future. Can one use redundancies between channel data to further reduce the bit rate so that it will not be directly proportional to the number of channels?

If one looks at the time data, there is generally little correlation between channels. There is more correlation in the frequency spectra of the different channels. This can be used to reduce bit rates.

Human binaural hearing behaves differently above and below 2 kHz, as discussed earlier. Below 2 kHz, phase differences between signals at the ears account for the direction finding, whereas above this frequency, the direction finding is based essentially on the time of arrival and intensity cues. The cocktail party effect allows us to separate sound sources in complex sound fields.

Simply coding the stereo as two single channels may result in unmasking of, for example, quantization noise since the perceived stereo sound field becomes more *transparent* to us. If two sounds sum up to zero because of phase differences, quantization noise may become audible, requiring an increase in resolution.

A different approach to stereo coding uses finding the direction of incidence of each sound component. One can then code the maximum direction signal together with a direction indication.

Audio Quality of Perceptual Coding/Decoding

The perceived transmission quality of any audio coder/decoder system is determined not only by the coder/decoder but also by the electroacoustic quality of the entire audio transmission chain, particularly the loudspeakers. These are the final link in the audio reproduction chain, and the better the loudspeakers, the more likely are listeners to hear whatever artifacts are introduced by the digital sampling and processing. Table 16.1 shows the specifications of the loudspeakers and listening room as required by the ITU-R BS.1116 recommendation for listening tests.

Typical unpleasant characteristics of codecs are quantization noise, pre-echo (spreading of quantization noise to before the arrival of a transient), aliasing (because of the side lobes of the filters), chirps (high-frequency spectral content suddenly appearing and disappearing), speech reverberation (where there is none), and multichannel artifacts (such as loss or shift of the phantom source direction).

Listening tests using trained listeners are an essential component in any development of digital audio systems, if perceptual coding is used. It is important to choose appropriate test routines, test audio, and test subjects. The statistical spread in any routine using human listeners as measurement instruments is likely to be higher than using physical instruments. Different psychometric methods will have different sensitivity to various faults and shortcomings. In addition, prolonged listening, over weeks and months, may show sensitivity to characteristics missed in the conventional laboratory listening test where the test subject is listening to audio files that are only 10 to 20 s long. Some problems in subjective testing are discussed in Chapter 17.

Table 16.1 The specifications of the loudspeakers and listening room as required by the ITU-R B5.1116 recommendation for listening tests. [Ref 17.20]

Parameter	Specifications
Reference loudspeaker monitors amplitude vs. frequency response	40 Hz to 16 kHz ± 2 dB (noise in 1/3 octave bands, free-field) ± 10° frontal axis ± 3 dB re 0° ± 30° frontal axis ± 4 dB re 0°
Reference loudspeaker monitors' directivity index	6 dB ≤ DI ≤ 12 dB 0.5 kHz to 10 kHz
Reference loudspeaker monitors nonlinear distortion at 90 dB SPL	< –30 dB (3%) for $f < 250$ Hz < –40 dB (1%) for $f \geq 250$ Hz
Reference monitors' time delay difference	< 100 μs between channels
Height and orientation of loudspeakers	1.10 m above floor reference axis at listener's ears, directed towards the listener
Loudspeaker configuration	Distance between loudspeakers 2 to 3 m Angle to loudspeakers 0°, ±30°, ± 110°, Distance from walls > 1 m
Room dimensions and proportions	20 to 60 m² area for mono/stereophonic reproduction 30 to 70 m² for multichannel reproduction $1.1\ w/h \leq l/h \leq (4.5\ w/h\text{-}4)$ $l/h < 3\ w/h < 3$ where: l is length, w is width, h is height
Room reverberation time	$T = 0.25(V/100)^{1/3}$ for 200 Hz ≤ f ≤ 4 kHz where: V = volume of the room The following limits apply: + 0.3 s @ 63 Hz to + 0.05 s @ 200 Hz ± 0.05 s @ 200 Hz to 4 kHz ± 0.1 s @ 4 kHz to 8 kHz
Room early reflections	< –10 dB for t ≤ 15 ms
Operational room response	≤ + 3 dB, – 7 dB @ 50 Hz to ≤ + 3 dB @ 250 Hz ≤ ± 3 dB, @ 250 Hz – 2 kHz ≤ ± 3 dB @ 2 kHz to ≤ + 3 dB, – 7.5 dB @ 16 kHz
Background noise (equipment & HVAC on)	≤ NR15 in the listening area ≤ NR10 in the listening area recommended

Audio Systems and Measurement

17

17.1 INTRODUCTION

In audio engineering, the term *high-fidelity* means the faithful reproduction of sounds of voice and classical nonelectronic musical instruments—sounds that are recorded by way of microphones and then played back using loudspeakers, headphones, earphones, or even shakers. Sounds of music and voice can also be generated electronically either by way of special synthesizers or by computer software. Such sounds can have waveforms and frequencies beyond those common to conventional musical instruments.

The properties of hearing were discussed in Chapter 3. When we listen to sounds, hearing converts the acoustic pressure history in the sound field that surrounds us to generate the percept of sound in our brain. Audio systems may be designed for reproduction sounds for maximum intelligibility (with or without noise), for maximum enjoyment or subjectively perceived quality, or for maximum similarity to what would have been heard if the listener had been present at the recording.

Four commonly accepted conditions for high-fidelity sound reproduction are (see Reference 17.1):

1. Full frequency range reproduction with negligible level and time delay differences between various parts of the audio spectrum
2. Noise and distortion free sound reproduction at all sound pressure levels
3. Reverberation characteristics of the original sound field must be retained
4. Spatial properties of the original sound field must be retained

In practice, the complete fulfillment of these conditions remains an unattainable goal. Engineering compromises are always necessary, and the listener must accept that, without the cues offered by vision and other modalities, the experience will, in any case, not be that of attending the live event. In many cases there is no "original" sound event or sound field, and for artistic reasons the recorded sound may have been processed in various ways. This chapter will give an introduction to some common sound recording and playback techniques, an introduction to physical metrics for audio quality, and finally a short discussion of measurement techniques using humans as measurement tools.

17.2 AUDIO CHANNELS AND LISTENING MODES

Telephones use a single channel pickup and sound reproduction system. Most other sound carriers (except for telephones), however, are designed to record and reproduce audio in formats using more than one channel. For stereo, at least two channels are usually used for recording and playback to render the spatial properties of sound in a better way. The audio contained in the two channels is usually reproduced by way of two headphones or loudspeakers but may be recorded using many microphones or pickup devices. Other carriers, such as radio and TV transmissions, CDs, DVDs, Internet radio (and other streaming audio), and others may be configured from one to several signal channels, two and five channels being the most popular. Because of the many ways that audio signals can be recorded, transferred, stored, and reproduced, it is important to understand how audio systems are characterized and how they differ.

17. 3 MONOPHONIC AND MONAURAL

The word ending *-phonic* is used to describe the number of channels that are used in the sound carrier. Old gramophone records and most AM broadcasts that are single channel carriers are said to be *monophonic*. In spite of telephone systems being essentially monophonic, they are seldom referred to in that way. Many people use headsets to communicate using telephone systems and the headsets usually use one transducer for each ear. Presenting the same channel to both ears is called dichotic presentation.

The number of ears used in listening is referred to using the word ending *-aural*. We can listen monaurally or binaurally (using one or two ears). The principles of mono- and stereophonic audio systems are shown in Figures 17.1 and 17.2.

17.4 STEREOPHONY AND PHANTOM SOURCES

Vinyl records, albums, CDs, DVDs, and Internet radio are generally transmitting audio using two carrier channels and are called *stereophonic*, see Figure 17.1b. In stereophonic sound reproduction, the listener and the two loudspeakers are assumed to be at the corners of triangle as indicated in Figure 17.1b.

Typically, the loudspeakers need to be close so that the triangle will have a smaller apex angle ϕ at the listener—40° to 60°. The line between the loudspeakers is called the baseline. If the recording is correctly done, the baseline will be filled with appropriate *phantom sources*. Listening off-center, in principle, results in the phantom sources tending to collect in the loudspeaker closest to the listener. By using loudspeakers that have a controlled directivity pattern, it is possible to have phantom sources between the loudspeakers also for off-center listening positions. Of course the presence of room generated early reflections and reverberation will affect the perception of the phantom sources. Using crosstalk cancellation techniques discussed later in this chapter one can also have phantom sources outside the baseline.

Figures 5.9 and 5.10 show how the perceived angle of sound incidence depends on the relative time difference. Because recordings can use microphones with different directivity patterns and because of the way microphones may be positioned and directed relative to the sound source, different time-level patterns will occur in the reproduced audio. Figure 17.3 shows the effects of combinations of such recording induced time and level differences.

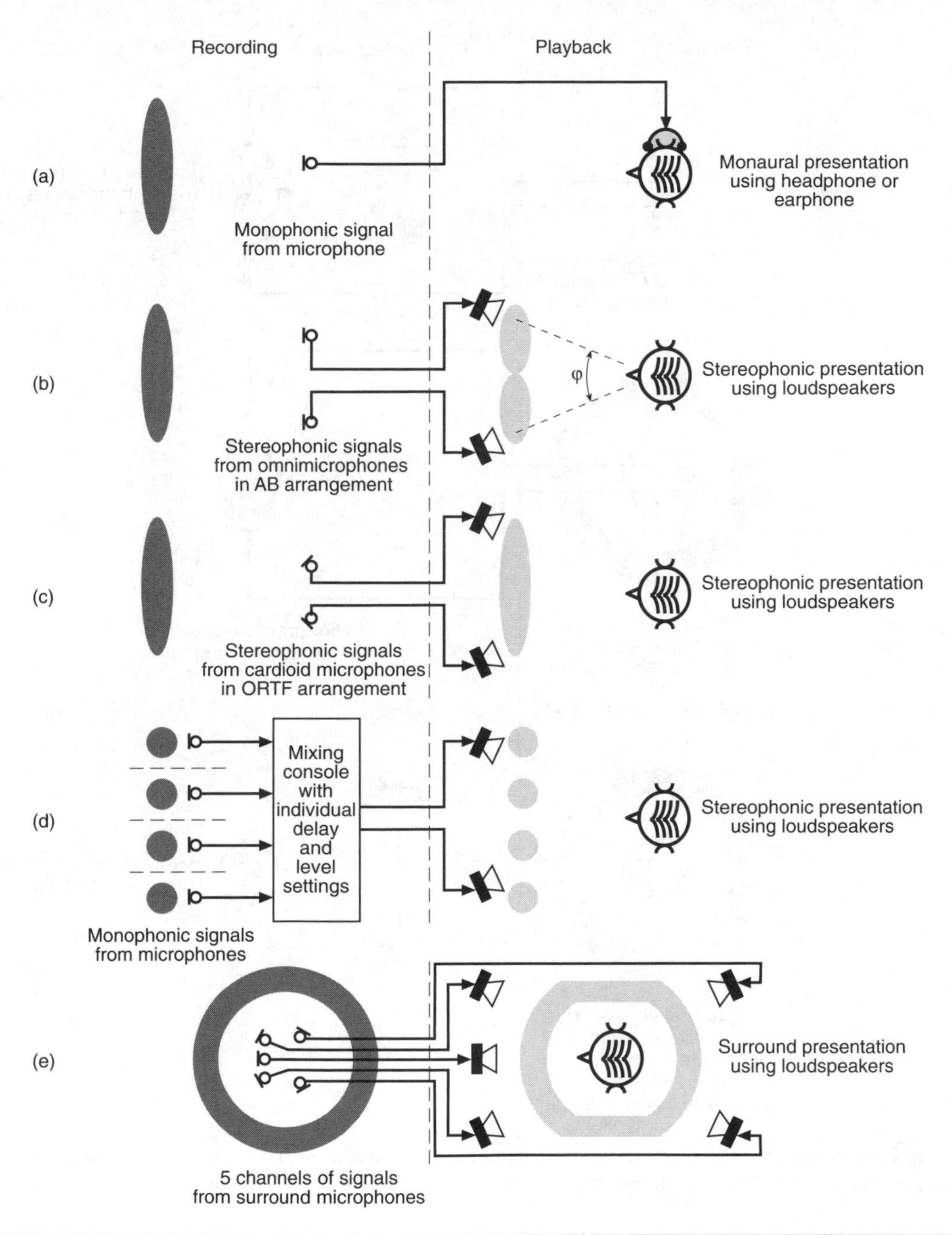

Figure 17.1 Some sound recording/playback systems: (a) monaural using a monophonic source, (b) stereophonic using AB microphones, (c) stereophonic using a ORTF microphone arrangement, (d) stereophonic using a multimicrophone technique, and (e) surround sound recording & playback. Many more are commonly used. Amplifiers, filters, delays, and other signal processing equipment are not shown.

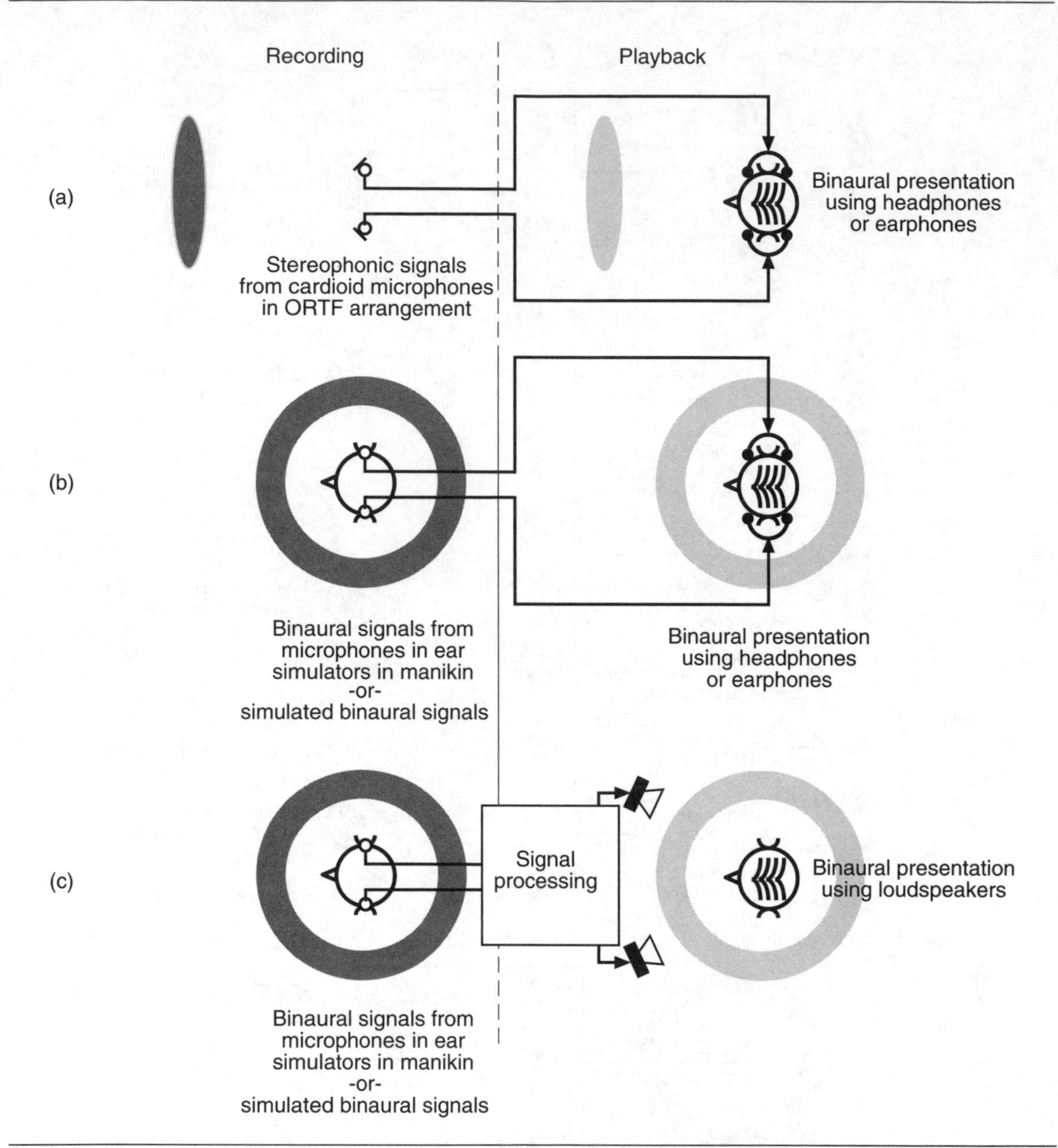

Figure 17.2 Some sound recording/playback systems for binaural playback using: (a) spaced cardioid microphones, (b) a recording manikin, and (c) cross-talk cancellation and other signal processing (see Figure 17.7).

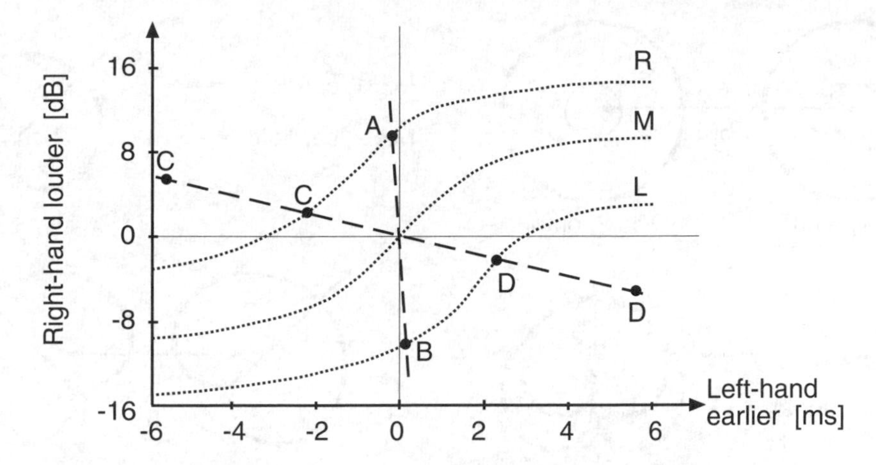

Figure 17.3 Lateral position of the phantom source in the case of mono sound source reproduced by two loudspeakers at $\varphi = \pm 30°$ with simultaneous variation of delay and level. Dashed lines explained in text. (L = phantom source in left-hand loudspeaker; M = phantom source in the middle between loudspeakers; R = phantom source in right-hand loudspeaker. (After Ref 5.6)

Figure 17.4 shows some popular ways of arranging the microphones for stereo and binaural sound recording. Originally, AB recording was used to record stereo sound. In AB recording, two spaced omnidirectional microphones provide the sound signals that are transferred to the stereo loudspeakers. In AB recordings, there will be both time and level differences due to the way music signals reach the microphone. Sometimes support microphones are used to provide more clarity for certain instruments or instrument groups.

Later, XY technology was developed in which two-coincident cardioid or bidirectional microphones pick up the stereo sound signals. In MS stereophony, coincident cardioid and bidirectional microphones are used, and their signals added and subtracted to obtain signals similar to those obtained by XY stereophony. In both XY and MS recording, there are mainly level differences between the stereo signals; these methods are called intensity stereophony.

Characteristic for all multichannel playback methods is the creation of phantom sources in listening. The curves marked R, H, and L in Figure 17.3 were obtained in listening tests and mark combinations that result in phantom source placement at the right hand loudspeaker, between the loudspeakers, and at the left hand loudspeaker respectively. Figure 17.3 shows two lines that indicate how AB and XY microphone techniques differ in stereo rendering capability. Line AB shows the typical behavior of MS recording technique signals, whereas line C′ C D D′ shows rendering using poor AB microphone placement. In the latter case, most of the phantom sources will be located in either loudspeaker. The effect is sometimes called *hole in the middle*, no sound sources are convincingly perceived as being in between the loudspeakers. The ORTF technique shown in Figure 17.4d has good loudspeaker/headphone compatibility.

The perceived depth localization is mainly due to level and spectral differences in the recorded reverberant and direct sound.

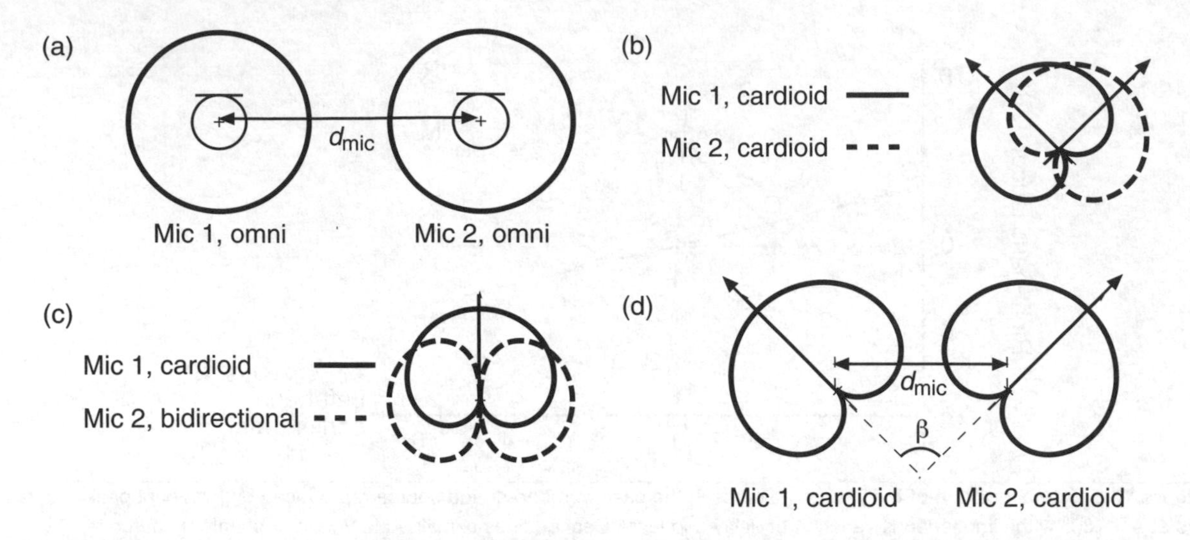

Figure 17.4 Some microphone arrangements for stereophonic recording of sound. Microphones 1 and 2 feed left- and righthand channels unless otherwise noted: (a) AB (time delay-based) the distance d_{mic} between the microphones is usually quite large, (b) XY (level-based), (c) MS (level-based) note that the microphone signals must be summed and subtracted to give left- and right-hand signals, and (d) ORTF microphone pair where the distance d_{mic} and the angle β between the two microphones are adjusted for optimum stereo sound reproduction with minimal sound stage distortion. Often d ≈ 0.2 m and β ≈ 110°.

In multimicrophone recordings, one can position the sound source more freely; there are now commercially available sound mixing desks that allow direct positioning of the phantom sources anywhere around the listener. These effects are often used in surround sound reproduction and to some extent, also in film sound. By adding simulated reverberation and early reflections, the phantom sources can be positioned inside a virtual space as well. This auralization-based technique is often used for recreating reverberation and *ambience*. The reverberation is usually generated by digital signal processing using either *finite* or *infinite impulse response filters* that can simulate the acoustic properties of real and imaginary environments.

17.5 STEREO USING HEADPHONES

Many listen to stereo recordings using headphones rather than loudspeakers. Since the two-channel recording does not contain the correct binaural cues (time and frequency response differences between the ear signals), at least two phenomena occur. One is that the sound recorded to appear mainly toward the right and left loudspeakers now appears in each headphone and is not externalized; the other is that the sound recorded to appear between the loudspeakers now appears inside the head, so called *in-head localization*. Several methods have been proposed to eliminate these problems. Commonly, binaural cues (time and level differences) are introduced to try to fool hearing that the sounds are actually emanating far from the head.

Figure 17.5a shows a hardware solution where the stereo loudspeaker signals are fed to an electronic network that introduces approximately correct cross-talk signals mimicking those that would be obtained in free-field listening, so that the stereo recording is better suited for headphone listening. A problem

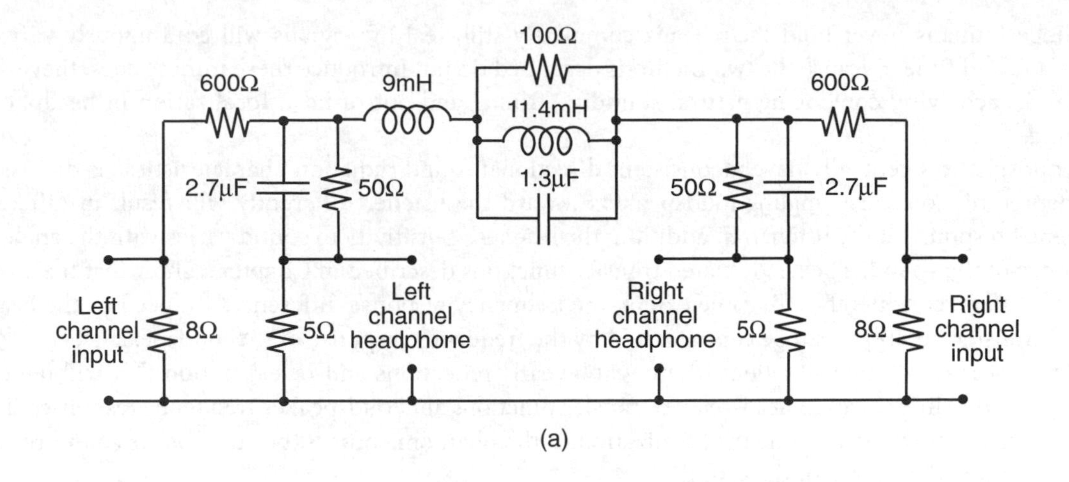

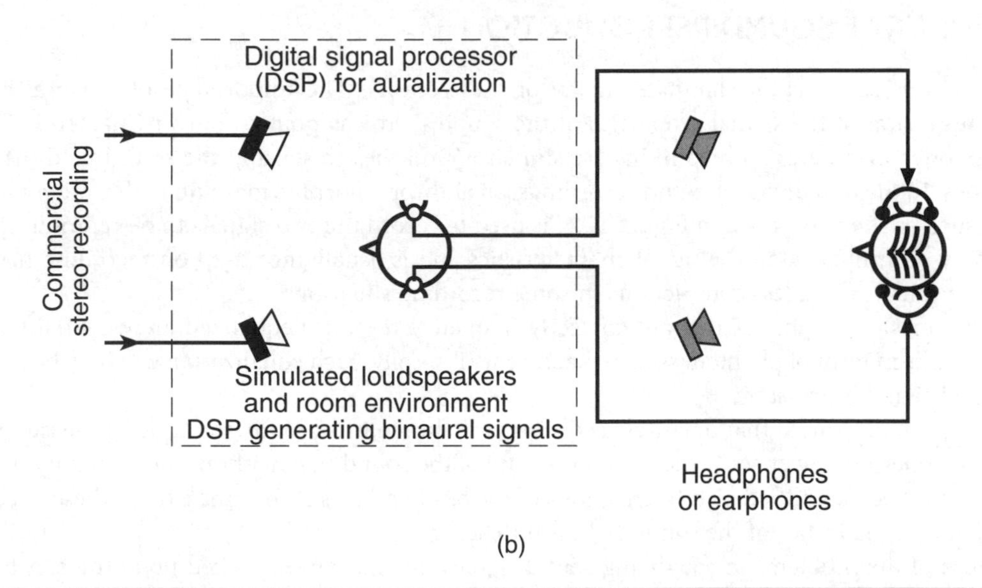

Figure 17.5 By introducing appropriate transfer functions and cross feed of signals one can simulate stereo and multichannel listening using headphones. Analog (a) or digital (b) signal processing techniques may be used. The circuit in (a) was proposed by B. B. Bauer. Using digital techniques as in (b) the room environment for stereo listening may be simulated as well.

with this approach, leading to an unsatisfactory sound, is that people listen to loudspeakers in reverberant rooms, not in anechoic chambers. The audio will not sound natural.

A software solution now available is using real-time filtering to introduce also the room cues into the audio signals as shown in Figure 17.5b. This is an application of auralization, as discussed in Chapter 6. Such systems are in common use in headphone sound reproduction systems used for aircraft passengers.

Since humans never hold their heads completely still, real-life signals will continuously vary in amplitude and time. Because the two methods described do not introduce these minute cues, they usually fail in achieving convincing natural sounding frontal and out-of-head localization in headphone listening.

Loudspeakers generally have inconsistent directional sound radiation characteristics, as discussed in Chapter 14, so clearly angling loudspeakers toward the listener differently will result in different frequency response at the listener. In addition, the listener's sensitivity to sound varies with the angle of incidence of the sound. The head-related transfer functions described in Chapter 3 show that these differences can be considerable. To some extent, the frequency response differences induced by the head-related transfer functions can be compensated by the frequency response of the loudspeakers.

In most real listening situations, there will be early reflections and reverberation that will tend to average out the differences in head-related transfer functions and loudspeaker frequency responses. The precedence effect will determine the localization of the phantom sources, but the room's sound reflections will affect the timbre of the sound.

17.6 BINAURAL SOUND REPRODUCTION

Binaural sound reproduction eliminates many of these effects. The importance of binaural hearing for our impression of the spatial properties of the sound field was pointed out in Chapter 3. The use of microphones in spheres and manikins are similar approaches to sensing the sound field the way a human does. In the best binaural sound recordings, an anthropomorphic manikin with a microphone in each ear, such as the two shown in Figure 17.6, is used to record the two signals to be reproduced. Hair has virtually no influence on the signal characteristics and is usually not used on recording manikins other than to make them less conspicuous in some recording situations.

Headphones or earphones that are correctly frequency response equalized are essential for good listening results in term of phantom source placement. Typically, such equalization can only be achieved using digital signal processing.

It is important to note that binaural recordings are difficult to use for sound level measurements. The microphones no longer sense the sound pressure of the sound field without the manikin but rather sound that has been amplified by the reflections of the head and torso. To some extent, these effects can be removed by equalization if the sound field is diffuse.

Because of the problems in equalizing sound signals that have been picked up at the concha or at the eardrum positions in a manikin, some prefer to use a hard sphere with diametrically positioned pressure sensing microphones at the surface of the sphere. Note that the reflection of the sound field at the surface of the sphere causes frequency response irregularities in any case.

Initially, binaural recording may seem straightforward and problem free. However, we are used to listening to our own ears, head, and torso and the specific time/frequency response patterns that those create in the cochlea. When using someone else's ears, these patterns are no longer the ones we are used to, and the binaural illusion falls apart, typically positioning frontal sound sources to the back or in the worst case causing in-head localization of the phantom sources. The absence of the minute response cues due to head movement discussed earlier contributes seriously to these effects. In experiments where the listener's head movements have been transferred to the manikin's head using a servo system,

(a)

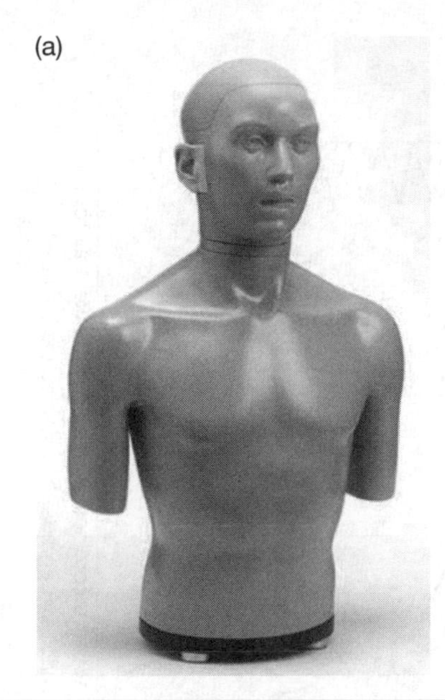

(b)

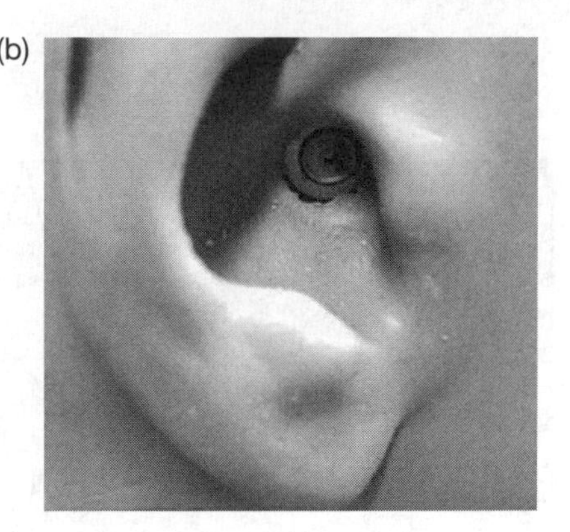

Figure 17.6 (a) The anthropometric head and torso simulator KEMAR manikin for binaural sound recording and measurement is based on worldwide average human male and female head and torso dimensions (Photo courtesy of G.R.A.S., Denmark). (b) Microphone mounted at the simulated ear canal entrance. (Photo by Mendel Kleiner.) (Compare to Fig 5.22)

it has been shown that the presence of these movement cues can largely restore the correct placement of phantom sources in front of the listener.

By using loudspeakers to reproduce the binaural signals, sometimes called *transaural sound reproduction* or *cross-talk compensation* (XTC), one can reduce the problems mentioned in the previous paragraph. Because the loudspeaker signals are affected by the listener's head-related transfer function, they need to be filtered (equalized) to remove this influence. The filtering also needs to include cross-talk cancellation to remove the left and right signals reaching the right and left ears respectively. The principle of cross-talk cancellation for binaural systems is shown in Figure 17.7.

17.7 LOUDSPEAKER-HEADPHONE COMPATIBILITY

Another disadvantage of binaural recording is its lack of compatibility with conventional stereo listening using loudspeakers. It is important for radio, television, and film that recordings sound good when listening using headphones or loudspeakers.

The ORTF recording technique addresses this goal by using a microphone arrangement according to Figure 17.4e. In the ORTF approach, two cardioid microphones are used. These are placed about

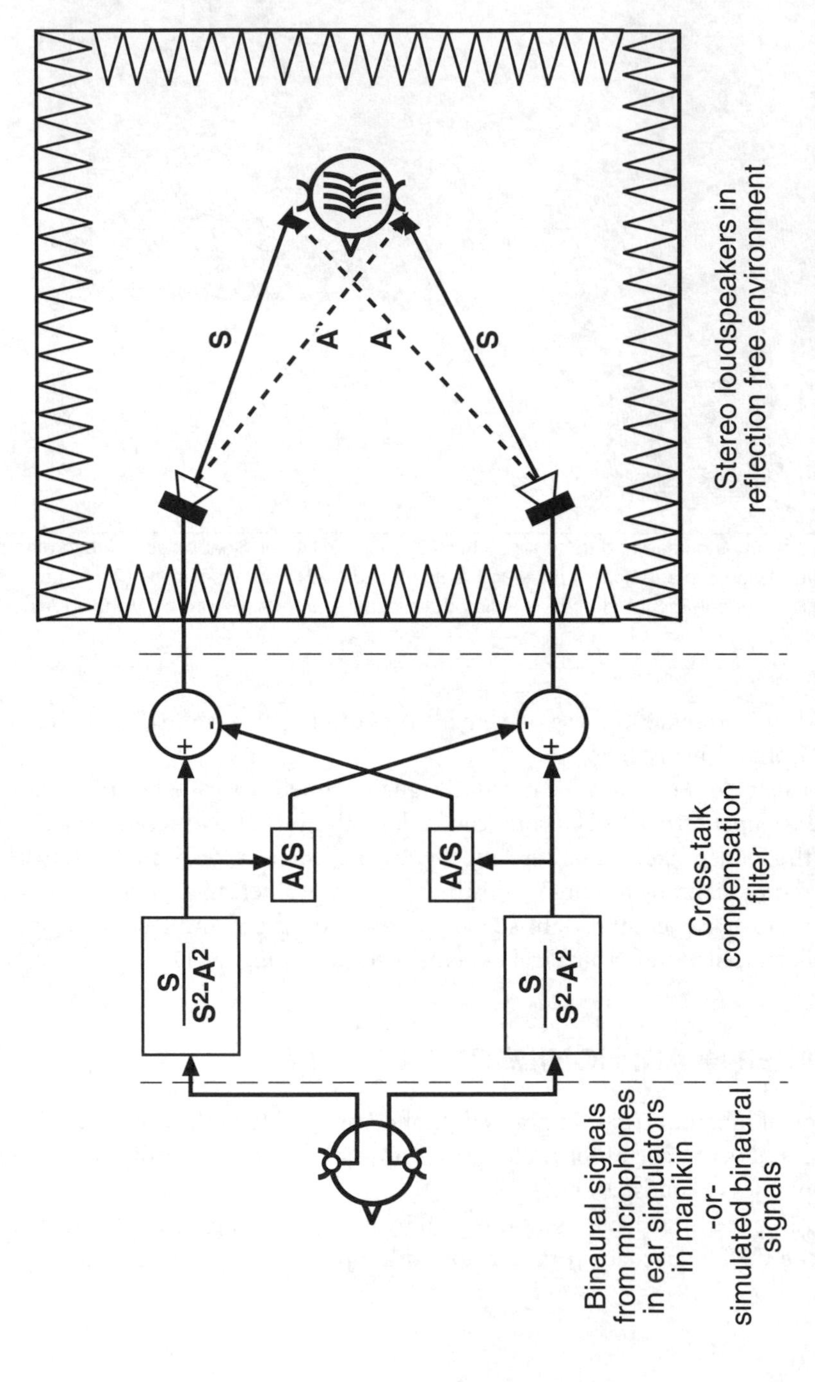

Figure 17.7 The principle of cross-talk compensation used for listening to binaural sound recordings using a loudspeaker pair for stereo. Note that off-center listening and lateral sound reflections from room surfaces seriously degrade the listening experience.

17 cm apart and angled away from one another so that their pickup angle covers the distribution of musical instruments. Figure 5.3 shows that maximum sensitivity of hearing to reflections is for sound that arrives at an angle of 45° to the median plane. If the ORTF pair microphones have their sensitivity lobes at right angles, their sound pickup will approximate this sensitivity characteristic.

The cardioid directivity pattern ensures effective loudspeaker function and the angling and intermicrophone distance ensures good binaural properties. The distance between the microphones results in approximately the same time delays as those that appear at human ears for different angles of incidence relative to the median plane.

17.8 MULTICHANNEL SOUND REPRODUCTION

By introducing more sound reproduction channels, one can obtain a more ideal sound reproduction involving 1) better phantom source localization, 2) phantom sources out of the horizontal plane, and 3) phantom sources that remain in place even when moving around in the sound field. Extra support microphones may be used as well as mentioned earlier.

So-called *5.1 surround sound* recording addresses the need for better phantom source localization. Surround recordings are intended for loudspeakers placed in the horizontal plane around the listener as shown in Figure 17.8 and aim at reproducing both low-frequency sound effects (the .1 channel), better front phantom source localization (the three front channels), and better effects and reverberation (the two rear channels).

Ambisonics is a surround sound recording technique that uses a tetrahedron of cardioid microphones (sound field microphone), such as the one shown in Figure 17.9, to pick up sound in four primary directions all separated by a 120° angle.

The tetrahedron microphone technique addresses the need to achieve phantom sources out of the horizontal plane and scalability. The signals picked up by the cardioid microphones are converted to four signals that represent those that would have been obtained if one had used one omnidirectional and three orthogonal bidirectional (cosine pattern) microphones. These four signals can then again be rematrixed to represent those that would be obtained for any other cardioid microphone direction. In this way, loudspeakers can be placed around the listener, and their position measured and entered into the software matrix so that there is a virtual microphone in each of these directions.

Wave field synthesis (WFS) is an approach to sound-field generation putting Huygen's principle to use and addresses the need for phantom sources that remain in place even when moving around in the sound field. Usually used for virtual sound sources, the signals are processed so that they appear at the loudspeakers with the time and amplitude that would have been obtained for a real sound field at the corresponding position. The distance between the loudspeakers should be smaller than one-half of the highest wavelength to be reproduced to avoid serious lattice effects, but this is currently not practical. Ideally, of course, a surface array of loudspeakers should be used. Since it is fairly easy to fool hearing into believing that there are out-of-plane signals, this is not a large limitation. A typical WFS array uses about 250 loudspeakers surrounding the listener in a plane. Figure 17.10 shows an example of a large experimental WFS loudspeaker array. An experimental cinema installation has shown that the principle can be used successfully for cinema sound reproduction.

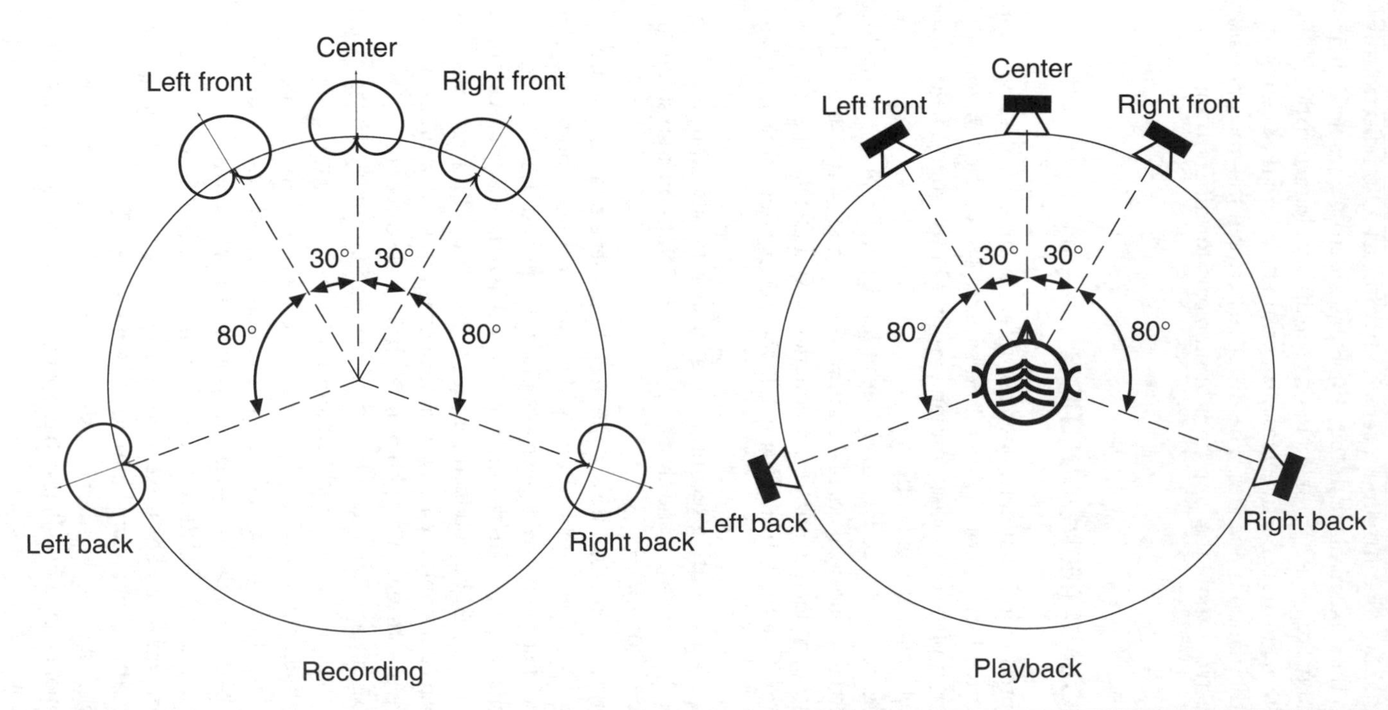

Figure 17.8 The principle of surround sound recording playback. There are many suggested recording setups, the one shown here is similar to the Wide Cardioid Surround Array suggested by DPA Microphones, Denmark. Five cardioid microphones directed radially are used at the various locations on a circle having a radius of 0.60 to 1 m. The array configuration shown here tends to give a wider "sweet spot" in listening than does the coincident array of cardioid microphones. For playback according to the ITU-R BS.775 recommendation five identical full frequency range loudspeakers are used. The positioning of the reference listening point is at the center of an imaginary circle having a radius between 2 m and 4 m (minimum and maximum radii defined by the ITU-R BS.1116–1 recommendation). Loudspeakers should be at a height of the listener's ears. A low-frequency effects channel and loudspeaker is usually added, hence the "5.1" designation.

Figure 17.9 A microphone using four cardioid microphones in a tetrahedron arrangement for Ambisonics recording. (Photo by Mendel Kleiner.)

Figure 17.10 An experimental rectangular array for WFS sound reproduction using 256 loudspeakers (photo by Mendel Kleiner.

17.9 METRICS FOR AUDIO CHARACTERISTICS

Audio and acoustic measurements are used to describe sound, typically with numerical values, expressed in metrics, determined using ad-hoc or predefined methods. Measuring audio quality is similar to the measurement of room acoustic quality in that it is necessary to measure transmission system properties that can truly only be described by listeners. Because of the complexity of audio signals and systems, one usually finds a large variation in the way listeners describe the system properties.

An audio transmission system starts in the recording venue, a concert hall, a stage outdoors, or recording studio, for example. At the recording venue, the venue's background noise, its acoustic properties, the microphones and their placement and directional characteristics, as well as the recording system (analog-to-digital [A/D] converters, sample word length, sampling frequency, codec, etc.), all influence the information that is uploaded to be stored or transmitted to the listener.

The system is reversed at the receiving end of the transmission system. Here the signals are decoded, amplified, reradiated by loudspeakers, and subject to the acoustics of the listening room. The frequency-dependent directivity of the loudspeakers and the sound absorption and scattering properties of the room will make each listening system have its own individual characteristics that are difficult to replicate and that will, in any case, be quite different from those of the recording venue.

One way of regarding the situation is to say that the best that can be expected is that the sound field at the listener is that assumed and strived for by the recording engineer or Tonmeister. A different way is to say that the best that can be expected is that the listener's audio expectations are fulfilled. In either case, one can say that the system has high audio quality. In practice, it is the latter view of what constitutes high audio quality that is the common one.

17.10 PHYSICAL AUDIO METRICS

The audio sound character descriptors can, to some limited extent, be mapped onto a set of audio metrics such as noise level, loudness, amplitude frequency response (usually called just frequency response), group delay, nonlinear distortion, modulation damping, directivity, and interaural cross-correlation, for example. Electronic devices such as amplifiers will require a set of metrics somewhat different from those suitable for characterization of electroacoustic devices such as loudspeakers. Note that the various audio sound character ratings are interdependent, so it is not possible to achieve a 1:1 mapping, of distortion onto roughness, for example. No audio metrics exist to characterize the spatial agreement between recording and playback sound.

Derived metrics, such as dynamic range, are also important in many practical situations. Dynamic range is usually taken to be the level difference between overload and background noise, so it will depend on the criteria used for determination of noise level and nonlinear distortion. The dynamic range will be different in different parts of the frequency spectrum.

Since audio metrics have been of great interest for a long time, many metrics have been introduced along with those discussed. Audio equipment manufacturers typically need to measure many more properties of their products as part of design, manufacture, and quality control. Many standards are available for example those issued by the International Electrical Commission (IEC) and the Audio Engineering Society (AES). The AES also issues *AES Recommended Practice* documents. Note that many metrics need the equipment to be properly calibrated to a reference. The ITU-R BS.1116 recommendation is often used to specify desirable playback conditions. An abbreviated version showing the main points can be found in Table 16.1. Other relevant recommendations are ITU-R BS.1283, 1284, 1285 and 1286 (see Reference 17.21).

17.11 NOISE LEVEL AND SIGNAL-TO-NOISE RATIO

Various metrics for noise were discussed in Chapters 2 and 6. Metrics such as octave and third octave band levels, as well as A-, B-, and C-curve frequency-weighted levels, were introduced in Chapter 2 and noise criteria (NC) usually used in room acoustics, such as NC curves, in Chapter 6. Additionally, there are many other weighting curves recommended by various standardization bodies.

It is important to remember that many types of noise have a recognizable character (for example, *popcorn* noise) and such noise should be rated differently from thermal (Gaussian) noise, (the previously mentioned weighting curves are intended for the evaluation of thermal noise). It is easy to convince oneself of the importance of noise *intelligibility* by listening to a time-inverted music sound file.

17.12 AMPLITUDE FREQUENCY RESPONSE

The transfer function H of a system is the complex ratio of the spectrum of the output signal S_0 divided by the spectrum of the input signal S_i:

$$\underline{H}(f) = \frac{\underline{S}_0(f)}{\underline{S}_i(f)} \tag{17.1}$$

The level of the transfer function L_H can be written:

$$L_H(f) = 20\log\left(\left|\underline{H}(f)\right|\right) \tag{17.2}$$

In audio engineering, the amplitude frequency response is usually taken relative to its level at 1 kHz. Other reference frequencies may be chosen, for example, when describing the frequency response of filters, low- and high-frequency loudspeakers, and other devices that may not have or be intended to have any useful response at 1 kHz. In such cases, the response reference frequency is typically taken as approximately the geometric mean of the low and high pass frequencies f_L and f_U where the transfer function level is –3 dB down relative to the level at the reference frequency f_{ref}. Some of the terms used in characterizing frequency response curves are shown in Figure 17.11.

$$f_{\text{ref}} \approx \sqrt{f_L f_U} \tag{17.3}$$

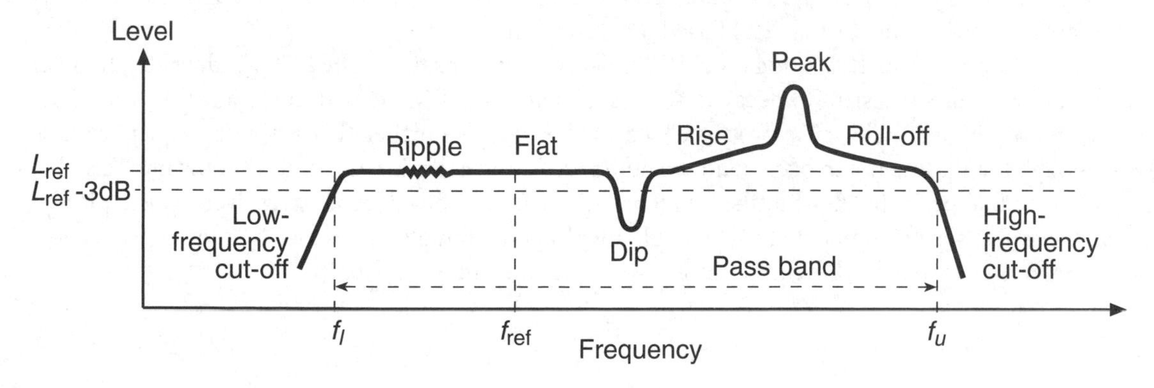

Figure 17.11 Various characteristics of the FR.

The amplitude frequency response function *FR* (usually just called the *frequency response* (FR), or simply the response) can be written:

$$FR(f) = \frac{L_H(f)}{L_H(1000)} \tag{17.4}$$

The *band limits* are usually taken as the frequencies where the response is –3 dB down relative to the level at some reference frequency, often 1 kHz. Small variations in the FR within the passband are usually called *ripple*. In most cases, one will want amplifiers, loudspeakers or microphones to have a FR that is as flat as possible—that have a minimum of ripple in the pass band. One also usually wants the FR to extend far beyond the pass band with slow decay so that there are only small deviations in phase in the transfer function. This is discussed further in the following section.

Strictly, the measurement of the FR should be done one frequency at a time, but a comprehensive measurement using this technique requires considerable time, so two other techniques are usually used. Using analog or digital techniques, the response can be measured using a sweep sine-wave generator set for slow sweep speed and a level recorder. In these measurements, the sweep speed (frequency increment per time unit, Hz/s) usually is set to increase proportionally to the signal frequency. Typically, the time necessary for a sweep measurement depends on the accuracy by which one wants to measure peaks or dips in the response. Assuming the resonant system to have a quality factor Q (or the filter used for the analysis) and that the resonance frequency is estimated as f_0, one can write the maximum sweep speed:

$$S_{\text{max}} < 0.9\left(\frac{f_0}{Q}\right)^2 \quad \text{Hz/s} \tag{17.5}$$

for linear sweeps (constant sweep rate in Hz per second). For logarithmic sweep rates (see Reference 17.1):

$$S_{\text{max}} < 1.3\frac{f_0}{Q^2} \quad \text{Octaves/s} \tag{17.6}$$

It is common to see commercial equipment measured at much too high sweep rates that make the equipment seem to have a *flatter* FR than it really has.

Fast sweep sine-wave signals are usually called *chirp* signals because of their birdsong-like sonic character. Note that such signals are unsuitable for analog measurement, but may be used successfully with digital fast Fourier transform (FFT) based measurement.

For digital frequency analysis using the FFT, similar relationships hold. The FFT window length must be large enough for any transient to decay to some sufficiently small value within the analysis time. The window function should be chosen appropriately for the signal type. Further, the sampling frequency must be high enough for aliasing not to occur, and the number of samples within the time window must also be large enough so that peaks (or dips) in the frequency spectrum can be adequately resolved. This typically requires at least five spectrum lines within one-half power bandwidth of a resonance frequency peak. Dips caused by interference may be difficult to resolve since they are often very sharp.

17.13 PHASE RESPONSE AND GROUP DELAY

Any audio system will be characterized by unavoidable time delays such as the delay of sound due to its travel from instruments to microphones and from loudspeakers to listeners. Characteristic for these time delays (for point sources and receivers) is that the delays are frequency-independent. Only at large distances (such as out-of-doors) will the influence of humidity on the velocity of sound be of interest.

Because of the resonant characteristics of audio transducers that cause frequency roll-off at spectrum extremes and peaks and dips in the FR, there will be additional delays because of the time it takes for energy to be stored and released by the resonant systems, no matter whether these are electrical, mechanical, or acoustical. Many physical filters are characterized by a minimum-phase response. For minimum-phase filter systems, the phase response is linked to the magnitude of $\underline{H}(f)$ via the Hilbert transform.

The rate of phase shift in the transfer function (at some frequency) as a function of frequency can be shown to be coupled to the group delay of the signal. The group delay is:

$$\tau_g = -\frac{1}{360}\frac{d\varphi}{df} \tag{17.7}$$

where $d\varphi$ is given in degrees and df in Hz.

The group delay is a function of frequency and is of considerable interest in the design of loudspeaker systems. The more resonant the system, the larger will be its group delay deviations from the purely geometrically caused delay because of the travel time for sound. The use of loudspeaker system crossover filters results in large group delays that need to be controlled between loudspeakers in a stereo or multichannel audio playback system.

A system having *constant group delay* has *linear phase*; that is, the phase of the transfer function increases proportionally to frequency. This preserves the shape of the wave. Mathematically, it means that all frequency components are time-synchronized exactly as they were in the original signal.

In practice, group delay is best measured using digital techniques—for example, using maximum-length sequence analyzers, such as MLSSA, or by FFT-analyzers using chirp signals. In these systems, the phase response is unwrapped and the phase versus frequency slope calculated to find the group delay.

17.14 NONLINEAR DISTORTIONS

One characteristic of nonlinear signal distortion is that not only the shape of the signal is changed but also that energy is shifted away from the excitation frequency to another frequency. Some types of nonlinear distortion characteristics are shown in Figures 17.12 and 17.13.

Figure 17.12 shows some important distortion generation mechanisms: hard clipping, soft clipping, hysteresis, frequency/amplitude modulation, and slew rate limiting. All of these distortion generating mechanisms may be present both in electronic and electroacoustic audio equipment and are typical for electrodynamic loudspeaker drivers. Some of the mechanisms will be acting the same way on both steady-state and transient signals; the action of others, such as hysteresis, may need to be analyzed separately for the two signal types. The distortion effects of resolution limiting in digital systems was discussed in Chapter 16.

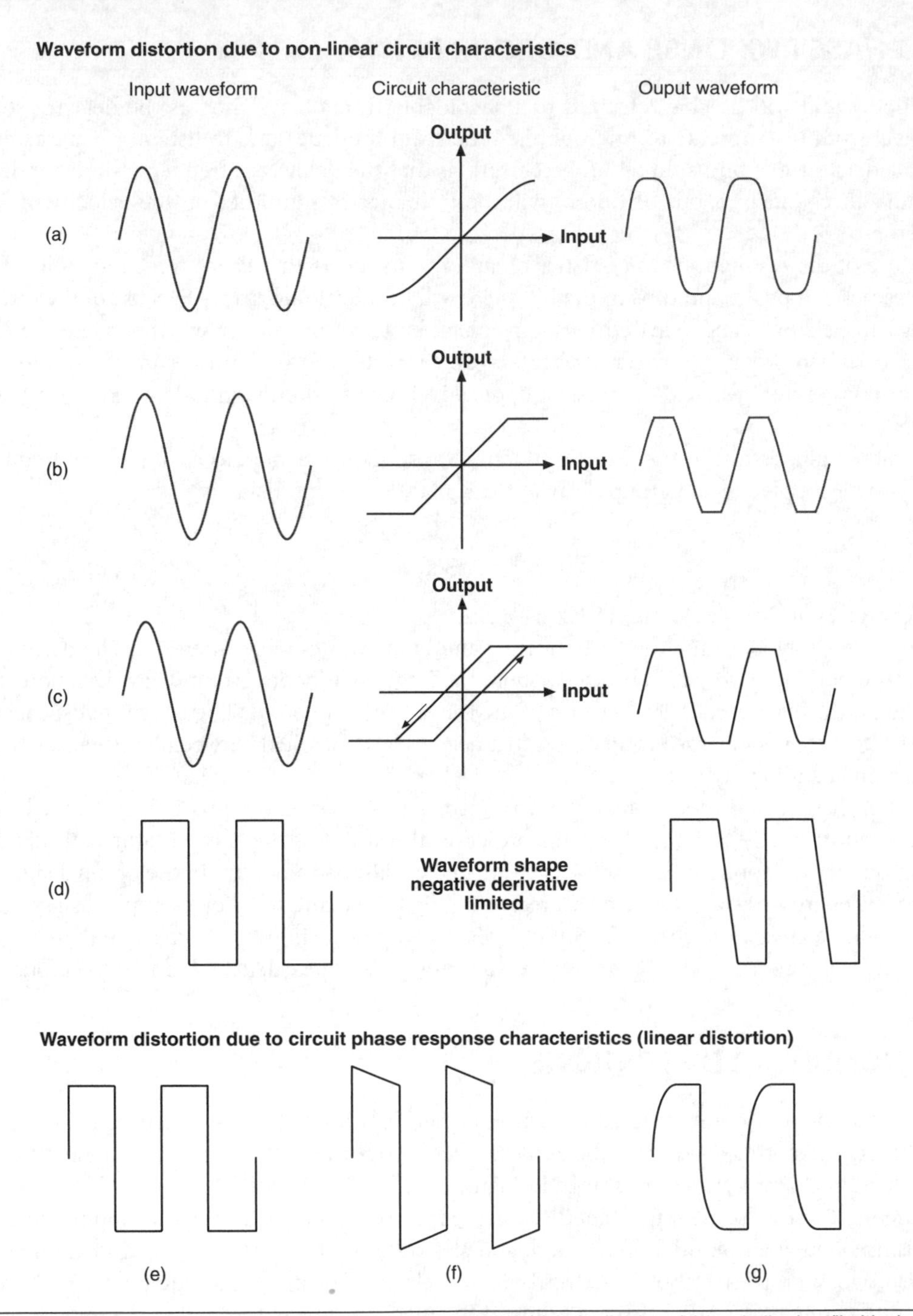

Figure 17.12 Waveform distortion characteristics for sine waves by: (a) soft clipping, (b) hard clipping (limiting), and (c) hysteresis. For square waves due to (d) slew rate limiting. Original square wave shown left (e). Square wave distortion shown for (f) phase distortion at low frequencies, and (g) *gentle* low-pass filter at high frequencies. Many other waveform distortions exist.

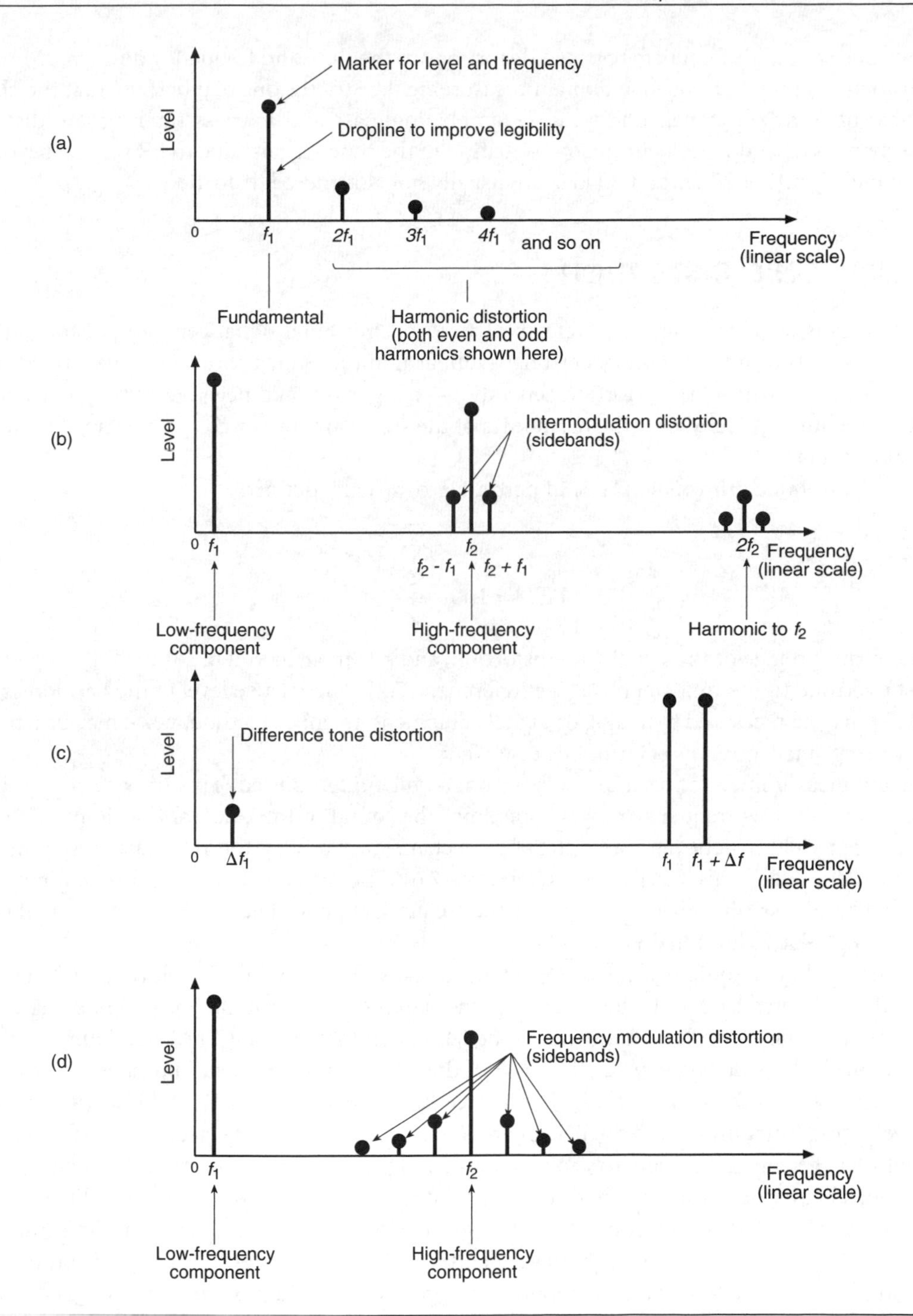

Figure 17.13 Examples of new spectral components generated by nonlinear distortion: (a) harmonic distortion (both odd and even harmonics shown in this example), (b) IMD, and (c) difference tone distortion, and (d) frequency modulation distortion.

Ideally, one should differentiate between distortions in the time and frequency domains. Most metrics for distortion apply to frequency domain measurements. Strictly, one could argue that the changes in waveform of transient signals due to, for example, highpass and lowpass filtering, are distortion. Moderate changes (that do not incur major oscillation in the time signals) due to FR roll-off beyond the audio frequency limits at 20 Hz and 20 kHz are usually not considered distortion.

17.15 HARMONIC DISTORTION

A system that causes a change in the waveform of an ideal sinusoidal signal generates distortion in the form of overtones (harmonics) and undertones (subharmonics). Subharmonics are generated by systems that feature nonsymmetric distortion generation—the positive and negative parts of the signal are distorted differently. Typically, only the harmonics of the signal are measured as a metric of nonlinearities in audio systems.

The *total harmonic distortion* (THD) in percentage is usually defined:

$$THD = 100\frac{\sqrt{\sum_{2}^{n} \tilde{e}_n^2}}{\tilde{e}} \tag{17.8}$$

Here $\tilde{e}$ is the rms voltage of the signal with distortion, and $\tilde{e}_n$ is the rms voltage of the n:th harmonic ($\tilde{e}_2$ is the first overtone to the fundamental) (see Reference 17.2). The relative level of the harmonics to the fundamental may be measured by analog or digital equipment. In either method, noise may be a problem when measuring small amounts of harmonic distortion.

In analog measurement of harmonic distortion, a *notch* filter is used. This filter rejects frequency components in a narrow frequency range. Sometimes the sound of the residual (harmonic distortion) components can yield insight into the distortion mechanism. Studying the residuals versus the signal on an X-Y oscilloscope is also useful (see Reference 17.6). The measurement can also be done using a narrow band bandpass filter that is swept over the frequency range. The signal spectrum is then best plotted showing relative level in dB.

When using digital equipment, it is important to use sufficiently high resolution A/D converters as discussed in Chapter 16. For the best results in measurement of harmonic distortion using FFT, the sampling frequency and window length should be chosen so that spectral line broadening due to frequency domain *leakage* does not occur. This is best done by having an integer number of cycles of the sine wave signal within the FFT analysis window (see References 17.4 and 17.5). Otherwise, a modern window weighting function such as the Blackman-Harris window should be carefully used. It must also be stressed that the sampling frequency must be much higher than any important harmonics, at least twice as high. The reason is that high harmonics may otherwise be mirrored into the audible frequency range due to aliasing. The test equipment must have suitable anti-aliasing filters. Built in sound cards, particularly in laptop computers, seldom have such filters and will also pick up noise from the digital circuits and power supply feeds of the computer. A high resolution, high sampling frequency outboard sampling unit is recommended.

Small amounts of harmonic distortion may be obscured by noise in the system. It may then be necessary to analyze the audibility of the distortion frequency components concerning their relative power in the critical bands where they occur.

It is difficult typically to see the presence of small amounts of low order harmonic distortion, less than 5%, in the waveform of a sine-wave signal. Due to the influence of frequency, domain masking in hearing the harmonic distortion may or may not be a suitable metric for nonlinearities. The audibility of the generation of signal harmonics will depend on the signal complexity. If the signal has multiple frequencies, intermodulation will occur, creating new nonharmonically related frequencies as discussed.

An important effect in the audibility of harmonic distortion is that the overtones of most musical instruments and voice are not exactly harmonically related to their fundamentals. When major harmonics are created by the nonlinearities of the audio system (so that they are about the same amplitude as those of instruments and voice), hearing will perceive the interaction between the two as the result of modulation. For modulation frequencies below 150 Hz, such modulation is called *roughness* (see Reference 17.3).

The audibility of harmonic distortion may depend on the frequency filter characteristic of the audio system component studied. Harmonic distortion generated by low- and mid-frequency range loudspeakers is likely to be more audible than that generated by high-frequency loudspeakers.

Here, only steady-state or quasi-steady-state measurements were considered. Recent advances in signal processing have made it possible to use sweep signals to considerably reduce the time needed for full frequency range measurement of harmonic distortion (see Reference 17.7).

17.16 DIFFERENCE FREQUENCY AND INTERMODULATION DISTORTION

Any nonlinear system will cause intermodulation between two or more frequency components. Difference tones will be created along with harmonics as shown in Figure 17.13. Two types of test procedures are common, the SMPTE and CCIF methods, named after the respective standardization bodies. In the SMPTE method, *intermodulation distortion* (IMD) is measured using a system input consisting of a mix of a low (70 Hz) and a high-frequency (6 kHz) signal where the voltage of the 70 Hz signal is 4 times that of the 6 kHz signal (Figure 17.13b). In the CCIF method, *difference frequency distortion* (DFD) is measured in which a mix of two high-frequency signals 10 and 11 kHz of equal voltage are supplied to the system under test (Figure 17.13c). These signals are used to test wide range and high-frequency systems respectively. Since the purpose of distortion should be testing the system in a way that reflects its behavior with *real* audio signals, the two frequencies should be chosen with this in mind.

The IMD and DFD distortions in percentage are typically defined in analogy to harmonic distortion as, for example:

$$IMD = 100\frac{\sqrt{\sum_{1}^{n} \tilde{e}_{IM,n}^{2}}}{\tilde{e}} \tag{17.9}$$

Here $\tilde{e}$ is the rms voltage of the modulated signal and $\tilde{e}_{dist}$ are the rms voltages of the respective sideband or DFD signals (see Reference 17.2).

Results of IMD measurement are sometimes claimed to be more accurate than harmonic distortion in corresponding to the subjective perception of nonlinearities in audio sound reproduction. For wide band electronic systems that have fairly similar distortion characteristics over the audio range, there will be little difference between the two in characterization of the leveled system's distortion properties. For frequency limited systems, such as loudspeakers, properly chosen mixed frequencies that generate IMD with suitable frequencies may be more relevant. Note that there will also be harmonic distortion generated by the mixed frequencies, so it is important to define which distortion frequency components are measured.

Even more than in the measurement of harmonic distortion, it is important in IMD measurement to choose analysis window length, sampling frequency, and number of samples within the analysis window wisely. The IMD and DFD test frequencies should be chosen appropriately to the systems bandwidth and other relevant characteristics. Since the measurement is not particularly sensitive to the exact frequencies used, it is clearly advisable to chose IMD and DFD test frequencies. These frequencies and the distortion frequency components will then correspond to the FFT analysis frequencies so that frequency domain leakage is minimized, unless measurement standards require differently.

Another factor that one might want to take into account is the loudness of the distortion frequency components. These might be masked and an evaluation procedure similar to that used in the determination of Zwicker's Loudness could then be appropriate see Chapter 3 and (Reference 17.3).

17.17 MULTITONE DISTORTION MEASUREMENT

As briefly mentioned in the previous section, it might be advantageous to choose test frequencies that are similar in behavior to *real* audio signals. Real audio signals are characterized by their statistical properties, such as positive and negative means, peak values, peak to rms values, and their modulation properties. The signal generated for Speech Transmission Index testing (see Chapter 3) is an example of a signal generated to simulate speech in audio testing. Similar signals have been created for music simulation, *Music Transmission Index.*

In practice, however, it might be wiser to chose signals having line spectra rather than noise signals since the frequency components can then be adjusted concerning phase as well so that a suitable peak to rms ratio can be obtained. A suitable input signal, without frequency domain leakage, can be obtained by using 5 low- and 5 high-frequency components to simulate popular music audio (see Reference 17.4 and general references for this chapter).

Since the signal consists of many different tones, the resulting distortion components will be spread over many frequencies, generating a distortion *noise*. Because leakage was eliminated by the choice of signal and analysis scheme, it is then straightforward to measure this distortion noise by simple elimination of the input signal frequency components. One can then define a frequency-dependent distortion density metric that can further be analyzed by taking the frequency domain masking of hearing into account.

17.18 FM DISTORTION

Because the signals radiated by a loudspeaker are from superimposed vibrations on the loudspeaker's diaphragm there will be intermodulation. This type of distortion is usually called frequency modulation (FM) distortion. The sound power radiated by a loudspeaker is proportional to the volume velocity generated by the diaphragm's motion. The volume velocity is proportional to the diaphragm excursion times the angular frequency. A small loudspeaker will thus need larger excursions at low frequencies to radiate the same power as a loudspeaker having a large diaphragm. This means that the smaller the loudspeaker the more FM-distortion will be generated. The FM distortion of the mix between a low- and a high-frequency tone will have similar characteristics to IM distortion but with many sidebands around the higher frequency as shown in Figure 17.13d.

17.19 MEASUREMENTS USING SPECIAL CHAMBERS

Two specialized types of rooms are used in audio testing, *anechoic chambers* and *reverberation chambers.* A special type of measurement room intended to simulate a living room environment is the *IEC 268-13 listening room* of which an example is shown in Figure 6.17 (see Reference 17.8).

Reverberation chambers are typically used to determine the power output (and sensitivity) of loudspeakers. The measurement is usually done using third octave band analysis with a random noise signal using equal energy per percentage bandwidth (that is having constant power in each third octave frequency band). Such a signal is often called a pink noise signal to differentiate it from a signal that has thermal noise characteristics. Since many loudspeakers such as studio monitoring loudspeakers are designed to have a flat, frequency-independent, on-axis FR the power response of the loudspeaker will typically drop with increasing frequency since the directivity of the loudspeaker typically increases with frequency. Figure 17.14 shows a reverberation chamber. It is important to

Figure 17.14 A typical reverberation chamber with diffusors. (Photo by Mendel Kleiner.)

note that reverberation chambers need to be large to support enough room modes for the sound field to be diffuse at low frequencies.

When measuring the power response in the reverberation chamber, it is important to position the loudspeaker relative to the chamber corner surfaces so that the placement corresponds to the recommended usage (see Chapter 4). Some loudspeakers are designed to be listened to in a situation where the direct sound of the loudspeaker is small compared to the reverberant sound; for such loudspeakers the reverberation chamber is ideal for measurement.

The rationale for using anechoic measurement chambers is to provide a sound reflection free measurement environment similar to out-of-doors conditions. Because of practicality, noise, weather, confidentiality, and such, it is more practical to use an anechoic chamber than to measure outdoors. Figure 17.15 shows such a room. For low frequency measurement an outdoor facility will be better since the lower frequency limit of anechoic chambers is determined by the sound absorption of the surface material and room size. Measurement of high directivity devices needs large rooms in order to be in the far-field of the device.

It is important to remember that an anechoic chamber still is a room with sound-reflecting walls and objects that create a diffuse sound field like that in a reverberation chamber. The relative level difference between direct sound and reverberant sound is much higher and the reverberation radius considerably larger.

The influence of the reverberant field in anechoic chambers is usually investigated by measuring the systematic and random deviation from the geometric attenuation, –6 dB per distance doubling, for a point source. In practice, it is difficult to build anechoic chambers having less than 1 dB standard deviation at 1 m distance.

Figure 17.15 An anechoic chamber having sound-absorptive wedges on the walls in use for HRTF measurements. The floor is made of sound transparent netting under which there are sound-absorptive wedges as on the walls. (Photo by Mendel Kleiner.)

Figure 17.16 shows the 67% confidence interval for pure tones levels in a room (see Reference 17.9). In addition to the distance between sound source and receiver, the confidence intervals are determined by the directivity of the sound source and microphone as well as by the reflection coefficient of the walls. Figure 17.17 shows the measured deviation from the distance law at 1 m distance for a small point source in the anechoic chamber shown in Figure 5.1.

The inside surfaces of anechoic chambers typically are covered by porous, wedge-shaped sound-absorptive wall elements. These quickly lose their sound-absorptive properties when their length becomes shorter than one quarter of the wavelength of sound. The modal behavior of the room is easily noticeable below the chamber's lower limiting frequency.

To obtain negligible reverberation and low-frequency modes at the low-frequency end of the audio range, the chamber and its wedges must be very large as indicated previously. Since building

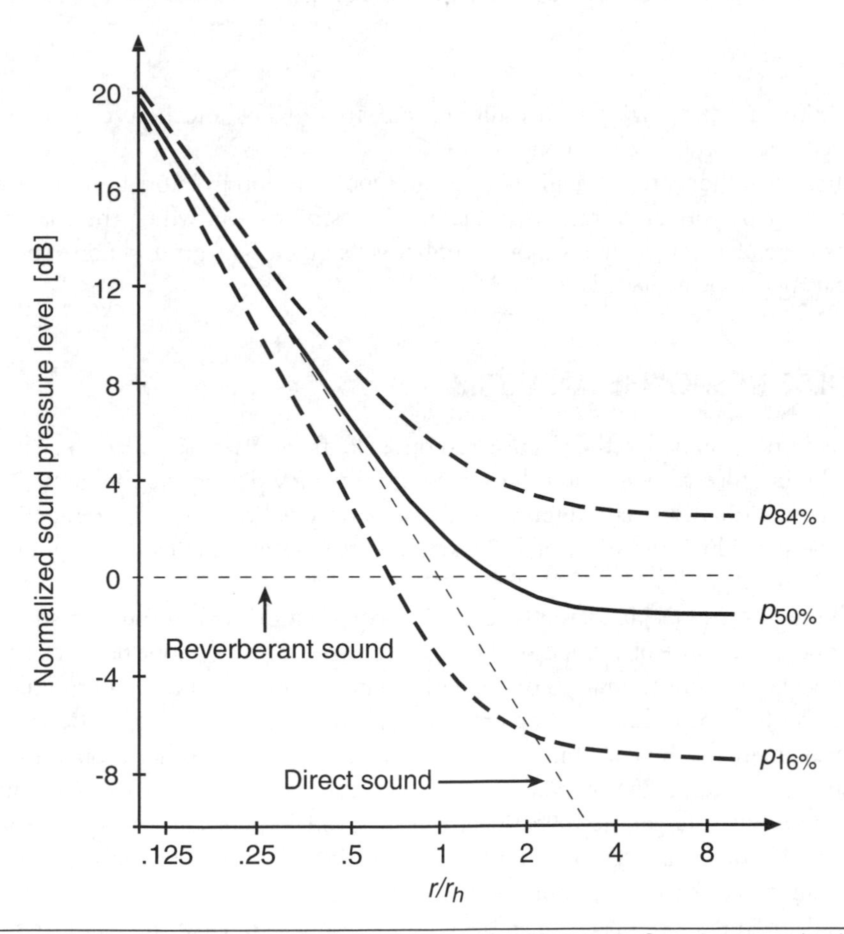

Figure 17.16 Expected spread in measurement of sound pressure level in a room. Sound pressure level distribution: 16, 50, and 84 percentiles shown. (After Ref. 17.9)

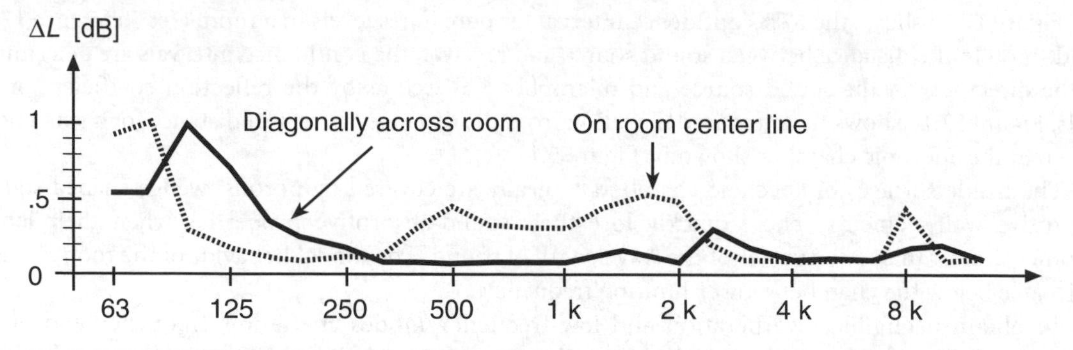

Figure 17.17 Measured deviation ΔL from the geometrical distance attenuation from a small sound source at 1 m distance in the anechoic chamber shown in Figure 5.1. Measured without the loudspeaker array shown in that figure present.

costs are approximately proportional to built volume, this makes anechoic chambers for low-frequency measurement prohibitively expensive.

Most anechoic chambers are suspended on spring-type vibration isolators to minimize sound leaking into the chamber by way of vibrations in the building structure in which the chamber is housed. One major advantage of a well-built anechoic chamber is its quiet. Background noise spectra below the threshold of hearing can be achieved.

17.20 IMPULSE RESPONSE ANALYSIS

The transfer function is coupled to the impulse response via the Fourier transform. The length (number of samples) of the impulse response and the sampling frequency determine the number of frequency points in the transfer function. The impulse response is only valid as a measurement for linear systems but is typically valid for the study of audio systems that are well below the amplitude limit set by clipping.

Analysis of the impulse response was discussed in conjunction with room acoustics in Chapter 5. Typically, the impulse response of a modern laboratory condenser microphone has such good properties that it does not play a role in the analysis of the performance of loudspeakers and rooms.

One way of avoiding the influence of wall reflections and subsequent reverberation on the measurement of loudspeakers in anechoic chambers (and to some extent also in other rooms) is to use impulse response measurement and apply time-gating to the measurement. Figure 17.18 shows the principle of time gating. If the room is large, the reflection-free time available between the direct sound from the source and the arrival of the scattered sound from the walls will be long. Time-gating corresponds to having a perfect anechoic chamber within the time slot under analysis.

Since the high-frequency components in the impulse response of electrodynamic loudspeakers and microphones generally have a length of interest of a few milliseconds, it is often advantageous to use time gating techniques and to remove the part of the impulse response that occurs after 10 ms after the

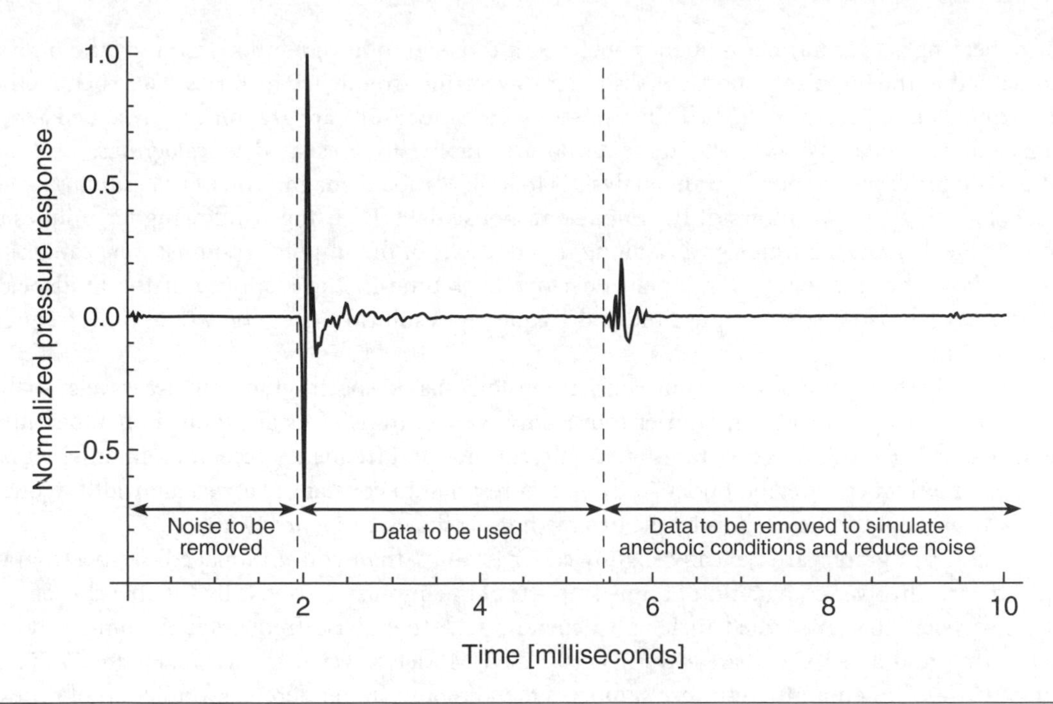

Figure 17.18 Application of time-gating to remove undesired reflection and noise from the measurement of the impulse response of a loudspeaker.

arrival of direct sound from the analysis. This corresponds to having the walls of the anechoic chamber more than 2 m away. This creates a virtually reflection-free acoustic environment. Some audio measurement software has built-in analysis routines for frequency-dependent time-gating.

The time-gating approach works quite well at medium and high-audio frequencies but gives problems in measuring the low-frequency properties of the loudspeaker unless one uses *zero padding* of the impulse response tail, extending its length. This approach causes errors in the measurement of the low FR, so it still makes sense to use a large chamber. One must remember that zero padding effectively assumes that the response of the loudspeaker is negligible after the zero gate time, so resonant systems characterized by strong long time extended oscillations will be incorrectly measured. Studying the sound pressure level envelope decay characteristics of the impulse response may give valuable guidance to the appropriateness of time gating.

17.21 FREQUENCY RESPONSE, SPECTROGRAM, AND WAVELET ANALYSIS

Once the impulse response has been trimmed by time-gating it can be analyzed to find interesting FR features of the loudspeaker or other device under test. Fourier transformation of the impulse response is trivial. Typically, the FR of the product such as a loudspeaker is one item of interest and is obtained using Equation 17.4.

Since hearing acts as a time/frequency analyzer, it is useful to have methods to study the information contained in the impulse response in a similar way as time-frequency histories. Two such methods are *short-time Fourier transform* (STFT) analysis for which the results are graphically presented as spectrograms and wavelet analysis for which the results are graphically presented as *scalograms*.

The basic principle of spectrogram analysis is to look at the FR magnitude in *time snippets*—gliding local analyzing time windows of the impulse response data. By using windowing techniques and moving the local analyzing time window along the time axis of the impulse response, one can graphically show how the frequency component develops over time in the response of the loudspeaker. Spectrogram analysis gives the same time and frequency resolution over the whole time/frequency region under study.

An alternative to the use of short-time Fourier transform-based spectrogram is to use wavelet analysis. While linear FR analysis is used in Fourier transforms, wavelet transforms are studied on a logarithmic frequency basis. The resulting scalograms offer different time and frequency resolution in different parts of the joint time/frequency region under study and correspond to constant relative bandwidth frequency analysis. The principle of the analysis by the two methods is shown in Figure 17.19.

The scalogram graphs are typically slightly easier to study than conventional STFT spectrograms. Wavelet analysis involves convolution of time snippets of the impulse response by suitably chosen short impulses, wavelets that are scaled in length depending on their base frequency. A commonly used wavelet is the modulated Gaussian wavelet, also called Morlet wavelet, shown in Figure 17.20. The energy in the resulting impulse response snippet is then graphically rendered as a function of time and frequency.

The advantage of wavelet analysis is clear when analyzing nonstationary signals, since it allows better detection and localization of transients than spectrograms. The compromise between good time or frequency resolution in the short-time Fourier transform analysis is eliminated in wavelet analysis because of the automatic rescaling of the resolution during the analysis of different frequency bands.

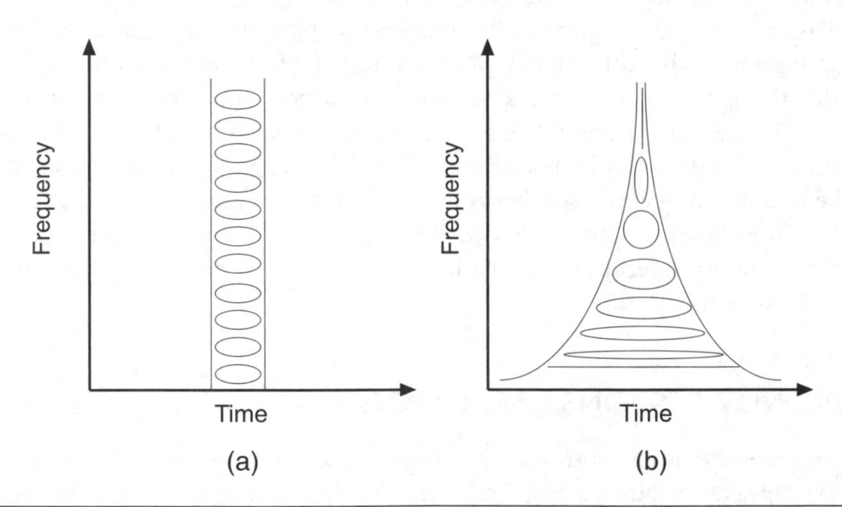

Figure 17.19 Time-frequency relationships in different octaves for: (a) the STFT-based spectrogram and (b) the wavelet transform-based scalogram). The spectrogram consists of a number of Fourier transforms of a signal multiplied by a *local analyzing window*, gradually shifted over the signal.

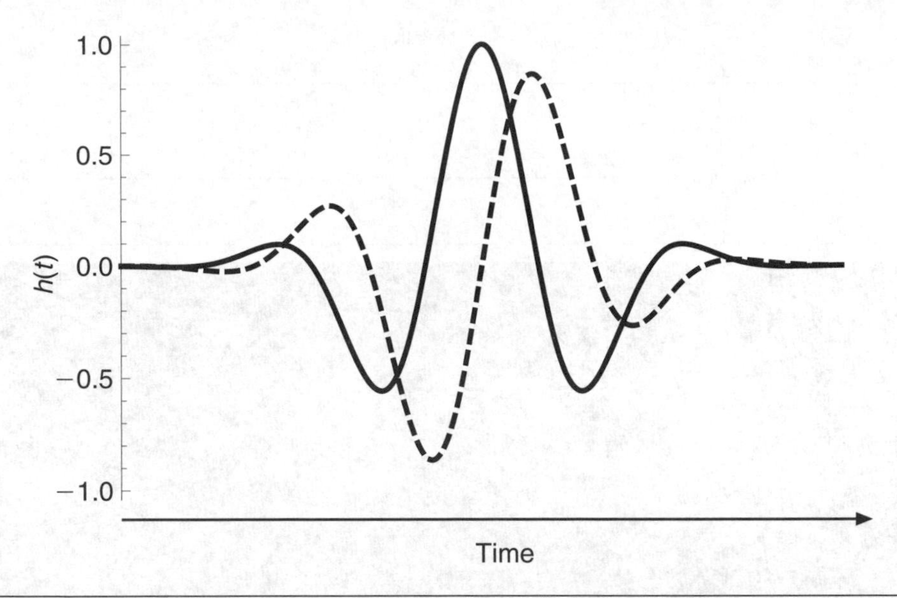

Figure 17.20 An example of the real (solid) and imaginary (dashed) parts of a Gaussian-type wavelet having the impulse response $h(t) = e^{-j\omega t/a}e^{-t^2a^2/2}$. The parameter *a* is called the scale factor and determines the time and frequency resolutions.

Since the critical bands of hearing are approximately one-third octave band wide (over 500 Hz) it is common to use at least 3 wavelets per octave (sometimes this is referred to as 3 voices per octave in wavelet literature). The use of ten wavelets (voices) per octave often gives a graphically pleasing representation of the signal in wavelet analysis as shown in Figure 17.21.

17.22 DIRECTIVITY

The directivity of loudspeakers and microphones are important properties for their correct use and was discussed in Chapters 12 and 14. The measurement of directivity for out of plane angles is quite difficult, because a *directivity balloon* will need a few hundred measurement points. Directivity measurements in the plane are comparatively easy to perform using a computer operated turntable in an anechoic chamber. Directivity measurement results are shown in either polar or linear formats. The latter has some advantages but is less intuitive. Directivity measurements are essentially the same as frequency or impulse response measurements except that the loudspeaker or microphone is turned or moved to a set of angles relative to the measurement device.

With highly directional transducers, the response will be weak at angles away from the main lobe. The reverberation of the room will then cause considerable random variation in the response unless time gating is used. In some types of directivity measurement, narrow band noise (such as 1/3, 1/6, or even 1/12 octave noise) or multifrequency signals (such as a frequency modulated sinewave) are used instead of pure sinewaves to achieve a frequency average. Note that the noise signals should have pink noise characteristics.

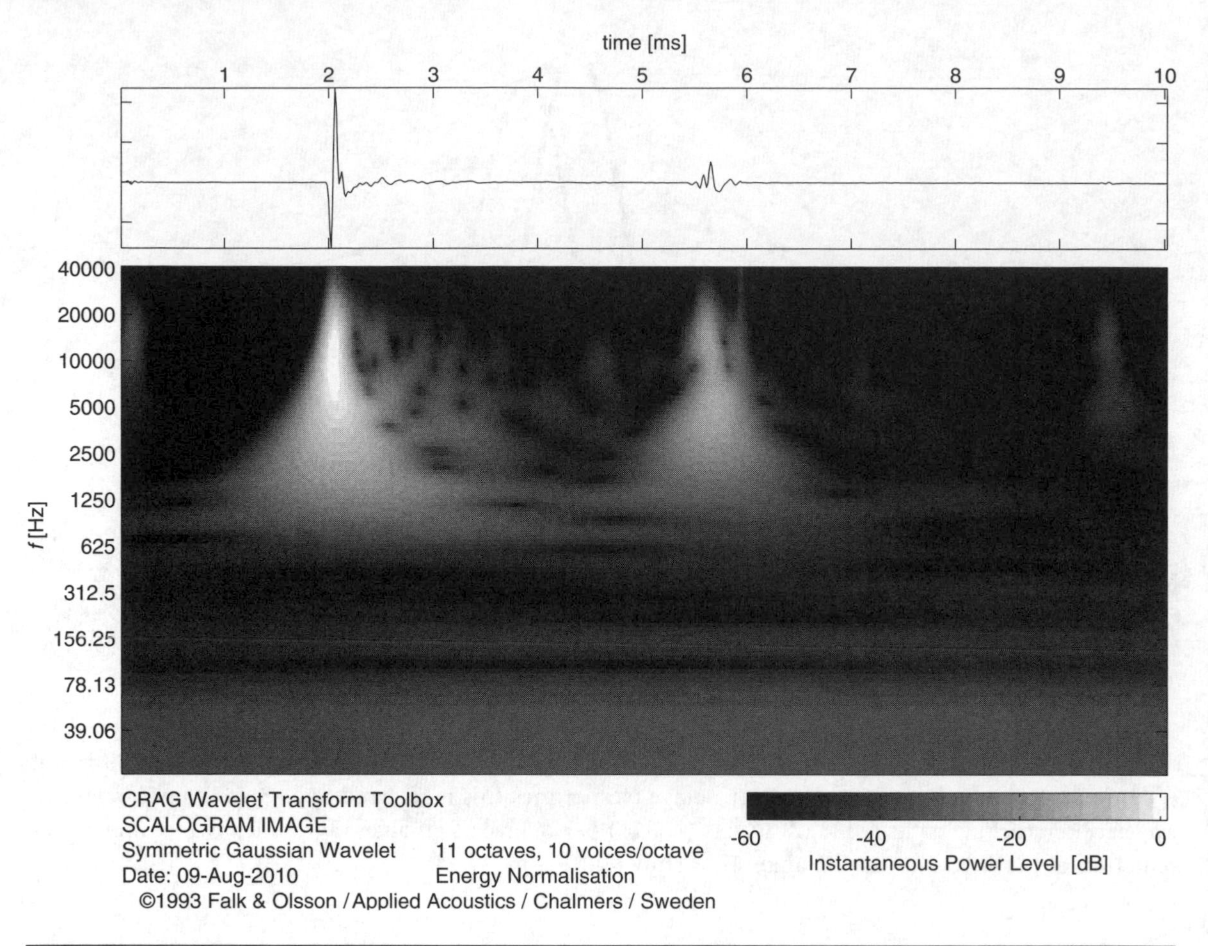

Figure 17.21 An example of the application of a wavelet transform-based scalogram to the loudspeaker impulse response shown in Figure 17.20.

A different way of measuring the directivity is to use a holographic approach, measuring the sound field at many points on a plane close to the diaphragm and then mathematically propagate the sound field to that at a large distance away from the loudspeaker.

17.23 SENSITIVITY

The sensitivity of loudspeakers and microphones (and gain of amplifiers) is comparatively easy to measure. Proper calibration of the measurement microphone is important. For loudspeakers, sensitivity is usually specified in sound pressure level for a certain input voltage, usually 2.83 volts (corresponding to 1 watt power input into a resistance of 8 ohms). For microphones the sensitivity is measured as the rms output voltage for the microphone placed in a plane wave sound field where the sound pressure level is 94 dB, that is, 1 Pa.

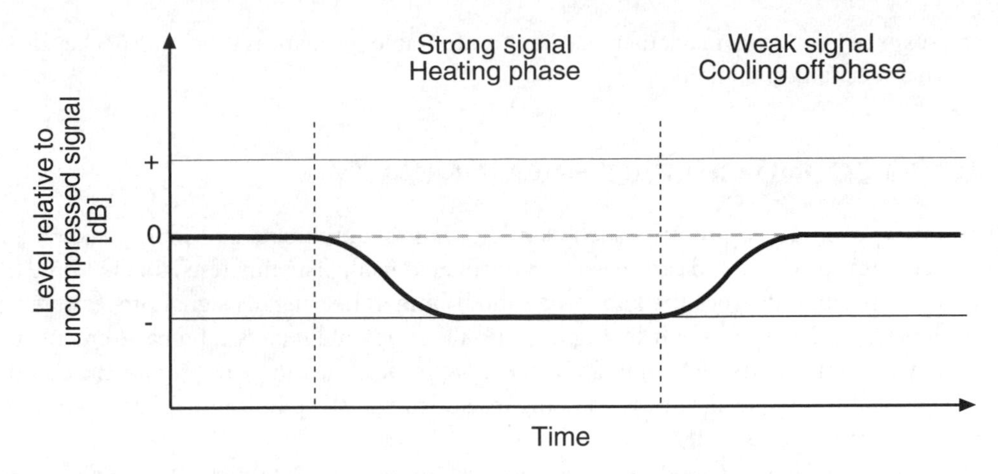

Figure 17.22 An example of the influence of compression effects on a loudspeaker's output.

17.24 COMPRESSION

A real audio signal such as voice and many types of music will be characterized by weak and strong parts—low-frequency modulation with modulation frequencies below 0.01 Hz. Since the voice coil of electrodynamic loudspeakers has limited thermal capacity, the voice coil will heat and cool with the signal envelope—the audio signal's low-frequency modulation. Since the electrical resistance of the voice coil is related to its temperature, the sensitivity of the loudspeaker will change causing a compression effect, resulting in a change in the *temporal envelope* of the radiated sound. This compression effect is important and is best measured using suitable gated/modulated tones or tone combinations or even music signals. An example of the influence of compression effects on the loudspeaker's output power is shown in Figure 17.22.

17.25 IMPEDANCE

The electrical input and output impedances of microphones, amplifiers, and loudspeakers are important properties. Typically, these are measured using small-signal methods, comparing the impedance of the measurement object to a reference using some form of electric series or bridge circuit. For loudspeakers, the FR of the electrical input impedance forms an important tool in tuning the loudspeaker enclosure to the loudspeaker driver. An example of the electrical input impedance characteristics of an electrodynamic loudspeaker was shown in Figure 14.11.

The traditional method of measuring loudspeaker impedance using a high value resistor in series with the loudspeaker gives erroneous results and should be replaced by a measurement using a low value series resistor in series instead. By measuring the voltage over this resistor, one can measure the current through the loudspeaker as it is driven by the amplifier output. This will let the loudspeaker operate as it is used in practice and give correct impedance values (although one is really measuring admittance that has to be inverted to yield impedance). This is particularly important if one is measuring a multidriver loudspeaker having a crossover network that may otherwise have undesirable resonances or a loudspeaker driver that has little internal mechanical damping.

17.26 AUDIO SOUND CHARACTER AND QUALITY

Many different types of audio products such as microphones, amplifiers, and loudspeakers, and also related products such as rooms and cars, need to be evaluated using listening tests, that is, using humans as measurement instruments since the knowledge about human hearing, its signal processing, and signal evaluation is weak. The results of listening tests usually supplement physical measurements of audio quality and may possibly be used for finding the relevant physical measure (metric) for the audio sound character by correlating the results of the listening tests with various physical metrics or combinations of metrics, as illustrated in Figure 17.23.

In audio product evaluation, we differentiate between *audio sound character* and *audio sound quality* (see References 17.10 and 17.11). Character refers to the sensory properties of the product's sound whereas quality refers to the adequacy of its sound to satisfy the listener's expectations. Figure 17.24 illustrates the differences between audio character and quality.

The term sound quality is often used in audio to describe what really should be called audio sound character. Audio sound character is the set of psychoacoustic dimensions of the sound at the listener such as the noisiness, loudness, FR, timbre, roughness, dynamic compression, and spatiality (for example auditory source width and envelopment) of the sound at a particular setting of the product. The terms are used here similarly to their use in product sound engineering and in room acoustic design. The sound character can be measured using listeners that are asked to rate the sound on semantic differential scales, magnitude scales, or by pair comparisons. It is obvious that the use of words introduces considerable possibilities for bias in such measurements. It is important to instruct and describe to the listeners how they should listen without bias (see References 17.12–17.14).

Perceived audio sound quality is a collection of features that confer the product's ability to satisfy stated or implied needs. Perceived audio sound quality determination is the evaluation of preference or goodness and refers to the adequacy of the product's sonic characteristics in evaluating the total auditory

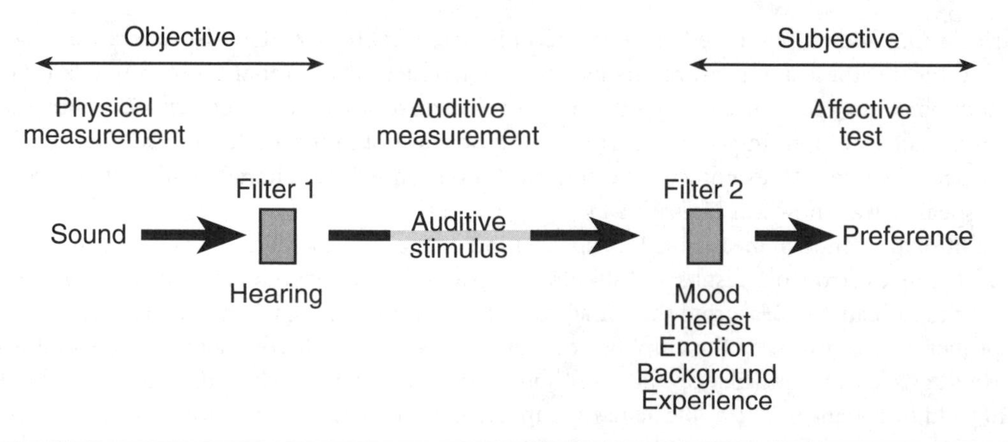

Figure 17.23 A simplified illustration of human sound perception and judgment.

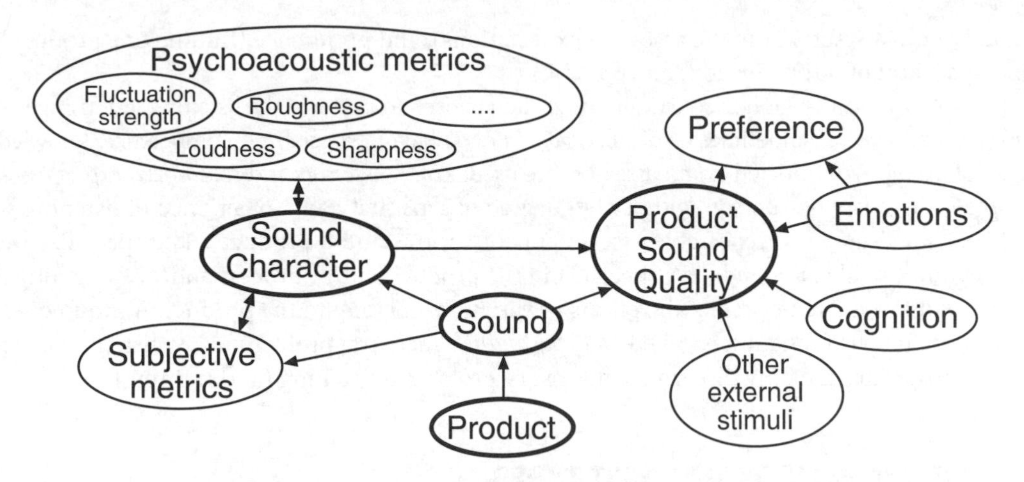

Figure 17.24 The difference between the audio sound character and sound quality concepts (inspired by Ref. 17.10).

characteristics of the sound, with reference to the set of desirable product features that are apparent in the user's cognitive and emotional situation. Any such evaluation is multidimensional and may be extremely dependent on distracters such as individual *taste* preferences, *cognitive factors* such as expectations (for example due to pretest information), *haptic* and *cognitive factors*, as well as *emotional* factors such as previous brand-experience and the audio program (sad or joyous music, for example).

Clearly, it is neither possible nor desirable to include these *distracters* into the set of audio metrics, but they need to be *neutralized* so that listeners/customers are able to account for their own biases and make informed decisions, for example, on recommendations for purchase. In other words, audio sound quality is not an inherent quality of the product but is a result of the perceptive and mental processes when a listener is exposed to the audio rendered by a product. Audio sound quality is not an absolute property but applies to a certain product and its customers. Audio sound quality tests must take acoustics, psychoacoustics, and psychology into account. By using proper test techniques and skilled listeners, the influence of distracters may be reduced.

17.27 LISTENING TESTS

Listening tests are *sensory tests* used in audio for the comparison of different products, situations and/or listening conditions (see Reference 17.15). In audio listening tests, one or more persons, *assessors*, are presented *audio samples* in a systematic way and asked to give their evaluations or other response in a prescribed manner. Listening tests may be either *objective* (perceptive—what do the assessors hear) or subjective (affective—what do the assessors prefer or dislike). The assessors may be naïve (no particular criterion) or initiated (have already participated in a sensory test). Typically, a group of assessors, a jury, is chosen to participate in a listening test. The test jury may be composed of engineers, marketing people, users, and/or buyers. The test jury is typically asked to judge not only the audio sound character but also the acceptability of a product in terms of its attributes, acceptance and quality (finding that a product answers satisfactorily to the assessors expectations), and preference (finding one product better than one or several others). (See Reference 17.16.)

Because of time and economic constraints, it is often chosen to perform the listening test, using *initiated assessors* (*expert listeners*, sometimes also called *golden ears*) having specialized competence, knowledge or experience. Two types of initiated *expert* assessors are used, *expert assessors* and *specialized expert assessors.*

An expert assessor is a person with a high degree of sensitivity and experience of listening who is able to make consistent and repeatable assessments of various audio products. The specialized expert assessor has additional experience as a specialist in the product and/or process and/or marketing and is able to listen and evaluate or predict the effects of variations relating to the product. A group of experts who participate in a listening test is called an *expert panel.* A central problem in any listening test is that some audio properties are only discernible after very prolonged listening (and training).

17.28 COMMON LISTENING TEST TYPES

Perceptive tests are objective tests where humans are used as measuring instruments. The characteristics of the perceived stimulus are *measured* in objective terms without asking the test persons for any preferences. The tests are usually made using a panel of expert assessors who have been trained for the tests and trained to express their sensory perception in terms that have been well defined in advance. Such an expert panel may be trained for a specific purpose. The tests may be perceptive or affective.

Perceptive listening tests are objective audio sound character measurements with speech and music as stimuli where the assessors express their perception of the sound characteristics in terms that describe the sound as, for example, *loud*, *soft*, *distorted*, *sharp*, *smooth*, *rough*, *muddled.*

Affective listening tests are audio sound quality measurements, normally performed with a group of naive assessors who are representative of the relevant user group, *a consumer jury*—for example, car audio buyers. As they may use other words than audio assessors or *experts* for the attributes of the product they hear, the relevant words for expression of the heard sound often have to be *found* before the answering forms for the listening tests can be made. This can be done by interviews or focus group discussions. The words may be *exclusive*, *efficient*, *powerful*, *acceptable*, for example. The immediate response of the persons' judgments is usually desired, so naïve listeners are preferred for these types of tests. Affective tests are used when the preferred characteristics or the *dislikes* of a product are sought (for example, in preference tests). The main purpose is to give information about *the product in relation to listeners in a given context.*

17.29 SOME COMMON LISTENING TESTS

In audio sound character determination, the two most common tests are the *paired comparison test* and the *semantic differential test.*

The semantic differential test gives an assessment with words for specific isolated sound characteristics. Listeners are asked to rate their impression on a scale (see Reference 17.16). Two typical graphical scales are shown in Figure 17.25. Such a scale does not restrict the test person to specific points and has the advantage of absence of any numerical value associated with the response plus the limited use

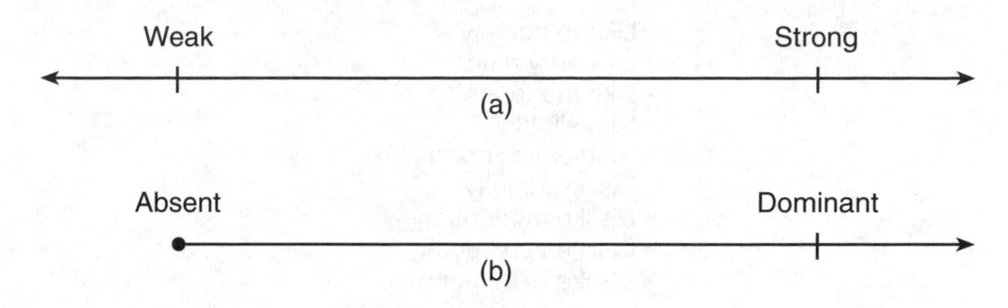

Figure 17.25 Examples of bipolar (a) and unipolar (b) scales for semantic differential tests. The upper scale has open ends, the lower scale has one closed end. Typically two anchors help determine the scale.

of words to minimize word bias. After the test, the answers are measured for example in mm from the left end of the line.

The test scale in Figure 17.25a is bipolar in the sense that the words at the ends describe opposite characteristics. The scales may also be unipolar as in Figure 17.25b and describe the intensity of a characteristic.

Numeric scales may also be used—for example, as seven- or nine-point scales. They are fast to read after the test but restrict the assessors to choose numbers between the fixed points.

Many other scales exist such as graphical and verbal scales. The verbal nine-point *liking* scale shown in Figure 17.26 is often used for audio sound quality measurements of product acceptance-preference. The words used for questions and/or to scale the responses must be familiar, easily understood, and unambiguous to the subjects. The words must be readily related to the product and the task, and they must make sense to the subjects as to how they are applied in the test.

Words that have specific and useful meaning to the person requiring the test and to the professional audio engineer but may be much less meaningful to subjects, especially if they are not trained in understanding the properties referred to. Words used for the scales may be found using focus groups (a guided discussion involving a small group of persons with small differences in background and age). Interviews are another method to obtain words that describe the audio characteristics and quality of the product. Words used for the characterization reproduced sound may be found in Section 17.8.

In *paired comparison tests*, the sounds are presented in pairs for comparison using defined attributes. Each sound is successively compared to all other sounds. At least two comparisons should be made for each pair (A, B): a comparison with sound A first and another comparison with sound B first. The number of comparisons will be N where N is the number of sounds. The paired comparison test gives the listeners the ability to distinguish between a specified characteristic of different sounds. The result will typically be of the type *higher* or *lower* without information about the absolute placement (in the *high* end or the *low* end) of the sounds. Pauses between sounds to be compared must be short, typically less than one second long. Analysis of preference using Thurstone's method is a convenient and quick way of presenting the data on a scale of subjective preference as shown in Figure 17.27 (see References 17.18 and 17.19).

- Like extremely
- Like very much
- Like moderately
- Like slightly
- Neither like nor dislike
- Dislike slightly
- Dislike moderately
- Dislike very much
- Dislike extremely

Figure 17.26 An example of a verbal nine-point *liking* scale.

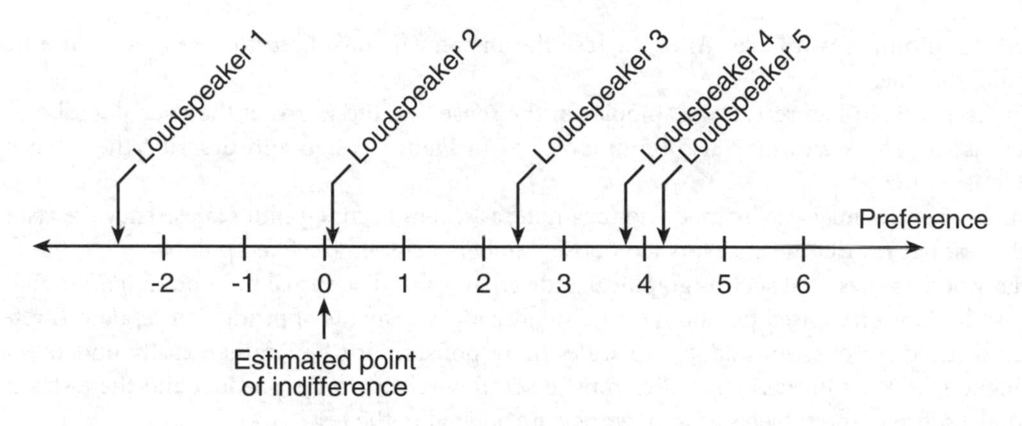

Figure 17.27 An example of a scale of subjective preference. The scale is given in units of normalized variance.

17.30 SELECTING AND TRAINING OF LISTENERS

The assessors for affective tests are normally a representative group of potential buyers, users, neighbors or other persons relevant for the actual situation. The persons should be qualified based on typical demographics (gender, age, education, cultural and ethnic background, income, etc.) physiological criteria (for example, hearing ability) and usage criteria.

In affective tests, one should in general avoid persons involved with the product, such as sales persons, technicians, designers, and employees at the company producing the product under test. If such persons are used, their responses should be compared with data from other groups to determine that the subjects are comparable, at least in their preferences toward a specific product or sound characteristic.

The *training* of the assessors for a test jury should be limited to training in the test procedure and to pre-tests with the purpose of ensuring that the words or other descriptors for the sounds presented are familiar and unambiguous to the subjects. The training will depend on the type of product and the accuracy and representability of the test in relation to the relevant market.

The *number of assessors* should be in the range from ten to several hundred. It is usually necessary to recruit two or three times the number of persons actually constituting the final panel. Successful tests are seen with 16 to 32 persons. The variance of the test results will give guidance for the reliability of the test.

17.31 EXPERT PANELS

The assessors in an expert panel are trained listeners. Expert listeners may be recruited from the office, plant, or laboratory staff. Still, as noted, one should avoid persons who are too personally involved with the products being examined, as they may cause bias. If such persons participate, their assessments should be compared with data from other groups in the panel to determine that the results are comparable for the specific sound characteristic under assessment. The expert listeners may also be recruited outside the organization by advertisement, among persons visiting the organization or among friends. The criteria for selection may be audiometric tests, reliability of judgments of pre-test sessions, and criteria for listening experience.

The main reason for training the expert assessors for a certain type of judgment is to make certain that different aspects of performance have reached an asymptotic level before the listening tests start. This will ensure that the assessments are precise as well as cost and time effective. The variability between selected and trained assessors is less than that between naïve listeners.

Ideally, it is not desirable to work with a panel of fewer than 10 selected expert assessors. For survey measurements, acceptable results may be obtained with fewer than 10 experts.

17.32 PLANNING AND PREPARATION OF A LISTENING TEST

It is difficult to avoid sources of bias in audio listening tests. Normally we want the listening test to be robust, unaffected by external variables. Planning, removal of nonacoustic sources of variance, selection of suitable assessors, randomizing the order of sample presentation can be used to minimize *bias.* However, in some tests regarding audio sound quality, knowledge of the product and the context are necessary to obtain relevant assessments. Multimodal sensory inputs are normal and must be taken into account during the planning of the test. Some modes of presentation are inherently difficult; for example, it is difficult to hide the tactile influence of headphone listening.

Bias is also a result from the choice of *listening level.* The listening level can, for example, be chosen as the natural level, loudness equalized level, or sound pressure equalized level. The equipment must be *calibrated* so that the playback level of the sound samples is known.

The presentation mode also may induce bias. Neither experimenter nor assessor should control the order of stimuli. So called *double-blind testing* is preferred—that is, randomized testing where the presentation order is out of the control of the test engineer.

Further bias may result from the *surroundings* and *context.* Are these real world conditions, laboratory, or simulated conditions? It is particularly difficult in an amateur listening test to avoid bias due to the *visual influence* of the listening room. Its room-acoustic properties, the placement of loudspeakers, and the background noise are difficult to hide. In most cases, one will want to avoid influence from nonacoustical parameters, such as visual and tactile cues, or artist recognizability.

If possible and practical, it is recommended that listening tests are performed with *custom recordings* of natural (live) sound sources. Special care should be taken to ensure that the audio is presented at calibrated sound pressure levels.

Audible noise caused by electric noise in the presentation set-up and other unwanted background noise from ventilation equipment and transportation should be avoided.

Announcements are usually necessary to be given in connection with presentation of the sound samples (announcement of sample numbers). These should be recorded at relevant levels and in similar acoustic surroundings as the sound samples, in order to avoid confusion or other effects of different acoustics for sound samples and announcements. The announcements should also be presented at a natural and calibrated level.

17.33 THE TEST SESSION

The listening test should start with a presentation of test organization and test leader, purpose, listening conditions, safety, the expected duration and pauses and other practical information. The assessors are then provided with aural and written instructions that describe the test situation and conditions, sounds and scenario (imagine that you are . . .), and their task.

It is advantageous to provide sample rating sheets, to perform a pretest, and to provide a questions and answers session ahead of the core listening test. After the test, there should be a debriefing session in which one discusses the assessors' observations and experience of the listening test. The test should take into account normal engineering ethics and the personal safety of the assessors. This will put a limit on the loudness levels that can be used in loudspeaker and headphone testing.

It is difficult to concentrate on a test for more than 20 minutes. After this time, a break for some 10 to 15 minutes is necessary. Products containing sugar, such as soft drinks and candy, may be used liberally to ensure that the blood sugar levels of the assessors remain high. Additionally, no listener should be engaged for more than two hours of listening per day and must be given regular breaks.

17.34 INFORMATION TO BE REPORTED

When reporting on listening tests, it is important that the type of equipment, its brand, serial number and other relevant data used is stated in the measurement report along with the technical data. The technical data should be given referring to the IEC 268 standards series or the ITU-R BS.1116 recommendation mentioned previously as applicable. The relevant acoustical and other characteristics of test environment should be specified and includes:

- Purpose of the measurements and involved parties
- Type of listening test (perceptive or affective)
- The test environment
- Descriptions of measuring objects/sound sources and their normal use
- The working conditions and stability of the source

- Recording, playback, and analysis equipment; type, make and model
- Description of the sound samples (recording technique, editing, announcements, length and pauses
- Criteria for selection of test persons, number of persons
- The method of the listening test, instructions to test persons, amount of information given to the test persons about the sound sources/situations of use or duration of the listening sessions, for example.
- The test conditions (surroundings, nonacoustical cues, individual or group test, etc).
- Results of listening tests, validation of results and test persons
- Graphical summary of data
- Mean standard deviations and confidence intervals of the results
- Statistical methods used
- Evaluation of listeners' qualifications from test results
- Generalization of test results
- Results of physical measurements
- Any correlation between physical measurement results and the results of the listening test, other types of statistical data processing
- Uncertainty and significance of the main results. Representativity of the results in relation to *non-test* situations. How representative are the tests?

17.35 PROBLEMS

17.1 The graph shows the measured output spectrum for a loudspeaker being fed a sine wave at 1 kHz.

Task: Calculate the THD.

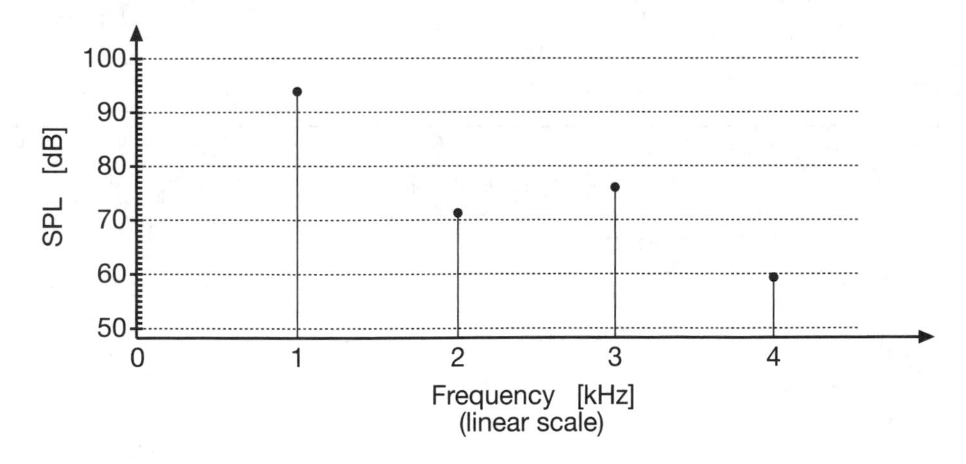

17.2 The graph shows the measured output spectrum of an amplifier for a signal consisting of 70 Hz and 6 kHz at a 4:1 ratio.

Task: Calculate the total IMD.

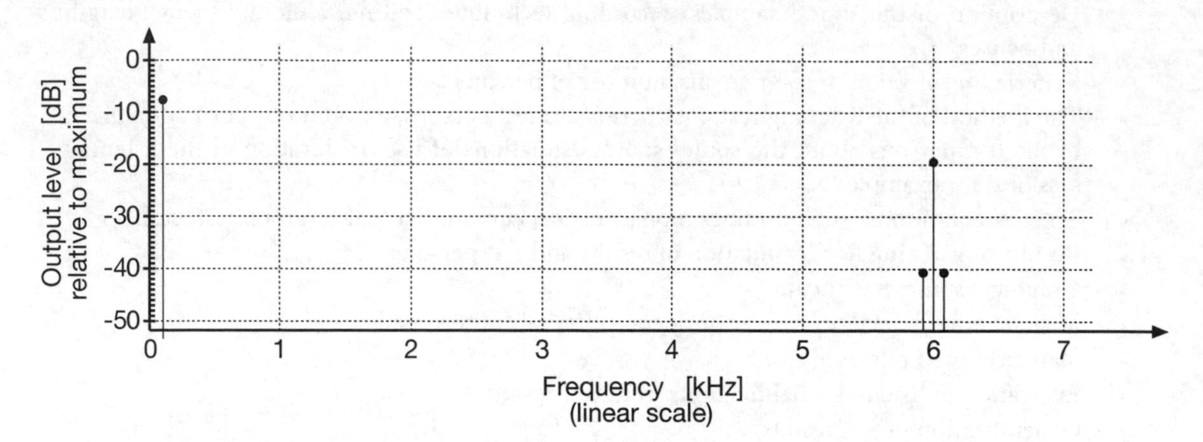

17.3 A loudspeaker was tested in an anechoic chamber using a two tone signal consisting of 4 kHz and 4.1 kHz sine waves. The test result is shown in the graph.

Task: Determine the harmonic and difference tone distortions. How can the test be improved to show greater clarity?

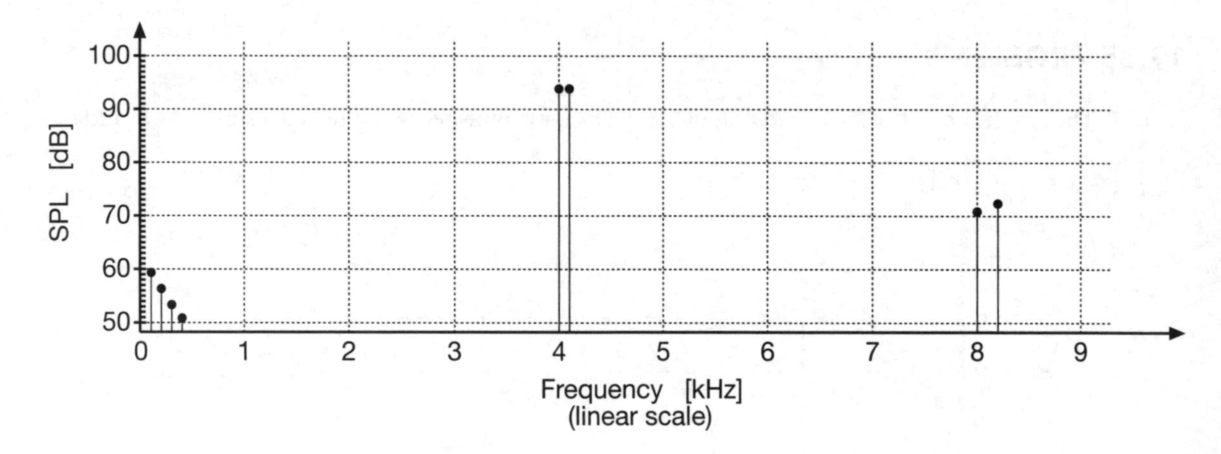

17.4 A loudspeaker was tested in an anechoic chamber with the FR result shown in the graph. How would you go about *improving* the measurement results (for example by redoing the test or by using signal processing) so that the loudspeaker becomes more attractive in the marketplace?

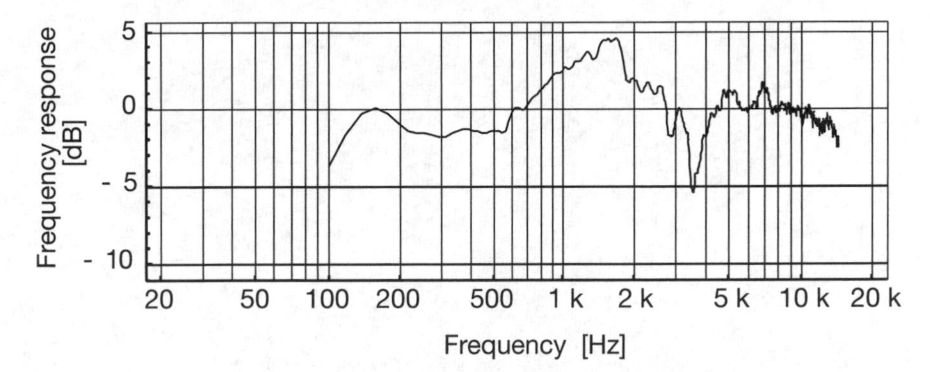

17.5 A listening test is to determine the relative timbre of 5 loudspeakers. How would you plan, design, and practically arrange the test and which difficulties do you foresee? Is the test sufficiently well specified?

References

Introduction

i.1 Beranek, L. L., *Acoustics*. New York: McGraw-Hill (1954), reprinted by the Acoustical Society of America (1986). ISBN-13: 978-0883184943.

i.2 Olson, H. F., *Acoustical Engineering*, 3rd ed. Princeton, NJ: D. Van Nostrand (1957). ASIN: B001HDVCMO.

i.3 Benson, B. K., *Audio Engineering Handbook*. New York: McGraw-Hill (1988). ISBN-13: 978-0070047778.

i.4 Ballou, G. M., *Handbook for Sound Engineers*, 2nd ed. Carmel, IN: Sams (1991).

i.5 Geddes, E., *Audio Transducers*. Northville, MI: Gedlee (2002). ISBN-13: 978-0972208505.

i.6 Borwick, J., *Loudspeaker and Headphone Handbook*, 3rd ed. Oxford: Focal Press (2001). ISBN-13: 978-0240515786.

i.7 Colloms, M., *High Performance Loudspeakers*, 6th ed. Chichester, UK: J. Wiley & Sons (2005). ISBN-13: 978-0470094303.

i.8 Gayford, M. L., *Electroacoustics: Microphones, Earphones and Loudspeakers*. London: Newnes-Butterworth (1970). ISBN-13: 978-0408000260.

i.9 Hood, J. H., *Audio Electronics*, 2nd ed. Oxford, UK: Newnes (1999). ISBN-13: 978-0750643320.

i.10 Woram, J. M., *The Recording Studio Handbook*. Plainview, NJ: Elar Publishing (1982). ISBN-13: 978-0914130017.

i.11 Ahnert, W., and Steffen, F., *Sound Reinforcement Engineering*. London: E & FN Spon (1999). ISBN-13: 978-0415238700.

i.12 Watkinson, J., *The Art of Digital Audio*, 3rd ed. Oxford, UK: Focal Press (2001). ISBN-13: 978-0240515878.

i.13 Raichel, D. R., *The Science and Applications of Acoustics*. New York: Springer Verlag (2000). ISBN-13: 978-0387989075.

i.14 Kinsler, L. E., et al., *Fundamentals of Acoustics*. New York: J. Wiley & Sons (1999). ISBN-13: 978-0471847892.

i.15 Beranek, L. L., ed., *Noise and Vibration Control*. New York: McGraw-Hill (1971). ISBN-13: 978-0070048416.

i.16 Kuttruff, H., *Room Acoustics*, 5th ed. London: Spon Press (2009). ISBN-13: 978-0415480215.

i.17 Ando, Y., *Concert Hall Acoustics*. New York: Springer Verlag (1985). ISBN-13: 978-3540135050.

i.18 Beranek, L. L., *Concert Halls and Opera Houses*, 2nd ed. New York: Springer Verlag (2004). ISBN-13: 978-1441930385.

i.19 Barron, M., *Auditorium Acoustics and Architectural Design*. London: E & FN Spon (2003). ISBN-13: 978-0419245100.

i.20 Doelle, L. L., *Environmental Acoustics*. New York: McGraw-Hill (1972). ISBN-13: 978-0070173422.

i.21 Cavanaugh, W. J., et al., *Architectural Acoustics: Principles and Practice*, 2nd ed. New York: J. Wiley & Sons (2009). ISBN-13: 978-0470190524.

i.22 Lindsay, R. B., "The Story of Acoustics." *Journal of the Acoustical Society of America*, vol. 39, pp. 629–644 (1966).

i.23 Pohlmann, K. C., *Principles of Digital Audio*, 6th ed. New York: McGraw-Hill (2010). ISBN-13: 978-0071663465.

Chapter 1 Sound

1.1 Beranek, L. L., *Acoustics*. New York: American Institute of Physics (1986). ISBN-13: 978-0883184943.

1.2 Kuttruff, H., *Physik und Technik des Ultraschalls*. Stuttgart, S. Hirzel Verlag (1988). ISBN-13: 978-3777604275.

1.3 Skilling, H. H., *Electrical Engineering Circuits*. New York: Wiley (1966). ASIN: B000V0M8OU.

General references

Kinsler, L. E., et al., *Fundamentals of Acoustics*. New York: J. Wiley & Sons (1999). ISBN-13: 978-0471847892.

Reynolds, D. R., *Engineering Principles of Acoustics*. Boston: Allyn and Bacon (1981). ISBN-13: 978-0205072712.

Seto, W. W., *Acoustics*. McGraw-Hill (1971). ISBN-13: 978-0070563285.

Chapter 2 Audio Signals

2.1 General Radio, *Handbook of Noise Measurement*. General Radio Division (1972). ASIN: B00314JQUW.

General references

Broch, J. T., *Principles of Experimental Frequency Analysis*. London: Elsevier Applied Science (1990). ISBN-13: 978-1851665549.

Hassall, J. R., and Zaveri, K., *Acoustics Noise Measurements*. Copenhagen, DK: Brüel & Kjær (1988). ISBN-13: 978-8787355216.

Randall, R. B., *Frequency Analysis*. Copenhagen, DK: Brüel & Kjær (1988). ISBN-13: 978-8787355070.

Chapter 3 Hearing and Voice

3.1 Blauert, J., *Spatial Hearing*. Cambridge, MA: The MIT Press (1996). ISBN-13: 978-0262024136.

3.2 Ando, Y., *Concert Hall Acoustics*. Berlin: Springer Verlag (1985). ISBN-13: 978-3540135050.

3.3 Zwicker, E., and Fastl, H., *Psychoacoustics*. Berlin: Springer Verlag (1990). ISBN-13: 978-3540526001.

3.4 von Békésy, G., *Experiments in Hearing*. New York McGraw-Hill (1960). ISBN-13: 978-0070043244

3.5 Möller, A. R., et al., *Man and Noise* (in Swedish). Lund, SE: Studentlitteratur (1978).

3.6 Spoor, A., "Presbyacusis Values in Relation to Noise-Induced Hearing Loss." *Audiology*, vol. 6, p. 48 (1967).

3.7 Brüel & Kjær, *Acoustic Noise Measurement*. Nærum, DK: Brüel & Kjær (1988).

3.8 Kleiner, M., "Problems in the Design and Use of Dummy-Heads." *Acustica*, vol. 41, pp. 183–193, (1978).

3.9 Fant, O., *Speech Sounds and Features*. Cambridge, MA: The MIT Press (1973). ISBN-13: 978-0262060516.

3.10 Flanagan, J. L., "Analog Measurements of Sound Radiation from the Mouth." *Journal of the Acoustical Society of America*, vol. 32, p. 1616 (1960).

3.11 Fletcher, H., *The ASA Edition of Speech and Hearing in Communication*. Acoustical Society of America (1995 reprint of 1961 edition). ISBN: 978-1563963933.

3.12 Mapp, P., "Measuring Intelligibility." *Sound and Communications*, pp. 56–68 (April 2002).

3.13 Houtgast, T., et al., "Predicting Speech Intelligibility in Rooms from the Modulation Transfer Function. I. General Room Acoustics." *Acustica*, vol. 46, pp. 60–72 (1980).

General references

Green, D. M., *An Introduction to Hearing*. Hillsdale, NJ: Lawrence Erlbaum Associates Publishers (1976). ISBN-13: 978-0470151884.

Kinsler, F. et al., *Fundamentals of Acoustics*, 2nd ed. New York: J. Wiley & Sons (1962). ASIN: B0012NMF3O.

Guyton, A. C., *Textbook of Medical Physiology*, 11th ed. London: Saunders (2005). ISBN-13: 978-0721602400.

Probst, R., et al., A Review of Otoacoustic Emissions. *Journal of the Acoustical Society of America*, vol. 89, pp. 2027–2067 (1991).

Chapter 4 Basic Room Acoustics

4.1 Cremer, L., et al., *Principles and Applications of Room Acoustics*. New York: Applied Science Publishers (1982). ISBN-13: 978-0853341130.

4.2 Kuttruff, H., *Room Acoustics*, 5th ed. London: Spon Press (2009). ISBN-13: 978-0415480215.

4.3 Harris, C. M., "Absorption of Sound in Air vs. Humidity and Temperature." *Journal of the Acoustical Society of America*, vol. 40, pp. 148–159 (1966).

Chapter 5 Spatial Sound Perception

5.1 Burgtorf, W., et al., "Untersuchungen zur Wahrnehmbarkeit verzögerter Schallsignale." *Acustica*, vol. 11, p. 97 (1961).

5.2 Seraphim, H. P., "Über die Wahrnehmbarkeit mehrerer Rückwürfe von Sprachschall." *Acustica*, vol. 11, p. 80 (1961).

5.3 Schubert, P., "Wahrnehmbarkeit von Einzelrückwürfen bei Musik." *Zeitschrift für Hochfrequenztechnik u. Elektroakustik*, vol. 78, p. 230 (1969).

5.4 Barron, M., "The Subjective Effects of First Reflections in Concert Halls." *J. Sound Vib.*, vol. 15, p. 4 (1971).

5.5 Meyer, E., and Schodder, G. R., "Über den Einfluss von Schallrückwürfen auf Richtungslokalisation und Lautstärke bei Sprache." *Nachr. Akadem. Wissenschaften*, Göttingen, Math.-Phys. Nr 6, p. 31 (1962).

5.6 Franssen, N. V., *Stereophony*. Eindhoven: Philips (1964). ISBN-13: 978-0333038444.

5.7 Haas, H., "The Influence of a Single Echo on the Audibility of Speech." *Journal of the Audio Engineering Society*, vol. 20:2, pp. 146–159 (1972).

5.8 Atal, B. S., et al., "Perception of Coloration in Filtered Gaussian Noise; Short Time Spectral Analysis by the Ear." *Proc. Int. Cong. Acoust.*, Copenhagen paper H3 (1962).

5.9 Cremer, L., et al., *Principles and Applications of Room Acoustics*. New York: Applied Science Publishers (1982). ISBN-13: 978-0853341130.

5.10 Atal, B. S., et al., "Subjective Reverberation Time and its Relation to Sound Decay." *Proc. 5th. Int. Cong. Acoust*, Liège, paper G32 (1965).

5.11 Damaske, P., "Subjektive Untersuchung von Schallfeldern." *Acustica*, vol. 19, p. 199 (1967–1968).

5.12 Reichardt, W., et al., "Raumakustische Nachbildungen mit elektroakustischen Hilfsmittein." *Acustica*, vol. 17, p. 75 (1966).

5.13 Beranek, L. L., *Concert Halls and Opera Houses*, 2nd ed. New York: Springer Verlag (2004). ISBN-13: 978-1441930385.

5.14 Kuttruff, H., *Room Acoustics*, 5th ed. London: Spon Press (2009). ISBN-13: 978-0415480215.

5.15 Kleiner, M., "Speech Intelligibility in Real and Simulated Sound Fields." *Acustica* (1978).

5.16 Kleiner, M., Dalenbäck, B-I., and Svensson, P., "Auralization—an Overview." *J. Audio Engineering Soc.*, vol. 41:11, pp. 861–875, (1993).

5.17 Schroeder, M. R., "New Method of Measuring Reverberation Time." *J. Audio Engineering Soc.*, vol. 35, pp. 299–305 (1987).

5.18 Kleiner, M., "Subjective Perception of Sound Field Simulation and Variable Acoustics by Some Passive and Electroacoustic Systems." *Applied Acoustics*, vol. 31, pp. 197–205 (1990).

5.19 Kleiner, M., and Svensson, P., "A Review of Active Systems in Room Acoustics and Electroacoustics." *Int. Symp. on Active Control of Sound and Vibration*. Newport Beach, CA (July 1995).

General references

Ando, Y., *Concert Hall Acoustics*. New York: Springer Verlag (1985). ISBN-13: 978-3540135050.

Burgtorf, W., et al., "Untersuchungen über die Richtungsabhängige Wahrnehmbarkeit verzögerte Schallsignale." *Acustica*, vol. 14, p. 254 (1964).

Mehta, M., Johnston, J., and Rocafort, J., *Architectural Acoustics*. Columbus, OH: Merrill Prentice Hall (1999). ISBN-13: 978-0137937950.

Meyer, E., and Schodder, G. R., "Über den Einfluss von Schallrückwürfen auf Richtungslokalisation und Lautstärke bei Sprache." *Nachr. Akadem. Wissenschaften*, Göttingen, Math.-Phys. no. 6, p. 31 (1962).

Chapter 6 Room Acoustics Planning and Design

6.1 Beranek, L. L. ed., *Noise and Vibration Control*. New York: McGraw-Hill (1971). ISBN-13: 978-0070048416.

6.2 Tocci, G. C., "Room Noise Criteria—The State of Art in the Year 2000." *Noise News International*, vol. 8, p. 3 (2000).

6.3 Beranek, L. L., *Concert Halls and Opera Houses*, 2nd ed. New York: Springer Verlag (2004). ISBN-13: 978-1441930385.

6.4 Kleiner, M., Klepper, D. L., and Torres, R. R., *Worship Space Acoustics*. Ft. Lauderdale, FL: J. Ross Publishing (2010). ISBN-13: 978-1604270372.

6.5 Newell, P., *Recording Studio Design*. Focal Press (2007). ISBN-13: 978-0240520865.

6.6 Rettinger, M., *Handbook of Architectural Acoustics and Noise Control: a Manual for Architects and Engineers*. New York: McGraw-Hill (1988). ISBN-13: 978-0830626861.

6.7 Toole, F., *Sound Reproduction: The Acoustics and Psychoacoustics of Loudspeakers and Rooms*. Focal Press (2008). ISBN-13: 978-0240520094.

6.8 Kleiner, M., Dalenbäck, B-I., and Svensson, P., "Auralization—An Overview." *J. Audio Engineering Soc.*, vol. 41:11, pp. 861–875 (1993).

6.9 Kleiner, M., "Sound Field Enhancement." *McGraw-Hill Encyclopedia of Science and Technology*, New York, (2001).

6.10 Kleiner, M., "Sound Reinforcement." *McGraw-Hill Encyclopedia of Science and Technology*. New York, NY: (2001).

6.11 http://www.comsol.com/products/multiphysics/.

6.12 Pryor, R. W., *Multiphysics Modeling Using COMSOL: A First Principles Approach*. Sudburg, MA, Jones & Bartlett Publishers (2009). ISBN-13: 978-0763779993.

6.13 SoundEasy and BoxCAB software. Information at http://www.interdomain.net.au/~bodzio/.

6.14 ABEC2. Information at http://www.randteam.de.

6.15 Pietrzyk, P., and Kleiner, M., "The Application of the Finite Element Method to the Prediction of Sound Fields of Small Rooms at Low Frequencies." paper 4423, *The 102nd Audio Engineering Society Convention* (1997).

6.16 Granier, E., et al. "Experimental Auralization of Car Audio Installations." *Journal of the Audio Engineering Society*, vol. 44, Issue 10, pp. 835–849 October 1996.

6.17 Veneklasen, P.S., *Design Considerations from the Viewpoint of the Professional Consultant*. ICA 1974, Architectural Acoustics, Edinburgh, UK (1974).General references

Ando, Y., *Concert Hall Acoustics*. Berlin: Springer Verlag (1985). ISBN-13: 978-3540135050.

Doelle, L. L., *Environmental Acoustics*. New York: McGraw-Hill (1972). ISBN-13: 978-0070173422.

Egan, M. D., *Architectural Acoustics*. Ft. Lauderdale, FL: J. Ross Publishing (2007). ISBN-13: 978-1932159783.

Klepper, D. L., ed., "Sound Reinforcement," *Journal of the Audio Engineering Society, Vol. 1*. (1980). ASIN: B0015O6RAM.

Klepper, D. L., ed., "Sound Reinforcement", *Journal of the Audio Engineering Society, Vol. 2*. (1980). ISBN-13: 978-0937803301.

Long, M., *Architectural Acoustics*. New York, NY: Academic Press (2005). ISBN-13: 978-0124555518.

Mehta, M., Johnston, J., and Rocafort, J., *Architectural Acoustics*. Columbus, OH: Merrill Prentice Hall (1999). ISBN-13: 978-0137937950.

Chapter 7 Absorbers, Reflectors, and Diffusers

7.1 Lindblad, S., and Gudmundson, S., *Compendium of Acoustics vol. V* (in Swedish). Lund University of Technology: Dept. of Applied Acoustics (1981).

7.2 Kuttruff, H., *Room Acoustics*, 5th ed. London: Spon Press (2009). ISBN-13: 978-0415480215.

7.3 Wack, R., and Fuchs, H., "On the Use of Micro-Perforated Sails in Assembly Rooms." *Proc. 30. Deutsche Jahrestagung für Akustik,* Strasbourg (March 2004).

7.4 Doelle, L. L., *Environmental Acoustics*. New York, McGraw-Hill: (1972). ISBN-13: 978-0070173422.

7.5 Beranek, L. L., and Hidaka, T., "Sound Absorption in Concert Halls by Seats, Occupied and Unoccupied, and by the Hall's Interior Surfaces." *Journal of the Acoustical Society of America*, vol. 104, pp. 3169–3177 (1998).

7.6 Maekawa, Z., "Problems of Sound Reflection in Rooms." *Auditorium Acoustics*, ed. Mackenzie, R., London: Applied Science Publishers, pp. 181–196 (1975).

7.7 Maekawa, Z., "Noise Reduction by Screens." *Applied Acoustics*, vol. 1, pp. 157–173 (1968).

7.8 Kleiner, M., Klepper, D. L., and Torres, R. R., *Worship Space Acoustics*. Ft. Lauderdale, FL: J. Ross Publishing (2010). ISBN-13: 978-1604270372.

7.9 Schroeder, M., *Number Theory in Science and Communication: with Applications in Cryptography, Physics, Digital Information, Computing, and Self-Similarity*. Springer (2009). ISBN-13: 978-3642099014.

7.10 Wu, T., Cox, T. J., and Lam, Y. W., "From a Profiled Diffuser to an Optimized Absorber." *Journal of the Acoustical Society of America*, vol. 108:2 (2000).

7.11 Cremer, L., et al., *Principles and Applications of Room Acoustics*. New York: Applied Science Publishers (1982). ISBN-13: 978-0853341130.

7.12 Pattersen, O.KR.Ø., Sound Intensity Measurements for Describing Acoustic Power Flow, *Applied Acoustics*, vol. 14, pp. 387–397 (1981).

7.13 Fasold, W., et al., *Bau-und Raumakustik*. (in German) VEB Verlag für Bauwesen, Berlin. (reprinted 1987 by Verlag Rudolf Müller as ASIN-B002TUURJS)

Chapter 8 Waves in Solids and Plates

8.1 Beranek, L. L. ed., *Noise and Vibration Control*. New York: McGraw-Hill (1971). ISBN-13: 978-0070048416.

8.2 Cremer, L., et al., *Structure-Borne Sound: Structural Vibrations and Sound Radiation at Audio Frequencies*, 3rd ed. Berlin: Springer Verlag (2005). ISBN-13: 978-3540226963.

8.3 Kinsler, L. E., et al., *Fundamentals of Acoustics*, 4th ed. New York: J. Wiley & Sons (1999). ISBN-13: 978-0471847892.

8.4 Kurtze, G., *Physik und Technik der Lärmbekämpfung*. Karlsruhe, GE: Verlag G. Braun (1964). ASIN: B0000BKML2.

8.5 Granhäll, A., Mechanical impedance and vibration measurement. Report F81/05 Applied, Acoustics, Chalmers Gothenburg (1981) ISSN 0348-288x.

Chapter 9 Sound Radiation and Generation

9.1 Beranek, L. L. ed., *Noise and Vibration Control*. New York: McGraw-Hill (1971). ISBN-13: 978-0070048416.

9.2 Gösele, K., "Schallabstrahlung von Platten die zu Biegeschwingungen angeregt sind." *Acustica*, vol. 3, p. 243 (1953).

9.3 Roozen, N. B., "Reduction of Bass-Reflex Port Nonlinearities by Optimizing the Port Geometry." *Proc. Audio Eng. Soc. Conv.*, vol. 104:4661 (1998).

Chapter 10 Sound Isolation

10.1 Brandt, O., *Acoustic Planning* (in Swedish). Statens Nämnd för Byggnadsforskning. Stockholm (1959).

10.2 Kleiner, M., Klepper, D. L., and Torres, R. R., *Worship Space Acoustics.* Ft. Lauderdale, FL: J. Ross Publishing (2010). ISBN-13: 978-1604270372.

10.3 Kurtze, G., *Physik und Technik der Lärmbekämpfung.* Karlsruhe, Verlag G. Braun (1964). ASIN: B0000BKML2.

General references

Vigran, T. E., *Building Acoustics.* London: Spon Press. (2008). ISBN-13: 978-0415428538.

Hopkins, C., *Sound Insulation.* Maryland Heights, MO, USA Butterworth-Heinemann (2007). ISBN-13: 978-0750665261.

Chapter 11 Vibration Isolation

11.1 TD150 Subchassis Concepts. www.theanalogdept.com (sampled 2011).

11.2 Muster, D., and Plunkett, R., "Isolation of Vibrations." In Beranek, L. L., ed., *Noise and Vibration Control.* Washington: Inst. of Noise Control Engineering (1988). ISBN-13: 978-0070048416.

General references

Harris, C. M., "Vibration Control Principles." In Harris, C. M., ed., *Handbook of Acoustical Measurements and Control.* Reprinted by the *Acoust. Soc. Amer.*, Woodbury, NY (1998). ISBN-13: 978-0070268685.

Racca, R. H., "Types and Characteristics of Vibration Isolators." In Harris, C. M., ed., *Handbook of Acoustical Measurements and Control.* Reprinted by the *Acoust. Soc. Amer.* Woodbury, NY (1998). ISBN-13: 978-0070268685.

Rivin, E. I., *Passive Vibration Isolation, American Society of Mechanical Engineers* (2003). ISBN-13: 978-0791801871.

Chapter 12 Microphones

12.1 Brüel & Kjær: *Measuring Microphones.* Nærum, Denmark (1971). ASIN: B000YT5TFI.

12.2 Dickreiter, M., *Mikrofonaufnahmetechnik.* Leipzig, Hirzel Verlag (2003). ISBN-13: 978-3777611990.

12.3 Wahlström, S., "The Parabolic Reflector as an Acoustical Amplifier." *Journal of the Audio Engineering Society*, vol. 33:6, pp. 418–429 (1985).

12.4 Olson, H. F., *Acoustical Engineering*, 3rd ed. Princeton, NJ: D. van Nostrand (1957). ASIN: B001HDVCMO.

General references

Abbagnaro, L. A., ed., "Microphones." *Journal of the Audio Engineering Society* (1980).

Ballou, G., ed., *Electroacoustic Devices: Microphones and Loudspeakers*. Focal Press Oxford, UK: (2009). ISBN-13: 978-0240812670.

Gayford, M. L., *Electroacoustics: Microphones, Earphones, and Loudspeakers*. London: Newnes-Butterworth (1970). ISBN-13: 978-0408000260.

Rossi, M., and Roe, P. R. W., *Acoustics and Electroacoustics*. Artech House Acoustics Library (1988). ISBN-13: 978-0890062555.

Woram, J. M., and Kefauver, A. P., *The New Recording Studio Handbook*. New York: Elar Publishing (1989). ISBN-13: 978-0914130048.

Chapter 13 Phonograph Systems

13.1 Olson, H. F., *Modern Sound Reproduction*. New York: Van Nostrand Reinhold. (1972). ISBN-13: 978-0442262808.

General references

Anderson, C. Roger, "Fifty Years of Stereo Phonograph Pickups: a Capsule History." *Journal of the Acoustical Society of America*, vol. 77:4, pp. 1320–1326 (April 1985).

Gravereaux, D. W., "Engineering Highlights of the LP Record. *Journal of the Acoustical Society of America*, vol. 77:4, pp. 1327–1331 (April 1985).

Roys, H. E., "Reminiscing—the Stereophonic Record." *Journal of the Acoustical Society of America*, vol. 77:4, pp. 1332–1334 (April 1985).

Stephani, O., "Geschichte von Schallplattentechnik." *Frequenz*, vol. 15, pp. 44–53 (1961).

Temmer, S. F., ed., "Disk Recording vol. 1: Groove Geometry and the Recording Process. *Journal of the Audio Engineering Society* (1980).

Temmer, S. F., ed., "Disk Recording vol. 2: Disk Playback and Testing, *Journal of the Audio Engineering Society* (1980).

Chapter 14 Loudspeakers

14.1 Beranek L. L., *Acoustics*. New York: McGraw-Hill (1954). ISBN-13: 978-0883184943.

14.2 Colloms, M., *High Performance Loudspeakers*, 6th ed. London: Pentech Press (2005). ISBN-13: 978-0470094303.

14.3 Olson, H. F., *Acoustical Engineering*. Princeton, NJ: Van Nostrand (1957). ASIN: B001HDVCMO.

14.4 Suzuki, H., and Tichy, J., "Radiation and Diffraction Effects of Convex and Concave Domes in an Infinite Baffle." Preprint 1675, *67th Convention Audio Engineering Society* (1980).

14.5 Tyrland S., *Construction of a Monitor Loudspeaker* (in Swedish). Chalmers University of Technology, Gothenburg: Dept. of Applied Acoustics (1974).

14.6 Waterhouse R. V., "Output of a Sound Source in a Reverberation Chamber and Other Reflecting Environments." *Journal of the Acoustical Society of America*, vol. 30, p. 1 (1958).

14.7 Button, D. J., and Gander, M. R., "The Dual Coil Drive Loudspeaker." *Proceedings of the 1st AES UK Conference* (1998).

14.8 Borwick, J., *Loudspeaker and Headphone Handbook*, 3rd ed. *Focal Press* Oxford, UK: (2001). ISBN-13: 978-0240515786.

General references

Ballou, G., ed., *Electroacoustic Devices: Microphones and Loudspeakers*. Focal Press (2009). ISBN-13: 978-0240812670.

Gayford, M. L., *Electroacoustics: Microphones, Earphones, and Loudspeakers*. London: Newnes-Butterworth (1970). ISBN-13: 978-0408000260.

Rossi, M., and Roe, P. R. W., *Acoustics and Electroacoustics*. Artech House Acoustics Library (1988). ISBN-13: 978-0890062555.

Cooke, R. E., ed., "Loudspeakers, vol. 1." *Audio Engineering Society* (1978). ASIN: B001ABN690.

Cooke, R. E., ed., "Loudspeakers, vol. 2." *Audio Engineering Society* (1984). ASIN: B001ABNACS.

Gander, M. R., ed., "Loudspeakers, vol. 3." *Audio Engineering Society* (1996). ISBN-13: 978-0937803288.

Gander, M. R., ed., "Loudspeakers, vol. 4." *Audio Engineering Society* (1996). ASIN: B001ABNDTI.

Chapter 15 Headphones and Earphones

15.1 Burkhard, M. D., and Sachs, R. M., "Anthropometric Manikin for Acoustic Research." *Journal of the Acoustical Society of America*, vol. 58, p. 1 (1975).

15.2 Kleiner, M., "Problems in the Design and Use of Dummy-Heads." *Acustica*, vol. 41, p. 3 (1978).

15.3 Gayford, M. L., *Electroacoustics: Microphones, Earphones and Loudspeakers*. London: Newnes-Butterworth (1970). ISBN-13: 978-0408000260.

General references

Ballou, G., ed., *Electroacoustic Devices: Microphones and Loudspeakers*. Oxford, UK: Focal Press (2009). ISBN-13: 978-0240812670.

Borwick, J., *Loudspeaker and Headphone Handbook*, 3rd ed.. Oxford, UK: Focal Press (2001). ISBN-13: 978-0240515786.

Chapter 16 Digital Representation of Sound

16.1 Blesser, B. A., "The Digitization of Audio." *Journal of the Audio Engineering Society*, vol. 26:10 (1978).

16.2 Blesser, B. A., et al., "Digital Processing in Audio Signals." In Oppenheimer ed., New York, NY: *Applications of Digital Signal Processing*. New Jersey: Prentice-Hall. (1978).

16.3 Blesser, B., et al., ed., "Digital Audio." *Proceedings of the AES Premier Conference* (1982).

16.4 Watkinson, J., *The Art of Digital Audio*, 3rd ed. Oxford, UK: Focal Press (2001). ISBN-13: 978-0240515878.

16.5 Pohlmann, K. C., *Principles of Digital Audio*, 6th ed. New York: McGraw-Hill (2010). ISBN-13: 978-0071663465.

16.6 Bosi, M., and Goldberg, R. E., *Introduction to Digital Audio Coding and Standards.* Boston: Kluwer Academic Publishers (2003). ISBN-13: 978-1402073.

16.7 Lehtonen, K., "T-61.246 Digital Signal Processing and Filtering: GSM Code." www.cis.hut.fi/Opinnot/T-61.246/Kutri2003/lehtonen_gsmdoc.pdf (sampled 2006).

16.8 http://www.iis.fraunhofer.de/amm/download/index.html (sampled 2006).

16.9 Kriz, J. S., "A 16-Bit A-D-A Conversion System for High-Fidelity Audio Research." *IEEE Trans. on Acoustics, Speech, and Signal Processing*, vol. 23, p. 1 (1975).

16.10 Lee, F. F., et al., "Floating-Point Encoding for Transcription of High-Fidelity Audio Signals." *Journal of the Audio Engineering Society*, vol. 25, p. 5 (1977).

16.11 Finger, R. A., "On the Use of Computer Generated Dithered Test Signals." *Journal of the Audio Engineering Society*, vol. 35, p. 6 (1987).

16.12 Vanderkooy, J., et al., "Resolution Below the Least Significant Bit in Digital Systems with Dither." *Journal of the Audio Engineering Society*, vol. 32, p. 3 (1984).

Chapter 17 Audio Systems and Measurements

17.1 Olson, H. F., *Modern Sound Reproduktion.* New York, NY: Van Nostrand Reinhold (1972). ISBN-13: 978-0442262808.

17.2 IEC 268-3 *Sound System Equipment. Amplifiers.*

17.3 Zwicker, E., and Fastl, H., *Psychoacoustics.* Berlin: Springer Verlag (1990). ISBN-13: 978-3540526001.

17.4 Kleiner, M., and Lindgren, C., *Objective Characterization of Audio Sound Fields in Automotive Spaces.* Paper 15-007. Audio Eng. Soc. 15th International Conference: Audio, Acoustics & Small Spaces (October 1998).

17.5 Metzler, R. E., *Audio Measurement Handbook.* Portland, OR: Audio Precision (1993). ISBN-13: 978-9994569885.

17.6 Cabot, R. C., "Fundamentals of Modern Audio Measurement." *Journal of the Audio Engineering Society*, vol. 47:9, pp. 738–744, 746-762 (1999).

17.7 Farina, A., "Simultaneous Measurement of Impulse Response and Distortion with a Swept Sine Technique." *Proceedings 110 AES Convention*, Paris (2000).

17.8 IEC 268-13 *Sound System Equipment, Listening Test on Loudspeakers.*

17.9 Diestel, H. G., "Zur Schallausbreitung in reflexionsarmen Räumen." *Acustica*, vol. 12, pp. 113–118 (1962).

17.10 Sköld, A., *Integrative Analyses of Perception and Reaction to Information and Warning Sounds in Vehicles*. Dissertation 2784, Chalmers University of Technology Dept. of Applied Acoustics, Gothenburg, Sweden (2008). ISBN 978-91-7385-103-9.

17.11 Blauert, J., and Jekosch, U., *Sound Quality Evaluation—A Multilayered Problem*. EEA-Tutorium Antwerp, Belgium: (1996).

17.12 Bech, S., "Selection and Training of Subjects for Listening Tests on Sound-Reproducing Equipment." *Journal of the Audio Engineering Society*, vol. 40:7/8 (1992).

17.13 Bech, S., and Zacharov, N., *Perceptual Audio Evaluation—Theory, Method and Application*. Wiley. ISBN-13: 978-0470869239. New York (2006)

17.14 Meilgaard, M. C., et al., *Sensory Evaluation Techniques*, 4th ed. CRC Press (2006). ISBN-13: 978-0849338397.

17.15 ISO 5492 *Sensory Analysis—Vocabulary*

17.16 Nordtest recommendation: *Measurements and Judgments of Sound in Relation to Human Sound Perception*. Copenhagen, Denmark (2001).

17.17 ISO 4121 *Sensory Analysis—Methodology—Evaluation of Food by Methods Using Scales.*

17.18 ISO 4121 *Sensory Analysis—Methodology—Paired Comparison Test.*

17.19 Guilford, J. P., *Psychometric Methods*. McGraw-Hill, ISBN-13: 978-0070993396. New York (1956).

17.20 ITU-R BS.1116 Recommendation: *Methods for the subjective assessment of small impairments in audio systems including multichannel sound systems* (1997).

17.21 ITU-R BS.1283 Recommendation: *Subjective assessment of sound quality—A guide to existing Recommendations (1997).*

General references

Scott, H. H., "Intermodulation Measurements," *J. Audio Eng. Soc.*, Volume 1, Issue 1, pp. 56–61; January 1953.

Czerwinski, E. et al., "Multitone Testing of Sound System Components. Some Results and Conclusion, Part 1: History and Theory." *J. Audio Eng. Soc.*, Volume 49, Issue 11, pp. 1011–1048; November 2001.

Cabot, R. C., "Comparison of Nonlinear Distortion Measurement Methods" in *Proc. of the 11th International Conference: Test & Measurement*, Audio Engineering Society, New York (May 1992).

Bohn, D., "Audio Specifications." RaneNote 145. *Rane Corporation*, Mukilteo, WA 98275-5000. (also at *http://www.rane.com/note145.html* [sampled May 2011])

Answers to Problems

1.1

r [m]	$\lvert p \rvert$ [Pa]	$\lvert u \rvert$[mm/s]	ϕ [°]
0.1	1	2.4	29
0.2	0.5	1.2	15
0.4	0.25	0.61	7.8
0.8	0.125	0.30	3.9

1.2

f_{mid} [Hz]	ΔL [dB]
63	6.0
125	5.8
250	5.1
500	1.6
1k	−5.2
2k	4.6
4k	−1.1
8k	1.8

1.3

f_{mid} [Hz]	L_W [dB] re 10^{-12} W
125	105
250	111
500	117
1k	121
2k	123
4k	124

1.4 a) $\alpha = 0{,}34 \pm 0{,}06$

b) Reflection coefficient phase angle $\phi = 2\,k\,x_{min} - \pi$

2.1 $L_{p,tot} = 66$ dB

2.2 $L_{pA,tot}$ = 60 dB. The energy in the 250 and 500 Hz octave bands contributes most to the sound level.

2.3 a) $p_{rms,tot}$ = 0.67 Pa
b) $p_{rms,tot}$ = 0.50 Pa
c) $L_{p,a}$ = 90.5 dB and $L_{p,b}$ = 88.0 dB

2.4 $L_{Aeq,\,T}$ (8 hours) = 83 dB

2.5 Fundamental + harmonics (60, 120, 180, 240, . . .) will be in the 63, 125, 200, 250 Hz, ... third-octave bands.

2.6 a) $L_{p,A}$ = 83 dB, $L_{pA,A}$ = 74 dB, $L_{p,B}$ = 64 dB, $L_{pA,B}$ = 62 dB.
b) $L_{p,A+B}$ = 83 dB, $L_{pA,A+B}$ = 74 dB

2.7

f_{mid} [Hz]	Third-octave band SPL	Octave band SPL
100	72 dB	
125	74 dB	79 dB
160	76 dB	
200	78 dB	
250	80 dB	85 dB
315	81 dB	
400	83 dB	

2.8 T = 3.3 s

2.9

Octave band [Hz]	$L_{p,signal}$ [dB]
125	34
250	36
500	36
1k	37
2k	40
4k	40

3.1 A filter having a 100 Hz bandwidth will not improve the audibility of a signal at 1 kHz since the critical bandwidth is approximately 60–80 Hz.

3.2 The threshold will be raised by 20 dB (Diagram B)

3.3 The 1 kHz-tone must have a sound pressure level higher than 72 dB to be audible.

3.4

f [Hz]	Perceived change [phon]
100	+ 38
400	+ 32
1 k	+ 30
5 k	+ 31
10 k	+ 32

3.5

f_{center} [Hz]	Δf_{cb} [Hz]	Comment
100	≈ 100	≫1/3 octave band
200	≈ 100	> 1/3 octave band
400	≈ 110	≈ 1/3 octave band
1 k	≈ 170	≈ 1/4 octave band
2 k	≈ 320	≈ 1/4 octave band
4 k	≈ 650	≈ 1/4 octave band
10 k	≈ 2500	> 1/3 octave band

4.1 The modes (3 0 0), (3 1 0), (1 2 0), (2 2 0), (2 0 1), (2 1 1), (0 2 1) will fit in the 44.5 Hz–56.1 Hz frequency band.

4.2 $\Delta N = \frac{4\pi V f^2}{c^3}\Delta f$

Third-octave band [Hz]	Number of modes, ΔN
100	3
125	7
160	14
200	27
250	53
315	≈ 110
400	≈ 220
500	≈ 420
630	≈ 840
800	≈ 1,700
1k	≈ 3,400

4.3 a) $\alpha = 0.10$ b) $\alpha = 0.09$ c) $\alpha = 0.09$

4.4 According to Schroeder's formula (Eq. 4.67), one can use the statistical room acoustics approach from a frequency of 160 Hz and upwards.

4.5

r [m]	$p_{rms,tot}$ [Pa]	Lp [dB]
0.5	1.3	96
1.0	0.81	92
2.0	0.64	90
4.0	0.59	89
8.0	0.58	89

4.6 a) $r = 0.7$ m b) $r = 3.4$ m c) $r = 5.3$ m

4.7 a)

Octave band	500	1k	2k	[Hz]
$A_{S,added}$	7.7–12.0	4.7–9.0	4.7–9.0	[m²S]

b) To reach $A_{S,added}$ 9 sheets of absorber are necessary.

c)

Octave band	500	1k	2k	[Hz]
T_{Eyring}	0.28	0.25	0.26	[s]

4.8 To obtain the desired added room absorption one can, for example, treat the rear wall by adding 24.5 m² of perforated board. In addition 40 m² of the membrane absorber will be needed. Using these extra absorbers one will reach the $A_{S,added}$ listed below.

Octave band [Hz]	Desired $A_{S,added}$ [m²S]	Achieved $A_{S,added}$ [m²S]
125	32.5–47.4	37.9
250	34.4–49.3	39.3
500	21.2–36.1	28.7
1 k	18.3–33.2	18.9
2 k	11.0–25.9	14.4
4 k	3.7–18.6	9.8

4.9 a) L_p at coordinate (4,1,2) m = 83.5 dB.
b) In a diffuse field L_p is the same at all positions, i.e. L_p = 75 dB also at coordinate (4, 1, 2) m.

4.10 The sound pressure level will be 15 dB higher in the corner position than in the middle of the room, i.e L_{pA} = 50 dBA at the corner.

5.1 $\delta = 6.91 / T$

5.2 $C_{80} = 10 \cdot \log\left(e^{\frac{1.1}{T}} - 1\right)$

$D_{50} = 100\left(1 - e^{\frac{-0.69}{T}}\right)$

5.3

θ [°]	Δ [dB]	A_V [dB]	A_H [dB]
24	15	4.6	−10.4
20	11	3.8	−7.2
16	8.5	3.2	−5.3
12	6	2.5	−3.5
8	4	1.8	−2.2
4	2	0.9	−1.1
0	0	0	0

6.1 The level difference between the direct sounds from source 2 and source 1 is −6.5 dB and the difference in arrival time is 59 ms. The number of people annoyed by the echo should then be less than 10%.

6.2 To achieve an arrival time difference of less than 20 ms, the time delay Δt must be within the intervals listed below:

Position	Δt [ms]
1	26–46
2	0–20
3	21–41
4	50–70

6.3

r [m]	$H_{without}$ [dB]	$T'_{without}$ [s]	H_{with} [dB]	T'_{with} [s]
10	−1.5	1.8	2.5	1.5
20	−7.5	2.0	−3.5	1.9
30	−11	2.1	−7.0	2.0

7.1

f [Hz]	$\alpha_{0,estimated}$
100	0.16
125	0.18
160	0.26
200	0.33
250	0.44
315	0.48

7.2 $$\alpha = \frac{1}{1+\left(\frac{\omega m''}{2Z_0}\right)^2}$$

7.3 The mass per unit area, $m'' \leq 0.12$ kg/m^2

7.4 Both a membrane absorber at a distance of 36 mm from the wall and the perforated board at a distance of 30 mm from the wall can give a 100 Hz resonance frequency. The perforated board is more suitable in this application because of its sharper and higher absorption peak.

7.5 3 – a, 4 – c, 5 – b, 2 – d, 1 – e

7.6 The perforated brick will act as a resonance absorber, peaking at 382 Hz. In addition there will be resonance peaks at 2.3 kHz, 4.7 kHz, . . . i.e. the frequencies at which the brick thickness is a multiple of half wavelengths. At these frequencies the brick is acoustically transparent because of impedance matching between the two sides of the brick if there is some absorption inside the air space. (Losses at the walls of the brick holes also help because of the half wave resonance.)

8.1 The modal density is $dN/df \approx 0.65$ Hz^{-1} which leads to $\Delta N \approx 15$ in the 100 Hz and $\Delta N \approx 120$ in the 800 Hz third octave bands respectively.

8.2 The rms value of the acceleration is approximately 0.012 m/s^2 at a sound pressure level of 70 dB.

8.3 a) The modal density is approximately given by

$$\frac{dN}{df} \approx \frac{4\pi V f^2}{c^3} + \frac{\pi S f}{2c^2} + \frac{L}{8c} \quad [\text{Hz}^{-1}]$$

f [Hz]	dN/df [Hz^{-1}]	Comment
100	0.34	Not audible
200	1.0	Not audible
500	5.4	Not audible
700	10	Not audible

b & c) The modal densities are $dN/df \approx 1.4$ Hz^{-1} for the sheet of steel (audible coloration) and $dN/df \approx 4.9$ Hz^{-1} for the gold foil (no audible coloration).

9.1

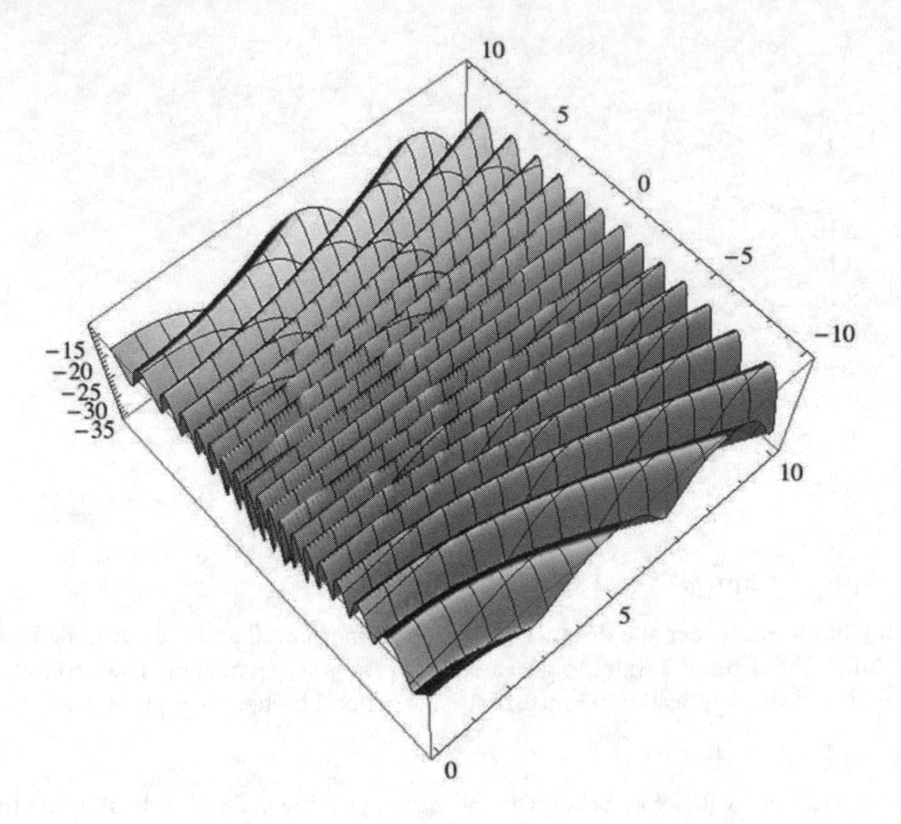
10
5
0
−5
−10
−15
−20
−25
−30
−35
10
5
0

9.2

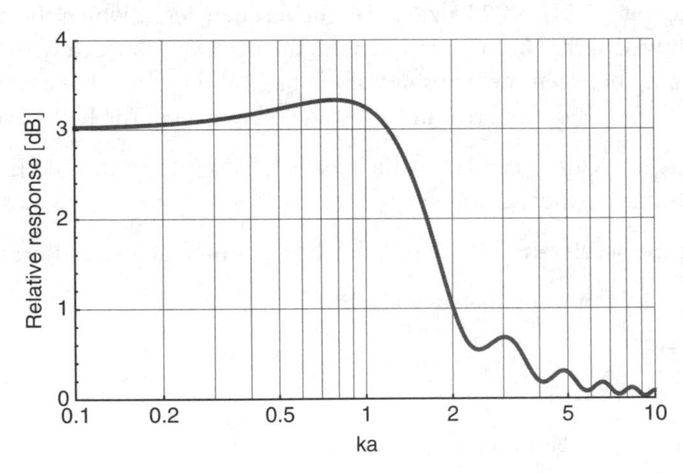
Relative response [dB]
ka
4
3
2
1
0
0.1
0.2
0.5
1
2
5
10

9.3

Octave band	0.5 k	1k	2k	4k	[Hz]
Radiation ratio	0.05	0.3	1	1	

9.4 $\Delta L_W \approx -20$ dB

9.5 Port power level ≈105 dB at 34 Hz and port flow noise level is ≈26 dB peaking at about 12 Hz.

10.1 $R_{tot} = 48$ dB

10.2 $L_{pA} = 27$ dB

10.3 a) Same L_p in the room without the TV set in both cases.
b) lower L_p in the room without the TV set if the TV set is in the small room.

10.4 The fundamental resonance frequency of the wall is 106 Hz. The critical frequencies for the two sheets are in the intervals 3.6–4.4 kHz (9 mm sheet) and 2.5–3.1 kHz (13 mm sheet). To increase the transmission loss one must lower the fundamental resonance frequency, for example by separating the two sheets. In addition it is advisable to have separate battens for each wall so that the wall will act more like a theoretical double wall.

10.5 The sound level in room 2 is $L_{pA} = 42$ dB.

10.6 The fundamental resonance frequency is 94 Hz. The critical frequencies are 1.2 kHz and 3 kHz respectively. The modes of the cavity are at the frequencies 95, 190, 200, 222, 275 Hz etc.

11.2 $f_{res} \approx 48$ Hz

11.3 $f_{res} \approx 4.7$ Hz

11.4 $f_{res} \approx 18$ Hz

12.1 a) The equation of motion: $F(x) = M\dfrac{d^2x}{dt^2} + R\dfrac{dx}{dt} + \dfrac{1}{C_m}x$

The mechanical impedance: $\underline{Z}_M(\omega) = j\omega M + R + \dfrac{1}{j\omega C_M}$

The resonance frequency: $f_{res} = \dfrac{1}{2\pi\sqrt{MC_M}}$

b) The output voltage is proportional to the sound pressure:
1) below f_0 (for example carbon and condenser microphones)
2) at f_0 (for example dynamic microphones)
3) over f_0

12.2 a) $\tilde{e} = \dfrac{BlS}{M\omega_0\sqrt{\left(\dfrac{\omega_0}{Q}\right)^2 + \left(\dfrac{\omega}{\omega_0} - \dfrac{\omega_0}{\omega}\right)^2}}\tilde{p}$

b)

f [Hz]	$\tilde{e}_{rms}$ [mV] $Q = 1$	$Q = 0.1$	$Q = 0.01$
75	1.4	0.52	0.055
150	3.1	0.55	0.055
300	5.5	0.55	0.055
600	3.1	0.55	0.055
1.2 k	1.4	0.52	0.055
2.4 k	0.7	0.43	0.055

12.3 a) $Q_1 = Q_2$

b) $Q_1 = -k\,Q_2$; k should be larger than 0 but smaller than 1.

c) $Q_1 = -Q_2$

12.4 Directivity index $DI = 4.8$ dB

12.5

f [Hz]	$\tilde{e}_{rms}$ [mV]
0.5 k	0.016
1 k	0.022
2 k	0.009

12.6 The signal-to-noise ratio is 42 dB.

14.1 a) The resonance frequency is 71 Hz.

b) The diaphragm velocity is $\tilde{u} = \dfrac{Bl}{\sqrt{R^2 + \dfrac{M}{C_M}\left(\dfrac{\omega}{\omega_0} - \dfrac{\omega_0}{\omega}\right)^2}}\tilde{i}$ and the displacement is $\tilde{x} = \dfrac{\tilde{u}}{\omega}$

c)

f [Hz]	$\tilde{u}_{rms}$ [m/s]	$\tilde{x}_{rms}$ [mm]
17	0.89	8.0
36	2.1	9.4
71	7.5	16.7
142	2.1	2.3
285	0.89	0.49

14.2 The resonance frequency is 78 Hz.

14.3 a)

f [Hz]	W [mW]
17	0.6
71	940
285	160

14.4 a) Region I—below f_0 the response is compliance controlled, L_p increases by 12 dB/octave.
Region II—the peak at f_0 is determined by the damping of the mass-compliance resonance.
Region III—above f_0 the response is controlled by the diaphragm mass and L_p will be constant except that, because of the increased directivity at higher frequencies, there will be a slight increase with frequency.
Region IV—interferences between the contributions from various parts of the diaphragm because of bending wave resonance mode patterns.
Region V—radiation resistance no longer increases with frequency (remains constant) so the radiated power decreases and thus also L_p.

b) Regions I–III—choice of diaphragm mass, suspension mechanical compliance and loudspeaker box acoustic compliance, and losses will determine the frequency and Q-value of the fundamental resonance.
IV—damping bending wave resonance for example by using a diaphragm having a viscoelastic layer.
V—a smaller diaphragm will lead to a higher radiation resistance cut-off.

14.5 The resonance frequency is 44 Hz.

14.6 The resonance frequency is $f_0 = \frac{1}{2\pi}\sqrt{\frac{\frac{1}{C_M} + \frac{\rho c^2}{V} S^2}{M}}$

The absorption coefficient is $\alpha = \frac{4a}{(a+1)^2 + b^2}$

where $a = \frac{R_M}{SZ_0}$

and $b = \frac{1}{SZ_0\left(\omega M_M - \frac{1}{\omega}\left(\frac{1}{C_M} + \frac{\rho c^2}{V} S^2\right)\right)}$

14.7 At frequencies above approximately 74 Hz: $W_{max} = 600$ mW

At frequencies below approximately 74 Hz : $W_{max} \approx 0.23\frac{\omega^4}{\omega^2 + 1.8 \cdot 10^7}$

14.8 The sound pressure level is 63 dB.

15.1 $\underline{p} = \frac{S\rho c^2}{j\omega V R_{VC}} \frac{Bl}{j\omega M_M + \frac{1}{j\omega C_M} + R_m} \underline{e}$

15.2 The output voltage will be about 0.05 mV.

15.3 The frequency response will have a +6dB/octave characteristic if the near field pressure close to one side of the headphone was adjusted for flat frequency response.

15.4 The frequency response at the eardrum will be flat up to about 2 kHz.

17.1 THD ≈ 14%

17.2 THD ≈ 13%

17.3 THD ≈ 7.5% and DFD ≈ 1.8%

Index

B

C

D

E

N

O

P

Q

R

T

U

V

W

Y

Z